PUBLIC POLICIES
AND FOOD SYSTEMS
IN LATIN AMERICA

Jean-François Le Coq, Catia Grisa,
Stéphane Guéneau, Paulo Niederle, editors

Éditions Quæ

Update Sciences & Technologies Collection

Méthodes d'investigation
de l'alimentation et des mangeurs
O. Lepiller, T. Fournier,
N. Bricas, M. Figuié, coord.
2021, 244 p.

La santé globale au prisme de l'analyse
des politiques publiques
S. Gardon, A. Gautier, G. Le Naour,
O. Faugère, R. Payre, coord.
2020, 248 p.

Eating in the city
Socio-anthropological perspectives
from Africa, Latin America and Asia
A. Soula, C. Yount-André, O. Lepiller,
N. Bricas, J-P. Hassoun, coord., D. Manley
(traduction)
2021, 158 p.

Manger en ville
Regards socio-anthropologiques d'Afrique,
d'Amérique latine et d'Asie
A. Soula, C. Yount-André, O. Lepiller,
N. Bricas, J-P. Hassoun, coord.
2020, 172 p.

This book was edited by James Calder, Lindsay Higgins, Teri Jones-Villeneuve,
Abisola Persem and Claire Walker from the original edition,
entitled *Políticas Públicas y Sistemas Alimentarios en América Latina*
and published by E-papers (http://www.e-papers.com.br; ISBN: 978-65-8706-526-7).

This book has received financial support from CIRAD.

To cite this book :
Le Coq J.-F., Grisa C., Guéneau S., Niederle P. (Eds), 2022. Public policies and food systems in
Latin America. Versailles, Quæ, 446 p.

Éditions Quæ
RD 10, 78026 Versailles Cedex
www.quae.com – www.quae-open.com

© Éditions Quæ, 2022

ISBN papier : 978-2-7592-3535-3
ISBN ePub : 978-2-7592-3537-7

ISBN PDF : 978-2-7592-3536-0
ISSN : 1773-7923

Table des matières

PART IV
BUILDING FOOD POLICIES AND ACTIONS IN REGIONS AND CITIES

PART V
THE CHALLENGES OF SPECIFIC FOOD POLICIES

Foreword

The profound transformation of sustainable food systems is a major challenge for agriculture and societies. We must not only solve the problems of malnutrition, but also address all the challenges of sustainable development in a comprehensive way. In addition to the recurrent food and nutrition crises that have marked previous decades, changing dietary patterns have created a situation where population supply challenges are compounded by nutritional and health challenges, such as obesity and its many related diseases. Although there is now near-universal abundance of food, malnutrition is the main public health problem.

Moreover, the environmental and climate crises and their impacts invite us to rethink food systems to reduce their footprint, support the renewal of resources and increase the resilience of societies and ecosystems. Finally, the Covid-19 pandemic both reflects the systemic dimension of the crises that follow one after the other and reinforces the need for profound structural transformations. This pandemic has generated an additional shock that urgently requires more rapid and systemic actions.

Society can respond to these challenges in various ways and at multiple levels. From international organizations to local civil society actors, social movements, and national and local governments, initiatives to promote food systems are emerging to promote sustainable food systems. However, although awareness is growing and consumers around the world are increasingly demanding healthy food that respects cultures and the environment, current solutions are limited given the scale of the challenge. Truly successful solutions will require actors from different sectors, at different scales and with different visions to all work together. This book highlights many initiatives and transformations, often developed and carried out in different places and within specific spaces. We know that such initiatives are rarely replicable, given the specificity of contexts, and that we cannot imagine remaking the world simply by reproducing them. We also know that local changes depend on actions undertaken in other spheres, which make new transformations possible and drive them forward. The aim is therefore to develop a new system of action, a framework that structures and enhances action to support sustainable development. In the face of this challenge, public policies play a central role. They educate populations and raise their awareness; adjust production and consumption standards; create new legal and regulatory frameworks; encourage new forms of production, marketing and consumption while prohibiting others; facilitate the creation of spaces for cooperation; promote and support local initiatives; address conflicts of interest, and more.

This book, proposed by the Public Policy and Rural Development in Latin America (PP-AL), contributes to this global reflection. The original chapters compiled here analyze food policies in Latin America, a heterogeneous continent where both food issues and innovative initiatives to address them abound.

With chapters covering the situation in ten Latin American countries, this book presents a broad overview of the region's food policy trajectories, illustrating the development of policies over the last two decades to provide access to food for the poorest and most vulnerable populations. These food and nutrition security policies and programs have integrated the health dimension in the last decade. However, they have only marginally integrated the environmental one necessary to establish sustainable food systems. Moreover, recent political changes in certain countries have affected the implementation of these programs, leaving vulnerable populations in a critical situation.

This book also takes an original look at emerging initiatives from urban and rural territories, through which the many different situations can be explored by substituting or complementing national public actions. Certain local initiatives and dynamics presented in this book reflect initial efforts to reconfigure local alliances between producers and consumers and the emergence of new actors working to promote alternative food culture and values. They also show the possible development of integrated policies that seek to address the various economic, social, health and environmental challenges in a coordinated manner, and to better ensure coherent sector-specific policies.

Additionally, this book explores the processes by which instruments and regulations are created and local initiatives emerge. The analysis reveals that the social movements of family farming and agroecology have played a key role in promoting inclusive regulations and environmentally friendly local food systems.

However, although promising, the initiatives presented in this book are recent and relevant to a specific moment or place. This book shows that innumerable constraints persist in fostering sustainable food systems in the region. The first constraint results from agrarian dynamics, competition between different models of agricultural development and centuries-old inequality in the way land and natural resources are distributed. As a result, the food situation continues to deteriorate very rapidly for certain rural population groups, particularly traditional communities. The second constraint highlighted in this book stems from the difficulties of coordinating different actors and levels of government to build integrated policies for sustainable food systems. In fact, although there has been a shift towards more integrative *référentiels* for a new generation of policies, there are still no integrated national food policies that include all the dimensions required to develop sustainable food systems and provide the necessary framework for action for the local production of public goods. To create such policies, policymakers would need to address intersectoral challenges, which inevitably emerge with multifaceted issues such as sustainable food systems. There would also need to be greater dialogue and integration between rural social movements and consumer movements and urban civil society organizations, on the one hand, and environmental movements, on the other, to advocate for sustainable and equitable food system models. This book does not propose a normative vision of sustainable food systems, but instead presents historical and recent policy trajectories along with their current limitations, and provides an overview of innovative perspectives. Through these contributions, it offers food for thought for the scientific community, as well as for public decision-makers and social movements, to consider and promote policies for sustainable food systems. It also offers suggestions and perspectives from which to tackle an important challenge for the design of our future: coordinating local initiatives, regional and national public policies and international frameworks for action. In a context of increasing political volatility and re-politicization of food issues, this book offers an important contribution to defining the political strategies that must now be implemented to navigate towards a sustainable world and mobilizing the lessons learned, alliances and power relations across different scales of action.

Patrick Caron
CIRAD,
Vice-President, University of Montpellier

Introduction

Jean-François Le Coq, Paulo Niederle,
Catia Grisa, Stéphane Guéneau

This book presents the results of a collective research effort coordinated by the Public Policy and Rural Development in Latin America network (PP-AL), an inter-institutional platform created in 2011 to enable regional academic and policy dialogue on the design and implementation of rural development policies. Motivated by the results of previous projects,[1] in September 2018 the network's coordination committee proposed a new research project on "Public policies and food security for sustainable food systems." Between March 2019 and March 2021, researchers from ten countries worked on this project seeking answers to a set of guiding questions posed by the coordinators of the research and this publication. In this introduction, we return to these questions and briefly describe the project's path to the publication of the chapters in this book.

We begin with the concept of food and nutritional security. This concept, used in Latin America since the 1990s, sums up a movement whereby the key food-related problems shifted from being focused on eradicating hunger (Chonchol, 1987) to return to the broader issue of the human right to healthy food, which incorporates the discussion on access to food in adequate quantity and quality, but also elements related to citizenship and environmental protection (FAO et al., 2011). In other words, the concept of food and nutritional security acquired new dimensions that were integrated into public policies in several Latin American countries in the mid-2000s (Almeida Filho & Ramos, 2010).

The discussions that underpinned the construction of food and nutritional security policies emphasized that the problems of undernutrition or malnutrition are not due to lack of food so much as to problems of access to safe and nutritious food (Maluf, Burlandy & Alexandre, 2020). This is one of the reasons why some Latin American countries have implemented policies to better distribute food or subsidize consumer prices.

---

1. See: Sabourin, E., Samper, M., Sotomayor, O. (Eds.). (2015). *Políticas públicas y agriculturas familiares en América Latina y el Caribe: nuevas perspectivas*. San José: IICA; Sabourin, E., Patrouilleau, M., Le Coq, J.F., Vasquez, L., Niederle, P. (Eds.). (2017). *Políticas públicas a favor de la agroecología en América Latina y El Caribe*. Porto Alegre: Red PP-AL; Goulet, F., Le Coq, J-F., Sotomayor, O. (Eds.). (2019). *Sistemas y políticas de innovación para el sector agropecuario en América Latina*. Rio de Janeiro: E-papers.

It was from this perspective that was developed, for example, the Brazilian Zero Hunger program, launched in 2003 to eradicate hunger and extreme poverty in Brazil (Graziano da Silva, Del Grossi & França, 2010), or the National Food Program (PNA)[2] in Nicaragua, to increase the supply of staple foods and improve access to healthy food (Freguin-Guesh & Cortes, 2018). In addition, countries such as Brazil, Colombia, Uruguay and Ecuador have also implemented policies for public procurement of food for redistribution in school canteens and public institutions for vulnerable people (Clark, 2016; Wittman & Blesh, 2017; Grisa, Perafán & Calderón, 2018; Schneider & Bohórquez, 2019). The common feature of these policies is that they aim to combat hunger and malnutrition not only through mechanisms that give vulnerable populations more affordable access to food, but also by stimulating the provision of affordable and quality food products.

Despite the spread of these policies in several Latin American countries (Sabourin & Grisa, 2018; Lopes Filho, 2018; Caldas & Ávila, 2018; Grisa & Niederle, 2019), recent data show that the problems of hunger and malnutrition have reappeared with a high level of prevalence (FAO et al., 2020, 2018). First, economic and political crises have led to a sharp rise in prices, which has directly affected net food importing countries as well as exporters. Today, even the richest countries of the continent, where food and nutrition insecurity was considered low by the FAO, are in a very difficult situation. This is the case, for example, in Brazil, where malnutrition once again became a problem for more than 5.2 million people in 2017 (FAO et al., 2018). By the end of 2020, already under the effects of the pandemic, 19 million Brazilians (9% of the population) were hungry (severe food insecurity) and another 43.4 million (20.5%) did not have enough food (moderate or severe food insecurity) (Rede Penssan, 2021).

In addition to the problems of famine and malnutrition, food safety issues have received increasing attention in Latin America. This is a consequence of a number of developments such as urban population growth, the restructuring of major food supply chains, the growing influence of supermarkets, the increase in food consumption outside the home, and the rise of fast food (Popkin & Reardon, 2018; Schubert, Schneider & Dias-Mendez, 2017). The biophysical-chemical and nutritional quality of food and consumer information on health risks became one of the central objectives of FSN policies because of repeated food scandals and crises (Schubert & Ávalos, 2020).

Large-scale food production has benefited from changes in food safety standards because, until now, these have focused on a hygienic and statistical view of contamination risks. This evaluation system, which has tended to standardize food according to essentially sanitary criteria, has effectively excluded a large part of non-industrialized products derived from family farming from sustainable food policies (Cintrão, 2017). Moreover, due to the adoption of food safety regulations, the supply of ultra-processed foods has strengthened the dual trend towards food industry concentration and food standardization.

Another effect is that people in urban and even rural areas of Latin America have experienced a rapid transformation of their diets. This shift, together with a decrease

---

2. These acronyms are in Spanish or Portuguese (see the list at the end of this book).

in physical activity, has led to a rapid increase in obesity in several countries where these problems were already alarming, such as Mexico, or have increased sharply since the 2000s after a decline in the 1990s, as in Brazil (Rivera et al., 2004; Monteiro et al., 2019; Popkin & Reardon, 2018; Fernandéz et al., 2017). Today, more than half of women and more than 20% of children are overweight or obese (Corvalán et al., 2017).[3] In some countries such as Chile and Mexico, these indicators are true for two-thirds of women and more than half of men (Popkin and Reardon, 2018). Of course, poor nutritional quality is also responsible for the high rates of diabetes, hypertension and other diseases that increasingly affect Latin American and Caribbean populations (Anauati, Galiani & Weinschelbaum, 2015).

Faced with this scenario, public action needs to address a dual food and nutrition problem, resulting from the coexistence of malnutrition and obesity in the same environment and often in the same households. In this sense, new dynamics of action, both public and private, are developing that contradict the food and nutritional security framework, which focused only on food safety (Triches, Gerhardt & Schneider, 2015). This new approach is especially evident in movements that support agroecology, sustainable food and local food supply chains, with a particular focus on public action to strengthen the contribution of new forms of small-scale agricultural production (agroecology, urban and peri-urban agriculture, family farming) to ensuring the food and nutrition security of the poorest households (Cabanãs, Nigh & Pouzenc, 2020; Preiss & Schneider, 2020; Portilho, 2020; Moura, Souza & Canavesi, 2016; Guarín, 2013).

Moreover, in urban areas, where more and more consumers are looking for healthier but also more authentic products with regard to certain regions, food issues go beyond the strict scope of nutrition. They increasingly involve cultural and social issues related to the social inclusion of the most disadvantaged producers and efforts to save certain types of food or recipes from rural or traditional communities (Cabanãs, Nigh & Pouzenc, 2020; Guéneau et al., 2017; Niederle & Wesz Jr., 2020). These issues also increasingly relate to environmental issues, given the desire of a swath of consumers to minimize the impact of production processes (Portilho, 2020). Moreover, in some Latin American cities, the growing demand for local food chains has favored the development of alternative food networks, resulting in the development of urban agriculture and the participation of consumers in food production, as part of the growing number of "agricultural support communities" (Gianella & Pinzás, 2019).

Food is thus a social symbol that expresses much more complex ways of being. Contemporary consumers increasingly make food choices as a political act (Paredes et al., 2020; Portilho, 2020), leading to a growing demand not only for a more diversified food supply, but also for new production, distribution and consumption practices. The shortening of supply chains, the emergence of "co-farming" consumers, the multiplication of restaurants offering dishes prepared from unconventional food plants or products derived from "socio-biodiversity," are some examples of the innovative dynamics underway in major cities (Duarte et al., 2021; Guéneau et al.,

---

3. According to data on obesity in Latin America, the percentage of children under 5 years of age who are overweight increased from 49.8% in 2000 to 59.6% in 2018, and the percentage of children aged 5-19 years who are overweight increased from 21.6% to 30.6% (FAO et al., 2019).

in chapter 2 of this book). Most of these innovations are based on the belief, real or not, that they contribute to sustainable development, such as the idea that local food production is ecological (Moustier, 2017).

This dynamic of transition to alternative food systems is part of a context of rapidly changing consumption practices, particularly in urban areas, especially in larger urban areas. The proportion of the population living in these areas is rising considerably in Latin America. In major Brazilian cities, consumers' food budgets are increasingly allocated to meals outside the home, where consumption patterns are becoming more diverse (fast food, food trucks, "kilo" restaurants, etc.) (Zaneti, 2017; Schubert, Schneider & Dias-Mendez, 2017). Moreover, new actors such as chefs are increasingly present in the media and on television programs dedicated to gastronomy, which are becoming increasingly popular. Latin American chefs such as the Peruvian Gastón Acurio and the Brazilian Alex Atala are playing a vital role as activist entrepreneurs to put the issue of food sustainability on the political agenda (Zaneti, 2017).

While these various factors do not yet seem to have led to a fundamental change in food systems, they do lead us to imagine a shift in the perspective of public action on food challenges. While policies have thus far been developed from a top-down, productive, rural-to-urban point of view, various groups of actors in urban settings have suggested adopting a new perspective: a bottom-up, consumption-driven, urban-to-rural approach that involves consumers and places the issue of food systems on the political agenda. Although some research focuses on the many innovative initiatives and new local policies aimed at transforming the food system and enabling consumers to regain control of their food (Paredes et al., 2020; Blay-Palmer, 2016; Preiss, Charão-Marques & Wiskerke, 2017), research has not yet captured the full importance of the ongoing dynamics in cities for sustainable food policymaking. For this reason, our research began with the attempt to shed light on the following questions:
– How do public policies take into account the new challenges of food security and the promotion of sustainable food systems? What are the initiatives aimed at designing public policies for sustainable food in rural and urban areas?
– What are the reconfigurations of stakeholder coalitions around the issue of sustainable food systems? How does public action renew the notion of sustainable food system?

How are food policies linked with agricultural, rural development, health, education, environmental and urban policies to address the food challenges of urban and rural populations?

To what extent can food policies constitute new vectors for changing agricultural production models towards greater sustainability and a profound transformation of the rural world?

The greatest difficulty of the project was that the research was conducted during a highly unstable period. In addition to the Covid-19 health crisis, which severely aggravated food insecurity conditions and made it exponentially more difficult for states to address food problems, there were strong political and economic changes throughout Latin America and the Caribbean over the two years of the project.

In 2018, when we began our research, the Latin American political scene pointed to a rise of conservativism, which had already set in motion an accelerated process of dismantling public policies for rural development, food security and family farming support (Sabourin et al., 2020). And it was precisely this reality that many researchers encountered during their work. However, in 2021, some countries were beginning to show signs that the conservative wave might not last as long as initially projected, with governments in countries such as Argentina and Bolivia once again under the leadership of progressive political coalitions. In short, over the last two years, Latin America has become a much more complex political mosaic.

In terms of the continent's political trajectory, the most direct effects of the uncertainties are reflected in the agenda and dynamics of the multilateral blocs. In the first two decades of the 2000s, these blocs were instrumental in the regional dissemination of food security policies (Sabourin & Grisa, 2019; Grisa & Niederle, 2019). MERCOSUR, and in particular its Specialized Meeting on Family Farming (REAF), was hollowed out by the Brazilian government of Jair Bolsonaro, who considers the bloc an obstacle to direct bilateral relations and market liberalization. Furthermore, because Bolsonaro's attempts to end the bloc come at a political cost and have encountered resistance from other countries, the Brazilian government has demanded the adoption of an agenda that seeks to open markets and trade agreements allowing agricultural product exports. Issues such as social participation, democracy and regional integration were withdrawn from the discussions, while the many forums for dialogue on developing food security policies were discouraged or terminated.

The situation of the Community of Latin American and Caribbean States (CELAC) and the Union of South American Nations (UNASUR) is even more dramatic. For the CELAC, the entity became the scene of strong political conflicts between countries with very different ideological stances. Once again, the Brazilian president Jair Bolsonaro took the lead and, in January 2020, suspended the country's participation in CELAC, stating that he could not communicate with governments that impose non-democratic regimes. Given Bolsonaro's own authoritarian bias, such a move was simply a rejection of left-leaning governments. UNASUR was dismantled after a series of withdrawals. At the beginning of 2018, Colombia, Argentina, Brazil, Chile, Paraguay and Peru announced they would be temporary stopping their involvement. However, by the end of 2018, Colombia withdrew definitively. In 2019, Ecuador and Bolivia followed suit, and in 2020, Uruguay also dropped out.[4]

These conflicts, compounded by the effects of Covid-19, have severely compromised the food security agenda. As a result, the social scourges so familiar to the Latin American population – poverty, hunger, inequality, authoritarianism – have made a comeback. Until the early 2010s Latin America was recognized as one of the regions that had made the most rapid progress in addressing these problems. However, in

---

4. To replace these entities, in 2017, foreign ministers from Argentina, Brazil, Canada, Chile, Colombia, Costa Rica, Guatemala, Honduras, Mexico, Panama, Paraguay and Peru created the Lima Group, stating that "We urge the international community to work together to support Venezuelans in finding a peaceful solution that urgently addresses the serious crisis they face and leads to the reestablishment of the rule of law and the constitutional and democratic order in Venezuela." However, in 2021, in a sign of support for the Venezuelan government, Argentina withdrew from the group.

recent years it has been pinpointed as one of areas facing the greatest setbacks. In 2019, nearly one-third of the world's moderate or severe food insecurity occurred in Latin America and the Caribbean (preceded only by Africa, where more than half of the world's population lives). If recent trends continue, hunger will continue to rise in Latin America and the Caribbean until 2030, taking this region farther away from the Sustainable Development Goals (SDGs) (FAO et al., 2020).

From an economic viewpoint, the scenario is just as complex. In the last two years, agricultural commodity prices have soared in international markets, contributing to rising exports from several Latin American countries (Argentina, Brazil, Chile, among others). Meanwhile, despite a loss of purchasing power among the population (resulting in reduced domestic demand), growing international demand, the national currency devaluation crisis (against the dollar) and a lack of market regulation have led to increasing domestic food prices, thereby aggravating food crises, especially for the most vulnerable urban and rural populations. These processes also impact the accelerated expansion of neo-extractivist practices and related social processes, such as agricultural land grabs, deforestation and violence in the countryside (Sosa Varroti & Gras, 2020; Sauer, 2018).

Rising food prices, the unemployment crisis, and falling incomes for the poorest population (net food buyers) have increased purchases of industrialized and ultra-processed foods. Such foods are cheaper and easily accessible because they are widely distributed through large supermarket chains. Thus, despite the progress made in the region with regard to discussions on producing and consuming healthy foods, much of the population has access to them only with the support of public policies. Unfortunately, state fiscal crises and the rise of governments aligned with the interests of the food industry have weakened such public policies.

These difficulties also matter when discussing food and nutritional security policies aiming to support the development of sustainable food systems. Rising agricultural commodity prices alone are already a factor driving the monoculture expansion, pesticide use and biodiversity losses. However, some Latin American governments have also spoken out against the global movement to curb the effects of climate change and attempts to build more sustainable production, distribution and consumption models. Thus, rather than working to strengthen policies that address these concerns, civil society, certain segments of academia and the media in many countries have been unable to do much more than pressure governments to reduce the environmental damage caused by agribusiness exports (Sabourin et al., 2020).

It was in this climate of controversy and abrupt economic and political changes, aggravated by the pandemic, that we produced the analyses presented in this book. Before making them public, we held seminars and debates among our team. The aim was to produce a coordinated interpretation that could explain the hegemonic food dynamics on the continent. In the end, as expected, what emerges from the compiled chapters is that the political and economic kaleidoscope of Latin America reflects a variety of food systems and policies. This variety accounts for the different ideological stances that have prevailed in each country – in governments as well as in interest groups, social organizations and academia – and the different dynamics of how food is approached a political problem in each context, referred to as the "politicization of food" (Grisa et al., chapter 1).

This book is divided into five parts. The first comprises three articles that present a cross-cutting look at regional dynamics, focusing on the diversity of reference points that guide food policies in Latin America (chapter 1), an analysis of foresight studies on food policies that were produced in the region (chapter 2), and a review of the role of international organizations in the dissemination of those policies (chapter 3). The second part presents a historical overview of national food policies based on the experience of four countries: Nicaragua (chapter 4), Paraguay (chapter 5), Mexico (chapter 6), Bolivia (chapter 7) and Chile (chapter 8). The third part then looks at recent changes in national food policies, examining the cases of Argentina (chapter 9), Peru (chapter 10) and Paraguay (chapter 11). The fourth part explores innovative experiences at the local level, analyzing policies developed in the cities of Brasilia (chapter 12), Cali (chapter 13), San Ramón (chapter 14) and Antioquia (chapter 15). Finally, the last part of the book discusses the challenges related to developing specific policies: the food procurement programs of Brazilian state governments (chapter 16), food safety regulations in the Argentine context (chapter 17), public procurement in Uruguay (chapter 18) and the intersection between food policies and the food cultures of the indigenous peoples of Paraguay (chapter 19).

In chapter 1, Catia Grisa, Paulo Niederle, Stéphane Guéneau, Jean-François Le Coq, Clara Craviotti, Graciela Borrás, Daniel Campos RD, Héctor Ávila-Sánchez, Sandrine Freguin-Gresh, Junior Miranda Scheuer and Jorge Albarracin review the evolution of benchmarks used in Latin America to address food challenges. Based on an analysis of policies in the countries of the region, they identify eight main *référentiels*, which promote different values and norms and are reflected in different policy instruments. Chapter 1 also shows that policies have evolved from a "food-for-market" *référentiel* to more "integrated" *référentiels* that seek to achieve adequate, healthy, sustainable and responsible consumption through a broad set of policy tools.

In chapter 2, Maria Mercedes Patrouilleau, Diego Taraborrelli and Ignacio Alonso analyze foresight studies carried out in the Latin American region. This chapter shows a strong increase in foresight studies since the 2000s, and a revaluation of food issues in recent years.

In chapter 3, Fernanda França de Vasconcellos analyzes how the narratives and *référentiels* used by international agencies were translated into food and nutritional security policy recommendations in Latin America and the Caribbean. Based on a literature review focusing on four organizations (Food and Agricultural Organization – FAO, World Food Programme – WFP, Inter-American Institute for Cooperation on Agriculture – IICA and International Fund for Agricultural Development – IFAD), the author shows consistent coordination of public policy reference points to prevent conflicting proposals, even when different points coexist in the narratives. It is also evident that the narratives have become increasingly catastrophic, reflected in the rising numbers of hungry people and increasing rates of obesity, overweight and diseases caused by malnutrition. Policy proposals have become more inclusive, in line with the premises of the international agenda comprising the SDGs.

The second part of this book presents a historical overview of national food policies. In the case of Nicaragua, Sandrine Freguin-Gresh and Geneviève Cortes discuss a critical period in the late 2000s that led to a gradual institutional change of policies. From this point, policies began prioritizing family farming and the fight against poverty and hunger, which resulted in the creation of two flagship programs, "Zero Hunger" and "School Meals." However, despite the progress made, these programs have encountered significant limits in terms of governance, particularly when it comes to selecting beneficiaries, which calls into question the universality of these programs.

In the case of Paraguay, Daniel Campos and María Benavidez conclude that the development of a food and nutritional security system that fully guarantees the right to food is still lagging far behind. Several laws and policies do support family farming, and various programs attempt to address food security as a social protection issue. However, no profound changes have been undertaken that could promote strategies for a national policy on food security and sovereignty, and there is no coordinated participation among social, private and public sectors, including vulnerable groups, most of whom are from the rural and peri-urban sector.

In the case of Mexico, Héctor Ávila-Sánchez discusses the strong state interventionism in agrarian issues throughout history. The author highlights a recent shift in policies under the current government, which takes up the previous struggles and social movements, and asserts the role of small and medium-sized producers. This shift suggests an intention to link rural programs and policies to food sovereignty. While presenting a broad set of policies and programs to promote food and nutritional security, the author concludes by questioning the real capacity to implement these policies focused on small and medium-scale producers in an environment dominated by the conventional production of large companies and the monopolistic expansion of transnational agribusinesses.

In the case of Bolivia, Jorge Albarracin highlights a dual and conflictive agrarian history between agroindustry and agribusiness on the one hand, and family and indigenous agriculture on the other. Bolivia has maintained and reinforced a food system focused on the production of products and food that follows the structure of the agroindustrial and agribusiness model, positioning itself as a supplier of raw materials to the world. Meanwhile, the more traditional production systems suffer from deficient public investment in both production support and the development of local markets, which both maintains and aggravates the problems of malnutrition in the Andean and valley areas.

In Chile, Michel Leporati Néron and Pablo Villalobos Mateluna propose an institutional, political and economic analysis of the public policies that have shaped the Chilean food system from its origins to the present day. Initially, the authors discuss changes in social protection policies, healthy eating and modernization of the food safety and quality management system. They then provide a historical overview of shifts in agricultural development and promotion policies. Based on these historical components, they close out the chapter by discussing the performance of the Chilean food system with a forward-looking view of present and future challenges.

The third part of this book explores recent changes in food policies. In Argentina, Cecilia Aranguren, Ana María Costa, Susana Brieva and Graciela Borrás analyze the agendas and processes of social construction of food policies since 2001. The authors show that, since the crisis of the early 2000s, food issues have become a key public concern within the country's socioeconomic policy agenda. After a detailed analysis of two programs, the National Food Security Plan (PNSA) and the National Plan for a Healthy Argentina (PNAS), they conclude that there was little citizen participation in the development of these programs, and call for a greater alliance between the public sector and civil society to create sustainable and healthy food systems that integrate the interests of the actors involved in all stages of the food chains.

In Peru, Carolina Trivelli and Carlos Urrutia analyze the evolution of food programs and changes in consumption patterns from 2004 to 2018. The authors show that the main public policies on food security are deeply disjointed with regard to their specific objectives. On one side are food policies with a productive approach that focus on access to food, while on the other, social policies target health problems. In addition, the authors highlight the ways these programs changed from 2004 to 2019, marked by inconsistent public intervention. Finally, they show that shifts in consumption were driven by the sustained increase in the cost of the food basket over the last few years. For this reason, the authors call for greater consideration of food quality in studies on these food security programs.

In Paraguay, Silvia Zimmermann and Noelia Riquelme analyze the policy-politics-polity dimensions of food policies. National public policies aimed at combating hunger and malnutrition began in the 2000s, with the creation of three national programs: Tekoporã, the Comprehensive Nutritional Food Program (PANI) and Abrazo. The policies then evolved along with the successive governments, namely with the creation of the National Plan for Food and Nutrition Sovereignty and Security (PLANAL). This plan combined different actions and was an important attempt at institutional coordination between 2009 and 2012. The policy changes are strongly linked to changes in government, with a recent tendency to treat the food problem as an issue of social assistance and protection, rather than questioning the economic growth model and its implications for food security and sovereignty.

The fourth part of the book analyzes innovative food policy initiatives and actions in place in different regions and cities. Stéphane Guéneau, Mauro Capelari, Janaína Deane de Abreu Sá Diniz, Jessica Pereira Garcia and Tainá Bacellar Zaneti investigate how the food issue is politicized in the city of Brazilia (Brazil). The authors show how local initiatives and policies are developed in response to a desire for change expressed by part of the urban population (especially those with higher education and income). They also discuss how such actions tie in with the desire for access to healthy food, while considering how to improve environmental and social conditions, enhance waste management and reduce food waste. The authors show how these actions have spurred local actors to become more involved as they attempt to tackle the challenge of regaining control of public problems managed at higher national and international scales, while battling strong interference from powerful private actors in the agrifood sector. The authors also highlight how the shift towards more sustainable food systems was initially initiated by actors from the alternative agricultural world, and that including such systems in policies is largely dependent on the action of a coalition focused on agroecological transition.

Ruby Castellano, Guy Henry and Sara Rankin analyze the construction of the urban food policy of Cali (Colombia). Based on the concept of sustainable food systems, the authors present the dynamics of the construction of the Municipal Public Policy

Proposal for Food and Nutritional Security and Sovereignty (PSSAN), which was created through collaboration between the municipal administration, research centers and universities, and local civil society actors as part of a food security roundtable. The authors highlight the dimension of trust between actors that was developed throughout the process as a key factor in the creation of this plan.

Jairo Rojas Meza, Pedro Pablo Benavídez and Francisco Chavarría Aráuz analyze the transformations of food systems in the rural municipality of San Ramón (Nicaragua). The authors show that the current food system in San Ramón is the result of a historical process of agricultural public policies and relations with the national and international market. The access of landless rural families to food products depends on the income generated, which is uncertain. The results of public policies promoted in programs seeking to capitalize peasant farms, such as the Food Production Support program (also known as Zero Hunger), have been questionable. The agroecological transformation of peasant farms has been mainly been achieved through efforts by trade and social organizations. Finally, the authors conclude that there is still insufficient coherence between national and local policies, coordination mechanisms and intersectoral cooperation.

José Anibal Quintero Hernández analyzes the evolution of and changes in food and nutritional security public policy in the Department of Antioquia (Colombia). The author outlines the process by which the pioneering, department-level Food and Nutrition Improvement Program of Antioquia (MANA), which began in 2001, was created and implemented. This food security policy aims to help people facing the greatest difficulties, especially children under 14, and has been based on public procurement programs for school meals. The authors describe the five successive stages that this policy has known, and emphasize that it has made great strides in bringing together large sectors, academics, politicians and economists around the issue of food security. Finally, the authors conclude that, while this policy has made it possible to improve food and nutritional security among children in the department, the problem has not been entirely solved. Doing so will require greater participation of social organizations.

The last part of the book provides a more focused analysis of specific food policy tools and innovations. First, Catia Grisa, Mario Avila and Rafael Cabral analyze the politicization of the 12 public food procurement programs for family farming of state governments in Brazil. The authors discuss the various issues that contributed to setting the public agenda and politicizing public food procurement, including the innovations and results of the Food Acquisition Program (PAA) and National School Feeding Program (PNAE), the financial and political crisis of the PAA (2013 onwards), the dismantling of public policies for family farming and food and nutrition security, as well as national and state political changes and the Covid-19 pandemic. Because the federal government has done little to address food challenges, various social actors have picked up the slack to strengthen the food policy agenda by promoting food activism through social movements and family farming unions, bringing their representatives to lobby the state government, and engaging in institutional activism by mid- and street-level bureaucrats. The authors highlight the predominance of family farming organizations among the actors involved in developing state programs. While these organizations pressure, interact and work with the bureaucracy and government actors, there has not yet been involvement from consumers, consumer organizations, other professional categories or public management areas in modifying food decrees in force at the level of the different Brazilian states.

Clara Craviotti looks at the process of regulations that support the production of food by family farming in Argentina, which began in the late 2000s. Based on the premise that health regulations are not only technical but also political objects, the author analyzes the formulation and incorporation of these regulations in the Argentine Food Code. She highlights a series of stages in the formulation process that address the inclusion of family farming in the institutional agenda when health standards are updated. The author discusses the role of a pro-family farming coalition in a negotiation process in which disputes were resolved and which led to the inclusion of a specific article in the Argentine Food Code. She also covers the various translation activities the pro-family farming coalition was required to undertake to negotiate this institutional innovation.

Junior Miranda Scheuer explores the public procurement programs for family producers and fishermen in Uruguay. The author shows that Uruguay's Public Procurement Act (LCP) emerged from regional debates promoted by the Specialized Meeting on Family Farming (REAF), which were then adapted by the Uruguayan state. Unlike in Brazil, where state programs to purchase foods produced by family farming have a strong rural development and sustainable food systems bias, the LCP plays a major role in the procurement of food products by the state. However, the author identifies several limitations of this law: it did not explicitly indicate the problems it intends to address; although the target audience is defined, the law allows non-family producers to access public procurement; it does not encourage sustainable (organic or agroecological) food production; nor does it provide a mechanism to differentiate the value of family producers working in a sustainable manner. In view of these shortcomings, the author stresses the importance of linking this law with other policies to promote a policy mix that generates more comprehensive solutions to strengthen sustainable food systems based on family farming.

Finally, Silvia Zimmermann, Diana Cohene and Noelia Riquelme investigate the food issue of indigenous peoples and public policies in Paraguay. The authors' detailed analysis looks at the reality of production, harvesting and consumption of indigenous peoples, their agrifood system, and the national public policy initiatives of food sovereignty and security that attempt to guarantee the *tekoporã* (which means "good life" in the Guarani language) of indigenous peoples. The authors highlight a very recent recognition of indigenous peoples in Paraguay's policies. Indigenous peoples are facing an increasingly degraded food situation. This problem is a result of structural problems, and especially the expansion of agribusinesses (mainly soybean production and cattle ranching), which are encroaching on their territories. The Census of Indigenous Peoples was fundamental for their visibility, and the recent construction of the National Program for Indigenous Peoples (PNPI), thanks to the mobilization of civil society, has been a significant step forward. However, the reduction of the investment funds of the Paraguayan Indigenous Institute (INDI) jeopardizes these initial advances. The authors conclude that not only is there a strong need to guarantee PNPI gets implemented, but also to rethink Paraguay's rural development, the advance of agribusiness, as well as its forms of production and exploitation of nature, in order to solve the food problems of the country's indigenous populations.

⏩ References

Almeida Filho, N., & Ramos, P. (Eds.). (2010). *Segurança alimentar, produção agrícola e desenvolvimento territorial*. Alínea.

Anauati, M. V., Galiani, S., & Weinschelbaum F. (2015). The rise of noncommunicable diseases in Latin America and the Caribbean: challenges for public health policies. *Latin American Economic Review, 24*(11). https://doi.org/10.1007/s40503-015-0025-7

Blay-Palmer, A., Sonnino, R., & Custot, J. (2016). A food politics of the possible? Growing sustainable food systems through networks of knowledge. *Agriculture and Human Values, 33*(1), 27–43. https://doi.org/10.1007/s10460-015-9592-0

Cabañas, A. A. G., Nigh, R., & Pouzenc, M. (Eds.). (2020). *La comida de aquí: retos y realidades de los circuitos cortos de comercialización*. Universidad Nacional Autónoma de México.

Caldas, E. L., & Ávila, M. L. (2018). Compras públicas e alimentação escolar no Paraguai: a disseminação da experiência brasileira e a adaptação do modelo. In Sabourin, E., & C. Grisa (Eds.), *A difusão de políticas brasileiras para a agricultura familiar na América Latina e Caribe* (pp. 171–188). Escritos.

Chonchol, J. (1987). O desafio alimentar: a fome no mundo. Marco Zero.

Cintrão, R. (2017). Segurança alimentar, riscos, escalas de produção – Desafios para a regulação sanitária. [Food security, risks, scales of production – challenges to sanitary regulation]. *Vigilância Sanitária em Debate, 5*(3), 3–13. https://doi.org/10.22239/2317-269x.00971

Clark, P. (2016). Can the State Foster Food Sovereignty? Insights from the Case of Ecuador. *Journal of Agrarian Change, 16*(2), 183–205. https://doi.org/10.1111/joac.12094

Corvalán, C., Garmendia, M. L., Jones-Smith, J., Lutter, C. K., Miranda, J. J., Pedraza, L. S., Popkin, B. M., Ramírez-Zea, M., Salvo, D., & Stein, A. D. (2017). Nutrition status of children in Latin America. *Obesity Reviews, 18*(S2), 7–18. https://doi.org/10.1111/obr.12571

Duarte, L. M. G., Guéneau, S., Diniz J. D. A. S., & Passos, C. J. S. (2021). Valorización de los patrimonios alimentarios y productos agroextractivistas del Cerrado brasileño en la gastronomía: un estudio sobre el Festival Gastronómico Cerrado Week. In Rebaï, N., Bilhaut, A-G., De Suremain, C-E., Katz, E., & M. Paredes (Eds.), *Patrimonios alimentarios en América Latina: recursos locales, actores y globalización* (pp. 109–136). IFEA/IRD.

FAO, IFAD, UNICEF, WFP, & WHO. (2018). *El estado de la seguridad alimentaria y la nutrición en el mundo 2018. Fomentando la resiliencia climática en aras de la seguridad alimentaria y la nutrición* [The state of food security and nutrition in the world 2018. Building climate resilience for food security and nutrition]. FAO. https://www.fao.org/3/I9553ES/i9553es.pdf

FAO, IFAD, UNICEF, WFP, & WHO. (2019). *El estado de la seguridad alimentaria y la nutrición en el mundo 2019. Protegerse frente a la desaceleración y el debilitamiento de la economía* [The state of food security and nutrition in the world 2019. Safeguarding against economic slowdowns and downturns]. FAO. https://www.fao.org/3/ca5162es/ca5162es.pdf

FAO, IFAD, UNICEF, WFP, & WHO. (2020). *El estado de la seguridad alimentaria y la nutrición en el mundo 2020. Transformación de los sistemas alimentarios para que promuevan dietas asequibles y saludables* [The state of food security and nutrition in the world 2020. Transforming food systems for affordable healthy diets]. FAO. https://www.fao.org/3/ca9692es/online/ca9692es.html

FAO, IFAD, & WFP. (2011). *El estado de la inseguridad alimentaria en el mundo 2011. ¿Cómo afecta la volatilidad de los precios internacionales a las economías nacionales y la seguridad alimentaria?* [The state of food insecurity in the world 2011. How does international price volatility affect domestic economies and food insecurity?]. FAO. https://www.fao.org/3/i2330s/i2330s.pdf

Fernández, A., Martinez, R., Carrasco, I., & Palma, A. (2017). *Impacto social y económico de la doble carga de la malnutrición: modelo de análisis y estudio piloto en Chile, el Ecuador y México.* CEPAL. https://www.cepal.org/sites/default/files/publication/files/42535/S1700443_es.pdf

Freguin-Guesh, S., & Cortès, G. (2018, June 20–21). *Politiques publiques et sécurité alimentaire au Nicaragua : Trajectoires socio-historiques et défis actuels* [Conference presentation]. SFER Symposium, Montpellier SupAgro, France. https://www.sfer.asso.fr/source/Coll-trajectoire-2018/articles/B31_FreguinGresh.pdf

Gianella, T., & Pinzás, T. (Eds.) (2019). Agricultura urbana en América Latina [Special issue]. *LEISA*, *35*(3). https://www.leisa-al.org/web/images/stories/revistapdf/vol35n3.pdf

Graziano Da Silva, J., Del Grossi, M. E., & De França, C. G. (Eds.) (2010). *Fome Zero: a experiência brasileira.* MDA. https://www.fao.org/3/i3023o/i3023o.pdf

Grisa, C., Valencia Perafán, M. E., & Giraldo Calderón, P. E. (2018). Transferência e tradução de políticas públicas do Brasil para a Colômbia: o caso das compras públicas da agricultura familiar. *Estudos Sociedad & Agricultura*, *26*(2) 353–375. https://doi.org/10.36920/esa-v26n2

Grisa, C., & Niederle, P. (2019). Transferência, convergência e tradução de políticas públicas: a experiência da Reunião Especializada sobre Agricultura Familiar do Mercosul. *Dados*, *62*(2), 1–37. https://doi.org/10.1590/001152582019175

Guarín, A. (2013). The value of domestic supply chains: producers, wholesalers, and urban consumers in Colombia. *Development Policy Review*, *31*(5), 511–530. https://doi.org/10.1111/dpr.12023

Guéneau, S., Diniz, J. D. D. A. S., Mendonça, S. D., & Garcia, J. P. (2017). Construção social dos mercados de frutos do Cerrado: entre sociobiodiversidade e alta gastronomia. *Século XXI-Revista de Ciências Sociais*, *7*(1), 130–156. https://doi.org/10.5902/2236672528133

Lopes Filho, M. A. (2018). Compras locais no Haiti: disseminação e interpretação em âmbito nacional dos modelos brasileiros. In Sabourin, E., & C. Grisa (Eds.), *A difusão de políticas brasileiras para a agricultura familiar na América Latina e Caribe* (pp. 141–170). Escritos.

Maluf, R. S., Burlandy, L., & Alexandre, V. P. (2020). Pesquisas em soberania e segurança alimentar e nutricional no Brasil: enfoques e conexões com as políticas públicas. In Preiss, P. V., Schneider, S., & G. Coelho De Souza (Eds.), *A contribuição brasileira à segurança alimentar e nutricional sustentável* (pp. 137–152). UFRGS. https://www.ufrgs.br/agrifood/images/2020/07-julho/001115755.pdf

Monteiro, C. A., Cannon, G., Levy, R. B., Moubarac, J-C., Louzada, M. L., Rauber, F., Khandpur, N., Cediel, G., Neri, D., Martinez-Steele, E., Baraldi, L. G., & Jaime, P. (2019). Ultra-processed foods: what they are and how to identify them. *Public Health Nutrition*, *22*(5), 936–941. https://doi.org/10.1017/S1368980018003762

Moustier, P. (2017). Short urban food chains in developing countries: signs of the past or of the future? *Natures Sciences Sociétés*, *25*, 7–20. https://doi.org/10.1051/nss/2017018

Niederle, P., & Wesz, Jr., V. J. (2020). *Agrifood System Transitions in Brazil: New Food Orders.* Routledge.

Paredes, M., Prado, P., Valero, Y., & Cole, D. (2020). Mensuração do consumo responsável de alimentos: um insumo para fortalecer a campanha "250.000 famílias saudáveis" no Equador. In Preiss, P. V., & S. Schneider (Eds.), *Sistemas alimentares no século XXI: debates contemporâneos* (pp. 325–336). UFRGS. https://www.lume.ufrgs.br/bitstream/handle/10183/211399/001115756.pdf

Popkin, B., & Reardon, T. (2018). Obesity and the food system transformation in Latin America. *Obesity Reviews*, *19*(8), 1028–1064. https://doi.org/10.1111/obr.12694

Portilho, F. (2020). Ativismo alimentar e consumo político – duas gerações de ativismo alimentar no Brasil. *Redes*, *25*(2), 411–432. https://doi.org/10.17058/redes.v25i2.15088

Preiss, P., Charão-Marques, F., & Wiskerke, J. S. C. (2017). Fostering sustainable urban-rural linkages through local food supply: a transnational analysis of collaborative food alliances. *Sustainability*, *9*(7), Article 1155. https://doi.org/10.3390/su9071155

Preiss, P. V., & Schneider, S. (Eds.). (2020). *Sistemas alimentares no século XXI: debates contemporâneos.* UFRGS. https://www.lume.ufrgs.br/bitstream/handle/10183/211399/001115756.pdf

Rede PENSSAN. (2021). Inquérito nacional sobre insegurança alimentar no contexto da pandemia da Covid-19 no Brasil. Rede PENSSAN. http://olheparaafome.com.br/VIGISAN_Inseguranca_alimentar.pdf

Rivera, J. A., Barquera, S., González-Cossio, T., Olaiz, G., & Sepúlveda, J. (2004). Nutrition transition in Mexico and in other Latin American countries. *Nutrition Reviews*, *62*(S2), S149–S157. https://doi.org/10.1111/j.1753-4887.2004.tb00086.x

Sabourin, E., & Grisa, C. (Eds.). (2018). *A difusão de políticas brasileiras para a agricultura familiar na América Latina e Caribe*. Escritos.

Sabourin, E., Grisa, C., Niederle, P., Leite, S., Milhorance, C., Ferreira, A., Sauer S., & Andriguetto-Filho, J. (2020). Le démantèlement des politiques publiques rurales et environnementales au Brésil. *Cahiers Agricultures*, *29*, Article 31. https://doi.org/10.1051/cagri/2020029

Sauer, S. (2018). Soy expansion into the agricultural frontiers of the Brazilian Amazon: The agribusiness economy and its social and environmental conflicts. *Land Use Policy*, *79*, 326–338. https://doi.org/10.1016/j.landusepol.2018.08.030

Schneider, S., & Bohórquez, N. V. (2019). A aquisição pública de alimentos como mecanismo de desenvolvimento na Colômbia. *Redes*, *24*(1), 81–105. https://doi.org/10.17058/redes.v24i1.13048

Schubert, M., & Ávalos, D. (2020). Sistemas alimentarios globales y ley de etiquetado de alimentos en Chile. *Redes*, *25*(2), 527–544. https://doi.org/10.17058/redes.v25i2.14850

Schubert, M., Schneider, S., & Díaz Méndez, C. (2017). O "comer fora de casa" no Brasil, Reino Unido e na Espanha: uma revisão das bases de dados estatísticos oficiais e perspectivas para comparação. *Estudos Sociedade e Agricultura*, *25*(2), 276–304. https://doi.org/10.36920/esa-v25n2-4

Sosa Varrotti, A., & Gras, C. (2021). Network companies, land grabbing, and financialization in South America. *Globalizations*, *18*(3), 482–497. https://doi.org/10.1080/14747731.2020.1794208

Triches, R. M., Gerhardt, T. E., & Schneider, S. (2014). Políticas alimentares: interações entre saúde, consumo e produção de alimentos. *Interações*, *15*(1). https://doi.org/10.1590/S1518-70122014000100011

Wittman, H., & Blesh, J. (2017). Food sovereignty and *Fome Zero*: Connecting public food procurement programmes to sustainable rural development in Brazil. *Journal of Agrarian Change*, *17*(1), 81–105. https://doi.org/10.1111/joac.12131

Zaneti, T. B. (2017). *Cozinha de raiz: relaciones entre chefs, productores y consumidores a través del uso de productos agroalimentarios únicos en la gastronomía contemporánea* [Doctoral dissertation, Universidade Federal do Rio Grande do Sul]. UFRGS Digital Repository. http://hdl.handle.net/10183/164708

Part I

Regional food policy analysis

Chapter 1

Food policies and the politicization of food: the Latin American experience

CATIA GRISA, PAULO NIEDERLE, STÉPHANE GUÉNEAU,
JEAN-FRANÇOIS LE COQ, CLARA CRAVIOTTI, GRACIELA BORRÁS,
DANIEL CAMPOS, HÉCTOR ÁVILA-SÁNCHEZ, SANDRINE FREGUIN-GRESH,
JUNIOR MIRANDA SCHEUER, JORGE ALBARRACIN

▸▸ Introduction

The politicization of food issues is not a recent phenomenon in Latin America. In fact, throughout the 20th century, access to food was considered one of the key public issues in the region, even becoming the trigger for paradigmatic changes in the development models that guided state action (Gilardon, 2016; Barquera, Rivera-Dommarco & Gasca-Garcia, 2001; Linhares & Silva, 1979; Linhares, 1979). However, as verified in other parts of the world, currently this phenomenon reveals new dynamics (Fouilleux & Michel, 2020; Portilho, 2020). The principal development is the proliferation of insights regarding the reasons why food has become a key contemporary public issue. In addition to the issue of hunger, which has predominated throughout Latin American history and has returned to the agenda due to the economic crisis, aggravated by the recent pandemic, discussions linking food practices to issues of health, food sovereignty, food rights, cultural heritage, sociobiodiversity, and climate change now abound (Swinburn et al., 2019). Moreover, politicization manifests itself, not only in the realm of production (Molina, García & Casado, 2017) and the public sphere (Fouilleux & Michel, 2020), but increasingly includes consumption and the private sphere (Portilho, 2020; Tanaka & Portilho, 2019).

The propagation of views regarding the role of food as an organizational practice of different dimensions of social life is reflected in the disputes between the actors involved in the construction of food policies[1] (Fouilleux & Michel, 2020). Depending on how they interpret and prioritize public problems, the actors project different solutions, propose and justify public policy instruments that do not have common objectives and often end up having contradictory effects: credit for grain production

---

1. Through food policies, states influence the way in which food is produced, processed, distributed and consumed; organize political structures and agreements for the supply of food; and define their forms of governance (Lang, Barling & Caraher, 2009). Thus defined, food actions and policies (or their absence) have an impact on agriculture, health, nutrition and local, territorial and national development.

leads to greater consumption of ultra-processed foods, aggravating health problems; support for meat exports promotes the expansion of livestock farming into environmental conservation areas; subsidies for the pesticide industry cause environmental and health effects; etc.

There are numerous particularities in the way in which the politicization of food is expressed in each territory. However, the chapters that make up this book also reveal that Latin American countries share a number of similarities, both in terms of the dynamics of food systems and the profile of the policies created to respond to the problems identified. This opening chapter presents an interpretation of the process of institutionalization of food policies in Latin America that extends beyond the analytical barrier imposed by national borders. Based on the knowledge provided on the reality of each country in the other chapters of this book and the discussions held by the research team, we will examine the different construction processes corresponding to public policies and associated instruments according to the way they are oriented by a cognitive and normative frames of reference called *"référentiels"* (Jobert & Muller, 1987; Muller, 2019). This interpretation is based on a dialogue involving two schools of contemporary social thought on public policy analysis. On the one hand, the so-called "cognitive approach" lends us the concept of *référentiels*, which manifest the ideas negotiated, selected, agreed and institutionalized within public policies (Fouilleux & Michel, 2020; Fouilleux & Jobert, 2017; Fouilleux, 2011; Muller, 2019). On the other, historical and sociological neoinstitutionalism provides the conceptual elements to understand the production of these *référentiels* as the result of strategies, cooperation and conflict between different actors and political coalitions (Mahoney & Thelen, 2010; Fligstein & McAdam, 2011). From these perspectives, we understand the state as a political and institutional space permeable to the pressures and demands of different actors, from which public policies with diverse and often contradictory *référentiels* emerge.

This analysis aims to make sense of the changes taking place with regard to food policies in Latin America, attempting to answer the following questions: What are the processes of construction and institutionalization of food policies in Latin America? Who are the actors behind them and the ideas that these actors advocate? And finally, do these food policies constitute new vectors of change in food systems for greater sustainability and inclusion between the urban and rural worlds?

The chapter is divided into four sections, in addition to this introduction. The following section briefly presents the analytical model that guides our reflections on the development of public policies. We then go on to identify the main food policy *référentiels* present in Latin America, constructed by different actors and ideas and manifested in different instruments. The section that follows analyzes how public policies guided by these *référentiels* have been defended/strengthened or challenged/dismantled by the actions of actors/coalitions contesting the control of food systems.

▸▸ The model for food policy analysis in Latin America

The central element of our analytical framework (figure 1.1) is public policy, understood as a set of actions constructed and implemented by the state for society, with a variable degree of participation and power of both parties (Sabatier & Jenkins-

Smith, 1999; Kingdon, 1984). Based on the sociological perspective of public action, we understand public policies not only as devices instituted to solve a given public problem, but also, and above all, as processes of interpretation and social construction of reality (Lascoumes & Le Galès, 2012).

More specifically, the starting point of our analysis is the identification of public policies that, over the past two decades, have set out to generate innovations that change the practices of production, processing, distribution, supply and consumption of food and waste treatment. These public policies will be analyzed mainly on the basis of their *référentiels*, which are linked to different actors, ideas and instruments. Indeed, according to the cognitive approach to public policy analysis, each policy involves the definition of objectives that are defined on the basis of a representation of the problem, its consequences and possible solutions. The definition of a public policy is always based on a representation of reality that constitutes the *référentiel* of that policy (Muller, 2019). Following Fouilleux and Michel (2020), Fouilleux and Jobert (2017) and Fouilleux (2003), we understand "food policy *référentiels*" to be the institutionalization of ideas and interpretations of food problems. The *référentiel* is "a snapshot of the policy at a given time," expressing its organization, its objectives and its image (Fouilleux, 2003, p. 43).

The construction of each *référentiel* implies a permanent negotiation between political coalitions and is susceptible to inclusions, cuts and transformations according to changes in power relations and in the political objectives of the different actors involved in the construction of public policies. Thus, a set of ideas, the product of controversies and hybridizations, constitutes a stabilized and institutionalized *référentiel* through the dissemination of shared representations (Fouilleux, 2000; Fouilleux & Jobert, 2017).

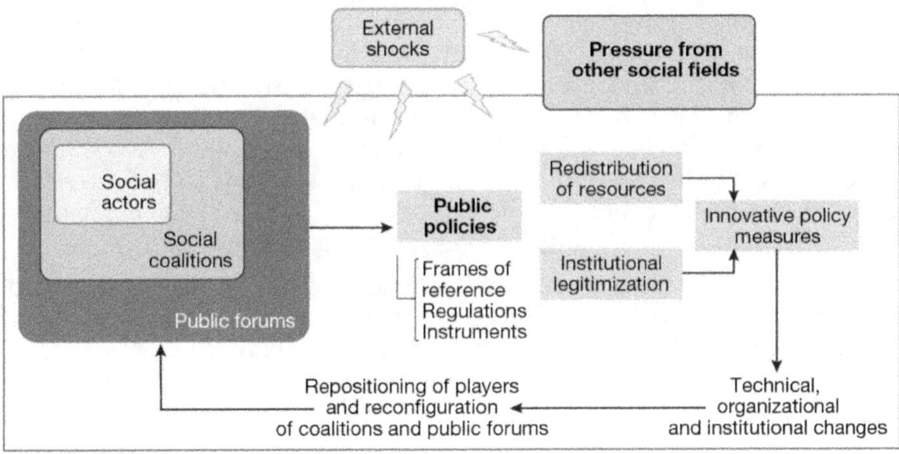

Figure 1.1. Elements and dynamics of the analytical framework for food policy analysis.

The identification of the *référentiels* of food policies in Latin America requires attention not only to the problems to be addressed, but also to variables such as the values that influence the positioning of actors in the construction of public problems and solutions, and the norms that guide public action (Muller, 1995). The concept of *référentiel* thus includes representations of the world and the means of acting on the world. Values can be defined as the fundamental representations of what is right or wrong, desirable or unacceptable. They define an overall framework for public action. Norms are the differences between the perceived reality and the desired reality. They define the principles of action, i.e., what must be done to achieve a desired situation. Each *référentiel* directs the development of specific public policy instruments, which can be defined as the socio-technical devices that organize the social relations between the State and policy recipients (Lascoumes & Le Galès, 2012).

The analysis then focuses on the upstream and downstream directions of the institutionalization of the policies. Upstream, the analytical framework focuses on the policy process of the construction of the *référentiel*. As such, we were interested in finding out about the relevant areas of negotiation and power relations, how they were shaped as public arenas governed by rules and hierarchies (Fouilleux, 2000). We also identified the political actors who dispute the construction of the *référentiels* that direct public policy, their interpretations of the world and the relational position of the political actors. To that end, our research focused on the relationships between collective state actors (ministries, departments, municipalities, universities, etc.) and non-governmental actors (social movements, trade unions, NGOs, companies, etc.) in producing policy policies.

To enhance their strategies and reinforce their positions in the public arena, actors create coalitions that allow them to share material and symbolic resources (meanings and identities) (Fligstein & McAdam, 2011). Inequalities exist within the coalitions in terms of access to these resources, so that there are, for example, those that are dominant and dominated among the dominant parties, as well as those that are dominant and dominated among the dominated parties. Moreover, the strategic use of resources should not be confused with purely rational or instrumental action. Choices are permeated by political interests, moral conceptions and fictitious expectations (Mahoney & Thelen, 2010). Moreover, the more a resource is used, the greater the actor's dependence on it may be, particularly if we are talking about the management of identities and meanings, which creates limits to the circulation of actors within the public sphere. As the literature shows, insofar as one organization adopts, for example, agribusiness as a political identity, and another defines agroecology as a signifier of its practices, the dialogue between them becomes more difficult (Sabourin, Craviotti & Milhorance, 2020; Niederle & Wesz Jr.).

In the downstream phase of the institutionalization of policies, the focus is, on the one hand, on their allocative role in terms of redistributing the resources that actors have at their disposal to act and, on the other, on their authoritative and legitimizing role, insofar as policies also act as institutional forces that favor or block the use of certain resources, also including identities and meanings. An example of this is the fact that some States have officially added terms such as "sovereignty" or "peasants" to give new meanings to the notions of "food security" and "family farming," respectively. In doing so, they not only set out a public policy *référentiel*, but also legitimize and strengthen the position of certain actors (in this case, peasants' social movements) in public arenas (which is always susceptible to power play and relations).

Public policies carry with them the assumption that they are capable of producing political innovations that will have repercussions on technical, organizational and institutional changes which, in turn, will have effects on the restructuring of public arenas (positioning of actors, reorganization of coalitions, alteration of hierarchies). In fact, this is what the most disadvantaged actors will expect from public policies. However, in the course of political disputes, the dominant actors can not only block the construction of policy, but also subvert its logic to ensure the maintenance of their relative positions of power. This does not mean that innovations or changes will not occur, but rather that they will not create fissures in the institutional structure that defines social and economic hierarchies. An example of this would be the reorientation of agroecology policy to promote the control of the organic food market by the dominant actors in the food system (Fouilleux & Michel, 2020).

Finally, it is also important to bear in mind that the dynamics of any policy will always be affected by changes in other social spheres, so that policies may be reinforced, blocked or reoriented due to events that do not directly involve the actors that most actively participated in their construction or implementation. Interference from other spheres can be perceived through incremental changes in policies, when, for example, the Ministry of Finance reduces resources or the Central Bank raises the interest rate, or in abrupt shocks when, in specific circumstances, external pressures generate significant ruptures capable of undermining the very existence of the policy.

As other chapters will show, external shocks that provoke abrupt changes happen on a recurring basis in Latin America. Some analysts compare the institutional scenario of many Latin American countries to a "perfect storm," that is, when an unfavorable event is dramatically aggravated by the occurrence of a rare combination of circumstances, turning into a disaster (Levitsky & Ziblatt, 2018). The main events of this confluence of crises include the political breakdowns associated with the removal of governments, the fiscal vulnerability of States as a result of the fluctuation of agricultural commodities on international markets, the failure of social protection systems and the explosion of unemployment and poverty, the impact of climatic events related to the increase in environmental devastation, and, finally, the aggravation of all the above by the Covid-19 pandemic.

▶▶ Food policy *référentiels* in Latin America

Various designs and objectives have been included in the food policies created by Latin American countries over the years. This is due, among other things, to the emergence of new *référentiels* that guide the actions of States. Based on the examples in the other chapters and other studies, eight public policy *référentiels* can be identified that guide the action of Latin American States, some of which, although classified according to the same set of ideas and interpretations, exhibit internal differentiations (what we call "secondary *référentiels*"). Following our analytical framework, these *référentiels* were defined on the basis of the identification of the public issues they focus on and the solutions they suggest to solve them; the values and norms associated with the social creation of these problems and solutions; the links to which they correspond within the food system (production, distribution, consumption); the main role conferred on the state and the type of social category that is favored (table 1.1). These *référentiels* are taken as "ideal types" and, in many public policies, it is possible to observe nuances or connections between two or more reference points.

Table 1.1. Primary and secondary *référentiels* related to food policies in Latin America and their main elements

Référentiel	Secondary *référentiel*	Constituent elements of *référentiels*			Positioning in relation to agrifood system issues			The role of the state
		Values	Norms		Production	Circuit type	Consumption	
Food-for-market (or liberal)		Neoliberal	Food supply regulated by the free market		Comparative advantages of each country	Extensive distribution/ conventional market	–	Market regulation
Urban food supply		Centralization, state power	Organization of food distribution centrally controlled and organized by the state		–	Specific routes managed or supported by the State	–	Organize the food supply in cities
Productivist	Productivist from agribusiness	Trust in technical progress	Application of the technological package of the green revolution (intensive and mechanized agriculture) to produce more food for the planet's growing population		Increase in agribusiness production	–	–	Boosting production
	Productivist based on family agriculture		Imitate the agro-industrial model to "modernize" family farming and produce more food (adopt intensive and mechanized techniques and forms of business management)		Increase in family farming production	–	–	Boosting production
Social food welfare		Right to food	Distributing food to food insecure populations		–	Specific state-controlled circuit	Access to food	Donate food

Référentiel	Secondary référentiel	Constituent elements of *référentiels*		Positioning in relation to agrifood system issues			The role of the state
		Values	Norms	Production	Circuit type	Consumption	
Technical food	Nutritionist	Public health (therapeutic food)	Supporting the provision of nutritious food	–	–	Improve the diet through a balanced nutrition, without deficiencies	Technical quality control of foodstuffs
	Hygienist		Ensuring a safe food supply	Ensuring food safety	Supply chain safety	Ensuring food safety	Food product safety control
Food sovereignty and security	Food autonomy	Sovereign right of peoples to define their food policies and practices.	Relocation of food supply, based on the organization of territories and rural communities	local production/ family production/ repositioned production	Short circuits	Consumption of local products	To value the production of the country/territory
	Heritage		Revival and valorization of food practices, products, ingredients, processes and knowledge.	Local and traditional production	Local circuits	Traditional foods and recipes	Valuing identities
	Gastronomic		Gastronomization of local products and popular culinary practices	Local/traditional/ sustainable production	Short circuits	Traditional/ local, ethical and sustainable foods and recipes	Valuing food as a comprehensive social event
Environmentalized food	Environmentalized food	Planetary boundaries	Production and consumption of food (diets) taking into account limited natural resources	Low environmental impact agriculture	Policy dependent variable	Waste reduction, healthy and sustainable diets	Protecting the environment to ensure food security in the future
Integrative	Integrative	Sustainability, integration and resilience	Profound transformation of current food systems into sustainable, integrated and resilient food systems that can provide healthy, safe, sufficient and environmentally friendly food.	Family, traditional, fair and sustainable agriculture production	Short circuits	Adequate, healthy, sustainable and responsible consumption	Creation of integrated food policies

The *food-for-market référentiel* is based on the liberal interpretation that supply should be organized on a global scale, taking advantage of the collective efficiency supposedly generated by the comparative advantages of each country. Claimed by political actors who criticize the "distortions" produced by state intervention in markets – while guzzling public resources to increase their comparative advantages (Bonanno, 2019) – this framework assumes that, provided the conditions for free trade are in place, global competition would guarantee the availability of food at an optimal price (i.e., Pareto efficiency). Since the late 1970s, the World Bank, the International Monetary Fund (IMF), the World Trade Organization (WTO) and other neoliberal organizations have been primarily responsible for the dissemination of this framework. The structural adjustment policies imposed on indebted developing countries advocated the liberalization of economies and trade as the main roadmap to follow, which led to a change of scale in the definition of global food security problems, moving from a national to a global (and, at the same time, micro, individual) approach. The primary focus for addressing the problem of hunger was supposed economic growth, with its implications in terms of poverty reduction (Jarosz, 2011). Thus, public policies would manage macroeconomic regulation, leaving the lead role to large corporations and distribution networks in the "global food supply" (McMichael, 2016).

In different degrees and configurations, this *référentiel* was present in the vast majority of Latin American countries, either during the 19th and early 20th centuries or, with the institutionalization of neoliberal measures and instruments, since the 1990s. Unlike other *référentiels* institutionalized in specific programs, this *référentiel* is featured, primarily, in trade agreements or treaties and in other normative instruments, measures and general directives.

Constructed by political and governmental actors identified with neoliberal ideas and demanded by corporations and competitive productive sectors in international markets, the free trade agreements for food in Mexico, Chile, Colombia, Peru and the Dominican Republic, among others, illustrate this *référentiel* (Maldonado, 2016; Baca, 2014). To a large extent, liberalization treaties were established based on the interpretation of their benefits in terms of expanding agricultural markets for farmers and production segments; price competitiveness benefits for consumers, who could also enjoy greater food diversity; and economic returns in terms of foreign exchange and income generation (Baca, 2014). In the case of Mexico, such treaties began with the North American Free Trade Agreement (NAFTA) in 1994 – which includes bilateral agreements between Mexico and the United States, the United States and Canada, and Mexico and Canada – that were soon followed by others with varying degrees of openness and tariff conditions (CEDRSSA, 2015; Baca, 2014). Although revised and readjusted over time, such treaties (mainly NAFTA) have been questioned, especially by peasants' movements and organizations defending food sovereignty due to the worsening of inequality and the productive and food changes induced (González, 2019; Baca, 2014). According to González (2019), before NAFTA, Mexico used to be self-sufficient in terms of corn, beans, rice and other commodities, but it is now dependent on imports from the United States, from where it sources 85% of its wheat, 73.7% of its rice and 38% of its corn.

Other policies that fit this *référentiel* are those that directed state reforms aimed at privatization, targeted regulations, reduced state intervention and free markets.

Food policies in Nicaragua, discussed by Freguin-Gresh and Cortes (see chapter 4), illustrate such actions. According to the authors, until the late 1970s, and defined by the agrarian elite and the commercial oligarchy, food policies were limited to supporting production for the foreign market (e.g., cotton, coffee, bananas, cattle) and regulating sanitary aspects of production, marketing and consumption (plant and animal health). Although the 1980s saw important political changes, in the following decade liberal perspectives returned to power, giving priority to macroeconomic stability, trade liberalization, privatization of enterprises and the reconstruction of a network of traders and distribution chains for private goods and services. Pressured by international events and organizations and by the ties between civil society and Sandinista parliamentarians, in the early 2000s, the government began to outline a number of food policies aimed, in particular, at improving the nutrition of children and schoolchildren.

The *urban food supply référentiel* gained ground especially in the second half of the 20th century due to the challenges posed by the growing food demand in large cities. Unlike the *food-for-market référentiel*, this *référentiel* is based on the power of the State and its planning capacity, and establishes as a principle of action the need to organize food distribution in a centralized manner. Specifically, this *référentiel* takes the form of a set of governmental actions that seek to ensure food supply in urban areas through the creation of structures administered or supported by the State. Its instruments focus, above all, on the organization of food supply to meet the demand of the service sector and private consumption.

Latin America's wholesale markets and food supply centers are emblematic cases in this regard. Encouraged by the FAO and boosted in the 1960s and 1970s by the French and Spanish experiences, these structures organized or supported by the States seek to offer a wide variety of fresh foods, not necessarily linked to the local territory and culture, through the coordination of a wide range of economic actors, especially intermediaries and commercial agents (Cunha, 2014). According to the FAO and the Latin American Federation of Supply Markets – FLAMA (2020), there are almost 300 centralized supply markets in Latin America, with different formats, sizes, dynamics and relationships with the States. These wholesale markets have functions that go beyond the "simple marketing of agricultural products," becoming "true food authorities" to the extent that they can regulate public supply, democratize information on food prices and quality, contribute to sanitary control, reduce losses and waste, and other food security actions (FAO & FLAMA, 2020).

As a complement to markets and supply centers, the promotion of temporary street markets (known as *ferias libres* in Chile) has been observed since the colonial period (Andrade, 2017; Tejada, 2013; Linhares, 1979). These initiatives assume different intensities and configurations, from those characterized by the presence of inter-mediaries and products circulating through long chains, to those marked by local supply chains, the presence of family farmers and peasants, and the promotion of agroecology and urban agriculture. In these cases, the street markets respond to the demands of urban supply and may also be related to other issues, such as concern for food and nutritional security, environmental issues, the promotion and marketing of family farming products and the strengthening of territories, intersecting with other *référentiels* discussed below.

According to the Observatorio de Ferias Libres (2013, p. 6), 933 street markets were organized in Chile in 2011, which "supply 70% of the fruit and vegetable market and 30% of the fish market (…) indicating the great importance of the channel both for food security and for the income of small producers and fishermen." Given this importance, the Chilean government and several municipalities have been mobilizing a set of actions to promote these markets, such as the Street Market Development Fund, the Street Market Registry and digitalization initiatives. Following Chile's example, several other national, regional and municipal governments have established similar actions (Ríos & Wesz Jr., 2020; Parrado Barbosa & Molina Ochoa, 2014; and see Guéneau et al., chapter 12).

Anchored in the technical-mercantile imaginary created by the modernization of agriculture in the 1960s and 1970s, the *productivist référentiel* interprets the problem of food security as the result of food shortages in the face of world population growth. From this perspective, the problem of hunger should be resolved, primarily, by increasing food supply. Although this *référentiel* has lost currency since the 1980s due to the availability of sufficient food to meet global demand – except in specific areas such as conflict zones or hard-to-reach areas – the 2007–2008 food crisis revived the interpretation, reinforcing increased agricultural production as the objective of many food policies in the international arena (Fouilleux, Bricas & Alpha, 2017).

With policies fundamentally aimed at promoting the agricultural sector (farmers and upstream and downstream industries), this reference framework can be divided into two secondary *référentiels*: *productivist based on agribusiness* and *productivist based on family farming*. Sharing ideas and interpretations of the world and, in several contexts, overlapping with the *food-for-market référentiel*, the *agribusiness-based productivist secondary référentiel* is based on measures focused on increasing production and competitiveness and, often, with the global market as a target. Neo-Malthusian arguments of food scarcity are at the heart of the discursive construction of this *référentiel* (Fouilleux, Bricas & Alpha, 2017), which have the potential to trigger various agricultural policy instruments (rural credit, agricultural research, dissemination of technological innovations, technical assistance, price guarantees, agricultural insurance, etc.).

One example of this is rural credit in Brazil. Instituted since 1965 and constituting the main agricultural policy in terms of allocated resources, this instrument offers the option of financing costs and investments for various agricultural products and activities, with subsidized credit conditions. Each year, in response to the interests of the sector's representative organizations and the interpretations of agricultural development by governmental and political actors, the Federal Government announces more financial resources and special conditions to boost production, especially that by medium and large producers, and to continue breaking agricultural production and export records. According to Wesz Jr. and Grisa (2017), almost 80% of the cost credit in 2015 was allocated to soybeans, corn, coffee and sugarcane, with soybeans receiving more than 40% of the resources. According to the authors, the characteristics of this policy "indicate the permanence of a style of state intervention that feeds an agro-export development model, which is guided by demand and remains a supplier of raw materials" (Wesz Jr. & Grisa, 2017, p. 103).

The Rural Change (*Cambio Rural*) program in Argentina is another example. Backed by international agencies and federal government organizations for agriculture,

the program was created in 1993 in a context of trade openness and globalized economies, aimed at "productive reconversion" through increased productivity and added
value, agricultural diversification and a shift towards crops with higher profitability
(Taraborrelli, 2017). Aimed at "capitalized producers, whose size and organization
allowed them to solve their problems with the support of the state (…) the program
consisted of technical assistance to producers with the objective of modifying their
productive structure and entrepreneurial capacity, as well as facilitating a connection
with different sources of financing" (Taraborrelli, 2017, p. 168). In the 2000s, due to
the political and institutional recognition of family farming, *Cambio Rural* was restructured under the name, Rural Change, Innovation and Investment (*Cambio Rural II*),
focusing on small and medium-sized agrifood and agro-industrial enterprises, cooperatives and family farmers, either capitalized and/or with capitalization potential, to
contemplate comprehensive territorial development (Patrouilleau, Taraborrelli &
Alonso, 2018). However, the political changes of 2016, with the formation of a new
government more closely aligned with neoliberal ideology, pushed back the proposal
for productive reconversion and technological incorporation with a productivist and
commercial bias, and focused on more consolidated farming operations (Jara et al.,
2019; Patrouilleau, Taraborelli & Alonso, 2018).[2]

In turn, the *family farming-based productivist secondary référentiel* emerged in the
1990s/2000s as a result of the actions of various social and trade union movements
of family farmers and peasants and/or government bureaucracies, considering the
importance of this social category in terms of food supply and the production of
affordable staples (given the nightmare of food price inflation being experienced
in Latin American). In light of the unequal treatment historically conferred by the
state on small-scale production, public policies guided by this secondary reference
often seek to imitate productivist policies aimed at the agro-industrial sector or large
agricultural establishments, giving priority to increasing agricultural production and
productivity. As in the case of agribusiness, these policies activate instruments that
rely on rural credit, price guarantees, production insurance, technical assistance and
rural extension, as well as the dissemination of technologies and innovations.

One notable example is the National Program for the Strengthening of Family
Farming (PRONAF), set up in Brazil in 1995. Given the historical marginalization of
this social category with regard to state actions (including the aforementioned rural
credit) and the difficulties imposed by trade liberalization (MERCOSUR), family
farming unions and social movements began to demand differentiated agricultural
policies. In the context of these demands, the creation of a credit policy that would
address their particularities became essential. Once the program was created, studies
and evaluations soon began to emerge indicating PRONAF's difficulties in breaking
away from elements that have been fundamental in the development of rural credit
in the country: incentives for commodity production based on conventional production models; difficulties in contemplating productive diversity and ways of relating
to land and nature; and limitations in incorporating the socioeconomic diversity of

---

2. Based on its organization into groups, *Cambio Rural* was developed in three stages: in the first year, the
costs of technical assistance would be absorbed by the State; in the second year, the costs of technicians
should be progressively assumed by the producer groups and other sources of financing; by the fourth
year, the group would be sufficiently mature to continue running autonomously (Taraborrelli, 2017).

the social category, being accessed mainly by the most capitalized family farmers in southern Brazil (Wesz Jr., 2020; Grisa, Wesz Jr. & Buchweitz, 2014). Despite these criticisms, the program is still a high priority on the agendas of the main family farming trade union organizations.

Another example is the Sustainable Modernization of Traditional Agriculture Program (MasAgro), set up in Mexico in 2010. The result of a partnership between the Ministry of Agriculture, Livestock, Rural Development, Fisheries and Food and the International Maize and Wheat Improvement Center (CIMMYT), the program was established as a ten-year cross-cutting project. The program aims to strengthen food security through research, capacity building and technology transfer in the field, and to help small and medium corn and wheat producers located in seasonal production areas to obtain high and stable yields, increase their income and mitigate the effects of climate change in Mexico (FAO/Sagarpa, 2012). The program operates through "regional innovation nodes," providing technical assistance to producers who "do not have access to modern technologies or functional markets" (FAO/Sagarpa, 2012). The program promotes: "i) the generation and distribution of suitable varieties; ii) the adoption of more efficient post-harvest technologies; iii) the use of conservation and precision farming practices; and, iv) the development of region-specific markets" (FAO/Sagarpa, 2012). Although it refers to food and nutritional security, several authors question this public policy (Martínez et al., 2017). For Chapela, Menéndez and Berlanga (2015, p. 248), this "program resumes the scheme of homogeneous technological packages, based on the use of industrial genetics owned by seed, fertilizer and agrochemical companies." Similarly, Sierra (2018, p. 4) states that this is a "new version of the Green Revolution," constituted in a neoliberal context marked by public-private alliance and at a time when the adverse effects of intensive agriculture on the environment are particularly evident.

Certain images are able to highlight food supply issues immediately, without any need for extended discourse. Such is the case of images of malnourished children that shock the general public and act as remarkable vectors for conveying the underlying values and norms that correspond to what we know as the *food social welfare référentiel*. Unlike the previous *référentiel*, the problem of hunger in the world is no longer approached as a question of global food insufficiency, but as a question of limited access to food for certain population groups. Thus, through some of the fundamental principles enshrined in the Right to Food,[3] this *référentiel* establishes access to food for the most vulnerable communities as a fundamental norm for public action. It is, in particular, mobilized in contexts and/or moments of economic, social or political crisis, or even as

---

3. The Right to Food has been incorporated into the national constitutions of several Latin American and Caribbean countries (Argentina, Brazil, Colombia, Bolivia, Dominican Republic, Ecuador, Nicaragua, Haiti). At regional level, the Protocol of San Salvador, annexed to the 1969 American Convention on Human Rights, explicitly recognizes the Right to Food as a binding provision for the States that have ratified it (Golay & Özden, 2012). In addition, in 2012, the Latin American Parliament (an entity that brings together the congresses and legislative assemblies of 23 countries in Latin America and the Caribbean) sanctioned the first Framework Law called "Right to Food, Food Security and Sovereignty," which recognizes the right to food at a supranational level, which strengthens international treaties. This opens up the possibility of legal action in the event that this right is violated. The law commits the state to "comply with its obligations, by virtue of the mandates incorporated in the political constitutions and its commitments implemented in international covenants and treaties."

a result of unexpected events that compromise access to income, as has occurred since the beginning of the Covid-19 crisis (Gurgel et al., 2020). In general, public policies directed by this *référentiel* focus on specific food distribution circuits controlled or organized by the state, including instruments such as food distribution, food baskets, cash transfers for food purchases, soup kitchens and community kitchens, and school canteens.[4] The main objective is to respond rapidly to problems of severe food insecurity (lack of food) by guaranteeing access to either free or subsidized supplies of basic foodstuffs. Although essential given the urgent food situations, it is not uncommon for these actions to be disconnected from a broader debate or from complementary actions for the promotion of healthy and sustainable food, or even to ignore concerns about the origin or nutritional quality of food.

Argentina's National Food Security Plan (PNSA), known as *El Hambre Más Urgente* ("Most Urgent Hunger"), based on the "National Nutrition and Food Program" (Act 25.724), and discussed by Aranguren et al. in chapter 9 of this book, represents an example of public policy related to the *food social welfare référentiel*. This plan was institutionalized in Argentina in 2003, as a response to the unprecedented economic, political and social crisis that began in 2001 (Gilardón, 2016). The crisis arose as a result of the neoliberal policies adopted in previous years that caused a rise in food prices in the domestic market, unemployment (21%) and poverty (affecting more than half of the population) and led, once again, to hunger and malnutrition becoming priority issues needing to be addressed. The plan "is intended to cover the nutritional requirements of children up to 14 years of age, pregnant women, the disabled and the elderly over 70 years of age living in poverty" (Argentina, 2003). Implemented by the Ministry of Social Development, the plan unified several programs and actions, and currently provides direct food and nutritional assistance (food distribution, vouchers or bank card[5] to buy food in markets), support for social, children's and community canteens, food services to canteens (breakfast, lunch, dinner or snack), distribution of milk and fortified milk for pregnant women and children, and support for kitchen garden production for household use (Aulicino & Langou, 2012). The latter dovetails with the ProHuerta program run by the National Agricultural Technology Institute (INTA) in Argentina, which, since 1990, has promoted access to healthy food, agroecological production practices for self-sufficiency (at family, school and community level),

------

4. Several of these instruments may take the form of other food policy *référentiels*, depending on their design and regulations. The case of school meals is most frequent in Latin America. As discussed by Vasconcellos (see chapter 3) and discussed later in this chapter, various international organizations and national efforts have tried to respond to different contemporary food challenges, such as the double burden of malnutrition (undernutrition and obesity), socioeconomic inclusion, and the promotion of biodiversity and agroecology through the implementation of school meal programs. In line with the suggestions made by Grisa, Avila and Cabral (see chapter 16) and Scheuer (see chapter 18), school meals and other public food procurement mechanisms could adopt provisions and policy instruments to address various contemporary food issues. We will return to this point later.

5. It is important to mention the existence of several cash transfer programs that seek to contribute to access to food for socially vulnerable families in Latin America. Although many of them are not limited to the purchase of food (the beneficiary can use the resource for other purposes), nor do they dictate the quality of the food purchased, these instruments are important for minimizing food insecurity. Examples include the Family Allowance (*Bolsa Família*) Program in Brazil, the Social Card (*Tarjeta Social*) Program in Uruguay, and the Social Credit (*Bono Social*) Program in Guatemala.

food education and the organization of alternative street and traditional markets with an inclusive approach for producer families (see Aranguren et al., chapter 9). Although the PNSA proposes new guidelines, such as the human Right to Food, the core of the social policy was based on targeted assistance, with the main benefits being the direct delivery of food, support to canteens and productive kitchen garden enterprises (Carrasco & Pautassi, 2015).

Long considered as a therapeutic remedy,[6] food was at the heart of discourse on health during the 2000s. In the mental representations of "eating well," food is associated with health from a very technical perspective (Mathé et al., 2008). This *référentiel*, which we can qualify as *technical food*, is characterized by a strong normative dimension, in particular through nutritional and health standards that are very present in public action (confronting traditional food practices), particularly in urban areas of Latin America (Martinez-Lomeli, 2020). This *référentiel* is divided into two secondary *référentiels*: *nutritionist* and *hygienist*.

Promoted by various actors – from civil society organizations, health professionals (especially nutritionists) and even segments of the food industry – the *nutritionist secondary référentiel* focuses on the problems of malnutrition and undernutrition, which are addressed through instruments focused on the composition of diets: (bio) enriched or fortified foods; supplements; control of the amount of sugar, sodium, fat, vitamins and calories. Although many of these actions are essential for nutritional adequacy, the promotion of health and consumer awareness, others are challenged by the fragmented, simplified and medicalized performance of food production and consumption (FBSSAN, 2016). In many situations, food policies align with a viewpoint that, according to Rocha (2020, p. 41), reduces "the nutritional value of foods to their individual nutrients, to the detriment of a broader understanding and more systemic solutions."

An example of food policy based on this *référentiel* is Chile's Food Labeling Act (Act No. 20.606/2016).[7] This law emerged from interactions initiated in 2007, between parliamentarians and academics seeking to address the problem of obesity, poor diet and associated diseases (Sánchez and Silva, 2018). Overcoming political and business resistance, among other measures, the law establishes that foods containing added sugars, sodium or saturated fats in proportions that exceed the limits established by the Ministry of Health must carry a warning label on the front, indicating, as appropriate, the high content of calories, saturated fats, sugars and/or sodium (Chile, Ministerio de Salud 2018). For Schubert and Avalos (2020, p. 535), the labeling law seeks to "guide consumers' decisions, by providing information about the ingredients the products contain and their nutritional value, and offering the choice to buy from another brand." In addition to labeling, the law establishes that such products may not be marketed or distributed free of charge in the country's schools, nor may advertising for them be aimed at children under 14 years of age

---

6. According to the famous quote by Hippocrates (460–356 B.C.), "Let thy food be thy medicine."

7. Food labeling is an instrument that has been gaining ground in Latin America as a result of actions by the health and consumer protection sectors. According to the South American Institute of Government in Health, several countries already have initiatives in place and others are in the process of implementation. Examples of cases already implemented include the "traffic light" model in Ecuador, and the octagonal model in Peru and Uruguay (ISAGS-UNASUR, 2019).

(Chile, Ministerio de Salud, 2018). Although the law has led to changes in consumer behavior, Schubert and Ávalos (2020) and Sánchez and Silva (2018) point out that other actions are needed to address malnutrition, such as offering alternative foods, increasing taxes on certain foods to discourage consumption, subsidizing and promoting the consumption of healthy foods.

Another example of a policy in this *référentiel* is Paraguay's Comprehensive Nutritional Food Program (PANI). PANI was originally set up as the National Assistance Program on Food and Nutrition (PROAN) in 2005, undergoing a change of name in 2011. Developed and implemented by the Paraguay National Institute of Food and Nutrition (INAN), linked to the Ministry of Public Health and Social Welfare, the program provides monthly nutritional supplements in the form of two kilos of whole milk powder fortified with iron, calcium, zinc, copper and vitamin C (see Zimmermann, Britez Cohene & Riquelme, chapter 19). The program was based on the "diagnosis of the problem of malnutrition in Paraguay, which is attributed to the lack of some or all of the nutrients that the body requires for adequate nutrition" (Ministerio de Educación y Ciencias, Paraguay, 2018, p. 3). With national coverage, the program targets children under five years of age, pregnant women and mothers of babies less than six months old, who are in a situation of poverty, underweight, malnourished or in any situation of social vulnerability (Ministerio de Educación y Ciencias, Paraguay, 2018). Along with milk delivery, monthly anthropometric measurements and sensitization, "on feeding and hygiene practices, which include the promotion of breastfeeding and the Dietary Guidelines," are conducted monthly (Paraguay, UTGS, 2018, p. 3). Like PANI, nutritional supplementation programs are common in Latin America, especially for the maternal and infant population.[8]

Present in a more expressive form since the intensification of urbanization and industrialization in Latin American countries (1940s/1950s), the *hygienist secondary référentiel* has gained even more ground with the exacerbation of concerns arising from sanitary crises (mad cow, swine flu and, more recently, Covid-19). Advocated by technical professionals and endorsed by government bureaucracies and industrial

---

8. As examples, we also cite the experiences of Mexico with its Social Milk Supply Program (Lincosa) and Chile's National Complementary Food Program and Complementary Food Program for the Elderly. In the case of the Mexican program, a company with majority State participation processes the milk and sells it at a subsidized price. All milk is fortified with iron, zinc, folic acid and vitamins A, C, D, B2 and B12 (see Avila-Sanchez, chapter 6). The first Chilean program consists of the distribution of powdered food to prepare an instant semi-skimmed milk and cereals-based drink fortified with vitamins and minerals; powdered food to prepare an instant soup for infants, based on cereals and legumes and fortified with vitamins and cereals; or nutrient-rich formulas for infants with cow's milk protein allergies (Chile, División de Control de Gestión Pública, 2020). The second Chilean program consists of the distribution of a cereal and legume-based instant food fortified with vitamins and minerals, or a cereal-based powdered milk drink fortified with vitamins and minerals, low in lactose, fat and sodium (Chile, Dirección de Presupuestos, 2020). Another emblematic example is *Maná Infantil,* run in the Department of Antioquia in Colombia, which provides enriched powdered milk, sweet cookies and *Bienestarina* (33g), a food supplement composed of a blend of rice flour, soy and powdered milk (Castro, Castaño & Correa, 2013; Montoya & Giraldo, 2012). According to Castro, Castaño and Correa (2013, p. 85), "The three products should be consumed every day of the week to provide a daily intake of 371 kcal, 14.2 g of protein, 450 mg of calcium, 10.2 mg of iron, 5.5 mg of zinc, 0.3 mg of thiamine, 0.4 mg of riboflavin, 5.2 mg of niacin, 0.6 mg of vitamin B12 and 108 mg of folic acid."

sectors, this secondary *référentiel* focuses on reducing contamination problems and ensuring food safety, which implies proposing technical standards for the production, industrialization, distribution and marketing of foodstuffs and the treatment of waste. Aiming to avoid public health problems, this secondary *référentiel* contributed to "separate cities from the organic world" (Daviron et al., 2017), aggravating the processes of artificialization, standardization, scientization and rationalization of food. Seals of conformity, standardized good manufacturing practices and sanitary codes are some examples of the instruments legitimized by this sub-*référentiel*.

To illustrate the importance of this *référentiel*, we cite Craviotti's article (see chapter 17), which discusses the difficulties and controversies involved in making Argentine family farming correspond to the rules governing the country's food production and marketing of food in the country. In fact, this is not specific to Argentina, and can be seen in several Latin American countries (Requier-Desjardins, 1999). According to Pablo (2015, p. 13), "in Latin America the food market involves both formal and informal trade. A relevant proportion of peasant family farming falls into the category of informal trade because it does not comply with the safety and quality standards established in national legislation. Guided by international food and sanitary codes, these laws adopt standard criteria aimed at large-scale industrialization, the transportation of food over long distances and its conservation over long periods of time. As Craviotti (see chapter 17) and Chávez and Muller (2020) in the case of Argentina, Gazolla (2017) in the case of Brazil, and David (2016) in the case of Colombia, point out, it is not uncommon for family farms and smallholdings to adopt artisanal processing practices based on traditional knowledge and flavors, short circuits, and embedded in territorial dynamics of production and marketing that generally run counter to the rules of industrial production and marketing. While the *hygienist référentiel* is mainly advocated by technical professionals, governmental organizations and the industrial food sector, family farming and peasant organizations, as well as certain government and academic actors, have promoted changes that take into account the specificities of the traditional food practices of family farmers and indigenous populations.

It is also important to note that within this *hygienist référentiel*, food contamination is mainly biological. The problem of chemical contamination, linked to the increased use of pesticides and other chemical products, is minimized by agribusiness actors who focus on the need to combat plant diseases, which is reflected in the common use of the word "*defensivos*" (crop protection agrochemical products) in the names of these products. However, others use the words "*agrotoxicos*" (toxic agricultural chemicals) or even "*venenos*" (poisons), to describe these products, thus reflecting the politicization of the debate raised mainly by agroecological organizations.

The *food sovereignty and food security référentiel* was constructed in the mid-1990s largely as an alternative to the *food-for-market* and *productivist food référentiels*. This *référentiel* questions the view that food security must be achieved through increased productivity or global trade, which would imply the transformation of peasant agricultural systems (considered archaic) into "modern" agricultural systems based on mechanization and the use of off-farm inputs. In contrast, from the point of view of food sovereignty, it is not peasant agricultural production that causes poverty and hunger, but its disorganization (Marques & Moal, 2014). Thus, according to the actors whose ideas are included in this *référentiel*, food sovereignty manifests "the right of communities to define their own food and agricultural policy; to protect and regulate national agricultural production and trade to achieve sustainable development objectives; to determine their degree of food autonomy and to eliminate dumping in their markets." (Paré, 2012, p. 88). The *food sovereignty and security référentiel* comprises three secondary *référentiels* with different but complementary emphases in relation to the valuation of food habits, territorial dynamics and "good living" (Acosta, 2016): the *food autonomy, heritage* and *gastronomic référentiel*.

Mainly proposed by rural and agroecological social movements, the *food autonomy sub-référentiel* seeks to guarantee food supply through the organization of territories and rural communities. Despite challenges in relation to its institutionalization within public policy, this secondary *référentiel* constructs discourses and calls for instruments that place value on the knowledge of peasants and rural women (incorporating the feminist critique of the centrality of women in food practices), promote healthy eating through guaranteed access to "real food" (FBSSAN, 2016) based on traditional, local and agroecological foods, and defend the autonomy of communities to manage common goods, especially seeds.

Although mainly developed by networks of family farmer and peasant organizations, agroecological organizations, non-governmental organizations and other social mediators (Vernooy, Sthapit & Shrestha, 2016), actions to promote and conserve traditional seeds are also supported by some public policies and serve as an example of this secondary *référentiel*. As Vernooy, Shrestha and Sthapit (2016, p. 16) point out, "there is a growing interest on the part of national and departmental governments in establishing and supporting community seed banks," as in Bolivia, Costa Rica, Mexico, the Central American countries and the subnational governments in Brazil (Paraíba, Alagoas and Minas Gerais). Although marked by discontinuity in terms of actions, Pinto, Ticona and Rojas (2016) give an account of the Bolivian government's support for conservation and community seed banks, particularly with the Community Agrobiodiversity Banks and the Community Quinoa and Canihua Banks. In the case of Costa Rica, Porras et al. (2016) report the establishment (2004) of the Bean Agricultural Research and Technology Transfer Program (*Pitta feijão*) which, organized by a group of governmental organizations, introduced a quality control protocol for local bean seed production and established technical committees for participatory plant breeding and seed production. These technical committees, formed by farmers' associations, culminated in the creation of the Southern Seed Production Union *(Unión de Semilleros del Sur)* in the Brunca region. In Mexico, the first seed banks were established in 2005 as part of a national strategy for in situ conservation and to support farmers in areas vulnerable to natural disasters (Sánchez et al., 2016). In 2015, the country's 25 community seed banks were integrated into a network of conservation centers of the National System of Plant Genetic Resources for Food and Agriculture (SINAREFI) (Sanchez, Santos & Aragón-Cuevas, 2015). The objectives of these and other efforts include the conservation of biodiversity, promotion of diversity and food self-sufficiency of local communities.

Another example of policy under this *référentiel* is the Basic Food Commodity Price Guarantee Program, implemented by the Mexican Food Security Agency

(SEGALMEX), a decentralized agency linked to the Mexican government's Secretariat of Agriculture and Rural Development (SADER). Although the Price Guarantee is not new in Mexican public policies, the program, established in 2019, presents important changes in relation to previous actions (Fierro, 2019; Ávila-Sánchez, chapter 6). According to the government website, the program aims to: "increase the income of small agricultural producers, with the objective of compensating their efforts, and contribute to improving their quality of life; stimulate the national production of basic grains (and fresh milk) seeking to reduce dependence on imports; ensure the availability of food for the basic food basket, whose main distribution is in rural areas."[9] The program guarantees small and medium-sized producers minimum prices for corn, beans, rice, wheat and milk, staples in Mexican food culture; prioritizes its application in the areas of greatest social vulnerability and in indigenous communities; and is implemented alongside other actions aimed at promoting food self-sufficiency. In other words, with the goal of ending the dependence on imports and the fragility of the *food-for-market référentiel*, the program was designed to promote self-sufficiency and combat rural poverty (Fierro, 2019).[10]

In a way, the *food autonomy secondary référentiel* also seeks to overcome the limitations of the *urban food supply référentiel,* especially in relation to the way it delimits the public problem (the need to guarantee food for the urban population) and the way it is approached (the action of the State mainly through the "logistics" of supply). Indeed, centralized urban supply is characterized by the strong presence and action of intermediaries in urban food supply chains and by the sale of food from "anywhere," which travels long distances and has no local cultural or socio-economic links (FAO & FLAMA, 2020; McMichael, 2009). The *food autonomy secondary référentiel* has reoriented public urban supply policies through three types of responses triggered by family farmer, peasant and agroecological organizations: i) demands for guaranteed commercial spaces in supply centers, such as the Agricultural Producers' Trade Center (*Centro de Comercialización para Productores Asociados,* CECOPROA) in Asunción/Paraguay (Ríos & Wesz Jr., 2020) and sales space "at the Producer's Farm" in Brazil's Supply Centers (a direct sales space for farmers) (Eugênio, 2018); ii) the creation of Family Farming Marketing Centers or Farmers' Markets, such as the Family Farming Marketing Centers in the states of Minas Gerais, Ceará, Pernambuco and Rio Grande do Norte, Brazil (Moraes &

---

9. Available at: https://www.gob.mx/preciosgarantia

10. The program is part of the National Development Plan 2019-2024, which, among other objectives, seeks, "food self-sufficiency and the rescue of the countryside." The analysis expressed in a government document illustrates the search to break away from the *food-for-market référentiel*: "The agrarian sector has been one of those most devastated by neoliberal policies. Since 1988, mechanisms that were fundamental for agrarian development were destroyed, public support was oriented towards electoral manipulation, and the hollowing out of agricultural communities was promoted. The indigenous communities, which have experienced oppression, plundering and discrimination for centuries, were particularly affected by this offensive. Official policies have favored the establishment of agribusinesses and megaprojects and left community members, ejidatarios and small landowners high and dry. This has not only been disastrous for the peasants themselves but also for the rest of the country: Mexico currently imports almost half of the food it consumes, as well as most of the inputs, machinery, equipment and fuels used for agriculture." (SEGOB, 2019).

Pires, 2019); and, above all, iii) the strengthening of family and peasants' farming fairs and other traditional communities.

Associated with the previous sub-*référentiel*, but primarily highlighting the demands of ethnic groups, traditional communities, their mediators and specific territories, the *heritage secondary référentiel* more clearly articulates the components of ethnic, local and territorial food practices linked to identity. Once the problem of the loss of food and cultural heritage has become evident, this *référentiel* proposes solutions through policies for the rescue and valorization of food practices, products, ingredients, processes and knowledge.[11] Another noteworthy aspect is a discourse that values the connection between rural communities and urban consumers, who would be important agents for policies related to the recognition and valorization of food heritage. Certain certification mechanisms, intangible heritage registers, support for local fairs and festivals, and community tourism all serve to illustrate the instruments of this secondary *référentiel*.

An example of government actions in this secondary *référentiel* refers to the identification stamps developed for products supplied by family farms, peasants, indigenous peoples and other traditional groups. According to the Specialized Meeting on Family Farming (REAF) of Mercosur, family farming employs "production systems based on product diversification, preserving traditional food production, and the identity of the production communities and native populations, contributing both to the generation of a balanced diet and to the preservation of agrobiodiversity and the cultural values of rural communities" (MERCOSUR/REAF, 2014). In view of this, it is, "important to ensure the visibility of the work and products associated with family farming," and "develop instruments that benefit consumers by enabling them to identify products and services sourced from family farming" (MERCOSUR/REAF, 2014). In fact, several countries have already adopted instruments to that end, such as: i) Brazil, with the Family Farming Product Identification Seal (SIPAF), the Brazilian Quilombos Seal and the Brazilian Indigenous Seal; ii) Argentina, with the Produced by Family Farming Seal; iii) Chile, with the *Manos Campesinas* Seal; iv) Ecuador, with the Family Rural Farm Production Seal; v) Uruguay, with the Murú Rural Women's Production Seal; vi) Colombia, with the Peasant, Family and Community Agriculture Seal; and vii) Bolivia, with the Bolivian Social Seal (Cruz, Marques & Haas, 2020).

Another example is the designations of origin of food products, which recognize the distinctive character of a food product by virtue of differentiated qualities that are due exclusively or essentially to the geographical environment, including natural and human factors (Niederle, 2013). Cotija cheese in Mexico, Turrialba cheese in Costa Rica, Chuao cocoa in Venezuela, coffee in Colombia, Arriba cocoa in Ecuador, giant white corn from Cusco in Peru and Neuquén goat in Argentina, are some of the many examples of foods whose instrument seeks to protect and value as a result of collective ownership of the place and its inhabitants (Riveros et al., 2008).

---

11. In Brazil, the Ministry of Culture's National Historical and Artistic Heritage Institute (IPHAN), the agency responsible, since 2003, for the preservation Intangible Cultural Heritage, has inventoried and included several cultural assets related to food, in the "Celebrations" and "Knowledge" categories. In Argentina, in 2018, the Secretariat of Culture declared, "the uses and spaces of yerba mate" as cultural heritage in conjunction with the other MERCOSUR countries (Rebaï et al., 2021).

Claimed, for the most part, by actors (with different socioeconomic conditions) linked to production, designations of origin become mechanisms for promoting territories, providing access to markets and added value. Although appropriated by dominant actors in the agrifood sector who use them for other purposes, the core idea of geographical indications is that they convey the potential for, "a symbiosis between market recognition and the safeguarding of cultural heritage," so that "if consumers value products deeply rooted in local cultures, the greater the chances are of reproducing the knowledge and customs that make that product distinctive" (Niederle, 2013, p. 46).

Closely related to that above, the *gastronomic secondary référentiel* highlights the food issues generated by the processes of "gastronomization" – a term coined by Poulain (2011) to explain the recent and growing interest of the gastronomic world in regional and popular cuisine. This process goes beyond assigning local and traditional foods heritage status, as it more clearly incorporates an aesthetic dimension (Duarte et al., 2021). Moreover, more recently, due to the growing impact of socio-environmental criticism, this gastronomization process includes concern for the "ethicalization of aesthetics," seen in the affiliation of many chefs with the drive to support agroecology and family and peasant farming (Guilherme & Portilho, 2018; Barbosa, 2016). In this case, more than a vital necessity, food becomes a manifestation of pleasure and civic responsibility on the part of chefs, restaurants, solidarity economy companies and consumers.

Driven by chefs who sought to "revolutionize" gastronomy (Matta, 2012), this secondary *référentiel* is responsible for directing various actions implemented by Latin American States, albeit to a greater or lesser degree institutionalized or enduring. Although also observed in other contexts,[12] the projects developed in Peru provide the most emblematic showcase for the influence of this secondary *référentiel*. In 2007, the Peruvian government, through the National Institute of Culture of Peru (INC), declared the country's gastronomy as National Heritage, which triggered a gastronomic boom in the period (Matta, 2012). In this context, a variety of other actions were then implemented by civil society, private and governmental organizations seeking to build links between gastronomy and small farm production and the appreciation of local and regional foods. As mentioned by Zanetti (2017, p. 106), "In 2007, when traditional Peruvian gastronomy was beginning to carve out a space in the realm of contemporary national gastronomy, which until then was dominated by French cuisine, a group of Peruvian chefs, together with representatives from the State's agricultural sector, met at a round table session to discuss how the emerging gastronomic boom could leverage the country's rural development. As a result of this reflection, the Peruvian Gastronomy Association – APEGA was created with the objective of serving as an interlocutor to channel initiatives that promote the development of national gastronomy, reviving the appreciation for typical regional dishes and products." Since its founding in 2008, APEGA has developed a series of initiatives in collaboration with the Central Government and a number of local governments, such as

---

12. In Colombia, the Ministry of Culture developed a policy to promote traditional cuisine that encouraged regional diversity. In Argentina, the Secretariat of Culture initiated a "food and gastronomic cultural heritage" program (Rebaï et al., 2021).

the Mistura gastronomic festival in Lima, aimed at building links between chefs and small farmers; the publication of "Peruvian Gastronomy for 2021: Guidelines for a Development Program of Peruvian Gastronomy within the Framework of the Bicentenary Plan," which sets out medium-term guidelines for gastronomy, among them the need to, "connect small producers to the growing national and international gastronomic market" (Peru, CEPLAN, 2012, p. 12); and a wide range of regional and local gastronomic festivals.

Another example of this secondary *référentiel* involves school canteens in Brazil. Seeking to "recognize the role of school *merendeiras*[13] in promoting healthy and adequate food in the school environment and promote the mobilization of the school community on the issue of food and nutrition education," the Federal Government has been developing the, "Best School Meal Recipe Competition" (Brazil, FNDE 2017). The recipes must "preferably contain important ingredients for healthy eating, limiting the amounts of unhealthy foods, such as canned foods, sausages, powders, concentrates and dehydrated foods"; be linked to some kind of educational action on food; and include foods produced through family farming in their preparation (Brazil, 2017). Innovative and original recipes that promote the use of regional foods are also more favorably regarded in the judging process. At national level, the National Commission is composed, among other actors, of a student, a prestigious chef and a nutritionist responsible for the program. Guéneau et al. (see chapter 12) discuss a complementary program developed by the government of the Federal District in Brazil. Through the Chef and Nutri at School Project, the government offers the opportunity for a chef to visit schools, valuing the work of the *merendeiras* (school cooks) and encouraging adequate and healthy nutrition. Driven mainly by government actors, these initiatives promote the gastronimization of school food, in its aesthetic and ethical dimensions, seeking to qualify school meals and sensitize those who prepare them about the multiple functions of the food provided.

The *environmentalized food référentiel* is based on a representation of the food problem largely linked to the interactions between the food system and the *Earth* system. The global food system is understood to be one of the main drivers and, at the same time, the principal victim of environmental degradation (Rockström et al., 2020). Some of the scientific actors involved in environmental sciences and NGOs support the concept of *planetary boundaries* that the food system cannot cross (Campbell et al., 2017; Gerten et al., 2020; Springman et al., 2018). Thus, the *environmentalized food référentiel* includes standards that seek to limit or minimize the impacts of food systems on the environment and the availability of natural resources (water, land, biodiversity, biomass, fossil oil and minerals), the stakes being meeting future food demand with current natural resources (Davis et al., 2016). The process of constructing this *référentiel* also largely stems from the actions of "consumer-actors" (Stassart, 2010) that have an impact both on companies' revenues (boycott and *buycott*) and on their political choices. Thus, govern-

---

13. Unlike in other countries, in Brazil school "merendeiras" are cooks hired by the State to prepare the meals served in school canteens. "A category composed mostly of women, with a low level of education, the *merendeiras* perform their work in school kitchens, preparing the food that will be offered to students during their time at the educational institution" (Cardilho, 2018, p. 11).

ments and companies have paid increasing attention to the creation of policies that engage with this *référentiel*, which is expressed, for example, in instruments related to environmental responsibility (e.g., the Soy Moratorium[14]), the reduction of losses and waste (Webb et al., 2020), the adoption of clean technologies, environmental certifications, Food Miles, the reduction of meat consumption and transformation of diets, etc.

One example that is evident in several Latin American countries refers to actions to promote low-carbon agriculture and/or livestock farming. In response to evidence and criticism regarding the amount of greenhouse gas emissions caused by agricultural and livestock practices, and often accompanying the establishment of national climate change mitigation and adaptation plans, several national governments have established sectoral plans for agriculture and livestock based on the reduction of carbon emissions. Examples include the Low Carbon Development Strategy (EDBC) for the livestock sector in Nicaragua (Canu, et al., 2018), the Low Carbon Livestock Strategy in Costa Rica (Costa Rica, Ministerio de Ambiente y Energía, 2015), the National Strategy for Low Emissions Sustainable Bovine Livestock (Guatemala, 2018), the Colombian Low-Carbon Development Strategy (ECDBC) with Nationally Appropriate Mitigation Action (NAMA) for Sustainable Livestock, and the Low-Carbon Agriculture (ABC) Plan and Program in Brazil (Brazil, 2012). The actions promoted by these programs and actions include: pasture rotation and division associated with crop-livestock-forest integration systems, live fences and dispersed trees in pastures; manure management through biodigesters; biofertilizer production and use; pasture recovery and productive intensification; promotion of livestock farming in areas more conducive to more intensive and profitable systems; technological innovation; and strengthening of value chains, access to markets and low carbon or carbon neutral certifications. In general, in response to consumer pressures and international environmental agreements, these government programs address the need to bring about changes and reduce greenhouse gas emissions in their food production systems, not infrequently through the intensification of production and technological innovations.

Another action that is on the governmental agenda of several countries, despite challenges in terms of its institutionalization, refers to the creation of programs or policies to reduce food losses and waste. In addition to the economic costs and damage to food and nutritional security, losses and waste represent significant environmental costs. As stated by the Government of Argentina, "producing food that no one will consume means using natural resources, services and other goods inefficiently. Added to this is the environmental cost of greenhouse gas emissions generated along the food chain, which unnecessarily contribute to global warming and climate change. The environmental consequences are greater as waste is produced at more advanced stages of the chain. In other words, the higher the degree of transformation of a food, the

---

14. Following evidence provided by an international NGO on the expansion of soybean production in the Amazon and its relationship with increased deforestation and its impact on European consumers, organizations representing the agribusiness chain signed the Soy Moratorium in 2006 committing not to purchase soybeans from deforested areas as of that date. In 2008, the Brazilian government joined the initiative, legitimizing it, contributing to the geospatial monitoring of the commitment and establishing other incentive actions (IMAFLORA, 2017).

greater the environmental impact. Finally, food waste increases the volume of waste and aggravates the problem of urban solid waste treatment" (Argentina, Ministerio de Agroindustria, 2017). To address this, the Argentine government created its National Program to Reduce Food Loss and Waste in 2015 and, in 2018, institutionalized Act 27,454 that creates the National Food Loss and Waste Reduction Plan, which was regulated in 2019. These initiatives include the implementation of information and communication campaigns to raise awareness among all actors in the food chain; improvements in transportation infrastructure, energy and market facilities; development and facilitation of access to equipment and new technologies/innovations; and governance between producers, intermediaries, food bank[15] associations, non-governmental organizations, educational establishments, among others, to make agreements to reduce losses and waste through donations (Argentina, Segretaría de Alimentos y Bioeconomía & Segretaría de Agroindustria, 2018).

In terms of consumer-actor actions, it is important to stress that the *environmentalized food référentiel* is largely focused on the issue of "sustainable diets," pointing to, "More environmentally-friendly dietary patterns, such as, for example, reducing the consumption of animal products, especially beef" (Takeut & Oliveira, 2013, p. 200). As a general rule, there is a certain convergence between sustainable diet and healthy diet (Triches, 2020), although the environmental issue is still little addressed in food guides, a public policy tool widely used to inform and raise consumer awareness about healthy and sustainable eating (Martinelli, Cortese & Cavalli, 2020). This environmental vision is beginning to direct several local and national food policies, such as food purchase programs for schools and other public institutions (prisons, hospitals, etc.), although a predominance of family farming and rural issues is still observed in these programs (see Grisa et al., chapter 16). Although incipient at municipal level, in recent years several public actions based on this *référentiel* have emerged in Brazil, including, as a priority objective, the reduction of animal protein consumption and the diversification of school meals, such as the Meatless Monday Program in São Paulo state and São Paulo municipality, the Innovative Tastes (*Inova Sabores*) Project in Jundiaí (São Paulo state), the Brazilian Food Awareness Program in Niterói (Rio de Janeiro state) or the Sustainable Schools Program in four municipalities of Bahia state (Cheib, 2020).

Finally, a new type of *référentiel* has emerged in recent years, which we call *integrative*. Although inspired by the values of sustainability, this *référentiel* goes beyond the purely socioecological dimensions of food, integrating other aspects related to nutrition and health (Ribeiro & Jaime, 2017). Thus, in recent years, the food problem has become a problem of integration between these different dimensions within a systemic

---

15. Food Banks are initiatives aimed at using food that, even if suitable for consumption, would otherwise be thrown away or wasted. Operated in many Latin American countries, these initiatives are usually promoted by civil society organizations, in addition to government initiatives. In the case of Brazil, the issue entered the governmental agenda in the early 2000s in the context of food and nutrition security policies. In 2000, the Municipal Government of São André (SP) created the Santo André Municipal Food Bank and, in 2003, the issue gained a foothold on the federal agenda, being supported by the Extraordinary Ministry of Food Security and Fight against Hunger (MESA). In 2017, the federal government estimated that there were 107 Food Banks being publicly managed by state and municipal governments (Grisa & Fornazier, 2018).

approach to food. In addition, the reference to the "sustainable food system,"[16] which has become dominant in the literature and the discourse of international institutions (Béné et al., 2019), has been evolving in recent years. The terms "integrated" and "resilient" have been added, firstly to reflect the integration between the different components of food systems, including the consideration of climate change and rural territories, and, secondly, to highlight the imperative for these systems to recover and adapt in the face of external shocks or stresses (Blay-Palmer, 2020). This *integrative référentiel* also emphasizes the need for "whole system change" to achieve the Sustainable Development Goals by changing food consumption patterns (sustainable consumption and waste reduction), transitioning to sustainable forms of production and value chains that ensure nutrition, mitigating climate change and improving rural territories (Caron et al., 2020).

Thus, public policies inspired by this new *référentiel* seek to systemically respond to various food problems, such as obesity, malnutrition, climate change, poverty and social exclusion (Sonnino et al., 2020; Swinburn et al., 2019; Willet et al., 2019; IPES Food, 2019). This *référentiel* legitimizes policies to promote sustainable production from family and peasant farming, with the promotion of adequate, healthy and responsible food, highlighting the importance of short marketing circuits. In addition, another innovative aspect of this reference framework is the requirement for greater intersectoral cooperation in public action, insofar as the promotion of sustainable food systems depends on the coordination of policies managed by ministries, secretariats and departments across different areas (agriculture, health, nutrition, social development, education, etc.).

An example of government action under the *integrative référentiel* is the Public Policy for Food and Nutritional Sovereignty and Security in Cali, Colombia, analyzed in this book by Castelhano, Henry and Rankin (see chapter 13). Stimulated by national advances,[17] in 2009 the Municipality of Cali initiated the construction of a municipal policy in a participatory manner, through the Food and Nutritional Security Roundtables, culminating with the regulation of the policy through Agreement No. 0470 of 2019. The policy seeks to "improve the quality of life of the population of Santiago de Cali through constant, coordinated and intersectoral action on food and nutritional security" (Concejo de Santiago de Cali, 2019). The policy is governed by 11 principles, namely integrality, considering that all food sovereignty and food security strategies and actions are elements that act in a coordinated and integrated manner in the food management process; universality, seeking to guarantee the

---

16. Sustainable food system is a food system that provides food security and nutrition for all in a way that does not compromise the economic, social and environmental basis for generating food security and nutrition for future generations (HLPE, 2014).

17. Following Castelhanos, Henry and Rankin (chapter 13), we refer to the National Food and Nutrition Plan (PNAN) established in 1999, and the National Food Security Policy, in 2008, with its respective National Food Security and Nutrition Plan (2012-2019). In fact, several countries have established national food and nutrition policies or strategies, examples of which include the National Food Security and Nutrition Strategy 2013-2021 (Peru, 2013), the National Food and Nutrition Policy of Chile (Chile, 2017), the National Food Security and Nutrition Policy 2011-2013 of Costa Rica (Costa Rica, 2011) and the National Food and Nutrition Policy of Brazil (Brazil, 2013). Often developed by Ministries of Health and/or Agriculture, these plans often face the challenges of implementation, broader intersectoral dialogue, and continuity given changes in government.

right to food for all people; diversity and interculturality; intersectoral cooperation and transversality, linking different social and institutional actors; community and citizen participation; food equity through the inclusion of socially vulnerable groups generally excluded from markets; solidarity through individual and collective actions; regional integration; agroecological sustainability; community strengthening; and support for young peasants (Concejo de Santiago de Cali, 2019). In addition, the policy is organized into six axes, each of which involves a set of actions, namely: availability, access, adequate and sufficient consumption, quality

and safety, biological use of food, and management and promotion of knowledge. Thus organized, it aims to address a wide range of contemporary food challenges, such as hunger, obesity, climate change and food democracy, seeking to transform the food system to be "sustainable over time and nutrition-sensitive" (Concejo de Santiago de Cali, 2019).

The performance of other programs corresponding to the *integrative référentiel* are also worthy of note, such as the National Urban, Suburban and Family Agriculture Program in Cuba and the configuration of the National School Feeding Program in some Brazilian municipalities. The Cuban program, which has been operating since 1997, is configured as a system from national to municipal level, bringing together several subprograms with the objective of "maximizing the production of diverse, fresh and healthy food in the available, previously unproductive areas. This production is based on ecological practices that do not pollute the environment, rational use of resources in each territory and direct marketing to the consumer" (FAO, 2016, p. 42). The program combines the landscaping of urban spaces, with the promotion of food supply, healthy eating, employment and income generation, socioeconomic inclusion, environmental preservation and social participation (FAO, 2016).

Regarding Brazilian school canteens, the program contributes to minimizing situations of hunger and food insecurity, and features a set of nutritional guidelines aimed at combating malnutrition, excess weight and obesity (e.g., controlling the supply of processed and ultra-processed foods) (Brazil, FNDE, 2020). In addition, since 2009, it has incorporated and guaranteed a market (at least 30%) for food produced by family farming, agrarian reform settlements, indigenous peoples and quilombola communities, giving priority to organic or agroecological and locally produced food (Brazil, FNDE, 2009). In addition to these federal regulations, some states and municipalities (e.g., the municipal government of Marechal Cândido Rondon and the state governments of Paraná and Santa Catarina) have adopted complementary laws, establishing minimum percentages or even total purchase of agroecological or organic food, in order to increase the supply of healthy food and contribute to more sustainable production practices. These actions are also complemented by a set of food and environmental education initiatives (composting, conservation of water and other natural resources, school gardens, etc.).

▸▸ Politics in food policy: from *référentiels* to coalitions

In the previous section we presented the different *référentiels* that frame and dispute public policies in relation to food problems in Latin America. In this section we analyze how actors-coalitions act to politicize certain issues, (de)construct and (de)legitimize certain *référentiels* in order to change the dynamics of social fields, i.e., food systems – thus also changing their positions in the structural hierarchy of these fields. As in the previous section, we do not intend to exhaust the subject. We have only mobilized elements that illustrate how, over time, actors-coalitions produce innovations that alter power relations within public arenas, leading to the (re)production of policies that, guided by *référentiels* more or less close to their interests/worldviews, stress continuity or changes in food systems.

It is worth noting at the outset that the dynamics of the fields (food systems) are not only influenced by public policies. In fact, there are fields in which the strategies of private companies and the actions of organized social movements are more relevant than the action of the State, and even fields in which there are still no public policies (or they are very specific and weak) with *référentiels* in line with the ideas of these actors. This is the case of the field that has been structured around alternatives to meat production and consumption. Despite fragmented actions (such as the inclusion of vegetarian and vegan alternatives in certain public canteens), it is rash to speak of public policies directed by a *référentiel* that institutionalizes ideas similar to those advocated by companies producing "vegetable meat" or by anti-speciesist and/or climacteric consumer movements. It is also the case of the movement for the ethicality of gastronomy that, although supported by public authorities (as discussed above), the level of activity from these channels is not comparable to the diversity and intensity of the actions and activism being driven by consumer movement, chefs and other actors. In these two examples, initiatives by non-state actors that empower and stress food systems predominate.

Secondly, it is important to ratify that the *référentiels* presented in the previous section refer to a set of ideas institutionalized, to a greater or lesser degree, in public policy. They are ideas, interpretations and regulations that come from different forums for the production of ideas, composed of individuals and groups that politicize and dispute the construction of public policies (Fouilleux, 2000). When entering the institutional and political game, many of the collective representations of individuals and groups have to adjust to and are confronted with the ideas of other individuals and groups, with their own and other institutional environments, and are influenced by external events. In other words, the ideas that exist in public arenas are reorganized, readjusted and molded as their process of institutionalization within public policy advances.

Third, it is important to note that, although manifested in legal and regulatory frameworks, *référentiels* present different degrees of vulnerability to change or, conversely, *path dependency* mechanisms (Pierson, 2000; Mahoney, 2001). These movements (vulnerability or institutional reinforcement) depend on the institutional fragilities and ambiguities of public policies and instruments, on the tensions that exist in agrifood systems, and on the power relations between groups of actors defending or challenging the representations of the world being brought into play. Throughout this book, we will see that while some *référentiels* have a long lifespan and are practically crystallized in institutions, others have an ephemeral life and do not withstand changes in mandates.

Moreover, it should be noted that the influence and relative importance of each *référentiel* in the dynamics of food systems depends, as we have said, on logics that go beyond the State's capacity for action, as well as the power conferred and legitimized by the political instruments that set them in motion. Although it is not a question of simply comparing budgets or numbers of beneficiaries, given that institutional and power logics are much more complex, it is to be expected that a billion dollar rural credit program with national scope and operated by financial organizations with a high degree of capillarity will have a much greater capacity to effect the way food production, supply and consumption is organized than local initiatives implemented by a municipal governments that are usually reliant on resources donated by charitable foundations.

Considering these political dynamics that determine the politicization of food issues and the construction, institutionalization and permanence (or not) of *référentiels*, certain inferences can be drawn regarding the general characteristics that influence the food policy production process in Latin American countries.

The new is not created from the rubble of the old

A well-known expression, attributed to Marx's early writings on the transition between different production systems, states that "the new is created from the rubble of the old." It is difficult to affirm that this maxim applies to food policies in Latin America. The results presented in this book suggest that the emergence of new *référentiels*, incorporating concerns about environmental issues, public health and cultural heritage, were not generated from the rubble of *référentiels* and policies on their way out or even in crisis. The "old" *food-for-market* and *productivist référentiels*, sustained by coalitions involving entities representing the more traditional agrarian elites and/or agrifood corporations, have been resurrected with each food supply or price crisis, legitimizing political spaces, institutional structures and some of the most important public policies on resources.

It is true that, to a greater or lesser extent, these *référentiels* have proven to be insufficient and have been harshly criticized in all countries, giving way to the emergence of others. Returning to the expression, we could say that the "new" is created, not from the rubble, but from the exacerbations and persistence of the old, shining the spotlight on its results and inadequacies. However, the level of institutionalization of other *référentiels* is highly variable. While the *food social welfare* and *technical food référentiels* are associated with historically legitimate public problems (malnutrition, food health, etc.), have the support of broad and even dominant coalitions, and are linked to relatively stable institutional structures, the *food sovereignty and security* and *integrative référentiels* guide incipient, fragile and contested public policies, which are generally produced and managed by peripheral actors, located in precarious governmental structures and face resistance from dominant coalitions.

In short, a striking feature of food policies in the Latin American context is the fact that they do not point unequivocally towards a process of transition between different food models. In international literature, especially that produced in the European context, it is common to find examples of policies that demonstrate attempts to replace the old extractive forms of agriculture, based on the plundering of natural resources, with models that respond to the principles of what have been called "sustainable food systems" (European Commission, 2020; IPES-Food, 2019).

There are examples of this type of public policy in Latin America, including experiences such as the Food Acquisition Program (PAA) and the National School Feeding Program (PNAE) in Brazil, which have become international public policy benchmarks for promoting territorial recognition and healthy eating, socioeconomic inclusion and more sustainable practices (WFP, 2013). However, a closer look at food policies in the region shows not only the coexistence of different *référentiels*, but also the predominance of institutional logic that supports *productivist* and *food-for-market référentiels* (Sabourin et al., 2020; Sabourin, Craviotti & Milhorance, 2020).

Three explanations can be highlighted to understand this phenomenon. First, the strength of the coalitions formed among the traditional agrarian elites, who control key spaces in the structure of the States, not only in the executive branch, but also in the legislative and judicial branches. In fact, even the progressive governments of the Latin American "pink tide" have had great difficulty innovating in the production of food policies and, when they have done so, they have generally had to yield to agreements that have also led to the strengthening of productivist and liberalizing policies (see Albarracin, chapter 7; Riella & Mascheroni, 2015; Delgado, 2012). Second, the public legitimacy of these *référentiels*. Characteristic of long-standing institutions, this legitimacy is manifested in the fact that, even among agrarian social movements that confront national elites, the demand for agrarian policies predominates, which, although aimed at specific actors, such as family farmers, are still basically directed by the *productivist référentiel* (Sabourin, Samper & Sotomayor, 2015). Third, the fact that this legitimacy is directly associated with the collective memory of the principal historical food problem in Latin America, namely the difficulty of access to food and its most violent expression in poverty and hunger. In the past two decades, the progress demonstrated by several countries in reducing phenomena such as these has been crucial in paving the way for other *référentiels* focused on other issues that are equally or more serious than hunger, such as obesity. However, when the economic crisis worsened and hunger returned to the public agenda, the *productivist référentiel* once again demonstrated why it continues to be hegemonic.

"External shocks" to reinforce "the old"

Another feature common to most Latin American countries, and which is directly associated with the explanations for the power of the old *référentiels*, is the existence of a series of "external shocks" produced in other arenas (Thelen & Mahoney, 2010; Mahoney, 2001) that, over time, influenced the design of food policies. Despite the potential for these events to lead to innovation and the emergence of new *référentiels*, we would argue that their effect has led to the reinforcement of the more consolidated *référentiels* discussed in the previous section.

In general, innovations in food policies were the result of relatively gradual, cumulative and transitory processes. Although progressive governments have occupied institutional political spaces over the years, they have encountered resistance from different orders and areas which, unlike the disruptions in food systems, have produced gradual changes in food policies amidst the preservation of power relations in certain spheres. Thus, based on the criticisms expounded by different social movements and any windows of opportunity that arose, progressive governments have been developing and incorporating new *référentiels*. This is particularly notable in the cases of policies based on the *integrative* and *food sovereignty* and *security référentiels,* which have been incorporated in some countries without compromising the institutional structure previously constructed for *productivist* and *food-for-market* based policies.

Given the changes (albeit gradual) that had been taking place with regard to food (and other broader areas), it was not long before conservative political coalitions attempted to respond by reestablishing hegemonic political paradigms in the region, either through democracy or institutional ruptures, particularly through the deposition of governments, a common scenario in the Latin American context. We are not claiming that the military or institutional coups that occurred in Latin America were exclusively or primarily caused by disagreements over the food agenda but would stress that, as they occurred, these events and ruptures strengthened the power of traditional coalitions, reinforced hegemonic dynamics in food systems and generated the right conditions for the dismantling of innovative policies.

Fragmentation and sectorization of public policy *référentiels*

In addition to the institutional instability mentioned above, the fragmentation and sectorization of public management can be highlighted as two other striking characteristics seen in Latin American states. In this chapter we have examined eight *référentiels* (with their corresponding secondary *référentiels*) that direct and challenge the construction of food policies in the region, and these *référentiels* are accommodated to a greater or lesser extent within certain state structures (ministries, secretariats, agencies, departments, public enterprises, etc.) responsible for managing food policies. States are arenas of power and dispute in which different coalitions, based on different understandings of food systems, are provided more or less political and institutional space to channel their demands and challenge the construction of public policies, reflecting the differences in power among them (as discussed in the first point). Even though all the *référentiels* form part of the governmental agenda, they do not command the same levels of resources and are usually dispersed across governmental structures.

We saw in the previous section, and in particular in table 1, that different problems (Kingdon, 1984; Cohen, March & Olsen, 1972) mobilize different actors (producers' organizations, family farmers, corporations, consumer organizations, chefs, professional categories and bureaucracies located in different government agencies), whose field of intervention and proposal of public policies is limited by the interpretations of the world that influence these actors, or by institutional and power structures that prevent/block the construction of more intersectoral policies and *référentiels*. Thus, although caricaturesque, it is not uncommon for food production issues to fall under the competence of the Ministries of Agriculture and Livestock – which generally deal with production incentives; the food access problems are dealt with by the Ministries of Social Support or Development – mainly concerned with ensuring access to food for the poorest; and nutritional and food security problems are dealt with by the Ministries of Health, which work to address nutritional deficiencies or imbalances and food risks. Although a large number of state agencies are involved and participate in the construction of food policies, difficulties exist in terms of interaction and coordination, which may even result in the development of contradictory actions.

Although this is not specific to the food issue (Repetto, 2005; Aureliano & Draibe, 1989), this sectorization and disjointed approach (regardless of the topic) produces fragmentation, overlapping and repetition with regard to state actions, and a reduction in the efficiency and effectiveness of financial and human resources. Moreover, as mentioned by Repetto and Fernandez (2012), this lack of streamlining minimizes the potential for the coherent and aligned inclusion of state interventions in long-term development objectives and plans through the process of consultation of the perspectives and views of actors. And this is particularly important in the contemporary context, when systemic crises (climate change, the triple burden of malnutrition, the economic crisis and the Covid 19 health crisis) interact and leverage each other (Pretty, 2020; Blay-Palmer et al., 2020; Preiss & Schneider, 2020; Swinburn et al., 2019).

In line with this critique, several authors and international organizations have highlighted the importance of building food policies that are integrated (Sonnino et al., 2020; European Commission, 2020; IPES-Food, 2019) or guided by the *integrative référentiel*. In the words of Sonnino et al. (2020), integrated food policies are those that are based on the interdependence (synergies and feedback) between environmental, socioeconomic, nutritional and health issues, and target the entire population, whether they are middle- or upper-class consumers, or vulnerable groups that often experience a number of food-related issues at the same time (poverty, malnutrition or undernutrition and environmental threats). These food policies require the participation of multiple stakeholders to achieve vertical coordination between different levels of government (federal, state and municipal) and horizontal coordination between different government sectors and actors (Sonnino et al., 2020). Rather than centralizing or aggregating a set of actions from different agencies and levels of governance – as is often observed in National Plans and Strategies – it is important that public actions and policies are constructed that seek to address different food issues simultaneously and in a complementary manner.

"New" actors constructing new reference points

We saw in the previous section that several *référentiels* and innovations in food policies were built on the ideas of civil society organizations and the interaction between the State and society, either through institutional fissures or pressure mechanisms in public arenas. We have also seen that, increasingly, there are "new" actors involved in the politicization of food and food issues, and seeking to bring about changes in food systems, whether from the sphere of private consumption, or from organizational practices (cooks' associations, consumer cooperatives, communities that support agriculture, etc.), or even through public policies. Where in the 1970s to 1990s, criticism of agrifood systems emanated mainly from agrarian and environmental actors (experiencing the effects of land inequality and the modernization of agriculture), in the 1990s to 2000s, the cause was taken up by the food activism of consumers, urban organizations and a host of professionals (nutritionists, urban planners, architects, engineers, chefs, etc.). Although not always organized into structured coalitions with the capacity to influence public spheres, this wide range of actors has been a decisive force in challenging food interpretations and practices.

Thus, in addition to the need for governmental actors to coordinate their efforts, we would also highlight the imperative need for participation from non-governmental actors (Sonnino et al., 2020) and the importance of the construction of spaces for consultation and social participation, such as the Food and Nutrition Security Councils operating in some countries and the Food and Nutritional Security Round-tables in Cali-Colombia. Rather than this social participation being purely the result of political pressure, we stress the importance of participation becoming common practice in public arenas and in the life span of public policies. In fact, this is also a dimension incorporated by the *integrative référentiel*, discussed in the previous point. The institutionalization of its ideas necessarily implies the participation of society in decisions related to food systems, considering that the construction of food democracy (Lang, 1999) is another element that structures this *référentiel*. As Hassanein (2003, p. 79) puts it, food democracy is "the idea that people can and should actively participate in shaping food systems, rather than remaining passive spectators on the sidelines." In other words, food democracy is about citizens being able to determine local, regional, national and global agrifood policies and practices. Changes in the asymmetry of power between coalitions and between *référentiels* also depend on the democratization of power spaces and opportunities for society as a whole to "have a voice" and be included in decision-making.

▸▸ Final considerations

The analysis set out in this chapter points to different ideas and interpretations that influence and question the construction of food policies in Latin America. Focusing our reflection on the different values and norms that guide the construction of public policies, as well as on the way in which actors interpret public problems, define solutions (where to intervene? How to intervene?) and position the role and instruments of the State, we map eight public policy *référentiels*, which are: i) *food-for-market*; ii) *urban food supply*; iii) *productivist* (with its variations between *productivist based on agribusiness* and *productivist based on family farming*); iv) *food social welfare*; v) *technical food* (with its manifestations in the *nutritionist* and *hygienist secondary référentiels*); vi) *food sovereignty and security* (with its differentiations between *food autonomy*, *heritage* and *gastronomic*); vii) *environmentalized food*; and viii) *integrative*.

As we have seen, these different *référentiels* are present in different countries, although with different degrees of incidence, institutionalization and stability. Such *référentiels* are present in different spaces and institutionalisms in each country, expressing, on the one hand, the fragmentation, disjointedness and contradiction that exists in public action and, on the other, the configuration of the State as a field of power and dispute. Some *référentiels* have long been institutionalized in public action (*productivist, food-for-market, urban supply référentiels*), expressing the power relations, influence and permeability of the agrarian elites, agroindustrial groups, market actors, international organizations; or even, the concern with historical and recurrent problems in Latin American agrifood systems (hunger, poverty, nutritional deficiencies, food security) (*food social welfare and technical food référentiels*).

Expressing their historical trajectory and their political and cognitive roots in societies, such *référentiels* are often reinforced by political, economic and social crises.

Other *référentiels* (*the food autonomy secondary référentiel, the heritage* and *gastronomic, environmentalized food* and *integrative référentiels*) have been challenging spaces in governmental arenas and agendas more recently and derive from the emergence/social construction of new public problems, the entry of new actors into the sphere of agrifood-related issues, and the emergence of new social movements and forms of collective action. Actors and social movements marginalized or negatively impacted by "old" *référentiels* (mainly family farming and peasant organizations), consumers and consumer organizations, different professional categories (chefs, cooks, nutritionists, urban planners), environmentalists, organizations in the agroecological field and more are some of the actors that began to form coalitions, dispute the state and create institutional fissures for the institutionalization of new ideas and interpretations of the world. This institutionalization, however, is susceptible to the influence of other coalitions of actors, external events and dynamics of other strategic fields of action.

Such public policy *référentiels* have different effects on the dynamics of agrifood systems. Several authors and international organizations highlight the way in which the actions of the State have contributed to and promoted the recurrent socioeconomic, health, food and environmental crises experienced by contemporary societies and, at the same time, emphasize the key and fundamental role of the State in the construction of and transition to sustainable food systems. The State has the resources and capacity to intervene, promote, coordinate, regulate and monitor healthy, sustainable and inclusive production and consumption. "(…) the State emerges as a powerful actor: its regulatory power, its huge budget and, not least, its mandate to act on behalf of the public interest, endow it with unique authority over both food supply and consumption" (Sonnino, Spayde & Ashe, 2016, p. 313). Thus, we highlight the importance of the State, in dialogue and co-construction with society, to advance in the institutionalization of public policy *référentiels* that, acting in different dimensions, contribute to a profound transformation of current food systems towards sustainable, integrated and resilient food systems. The *food sovereignty and security, environmentalized* and *integrative food référentiels* are particularly important in this regard.

▸▸ References

Acosta, A. (2016). *O bem viver: uma oportunidade para imaginar outros mundos*. (T. Breda, Trans.). Autonoma Literária, Elefante Editora. (Original work published 2008). https://rosalux.org.br/wp-content/uploads/2017/06/Bemviver.pdf

Argentina, Ministerio de Agroindustria. (2017). *Informe de avance Programa Nacional de Reducción de Pérdida y Desperdicio de Alimentos 2013–2015*. http://www.alimentosargentinos.gob.ar/HomeAlimentos/ValoremoslosAlimentos/imagenes/Informe_de_avance_PDA_2013_2015.pdf

Argentina, Programa de nutrición y alimentación nacional, Ley Nacional de Argentina 25.724 (2003). http://servicios.infoleg.gob.ar/infolegInternet/anexos/80000-84999/81446/norma.htm

Argentina, Segretaría de Alimentos y Bioeconomía & Segretaría de Agroindustria. (2019). *Informe de avance 2018. Programa nacional de reducción de pérdida y desperdício de alimentos*. Ministerio de Producción y Trabajo. http://www.alimentosargentinos.gob.ar/HomeAlimentos/ValoremoslosAlimentos/documentos/Informe_avance_2018_PDA.pdf

Aulicino, C., & Langou, G. D. (2012). *La implementación del Plan Nacional de Seguridad Alimentaria en ámbitos subnacionalies* (Documento de Trabajo n. 88). CIPPEC.

Aureliano, L., & Draibe, S. M. (1989). A especificidade do "welfare state" brasileiro. In Ministério da Previdência e Assistência Social (MPAS)/Comissão Econômica para América Latina e Caribe (CEPAL), *Reflexões sobre a natureza do Bem-estar* (pp. 86–178). Cepal. http://hdl.handle. net/11362/29505

Baca, A. S. (2014). *El patrón alimentario del libre comercio.* UNAM, Instituto de Investigaciones Económicas, CEPAL.

Barbosa, L. (2016). A ética e a estética na alimentação contemporânea. In Cruz, F.T., Matte, A., & S. Schneider (Eds.), *Produção, consumo e abastecimento de alimentos: desafios e novas estratégias* (pp. 95–123). UFRGS.

Barquera, S., Rivera-Dommarco, J., & Gasca-Garcia, A. (2001). Políticas y programas de alimentación y nutrición en México. *Salud pública de México, 43*(5), 464–477. https://www.saludpublica.mx/index.php/spm/article/view/6342/7639

Barton, D., Ring, I., & Rusch G. (2014). *Policyscapes*: nature-based policy mixes for biodiversity conservation and ecosystem services provision (Policy brief). POLICYMIX. https://policymix.nina. no/Portals/policymix/POLICYMIX%20Policy%20brief%202_web_1.pdf

Béné, C., Oosterveer, P., Lamotte, L., Brouwer, I. D., de Haan, S., Prager, S. D., Talsma, E. F., & Khoury, C. K. (2019). When food systems meet sustainability – Current narratives and implications for actions. *World Development, 113*, 116–130. https://doi.org/10.1016/j.worlddev.2018.08.011

Blay-Palmer, A., Carey, R., Valette, E., & Sanderson, M.R. (2020). Post Covid 19 and food pathways to sustainable transformation. *Agriculture and Human Values, 37*, 517–519. https://doi.org/10.1007/s10460-020-10051-7

Bonanno, A. (2019). State capitalism in neoliberal agrifood: the case of the Brazilian company JBS. In Bonanno, A., & J. S. B. Cavalcanti (Eds.), *State capitalism under neoliberalism: the case of agriculture and food in Brazil.* Lexington Books.

Brazil, Fundo Nacional de Desenvolvimento da Educação (FNDE). (2009). *Resolução/CD/ FNDE nº 38, de 16 de julho de 2009. Dispõe sobre o atendimento da alimentação escolar aos alunos da educação básica no Programa Nacional de Alimentação Escolar.* https://www.fnde. gov.br/index.php/acesso-a-informacao/institucional/legislacao/item/3341-resolução-cd-fnde-nº-38-de-16-de-julho-de-2009

Brazil, Fundo Nacional de Desenvolvimento da Educação (FNDE). (2017). *Edital nº 003/2017; Concurso Melhores Receitas da Alimentação Escolar – 2ª edição.* http://melhoresreceitas.mec.gov.br/docs/regulamento.pdf

Brazil, Fundo Nacional de Desenvolvimento da Educação (FNDE). (2020). *Resolução nº 6 de 08 de maio de 2020. Dispõe sobre o atendimento da alimentação escolar aos alunos da educação básica no âmbito do Programa Nacional de Alimentação Escolar-PNAE.* https://www.fnde.gov.br/index.php/centrais-de-conteudos/publicacoes/category/100-resolucoes?download=13857:resolução-nº-6,-de-08-de-maio-de-2020

Brazil, Ministério da Agricultura, Pecuária e Abastecimento (MAPA). (2012). *Plano setorial de mitigação e de adaptação às mudanças climáticas para a consolidação de uma economia de baixa emissão de carbono na agricultura.* https://www.gov.br/agricultura/pt-br/assuntos/sustentabilidade/plano-abc/arquivo-publicacoes-plano-abc/download.pdf

Brazil, Ministério da Saúde. (2013). *Política nacional de alimentação e nutrição.* https://bvsms.saude. gov.br/bvs/publicacoes/politica_nacional_alimentacao_nutricao.pdf

Campbell, B. M., Beare, D. J., Bennett, E. M., Hall-Spencer, J. M., Ingram, J. S. I., Jaramillo, F., Ortiz, R., Ramankutty, N., Sayer, J. A., & Shindell, D. (2017). Agriculture production as a major driver of the Earth system exceeding planetary boundaries. *Ecology and Society, 22*(4), Article 8. https://doi.org/10.5751/ES-09595-220408

Canu, F. A., Wrewtlind, P. H., Audia, I., Tobar, D., & Andrade, H. J. (2018). *Estrategia de Desarrollo Bajo en Carbono (EDBC) para el sector de Ganadería bovina de Nicaragua.* DTU Library. https://backend.orbit.dtu.dk/ws/portalfiles/portal/145121640/Nicaragua_Livestock_Spanish_2.pdf

Cardilho, V. H. (2018). *Análise do papel das merendeiras terceirizadas como atores implementadores no Programa Nacional de Alimentação Escolar: desafios na perspectiva do trabalho* [Master's dissertation, Universidade Estadual de Campinas, Faculdade de Ciências Aplicadas]. Unicamp.

Caron, P., Ferrero y de Loma-Osorio, G., Nabarro, D., Hainzelin, E., Guillou, M., Andersen, I., Arnold, T., Astralaga, M., Beukeboom, M., Bickersteth, S., Bwalya, M., Caballero, P., Campbell, B. M., Divine, N., Fan, S., Frick, M., Friis, A., Gallagher, M., Halkin, J.-P., Hanson, C., Lasbennes, F., Ribera, T., Rockstrom, J., Schuepbach, M., Steer, A., Tutwiler A., & Verbur, G. (2020). Sistemas alimentares para o desenvolvimento sustentável: propostas para uma profunda transformação em quatro partes (pp. 25–50). In Preiss, P. & S. Schneider (Eds.), *Sistemas alimentares no século XX*I: *debates contemporâneos*. UFRGS. https://www.lume.ufrgs.br/bitstream/handle/10183/211399/001115756.pdf

Carrasco, M., & Pautassi, L. (2015). Diez años del "Plan Nacional de Seguridad Alimentaria" em Argentina. Una aproximación desde el enfoque de derechos. *De prácticas y discursos*, *4*(5). http://dx.doi.org/10.30972/dpd.45805

Castro, M. A. C., Castaño, L. C., & Correa, L. M. M. (2013). Descripción de la ingesta de alimentos y nutrientes en niños expuestos y no expuestos al programa de complementación alimentaria Mana Infantil en el municipio de Envigado, Colombia 2006–2010. *Perspectivas en Nutrición Humana*, *15*(1), 83–96. https://revistas.udea.edu.co/index.php/nutricion/article/view/17907

Centro de Estudios para el Desarrollo Rural Sustentable y la Soberania Alimentaria (CEDRSSA). (2015). *El sector agropecuario de Mexico en sus tratados comerciales vigentes*. http://www.cedrssa.gob. mx/files/b/13/47El%20sector%20agropecuario%20de%20México%20en%20sus%20tratados.pdf

Chapela, G., Menéndez, C., & Robles Berlanga, H. (2015). México: políticas para la agricultura campesina y familiar: un marco de referencia (pp. 233–259). In Sabourin, E., Samper, M., & O. Sotomayor (Eds.), *Políticas públicas y agriculturas familiares en América Latina: nuevas perspectivas*. Rede PPAL, CIRAD, Cepal, IICA.

Chavez, M. F., & Muller, A. (2020). Procesos de innovación tecnológica en la agricultura familiar: análisis de dos modelos de salas queseras implementadas en Amblayo, Salta, Argentina. *Revista Americana de Empreendedorismo e Inovação*, *2*(1), 115–125. http://periodicos.unespar.edu.br/ index.php/raei/article/view/3257

Cheib, A. S. (2020). *As políticas públicas de redução do consumo de proteína animal e de substituição por uma alimentação à base de vegetais: uma análise de atores, instrumentos e objetivos* (Monograph, Graduação em Administração Pública, Escola de Governo Professor Paulo Neves de Carvalho da Fundação João Pinheiro). Fundação João Pinheiro. http://monografias.fjp.mg.gov.br/ handle/123456789/2691

Chile, Dirección de Presupuestos. (2020). *Evaluación Programa Nacional de Alimentación Complementaria y Programa de Alimentación Complementaria del Adulto Mayor*. División de Control de Gestión Pública. https://www.dipres.gob.cl/597/articles-205709_informe_final.pdf

Chile, Ministerio de Salud. (2017). *Política nacional de alimentación y nutrición*.

Chile, Ministerio de Salud. (2018). *Informe de evaluación de la implementación de la ley sobre composición nutricional de los alimentos y su publicidad*. https://www.minsal.cl/wp-content/ uploads/2018/05/Informe-Implementación-Ley-20606-febrero-18-1.pdf

Cohen, M. D., March, J. G., & Olsen, J. P. (1972). A garbage can model of organizational choice. *Administrative Science Quarterly*, *17*(1), 1–25. https://doi.org/10.2307/2392088

Concejo de Santiago de Cali. (2019). *Acuerdo n° 0470. Por el cual se adopta la política publica de soberanía y seguridad alimentaria y nutricional del Distrito especial de Santiago de Cali y se dictan otras disposiciones*. http://www.concejodecali.gov.co/descargar.php?idFile=18584

Costa Rica, Ministerio de Ambiente y Energía (MINAE). (2015). *Estrategia para la ganadería: Baja en carbono en Costa Rica. Informe final: estrategia y plan de acción*. http://www.mag.go.cr/bibliotecavirtual/L01-11006.pdf

Costa Rica, Ministerio de Salud. (2011). *Política nacional para la seguridad alimentaria y nutricional 2011–2021*. https://www.ministeriodesalud.go.cr/index.php/biblioteca-de-archivos/sobre-el-ministerio/politcas-y-planes-en-salud/politicas-en-salud/1106-politica-nacional-de-seguridad-alimentaria-y-nutricional-2011-2021/file

Cruz, F. T., Marques, V. P. M., & Haas, J. M. (2020). Os selos da agricultura familiar no Mercosul: implicações e desafios para valorização de produtos agroalimentares. *Revista Brasileira de Planejamento e Desenvolvimento*, 9(5), 757–798. http://hdl.handle.net/10183/224187

Cunha, A. R. A. A. (2014). Las centrales mayoristas de abasto y los circuitos cortos en America Latina. In UN/CEPAL, *Agricultura familiar y circuitos cortos: nuevos esquemas de producción, comercialización y nutrición.* (pp. 69–74). United Nations. http://hdl.handle.net/11362/44258

David, K. A. G. (2016). *Campesino paneleros higienizados por la biopolítica: sabers locales y biopolíticas en la molienda campesina panelera. Un estudio de caso de productores paneleros de la vereda San Miguel, municipio de Quebradanegra, Cundinamarca, 1990–2015* [Master's thesis, Universidad Pontificia Bolivariana]. Repositorio Institucional UPB. http://hdl.handle.net/20.500.11912/4253

Daviron, B., Perin, C., & Soulard, C. T. (2017). History of urban food policy in Europe, from the ancient city to the industrial city. In Brand, C., Bricas, N., Conaré, D., Daviron, B., Debru, J., Michel, L., & C.-T. Soulard (Eds.), *Designing urban food policies* (pp. 27–51). Springer. https://doi.org/10.1007/978-3-030-13958-2

Davis, K. F., Gephart, J. A., Emery, K. A., Leach, A. M., Galloway, J. N., & D'Odorico, P. (2016). Meeting future food demand with current agricultural resources. *Global Environmental Change, 39*, 125–132. https://doi.org/10.1016/j.gloenvcha.2016.05.004

Delgado, G. C. (2012). *Do capital financeiro na agricultura à economia do agronegócio: mudanças cíclicas em meio século (1965–2012).* UFRGS.

De Molina, M. G., García, D. L., & Casado, G. G. (2017). Politizando el consumo alimentario: estrategias para avanzar en la transición agroecológica. *Redes, 22*(2), 31–55.

De Pablo, M. (2015). Buenas prácticas: inocuidad de los alimentos y su relación con el acceso a mercado de la agricultura familiar. In A. Flores (Ed.), *Boletin de agricultura familiar para América Latina y el Caribe, 13* (pp. 13–15). FAO. https://www.fao.org/3/i5192s/i5192s.pdf

Duarte, L. M. G., Guéneau, S., Diniz J. D. A. S., & Passos, C. J. S. (2021). Valorización de los patrimonios alimentarios y productos agroextractivistas del Cerrado brasileño en la gastronomía: un estudio sobre el Festival Gastronómico Cerrado Week. In Rebaï, N., Bilhaut, A-G., De Suremain, C-E., Katz, E., & M. Paredes (Eds.). *Patrimonios alimentarios en América Latina: recursos locales, actores y globalización* (pp. 109–136). IFEA/IRD.

Eugenio, A. C. (2018). *Participação e caracterização da agricultura familiar na Pedra do Produtor da Ceasa-GO* [Master's dissertation, Universidade Federal de Goiás]. Biblioteca Digital de Teses e Dissertações. http://repositorio.bc.ufg.br/tede/handle/tede/8424

European Commission. (2020). *Comunicação da Comissão ao Parlamento Europeu, ao Conselho, ao Comitê Econômico e Social Europeu e ao Comitê das Regiões: Estratégia do Prado ao Prato para um sistema alimentar justo, saudável e respeitador do ambiente* [Communication from the Commission to the European Parliament, the Council, the European Economic and Social Committee and the Committee of the Regions: A Farm to Fork Strategy for a fair, healthy and environmentally-friendly food system] (COM/2020/381 final). https://eur-lex.europa.eu/legal-content/PT/TXT/?uri=celex:52020DC0381

FAO (Food and Agriculture Organization of the United Nations). (2016). *Análisis y diagnóstico de políticas agroambientales en Cuba.* https://www.fao.org/3/i5559s/i5559s.pdf

FAO & FLAMA (Federación Latino-Americana de Mercados de Abastecimiento). (2020). *Los mercados mayoristas: acción frente al covid-19* (Boletín nº 3, 13-07-2020). FAO. https://www.fao.org/3/cb0218es/CB0218ES.pdf

FAO & SAGARPA (Secretaría de Agricultura, Ganadería, Desarrollo Rural, Pesca y Alimentación). (2012). *Agricultura familiar con potencial productivo en México.* FAO. https://www.fao.org/3/bc944s/bc944s.pdf

FBSSAN (Fórum Brasileiro de Soberania e Segurança Alimentar e Nutricional). (2016). *Biofortificação: as controvérsias e as ameaças à soberania e segurança alimentar e nutricional.* https://fbssan.org.br/wp-content/plugins/download-attachments/includes/download.php?id=1277

Fierro, M. P. P (Ed.). (2019). *Los precios de garantía: avances y retos en la implementación* (Cuaderno de investigación nº 4). DGDyP/IBD, CDMX. http://bibliodigitalibd.senado.gob.mx/handle/123456789/4704

Fligstein, N., & McAdam, D. (2011). Toward a general theory of strategic action fields. *Sociological Theory, 29*(1), 1–26. https://doi.org/10.1111%2Fj.1467-9558.2010.01385.x

Fouilleux, E. (2000). Entre production et institutionnalisation des idées: la réforme de la politique agricole commune. *Revue Française de Science Politique, 50*(2), 277–305. http://www.jstor.org/stable/43119730.

Fouilleux, E. (2003). *La politique agricole commune et ses réformes*. L'Harmattan.

Fouilleux, E. (2011). Analisar a mudança: políticas públicas e debates num sistema em diferentes níveis de governança. *Estudos, Sociedade e Agricultura, 19*(1), 88–125. https://revistaesa.com/ojs/index.php/esa/article/view/337

Fouilleux, E., Bricas, N., & Alpha, A. (2017). "Feeding 9 billion people": global food security debates and the productionist trap. *Journal of European Public Policy, 24*(11), 1658–1677. https://doi.org/10.1080/13501763.2017.1334084

Fouilleux, E., & Jobert, B. (2017). Le cheminement des controverses dans la globalisation néo-libérale: pour une approche agonistique des politiques publiques. *Gouvernement et action publique, 6*(3), 9–36. https://doi.org/10.3917/gap.173.0009

Fouilleux, E., & Michel, L. (Eds.). (2020). *Quand l'alimentation se fait politique(s)*. Presses Universitaires de Rennes.

Gasselin, P., Lardon, S., Cerdan, C., Loudiyi, S., & Sautier, D. (Eds.). (2021). *Coexistence et confrontation des modèles agricoles et alimentaires: Un nouveau paradigme du développement territorial?* Quae. http://doi.org/10.35690/978-2-7592-3243-7

Gazola, M. (2017). Cadeias curtas agroalimentares na agroindustria familiar: dinâmicas e atores sociais envolvidos. In Gazolla, M., & S. Schneider (Eds.), *Cadeias curtas e redes agroalimentares alternativas: negócios e mercados da agricultura familiar* (pp. 175–194). Editora da UFRGS.

Gerten, D., Heck, V., Jägermeyr, J., Bodirsky, B. L., Fetzer, I., Jalava, M., Kummu, M., Lucht, W., Rockström, J., Schaphoff, S., & Schellnhuber, H. J. (2020). Feeding ten billion people is possible within four terrestrial planetary boundaries. *Nature Sustainability, 3*, 200–208. https://doi.org/10.1038/s41893-019-0465-1

Gilardon, E. O. A. (2016). Una evaluación crítica de los programas alimentarios en Argentina. *Salud Colectiva, 12*(4), 589–604. https://doi.org/10.18294/sc.2016.935

Golay, C., & Özden, M. (2012). *Le droit à l'alimentation. Un droit humain fondamental stipulé par l'ONU et reconnu par des traités régionaux et de nombreuses constitutions nationales*. Centre Europe-Tiers Monde (CETIM). http://agirledroit.clmayer.net/IMG/pdf/DroitAlimentation_CETIM.pdf

González, S. M. (2019). La seguridad alimentaria de México y la renegociación del TLCAN: oportunidad para una estrategia de desarrollo rural y de combate a la pobreza. *Portes, Revista Mexicana de Estudios Sobre la Cuenca del Pacífico, 13*(26), 27–60. http://www.portesasiapacifico.com.mx/revistas/epocaiii/numero26/2.pdf

Grisa, C., & Fornazier, A. (2018). O desperdício de alimentos e as políticas públicas no Brasil: entre ações produtivistas e políticas alimentares. In Zaro, M. (Ed.), *Desperdício de alimentos: velhos hábitos, novos desafios* (pp. 363–383). EDUCS. https://www.ucs.br/site/midia/arquivos/e-book-desperdicio-de-alimentos-velhos-habitos.pdf

Grisa, C., Wesz, Jr., V. J.; Buchweitz, V. D. (2014). Revisitando o Pronaf: velhos questionamentos, novas interpretações. *Revista de Economia e Sociologia Rural, 52*(2), 323–346. https://www.scielo.br/j/resr/a/FfGVnNCzjyTK6JgDCrqFfGg/?format=pdf&lang=pt

Guatemala, Ministerio de Agricultura, Ganadería y Alimentación. (2018). *Estrategia nacional de ganadería bovina sostenible con bajas emisiones*. https://www.maga.gob.gt/download/estrategia-ganado.pdf

Guilherme, N. O. S., & Portilho, F. (2018). Ecochefs, tapiocas e a gastronomização da agricultura familiar. In Garson M., & S. Torquato (Eds.), *Alimentação e ciências sociais: Perspectivas contemporâneas* (pp. 93–119). Autografia.

Gurgel, A. D. M., Santos, C. C. S. D., Alves, K. P. D. S., Araujo, J. M. D., & Leal, V. S. (2020). Estratégias governamentais para a garantia do direito humano à alimentação adequada e saudável no enfrentamento à pandemia de Covid-19 no Brasil. *Ciência & Saúde Coletiva*, *25*, 4945–4956. https://doi.org/10.1590/1413-812320202512.33912020

Hassanein, N. (2003). Practicing food democracy: a pragmatic politics of transformation. *Journal of Rural Studies*, *19*(1), 77–86. https://doi.org/10.1016/S0743-0167(02)00041-4

HLPE (High Level Panel of Experts on Food Security and Nutrition). (2014). *Food losses and waste in the context of sustainable food systems*. Committee on World Food Security. https://www.fao.org/3/i3901e/i3901e.pdf

IMAFLORA (Instituto de Manejo e Certificação Florestal e Agrícola). (2017). *10 anos da moratória da soja na Amazônia: história, impactos e a expansão para o Cerrado*. https://www.imaflora.org/public/media/biblioteca/IMF-10-anos-moratoria-da-soja-WB.pdf

IPES-FOOD (International Panel of Experts on Sustainable Food Systems). (2019). *Towards a common food policy for the European Union: the policy reform and realignment that is required to build sustainable food systems in Europe*. https://www.ipes-food.org/_img/upload/files/CFP_FullReport.pdf

ISAGS-UNASUR (Instituto Suramericano de Gobierno em Salud-Unión de Naciones Suramericanes). (2019). *Interferencia de la industria de Alimentos en las políticas etiquetado gráfico innovador de alimentos procesados en Suramérica*. https://www.rets.epsjv.fiocruz.br/sites/default/files/arquivos/interferencia-de-la-industria-en-etiquetado-isags.pdf

Jara, C. E., Sperat, R. R., Manrique, L. F. R., & Herrera, A. G. (2019). Desarrollo rural y agricultura familiar en Argentina: una aproximación a la coyuntura desde las políticas estatales. *Revista de Economia e Sociologia Rural*, *56*(2), 339–352. https://doi.org/10.1590/1806-9479.2019.191195

Jarosz, L. (2011). Defining world hunger: scale and neoliberal ideology in international food security policy discourse. *Food, Culture & Society*, *14*(1), 117–139. https://doi.org/10.2752/175174411X12810842291308

Jobert, B. Y., & Muller P. (1987). *L'Etat en action: politiques publiques et corporatismes*. Presses universitaires de France.

Kingdon, J. (1984). *Agendas, alternatives, and public policies*. Little, Brown.

Lang, T., Barling, D., & Caraher, M. (2009) *Food policy: integrating health, environment and society*. Oxford University Press.

Lascoumes, P., & Le Galès, P. (2012). A ação pública abordada pelos seus instrumentos. *Revista Pós Ciências Sociais*, *9*(18), 19–43. http://www.periodicoseletronicos.ufma.br/index.php/rpcsoc/article/view/1331

Levitsky, S., & Ziblatt, D. (2018). *Como as democracias morrem* [*How democracies die*] (R. Aguiar, Trans.). Zahar.

Linhares, M. Y. L. (1979). *História do Abastecimento: uma problemática em questão (1530–1918)*. Binagri.

Linhares, M. Y. L., & Silva, F. C. T. (1979). *História política do abastecimento (1918–1974)*. Binagri.

Mahoney, J. (2001). Path-dependent explanations of regime change: Central America in comparative perspective. *Studies in Comparative International Development*, *36*(1): 111–141. https://doi.org/10.1007/BF02687587

Mahoney, J., & Thelen, K. (2010). A theory of gradual institutional change. In Mahoney, J., & K. Thelen (Eds.), *Explaining institutional change: ambiguity, agency, and power* (pp. 1–37). Cambridge University Press. https://doi.org/10.1017/CBO9780511806414.003

Maldonado, E. F. (2016). *Impactos de los acuerdos comerciales internacionales adoptados por el Estado peruano en el cumplimiento y promoción de los derechos humanos en el Perú*. RedGE.

Marques, P. E. M., & Moal, M. F. L. (2014). Le Programme d'Acquisition d'Aliments (PAA) au Brésil: l'agriculture locale et familiale au cœur de l'action publique en vue de la sécurité alimentaire. *VertigO-la revue électronique en sciences de l'environnement*, *14*(1). https://id.erudit.org/iderudit/1027951ar

Martinelli, S. S., Cortese, R. D. M., & Cavalli, S. B. (2020). Contribuições de guias alimentares para uma alimentação saudável e sustentável. In Preiss, P. V., Schneider, S., & G. Coelho De Souza (Eds.), *A contribuição brasileira à segurança alimentar e nutricional* sustentável (pp. 53–68). UFRGS. https://www.ufrgs.br/agrifood/images/2020/07-julho/001115755.pdf

Martinez, J. Z., Ramírez, A. M., Zamora, R. G. L., & Rodríguez, M. L. H. (2017). Política agrícola neoliberal: el caso del programa de modernización sustentable de la agricultura tradicional en Tlaxcala. *Globalización: Revista Mensual de Economía, Sociedad y Cultura, Mayo*. http://rcci.net/globalizacion/2017/fg2955.htm

Martinez-Lomeli, L. (2020). «Manger dehors» dans les villes de Mexico et de Guadalajara: de quelques tensions entre les dimensions sanitaires et patrimoniales au Mexique. In Soula A., Yount-André C., Lepiller O., & N. Bricas (Eds.), *Manger en ville: regards socio-anthropologiques d'Afrique, d'Amérique latine et d'Asie* (pp. 57–69). Quae. http://doi.org/10.35690/978-2-7592-3091-4

Mathé, T., Pilorin, T., Hébel, P., & Denizeau, M. (2008). *Du discours nutritionnel aux représentations de l'alimentation* (Cahier de recherche n. 252). CRÉDOC, Centre de Recherche pour l'Étude et l'Observation des Conditions de Vie. https://www.credoc.fr/publications/du-discours-nutritionnel-aux-representations-de-lalimentation

Matta, R. (2012). *El patrimonio culinario peruano ante UNESCO: algunas reflexiones de gastro-política* (Working Paper Series n. 28). desiguALdades.net Research Network on Interdependent Inequalities in Latin America. http://dx.doi.org/10.17169/refubium-23227

McMichael, P. (2009). A food regime genealogy. *The Journal of Peasant Studies, 36*(1), 139–169. https://doi.org/10.1080/03066150902820354

McMichael, P. (2016). *Regimes alimentares e questões agrarias*. Editora da Unesp/Editora da UFRGS.

MERCOSUR/REAF (Reunión Especializada sobre Agricultura Familiar). (2014). *Sellos de identificación de la agricultura familiar* (Mercosur/CMC/Rec. Nº 02/2014). http://www.reafmercosul.org/biblioteca-reaf

Montoya, L. M. A., & Giraldo, F. A. F. (2012). Significado de la alimentación y del complemento alimentario MANA en un grupo de hogares de Turbo, Colombia. *Perspectivas en Nutrición Humana, 14*(2), 171–183. https://revistas.udea.edu.co/index.php/nutricion/article/download/16487/14311/

Moraes, J. G., & Pires, M. L. L. S. (2019). Agricultura familiar e mercados atacadistas: dinâmicas sociais da Central de Comercialização da Agricultura Familiar (Cecaf/Ceasa) em Recife–Pernambuco. *Revista de Economia e Sociologia Rural, 57*(2), 309–325. https://doi.org/10.1590/1806-9479.2019.181152

Muller, P. (1995). Les politiques publiques comme construction d'un rapport au monde. In Faure, A., Pollet, G., & P. Warin (Eds.), *La construction du sens dans les politiques publiques. Débats autour de la notion de référentiel* (pp. 153–179). L'Harmattan.

Muller, P. (2019). Référentiel. In Boussaguet, L., Jacquot, S., & P. Ravinet (Eds.), *Dictionnaire des politiques publiques* (pp. 533–540). Presses de Sciences Po. https://doi.org/10.3917/scpo.bouss.2019.01.0533

Niederle, P. A. (Ed.). (2013). *Indicações geográficas: qualidade e origem nos mercados alimentares*. UFRGS.

Niederle, P., & Wesz, Jr., V. J. (2020). *Agrifood system transitions in Brazil: new food orders*. Routledge.

Observatorio Feria Libre. (2013). *Características económicas y sociales de ferias libres de Chile: encuesta nacional de ferias libres*. Proyecto de Cooperación Técnica FAO-ODEPA-ASOF TCP CHI/3303. https://www.fao.org/3/as114s/as114s.pdf

Paraguay, UTGS (Unidad Técnica del Gabinete Social). (2018) *Programa Alimentario Nutricional Integral (PANI)*. https://gabinetesocial.gov.py/archivos/documentos/PANI-docu-eva_uhfqt18i.pdf

Paré, F. (2012). Pour la sécurité alimentaire: restaurer la responsabilité d'État. *Revue internationale de droit économique, 26*(4), 87-97. https://doi.org/10.3917/ride.258.0087

Parrado Barbosa, A., & Molina Ochoa, J. P. (2014). *Mercados campesinos: modelo de acceso a mercados y seguridad alimentaria en al región central de Colombia*. OXFAM, Universidad Nacional de Colombia.

Patrouilleau, M. M., Taraborrelli, D., & Alonso, I. (2018). La trayectoria de la "agricultura familiar" en la agenda agroalimentaria Argentina y las rigideces de la política nacional. *Raízes: Revista de Ciências Sociais e Econômicas, 38*(1), 22–35. https://doi.org/10.37370/raizes.2018.v38.36

Peru, Comisión Multisectorial de Seguridad Alimentaria y Nutricional. (2013). *Estrategia nacional de seguridad alimentaria y nutricional 2013–2021.* Ministerio de Agricultura y Riego. https://siteal.iiep.unesco.org/sites/default/files/sit_accion_files/estrategia_nacional_de_seguridad_alimentaria_2013-2021.pdf

Peru, CEPLAN (Centro Nacional de Planeamiento Estratégico). (2012) *Gastronomía peruana al 2021: lineamentos para un programa de desarrollo de la gastronomía peruana en el marco del Plan Bicentenário (2nd ed.).* https://www.ceplan.gob.pe/wp-content/uploads/files/Documentos/gastronomiaperuana.pdf

Pierson, P. (2000). Increasing returns, path dependence, and the study of politics. *American Political Science Review, 94*(2), 251–267. https://doi.org/10.2307/2586011

Pinto, M., Ticona, J. F., & Rojas, W. (2016). Bolivia: Bancos comunitários de semillas en el área del Lago Titicaca (A. Walter, Trans.). In Vernooy, R., Shrestha, P., Sthapit, B., & M. Ramírez (Eds.), *Bancos comunitarios de semillas: orígenes, evolución y perspectivas* (pp. 74–79). Bioversity International. (Original work published 2015). https://www.bioversityinternational.org/fileadmin/user_upload/BANCOS_COMUNITARIOS_DE_SEMILLAS_Vernooy.pdf

Porras, F. I. E., Villalobos, R. A., Fonseca, J. C. H., & Umaña, K. M. (2016). Costa Rica: Unión de Semilleros del Sur (A. Walter, Trans.). In Vernooy, R., Shrestha, P., Sthapit, B., & M. Ramírez (Eds.), *Bancos comunitarios de semillas: orígenes, evolución y perspectivas* (pp. 99–103). Bioversity International. (Original work published 2015). https://www.bioversityinternational.org/fileadmin/user_upload/BANCOS_COMUNITARIOS_DE_SEMILLAS_Vernooy.pdf

Portilho, F. (2020). Ativismo alimentar e consumo político – duas gerações de ativismo alimentar no Brasil. *Redes, 25*(2), 411–432. https://doi.org/10.17058/redes.v25i2.15088

Poulain, J. P. (2011). La gastronomisation des cuisines de terroir: sociologie d'un retournement de perspective. In Adell, N., & Y. Pourcher, *Transmettre, quel(s) patrimoine(s)? Autour du Patrimoine Culturel Immatériel* (pp. 239–248). Michel Houdiard Editions.

Preiss, P. V., & Schneider, S. (Eds.). (2020). *Sistemas alimentares no século XXI: debates contemporâneos.* UFRGS. https://www.lume.ufrgs.br/bitstream/handle/10183/211399/001115756.pdf

Pretty, J. (2020). New opportunities for the redesign of agricultural and food systems. *Agriculture and Human Values, 37,* 629–630. https://doi.org/10.1007/s10460-020-10056-2

Rebaï, N., Bilhaut, A-G., De Suremain, C-E., Katz, E., & Paredes, M. (Eds.). (2021). *Patrimonios alimentarios en América Latina: recursos locales, actores y globalización.* IFEA/IRD.

Repetto, F. (2005). La dimensión política de la coordinación de programas y políticas sociales: una aproximación teórica y algunas referencias prácticas en América Latina. In F. Repetto (Ed.), *La gerencia social ante los nuevos retos del desarrollo social en América Latina* (pp. 39–100). MagnaTerra Editores. https://publications.iadb.org/publications/spanish/document/La-gerencia-social-ante-los-nuevos-retos-del-desarrollo-social-en-América-Latina.pdf

Repetto, F., & Fernandez, J. P. (2012). *Coordinacion de políticas, programas y proyectos sociales.* Fundación CIPPEC. https://www.cippec.org/wp-content/uploads/2017/03/2425.pdf

Requier-Desjardins, D. (1999). Agroindustria rural y sistemas agroalimentarios localizados: cuales puestas. In Riveros, H., Salas, I., & S. Mazzotti (Eds.), *Revista conmemorativa por el X Aniversario de Prodar,* IICA, CIID, CIRAD, CIAT.

Ribeiro, H., Jaime, P. C., & Ventura, D. (2017). Alimentação e sustentabilidade. *Estudos avançados, 31,* 185–198. https://doi.org/10.1590/s0103-40142017.31890016

Riella, A., & Mascheroni, P. (2015). Gobiernos progresistas, politicas públicas y organizaciones agrarias en Uruguay. *Revista IDeAS – Interfaces em Desenvolvimento, Agricultura e Sociedade, 9*(2), 98–128. https://revistaideas.ufrrj.br/ojs/index.php/ideas/article/view/138

Rios, F. D. C., & Wesz, Jr., V. J. (2020). Agricultura familiar campesina y cadenas cortas agroalimentarias: la Feria Municipal de Yuty-Caazapá (Paraguay). *Interações (Campo Grande), 21*(4), 903–914. https://doi.org/10.20435/inter.v21i4.2815

Riveros, H., Vandecandelaere, E., Tartanac, F., Ruiz, C., & Pancorbo, G. (Eds.). (2008). *Calidad de los alimentos vinculada al origen y las tradiciones en América Latina: estudio de casos*. FAO-IICA. https://www.fao.org/3/au691s/au691s.pdf

Rocha, C. (2020). Impactos à saúde humana causados pelos sistemas alimentares. In Preiss, P. V., Schneider, S., & G. Coelho De Souza (Eds.), *A contribuição brasileira à segurança alimentar e nutricional sustentável* (pp. 27–52). UFRGS. https://www.ufrgs.br/agrifood/images/2020/07-julho/001115755.pdf

Rockström, J., Edenhofer, O., Gärtner, J., & DeClerck, F. (2020). Planet-proofing the global food system. *Nature Food, 1*(1), 3-5. https://doi.org/10.1038/s43016-019-0010-4

Sabatier, P. A., & Jenkins-Smith, H. C. (1999). The advocacy coalition framework: an assessment. In P. A. Sabatier (Ed.), *Theories of the policy process* (pp. 117–166). Westview Press.

Sabourin, E., Craviotti, C., & Milhorance, C. (2020). The dismantling of family farming policies in Brazil and Argentina. *International Review of Public Policy, 2*(2:1), 45–67. https://doi.org/10.4000/irpp.799

Sabourin, E., Grisa, C., Niederle, P., Leite, S., Milhorance, C., Ferreira, A., Sauer S., & Andriguetto-Filho, J. (2020). Le démantèlement des politiques publiques rurales et environnementales au Brésil. *Cahiers Agricultures, 29*, Article 31. https://doi.org/10.1051/cagri/2020029

Sabourin, E., Samper, M., & Sotomayor, O. (Eds.) (2015). *Políticas públicas y agriculturas familiares en América Latina y el Caribe: nuevas perspectivas*. IICA. https://agritrop.cirad.fr/578797/

Sánchez, K. S. V., Santos, R. G., & Aragón-Cuevas, F. (2016). Bancos comunitarios de semillas en México: una estrategia de conservación *in situ* (A. Walter, Trans.). In Vernooy, R., Shrestha, P., Sthapit, B., & M. Ramírez (Eds.), *Bancos comunitarios de semillas: orígenes, evolución y perspectivas* (pp. 248-254). Bioversity International. (Original work published 2015). https://www.bioversityinternational.org/fileadmin/user_upload/BANCOS_COMUNITARIOS_DE_SEMILLAS_Vernooy.pdf

Sánchez, V. S., & Silva, C. V. (2018). El impacto de la nueva ley de etiquetados de alimentos en la venta de productos en Chile. *Perfiles Económicos, 3*, 7–33. https://doi.org/10.22370/rpe.2017.3.1218

Schubert, M. N., & Ávalos, D. E. (2020). Sistemas alimentarios globales y ley de etiquetado de alimentos en Chile. *Redes. Revista do Desenvolvimento Regional, 25*(2), 527-544. https://doi.org/10.17058/redes.v25i2.14850

SEGOB (Secretaría de Gobernación). 2019. México. Plan Nacional de Desarrollo 2019-2024. http://www.dof.gob.mx/nota_detalle.php?codigo=5565599&fecha=12/07/2019

Sierra, P. L. (2018). *MasAgro: modernizando la agricultura tradicional en tiempos neoliberais*. CECCAM – Centro de Estudios para el Cambio en el Campo Mexicano. https://ceccam.org/sites/default/files/Folleto%20MasAgro%20digital.pdf

Sonnino, R., Callenius, C., Lähteenmäki, L., Breda, J., Cahill, J., Caron, P., Damianova, Z., Gurinovic, M. A., Lang, T., Laperriere, A., Mango, C., Ryder, J., Verburg, G., Achterbosch, T., den Boer, A. C. L., Kok, K. P. W., Regeer, B. J., Broerse, J. E. W., Cesuroglu, T., & Gill M. (2020). *Research and innovation supporting the farm to fork strategy of the European Commission* (Policy brief 3). FIT4FOOD2030. https://fit4food2030.eu/wp-content/uploads/2020/04/FIT4FOOD2030-Research-and-Innovation-supporting-the-Farm-to-Fork-Strategy-of-the-European-Commission-Policy-Brief.pdf

Sonnino, R., Spayde, J., & Ashe, L. (2016). Políticas públicas e a construção de mercados: percepções a partir de iniciativas de merenda escolar. In Marques, F.C., Conterato, M.A., & S. Schneider (Eds.), *Construção de mercados e agricultura familiar: desafios para o desenvolvimento rural* (pp. 311–330). UFRGS.

Springmann, M., Clark, M., Mason-D'Croz, D., Wiebe, K., Bodirsky, B. L., Lassaletta, L., de Vries, W., Vermeulen, S. J., Herrero, M., Carlson, K. M., Jonell, M., Troell, M., DeClerck, F., Gordon, L. J., Zurayk, R., Scarborough, P., Rayner, M., Loken, B., Fanzo, B., … Willett, W. (2018). Options for keeping the food system within environmental limits. *Nature, 562*(7728), 519–525. https://doi.org/10.1038/s41586-018-0594-0

Stassart, P. M. (2010). Le rôle des "consommateurs" dans la construction d'un accord entre agriculteurs et environnementalistes. In *Encontro da Rede de Estudos Rurais, 4, Anais eletrônicos*. UFPR.

Swinburn, B. A., Kraak, V. I., Allender, S., Atkins, V. J., Baker, P. I., Bogard, J. R., Brinsden, H., Calvillo, A., De Schutter, O., Devarajan, R., Ezzati, M., Friel, S., Goenka, S., Hammond, R. A., Hastings, G., Hawkes, C., Herrero, M., Hovmand, P. S., … Dietz, W. H. (2019). The global syndemic of obesity, undernutrition, and climate change: The Lancet Commission report. *The Lancet*, *393*(10173), 791–846. https://doi.org/10.1016/S0140-6736(18)32822-8

Takeuti, D., & Oliveira, J. M. (2013). Para além dos aspectos nutricionais: uma visão ambiental do sistema alimentar. *Segurança Alimentar e Nutricional*, *20*(2), 194–203. https://doi.org/10.20396/san.v20i2.8634597

Tanaka, J., & Portilho, F. (2019). Ambiguidades da politização do consumo: O caso do Movimento dos Pequenos Agricultores (MPA) na cidade do Rio de Janeiro. *Raízes*, *39*(2), 344–358. https://doi.org/10.37370/raizes.2019.v39.114

Taraborrelli, D. (2017). Políticas públicas rurales y modelos de desarollo em Argentina. El Programa Cambio Rural entre 1993 y 2015. *Estudios Sociales del Estado*, *3*(5), 164–188.

Triches, R. M. (2020). Dietas saudáveis e sustentáveis no âmbito do sistema alimentar no século XXI. *Saúde em Debate*, *44*(126), 881–894. https://doi.org/10.1590/0103-1104202012622

Vernooy, R., Shrestha, P., & Sthapit, B. (2016). Las valiosas pero poco conocidas crónicas de los bancos comunitarios de semillas (A. Walter, Trans.). In Vernooy, R., Shrestha, P., Sthapit, B., & M. Ramírez (Eds.), *Bancos comunitarios de semillas: orígenes, evolución y perspectivas* (pp. 1–9). Bioversity International. (Original work published 2015). https://www.bioversityinternational.org/fileadmin/user_upload/BANCOS_COMUNITARIOS_DE_SEMILLAS_Vernooy.pdf

Webb, P., Benton, T. G., Beddington, J., Flynn, D., Kelly, N. M., & Thomas, S. M. (2020). The urgency of food system transformation is now irrefutable. *Nature Food*, *1*, 584–585. https://doi.org/10.1038/s43016-020-00161-0

Wesz, Jr., V. J. (2021). O Pronaf pós-2014: intensificando a sua seletividade? *Revista Grifos*, 30(51). https://doi.org/10.22295/grifos.v30i51.5353

Wesz, Jr., V. J., & Grisa, C. (2017). O Estado e a soja no Brasil: a atuação do crédito rural de custeio (1999–2015). In Maluf, R., & G. Flexor (Eds.), *Questões agrárias, agrícolas e rurais: conjunturas e políticas públicas* (pp. 97–111). E-papers. http://oppa.net.br/livros/Questoes_agrarias_agricolas_e_rurais-Renato_Maluf-Georges_Flexor.pdf

Willett, W., Rockström, J., Loken, B., Springmann, M., Lang, T., Vermeulen, S., Garnett, T., Tilman, D., DeClerck, F., Wood, A., Jonell, M., Clark, M., Gordon, L. J., Fanzo, J., Hawkes, C., Zurayk, R., Rivera, J., De Vries, W., Sibanda, L. M., … Murray, C. J. (2019). Food in the Anthropocene: the EAT–Lancet Commission on healthy diets from sustainable food systems. *The Lancet*, *393*(10170), 447–492. https://doi.org/10.1016/S0140-6736(18)31788-4

World Food Programme. (2013). *El estado de la alimentación escolar a nivel mundo* [State of school feeding worldwide]. https://www.wfp.org/publications/state-school-feeding-worldwide-2013

Zaneti, T. B. (2017). *Cozinha de raiz: relaciones entre chefs, productores y consumidores a través del uso de productos agroalimentarios únicos en la gastronomía contemporánea* [Doctoral dissertation, Universidade Federal do Rio Grande do Sul]. UFRGS Digital Repository. http://hdl.handle.net/10183/164708.

Chapter 2

Futures studies and the food question in Latin America: a literature review

MARIA MERCEDES PATROUILLEAU, DIEGO S. TARABORRELLI,
IGNACIO ALONSO

▶▶ Introduction

As the precedents on the public policy cycle indicate, prior to the design and formulation of policies, there is an instance of policy analysis, sometimes presented as an instance of deliberation, of openness to think about alternatives. It is in this space of deliberation or discussion that studies that contemplate prospective approaches often come into play. Foresight or futures studies, as disciplines that seek to propose, analyze and evaluate probable, possible and desirable futures (Bell, 2003), are inserted into this hypothetical instance of the policy cycle, after the initial stage of problem definition, to contribute to defining the type of solution for the public issue in question (Subirats, 1989; Orlansky, 2007). A range of different technical organizations, research institutes, laboratories, research centers, international cooperation agencies and multilateral organizations contribute studies and analyses to direct and guide public policy with regard to the challenges and opportunities of the future.

Within this instance of policy analysis, futures studies are characterized by proposing a medium- or long-term view of the probable, possible or desirable futures corresponding to the issues in question. This long-term view seeks to anticipate certain processes, to facilitate the incorporation of different dimensions (with their different temporal rationales) as part of an interdisciplinary analysis, and to generate potential transformations that will make it possible to change the situation.

The field of futures studies includes a variety of different schools of thought, with certain differences in terms of methodological practices and epistemological conceptions of the future.[1] Some of these schools of thought are more closely linked

---

1. In Spanish, the term *"prospectiva"* is used for this discipline of futures studies. This term derives from the French *prospective*, which the French school used to define this discipline at the end of the 1950s. The French school was very influential, particularly in Latin America. In the English-speaking world, the term "foresight" is used (it would be the equivalent of *prospectiva*). Currently, at international level, the different schools are grouped around the area of "futures studies." The plural acceptation of the English term "futures" is intended to accommodate different approaches, as well as to differentiate it from classical scientific foresight by taking into account the multiplicity of possible futures, an open conception of the future, and defining it as indeterminate (Bell, 2003). Another more recent acceptation is that of "anticipation studies," a branch of futures studies that highlights the impossibility of foreseeing possible futures and that is proposed with an even more disruptive attitude than that of foresight, directly linked to action, i.e., to the design of ways of using the future to change the present (Miller, 2018).

to traditional scientific foresight (relying heavily on projections or models or on the analysis of the interaction between variables), while others emphasize the indeterminate condition of the future and the capacity of agency to forge desirable futures. In this paper we refer to foresight – or futures – studies as a broad spectrum of inquiries that seek to foresee, design, imagine or construct futures.

Different methods are used to address probable, possible and desirable futures. More traditional methods from the field of scientific forecasting (simulation, econometric, projections, evolutionary and dynamic models) can be used that make it possible to determine the occurrence probability of a given phenomenon, or methods generated in the field of futures studies, which integrate qualitative and quasi-structured aspects of reality, such as the Delphi method (Aichholzer, 2009), the scenario method (Börjeson et al., 2006), the normative scenario or backcasting (Robinson, 2003), and causal layered analysis (Inayatullah, 1998), among others.

This chapter sets out to analyze the way in which futures studies have addressed the food issue in Latin America. Given that futures studies are carried out to a large extent (not exclusively) to influence the positions of the state and the design of policies, it is interesting and pertinent to examine how these studies have addressed food systems, food or food security (from the different categories in which the food issue has been problematized).

Given that Latin America is a major food-producing region, a preliminary question to be answered is the extent to which foresight studies linked to the food production sector have considered food as a problem to be investigated in prospective terms. Thus, the principal questions that form the basis of the inquiry include: To what extent do the futures studies that have been developed incorporate the problem of food into futures analysis, and if they do, how do they approach it? Which approaches and categories problematize the food issue? Which approaches have prevailed at different moments in time? Which products are generated through these studies and which audiences are targeted for the use or discussion of these results?

A range of different approaches to the food issue were identified in the futures studies: a structuralist approach; a food security approach (in terms of access and availability); an approach that emphasizes food sovereignty; another that assumes that it is fundamentally markets, and especially incorporation into global markets, that can provide answers to food issues; another that highlights the "technical" attributes of food; one that prioritizes the environmental dimension and its relationship with food security; and, finally, an integral approach to sustainable food systems (the definition of each of these approaches can be found in the Appendix 2.1).[2] These different approaches, which are found more or less explicitly in the studies analyzed, emerge from the analysis of the argumentative content of the documents in their various parts: justification, diagnosis, analysis of variables, scenarios formulated or policy recommendations.[3]

------

2. Some of the categories used to define the different approaches are related to the *référentiels* used to analyze policies in the first chapter of this book. However, the set of categories does not correspond completely. Each chapter addresses a different object of analysis: futures studies in this case, and policies in the case of the former.

3. It should be clarified that not all papers contain the same components. They do not all generate scenarios or outline policy recommendations.

As a complementary objective, this chapter also analyzes how the field of futures studies has evolved, specifically in the agricultural sector and in rural and territorial development. It is also interesting to discuss how this discipline and practice of imagining possible futures to assist policy formulation was introduced into technical and academic circles. In this sense, this study aims to provide a literature review on the use of foresight in the strategic planning of companies, production chains, agricultural science and technology and rural and territorial development organizations and other government agencies.

The research is based on a survey of futures studies developed in Latin America from the 1970s to 2020. This survey includes 107 studies that address sectoral agricultural, food and rural, regional or territorial development issues. The survey included studies carried out in the region on a global scale, on a Latin American scale (approaches for all countries or for certain geographical areas, such as South America or the Southern Cone), and studies carried out in eleven countries: Argentina, Brazil, Chile, Colombia, Costa Rica, Cuba, Ecuador, Mexico, Peru, Venezuela and Uruguay (either on a national or subnational scale).

The criterion for considering the works included in the analysis was that they constituted a systematic foresight study on a specific problem, using a future perspective – whether or not a precise time horizon was specified – based on research activities and information analysis that applied certain methods linked to the prospective discipline and that the focus of the study, or at least a significant proportion of its analysis, was concerned with the agricultural sector, the agrifood system or the rural environment.[4]

As we are working on more than 100 studies, this chapter does not attempt to analyze each of the studies individually. Conversely, a global view of the survey enables us to determine how foresight studies have addressed food issues over time, by which agencies or organizations they have been conducted, with what scale of analysis, which approaches to futures studies have prevailed at different times, and what type of publications have been produced.

The chapter then goes on to highlight general aspects about the development of futures studies, and particularly agrifood futures studies, within the region, based on survey data from the 1970s and 1980s. The subsequent sections analyze the development of studies over the following three decades: the 1990s, the 2000s and 2010–2020. At the end of the chapter, a series of appendices follow the conclusions

---

4. A further clarification regarding the universe of studies addressed is that studies with a medium- or long-term time horizon (more than 5 years into the future) were considered for analysis. Projections for 1, 2 or 5 years were not considered. There are studies that are carried out periodically by multilateral organizations that follow the social and food impact based on the evolution of economic growth and certain other variables on a year-by-year basis. These studies were not considered as they generally look at a short-term period of between 1 and 5 years. Working in prospective terms requires a longer time horizon precisely to allow for possible and desirable transformations and for a longer-term perspective. The study also ruled out works that comment on other studies, futures studies manuals and publications that disseminate the discipline. In addition, since the unit of analysis is foresight studies and not publications, we have sought to only include the original (or, failing that, the most complete) study in each case, on which a number of publications can sometimes be based. Another practical criterion was also being able to access the studies, whether printed or digital publications.

providing information on the analytical categories used to examine the approaches applied in addressing the food issue (Appendix 2.1), a brief description of the foresight methods mentioned in the chapter (Appendix 2.2) and the complete list of the futures studies surveyed and analyzed (Annex).

▸▸ The origins of futures studies in Latin America and the food issue within the structuralist approach framework

The futures studies discipline emerged in the context of the second post-war period in the middle of the last century. From different fields of experience, approaches and methods were being forged to deal with the uncertainty of the future and to assist in future-oriented decision making. The first known developments were made from two different but connected fields: the French academic field and North American geopolitics (Durance, 2010). From these spaces, an epistemology for analytical work on the foresight and innovative methods, such as the Delphi and scenario method, began to be developed that provided a new way of producing knowledge, incorporating the future time horizon and motivating interaction and interdisciplinarity. Soon, futures studies also began to be developed in large companies, in the field of business management, through what was called scenario planning. Other conceptual frameworks were also incorporated into the new field of future studies: from sociology, humanities and post-structuralism (Masini, 1993; Inayatullah, 1998), systems theory (Gallopín et al., 1997), epistemology and critical realism (Bell, 2003), and post-normal science (Funtowicz & Ravetz, 1993), among other contributions. Towards the end of the 1980s and the beginning of the 1990s, foresight was consolidated as an area of study (futures studies at international level) with insertion and development at academic level as part of postgraduate programs, research and development centers and various government agencies.

The development of futures studies in Latin America suffered the difficulties of institutional instability, dictatorships and the exile of scientists and intellectuals in the initial stage of its development. The future-focused studies that began to be conducted were linked to the study of the economic and political viability of economic development, both at regional (Yero, 1993) and world levels (Herrera et al., 1977). The methodology of mathematical models and numerical experimentation methods had a strong influence in this early stage (Varsavsky & Calcagno, 1971). This primacy of mathematical methods, according to some authors, was a weakness for the exploration of alternative paths (Yero, 1993; Masini, 2005). The French school also had an influence in the Region, becoming an increasing presence over time, first through the influence of the Futuribles organization[5] and then through the set of methods and freely available associated software developed by Michel Godet.

The data from the survey conducted shows the pace of development of futures studies in the region. As can be seen in figure 2.1, the evolution in the number of studies

---

5. In Argentina, for example, the foresight approach disseminated by the Futuribles organization had been echoed. José Luis De Imaz and Agustín Merello had begun to develop it, and in 1969 they even published an issue of a magazine in association with that organization called: Prospectiva. Revista de Futuribles (Merello, 1969). See in this regard the work of Kozel and Patrouilleau (2016).

carried out has grown slowly since the 1970s and only in the last decade does there appear to be a trend towards the consolidation of this type of studies in the Region, with greater dissemination, with 62 studies identified between 2010 and 2020.

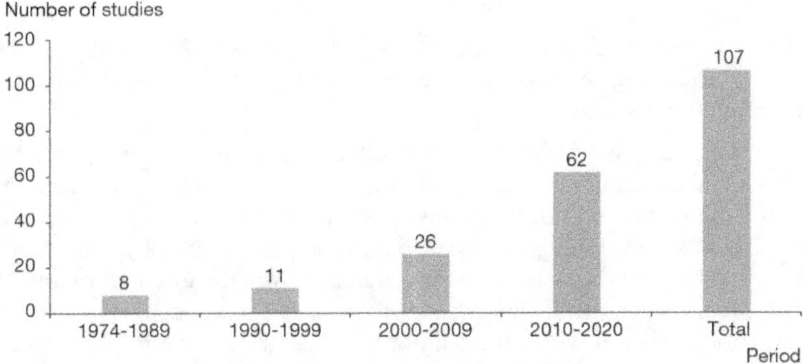

Figure 2.1. Number of studies surveyed per period.

During the 1970s and 1980s, with the incipient development of futures studies in the Region, eight studies were identified that in some way incorporate the long-term view in addressing development, agricultural sectoral and food issues. Only one of those eight studies fails to address the food issue, and another does so from a "market food" perspective. The rest of the studies are based on what we understand as a structuralist approach, although with nuances in the prioritization of certain variables, or in the explanatory models, but with common features that define them within this approach, which constituted an entire epochal climate in the region.

From the structuralist approach, the issue of food, poverty and quality of life deficits in Latin American countries are related to the structural problems of development, to the subordinate situation with regard to international trade in relation to the Central countries, to the structural heterogeneity between economic sectors within the economies themselves (Cimoli & Porcile, 2013), and even within the agricultural sector itself. The lack of access to land and to production technologies, the agrarian social structure and its forms of large estate and peasant farming, are considered among the fundamental causes of the lack of productive development and also of access to adequate food for the population (FAO & ECLAC, 1974).

Another noteworthy feature of the studies of this period is that, although few in number, most have been conducted at regional level. Of the eight studies conducted, five relate to Latin America, and, as a rule, have been financed and published by ECLAC or another UN agency. Two other studies refer to specific production chains: the pork chain in Venezuela and the citrus chain in Tucumán, Argentina. One other worldwide study is the high-profile Latin American World Model, coordinated by Amilcar Herrera (1977).

The regional and global studies are those that specifically focus on the food issue, in addition to dealing with production, energy and urbanization issues. Since few studies were collated for this period, it is worthwhile highlighting the regional and global studies that included observations about food and nutrition.

The study on the world model, coordinated by Herrera et al. (1977) was carried out by an interdisciplinary team at the Bariloche Foundation in Argentina. It was a very ambitious work that proposed an alternative hypothesis on the limits to world economic growth in response to the "The Limits to Growth" report presented years earlier by the Club of Rome (Meadows et al., 1972). It is the only world model carried out in the Region and had to be published at the time by a Canadian organization following the repression exercised by the military dictatorship on the institution and teams responsible.

The Latin American World Model proposed a normative approach to the material feasibility of a desirable future, concluding that the main economic and social problem of the future did not lie in the material lack of natural resources or in the overpopulation of developing countries, but in the consumerist style of the northern countries. It proposed a series of concepts that nurtured the discussions and agendas on development in the following decades, such as the concept of "basic needs," including food, together with other dimensions, such as demographics and health, housing and urbanization, education, renewable resources and other environmental aspects. This study criticized the distribution of world power as a conditioning factor of development, stating, in relation to food:

> The true causes of hunger have their roots in sociopolitical factors, at both the international and national levels. Especially important in underdeveloped countries are local factors of social and political organization that hinder the equitable distribution and production of food. (Herrera et al., 1977, p. 65).

The model also considered the levels of nutrition that could be achieved in the future, although in a very simplified way, considering the quantity of calories consumed.

The work of ECLAC and FAO (1974) represented a long-term projection, using simulation models, with a time horizon of 15 and 30 years (up to 2000). It is justified by reference to a possible growing demand for food at international level and in the context of food deficits in the Region. The study states that, in the future, the Region will have to respond not only to its unsatisfied food and nutrition needs, but also to growing international demand, and therefore aims to boost agriculture, the growth and specialization of production, the development of agro-industry, improved productivity, and encourages international cooperation to improve in all these areas.

Another regional study worth mentioning is that compiled by Martner (1986). This work, motivated by the approaching year 2000 milestone, and carried out by renowned Latin American social scientists and economists, set out to generate a regional discussion on the future of Latin America. Although in general terms signs of the structuralist approach can be identified in this work in the section that comprehensively explores the rural and agrarian issue in the region, the study compiled by Martner incorporates the report by Jacques Chonchol (protagonist of the agrarian reforms in Chile), which questions the future of capitalist modernization in agriculture and reveals the persistence of rural poverty. Chonchol proposed an ambitious set of measures for this work to modify the "bifurcation" between the economic and social objectives of Latin American economies, which included, for example, an agrarian reform program and goals such as food and energy self-sufficiency, a restructuring of social and cultural services, as well as modernization and the development of technological complexes. In this agrarian section, Martner's work examines the integral approach of sustainable food systems. Although it fails to include the environmental dimension present in the current approach to sustainable food systems, aspects such as renewable energies were considered in this study, as shown in the following excerpt:

> Faced with the problem of types of food, agrifood multinationals in Latin America have favored the development of a very expensive type of food in terms of the use of fossil fuels, which is only within the reach of the urban middle and wealthy classes. This is not conducive to the possibility of improving the food situation of the lower income sectors. In the future, it will be necessary to favor a new food model better adapted to the more efficient use of the natural resources of each country and the least costly possible in terms of the use of fossil fuels. (Martner, 1986, pp. 117-118).

In methodological terms, the study draws on a structured analysis of future possibilities. The scenario method had not yet been incorporated into Latin American studies when it was being written. As a policy (or rather strategy) recommendation, it emphasizes the need for integration among the economies of the region to overcome the difficulties of economic and social crises.

As mentioned above, during this initial period, Latin America was not the most favorable environment for the development of futures studies and their interaction with institutional and innovation systems, as the contexts of dictatorships, censorship and exile of scientists and professionals working in these disciplines made this type of innovation and any debate on possible and desirable futures a difficult task. It is possible that, because of these difficulties in the national spheres, futures studies were largely developed with the support of United Nations organizations, but that is also because regional integration operated as a possible horizon for overcoming the problems of development in Latin America. This wave of production focused on the regional scale was halted.

In methodological terms, projection and modeling studies prevailed. The scenario method had not yet been introduced in the Region, nor had other futures methods been formalized. The scenario method, for example, was just starting to be formalized in the field of strategic business planning and some other international organizations. Although some studies introduced the term scenarios, they did so by analyzing the most probable trends, without yet incorporating an express formalization of the method or the epistemology that this method contemplates today. In addition, the publications and projects were mainly developed through international organizations, with some participation from scientific and governmental organizations.

▸▸ Agrifood futures studies in the context of the transformations of the 1990s

During the 1990s, the development of the futures studies discipline and its insertion into innovation systems in the Region was still incipient, but a number of academic publications appeared, as well as mass publications produced by major publishing houses aimed at a wide audience (Siglo XXI and Trilce). This is interesting because it means that the goal was to raise awareness of the issue of the future of agriculture and food among a broader public, rather than solely targeting experts or decision-makers.

Eleven studies were identified for this period, in which a more focused interest in the agricultural sector (rather than in the broader terms of development studies) was observed, reprioritizing the sector within development strategies. There was a push for developments in biotechnology and the green revolution. Futures studies reflected the concerns defined by the new currents of globalization.

In terms of the scales of analysis, of the eleven studies identified, five were conducted at regional level and were promoted by IICA, ECLAC and FAO; and six were developed at a national or organizational or chain (subnational) scale. The countries that developed studies at a national or chain scale were Brazil (4), Chile (1) and Mexico (1).

As for specific topics, the foresight exercises were framed within the structural changes generated by the abandonment of import substitution industrialization and, in this context, agricultural activity was reprioritized and regained prominence in national economic strategies. The studies as a whole show a shift in concerns about the food issue. Sectoral approaches to the problems and prospects of agriculture prevail, focusing more on production, markets and new technologies and on the "technical" characteristics of food than on consumption, food or nutrition. Of the eleven studies, only three comprehensively address or link the food issue with the production issue. An incipient incorporation of environmental issues is also observed.

The studies that did not focus on the food issue concentrated on technological innovation, international competitiveness, environmental sustainability and the modernization of the institutional framework for agriculture for the new millennium. Figure 2.2 lists the number of studies according to their focus on addressing the food issue.

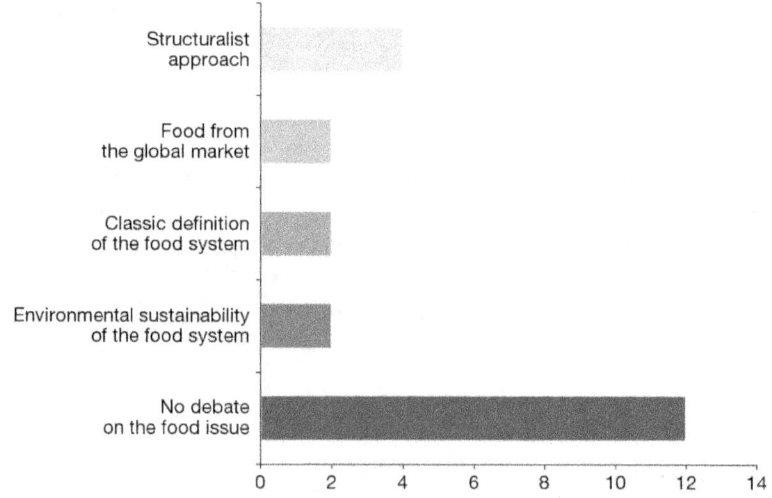

Figure 2.2. Approach to the food issue in the 1990s.

The agricultural developments that emerged during this decade were identified as part of the techno-economic revolution based on knowledge-intensive technologies. The new technological trends of the period heralded transformations in production processes and, therefore, also in the social structure of rural scenarios. Rodolfo

Quintero's (1993) foresight study on agrobiotechnologies, for example, pioneered the identification of agricultural biotechnology as a development in which the region, as a global geoeconomic space for food production, should develop institutional, scientific and technological capabilities. In this sense, Quintero's prospective exercise illustrates the strategic and normative nature of the policy recommendations that filtered into the future studies of this period. The "market food" approach is clear in this report, as it does not address concerns related to the food issue but is only concerned with the contribution of biotechnology to sectoral developments in the Region's agriculture.

Futures studies also focused on the development of certain production chains in view of the concern about the international integration of agricultural products due to the new quality and health parameters demanded by the markets of developed countries. This was the focus of a study conducted by the National Grape and Wine Research Center of the Brazilian Agricultural Research Corporation (EMBRAPA) in conjunction with the Future Studies Program of the Management Institute Foundation of the University of São Paulo (Wright et al., 1992). This study sought to guide the planning and development of the activities of the agents in the wine sector, identifying future scenarios and indicating paths and proposals for action in the international context, including MERCOSUR. Similarly, albeit using a different methodology, the foresight analysis of Chilean agro-exports carried out by ECLAC (Palma, 1990) identified the potential power of global certification bodies within the framework of the new dynamics of the international market.

One of the studies to have looked at the problems surrounding the food issue in-depth is that by Torres Torres (1990) that, although failing to define a specific prospective method, conducts a macro-historical analysis from a Latin American structuralist approach to analyze the technological revolutions that have been historically developed in agriculture, and particularly the green revolution in the Region that began to have an impact in the 1970s, followed by the era of the biotechnological revolution, analyzing its possible impact on production and agricultural markets in Mexico, as well as on consumption. It shows that the new scientific revolution will have an impact on the entire economy and on the agrarian social structure and analyzes its effect on food prices and on the redefinition of consumption patterns:

> These elements suggest that we are on the threshold of a type of agriculture in which the processes will be similar to those found in industry, with greater automation (even computerized in some of its stages) and without any link to the traditional production frameworks that currently persist. This will initially spark a major dispute over land and subsequently lead to an almost definitive elimination of small producers, although accompanied by an increase in the occupation of certain areas in the initial stages of technological incorporation (Torres Torres, 1990, p. 143).

The technological issue is also analyzed in the work of Herrera et al. (1994). The author and coordinator of the Latin American World Model of 1977, was, at that time, also coordinating a major technological futures study in Latin America, maintaining the normative approach of the former futures study but delving deeper into the logic of the scenario method by incorporating the concepts of "heavy trends," "seeds of the future," and identifying trend scenarios and a desirable scenario, with endogenous development. The study presents a normative scenario exercise, with analysis of possible trajectories.

This study by Herrera takes a different approach to the technological foresight studies that was generally encouraged by international organizations during this period. Technology is analyzed with an emphasis on the social and political dimensions that frame its development and appropriation by Latin American countries and focuses on developments in biotechnology, microelectronics and new materials technologies. The conditions of poverty among the population and food shortages are, for this study, overcome with structural reforms, such as agrarian reforms, but within the framework of other reforms, such as urban reform or regional integration. The environmental dimension is also considered, although not yet in terms of climate change. As such, this study is considered to correspond to the structuralist approach, although it contemplates some elements that now form part of the more contemporary sustainable food systems approach. The excerpt below illustrates the spirit in which the general issues are addressed, as well as those of food and agricultural production:

> Access to land for the poor, through agrarian reform, is a means of increasing food production in these countries. (...) However, land reform is not expected to address the food and other needs of the urban sector in the short term, as its immediate objective is to improve income distribution in rural areas, allowing a marginal population no access to the means of production. Its main impact will be to improve the supply of food to the rural population itself. (...) The agricultural sector should have a certain technological plurality, between a more productive segment, but with a low demand for employment, and a less productive sector with a higher demand for employment. This plurality should gradually be reduced. (...) The central guideline of economic policy will be to resume sustainable growth in the region. The region possesses a broad base of natural resources that can ensure the fulfillment of this objective, both in agriculture and in industry. There is sufficient agricultural land to feed the population and to support the other functions of agriculture, such as exports, production of raw materials and energy. However, since economic resources for agriculture are limited, the strategy will have to give priority, in the first stage, to domestic demand for food (Herrera et al., 1994, pp. 99-102).

In terms of the foresight methods used in the studies, the projections continued, but new methodological approaches more typical of the futures studies discipline were also being included, such as the scenario method – with horizons of between 10 and 15 years – the Delphi method and the technological roadmapping method. The latter two are related to the field of technology foresight.

In summary, in the 1990s, foresight studies in general were attracted by technological and agrifood market changes, although other voices appeared in the incipient field of futures studies, pointing out the structural, social and economic aspects, and the difficulties for the Region's countries to come out ahead in light of the persistent social and food shortages in Latin America.

▸▸ Regional agrifood foresight in the 2000s

During the first decade of this century, foresight studies in the region became significantly more widespread. The survey identified 26 studies during this period, more than doubling the number from the 1990s. For the first time, rather than multilateral organizations being the dominant force in terms of the production of studies, that position is now occupied by academic spaces and scientific journal publications. Figure 2.3 shows the distribution of studies by type of publication during the period.

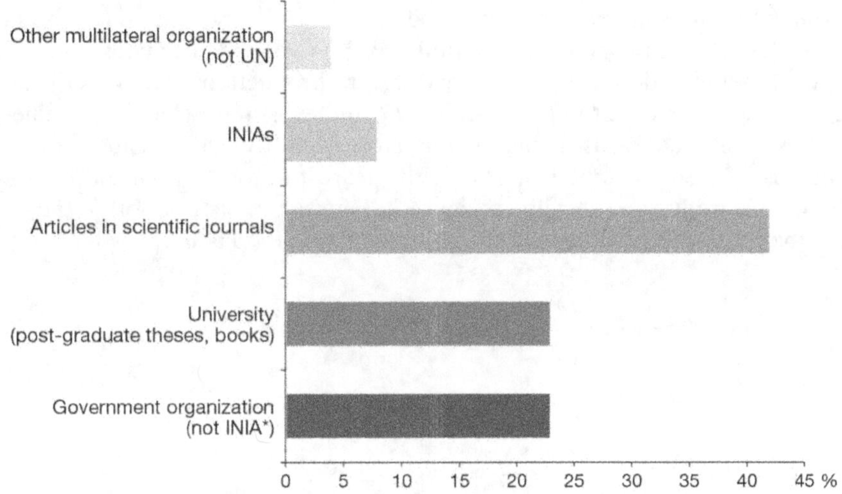

Figure 2.3. Agrifood foresight studies published in the period 2000–2010, by publication type. *INIA: National Institute of Agricultural Innovation.

In total, among academic publications, those produced by universities (graduate theses and books) and those in scientific journals, represent 65% of the studies for the period. These are largely case or chain studies carried out in different countries: Brazil (7), Mexico (4), Argentina (2), Peru, Colombia and Chile (1). These studies already echo the development of the scenario method, in particular by applying some of the methods disseminated by Michel Godet (Godet & Durance, 2011). Publications carried out in government agencies make up the second largest group, representing 23% of the total, with six studies conducted: one carried out in Brazil (the Brazil 3 Eras [*Brasil 3 Tempos*] project), two carried out within the framework of the Argentina's National Science and Technology Agency and three carried out by the Ministry of Agriculture and Rural Development in Colombia. Notable characteristics of these studies include the predominance of technology foresight studies, as well the increased diversity of methods used: scenarios, Delphi, technology roadmaps.

Two studies (8%) produced by National Agricultural Research Institutes (NARIs) were identified, one from Argentina and the other from Chile. The Argentinian project makes projections for agricultural production based on estimates of production gaps, land availability and technologies (Cap, 2002). The Chilean study also makes projections but introduces some elements relating to the scenario method: it includes elements of trend breaks and projects more than one future for each production chain or category (Odepa, 2004).

Following the trend of the previous decade, the works in this period focused on technology, generally understood as the axis of development, analyzing the role played by information and communication technologies and biotechnology. The agrifood sector presented an incipient concern for the problems of food security, incorporating the term and the concept of food security, as had been previously addressed in Europe, with emphasis mainly on access and availability (in five studies), and also incorporating the food sovereignty approach (in two studies). The issue of environmental sustainability linked to food appears to a lesser extent (in one study). The predominant emphasis of the period was, however, on productivity, efficiency and the global insertion of the region's different productions. Many of the studies did not even address the food issue at all (12 studies). In most of these studies, the analysis is oriented towards chains and the "closure of productivity gaps," considered parameters of global competitiveness, with regard to export opportunities and the possibility of producing specialties. Figure 2.4 shows the distribution of the studies during that period according to their approach to the food issue.

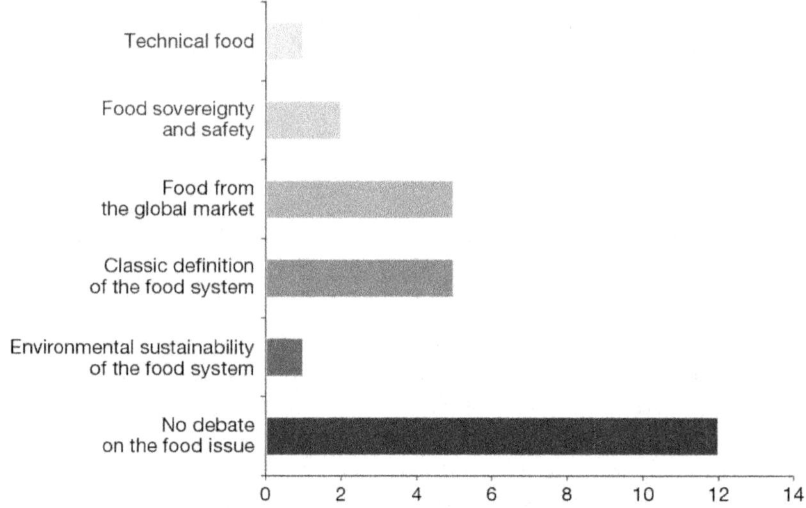

Figure 2.4. Approach to the food issue in the 2000–2010 period.

Halfway through these first ten years, the bioeconomy began to emerge as a conceptual umbrella for thinking about primary-based activities, as a way of redefining the concept of agroindustry in the light of the circular economy. In this sense, the terms of sustainability, with food security as an auxiliary concept, became installed in studies on agrifood systems. The "Biofuels and Food in Latin America and the Caribbean" scenario study published by the IICA (Gazzoni, 2009), presents an analysis of the correlation between food security and competition for productive resources, especially land, as a result of the increase in biofuel production. The study also highlights the comparative advantages of the region for the development of biofuels and concludes that there is no conflict with the destination of land for other uses (such as food) and warns about certain environmental risks of production, but also the opportunity to enter the carbon credit market. The main focus of the study on the food issue is that of "market food."

Another aspect to highlight regarding the studies in this period is the loss of the regional integration horizon as a strategy for overcoming food crises and deficits.

In general, these exercises are focused on national or smaller scales (subnational, regional, provincial, state or chain). Only one regional level study was identified (the aforementioned study on biofuels published by the Inter-American Institute for Cooperation on Agriculture [IICA]). Integration into world markets and global value chains is the main strategy envisaged in the studies. This aspect is discussed in more detail in the following section.

▸▸ Futures studies and agrifood problems in the last decade and general considerations on the period as a whole

The last decade shows a continued trend for the process of dissemination of futures studies. Again, the number of studies is more than double that identified in the previous decade. Sixty-two studies were identified for this period spanning 2010 to 2020.

As we saw in the previous section, the regional scale had lost relevance as a relevant scale for analyzing the possibilities for the future of Latin American economies during the 2000s, and also as a scale for developing viable strategies for overcoming economic difficulties. This is also reflected in the last stage, although regional studies once again feature within the larger set of studies.

During this period, national (23 studies) and sub-national (30 studies, including sub-national and organizational or chain scales) level projects predominated. It is also interesting to note the emergence of other regional scales, such as South America or the Southern Cone, which became relevant in light of the agriculturi-zation[6] processes developed with the increase in soybean cultivation and within the framework of the new integrationist strategies promoted by the governments repre-sentative of the shift to the left during this period. A total of eight studies consider supra-national regional scales (three from South America, two from the Southern Cone and three from Latin America). Figure 2.5 shows the scale chosen by all the studies for the entire period considered in the investigation.

During this period, futures studies were also disseminated in other areas. In particular, the Region's various national agricultural research institutes (INIAs) began to develop more sustained futures studies. The number of these studies in academic circles continues to grow, especially in scientific journals (where 18 articles on futures studies were identified). This data is compiled in figure 2.6, which shows the number of futures studies identified by type of publication and period.

With respect to addressing problems related to the food issue, although a large number of studies do not problematize the issue (16 studies, 25% of the studies of the period), this percentage is much lower than that of the previous decade, when a total of twelve studies failed to problematize the food issue, representing 46% of

---

6. The Southern Cone scale correlates to a process that has been developing since the new millennium and is currently gaining momentum throughout the region corresponding to the shifting of the agricul-tural frontier, displacing and reconfiguring other activities such as livestock, and of agricultural activity, with the preeminence of soybean cultivation. Argentina, southern Brazil, Uruguay, Paraguay and Bolivia are gradually becoming involved in this process and with the introduction of the technological package associated with soybean cultivation in the new millennium. For more information, see: Manuel Navarrete et al. (2005) and Manzanal (2017).

the studies of the period. Even fewer studies flagged food issues as a problem in the 1990s, as it did not feature in 55% of the studies reviewed. These relationships can be seen in figure 2.7.

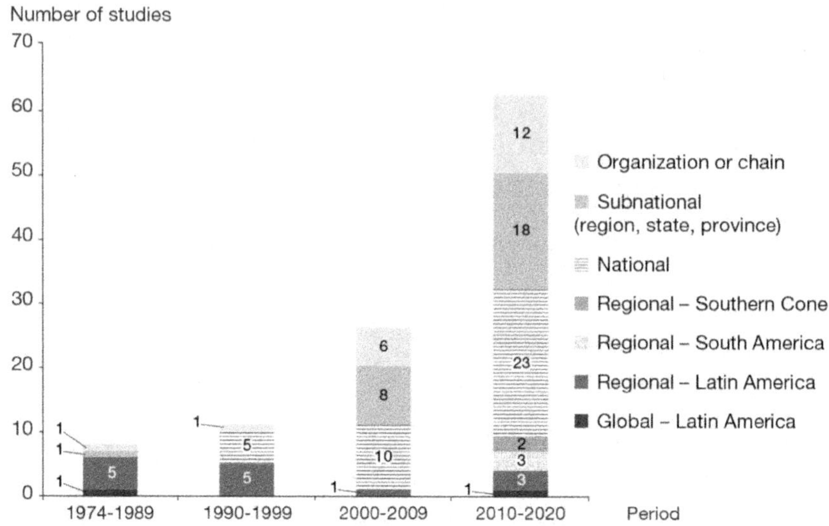

Figure 2.5. Number of studies per scale of analysis per period.

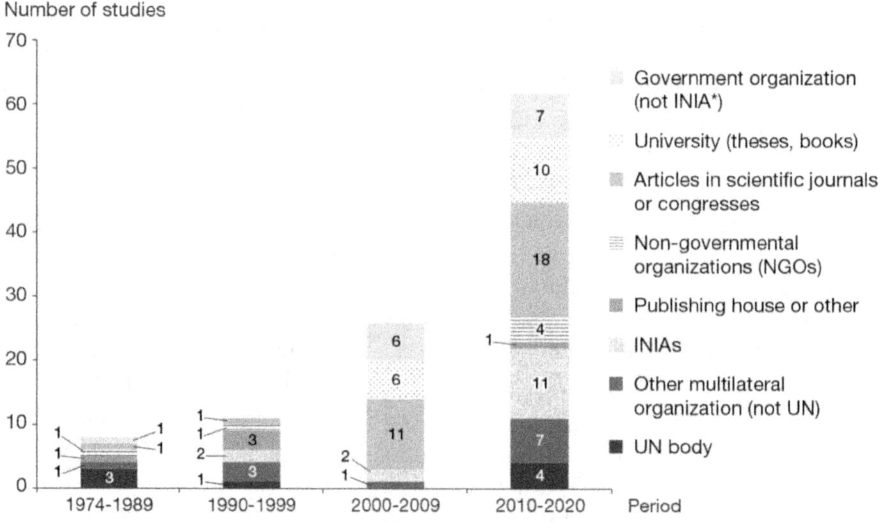

Figure 2.6. Number of studies by type of publication per period.
*INIA: National Institute of Agricultural Innovation.

Certain approaches that had already made a tepid appearance in the previous decade, such as food sovereignty and security, technical food and sustainability of food security, were consolidated during this period (between five and six studies for each approach). The structuralist approach, which did not feature during the previous decade, also makes a reappearance in three of the studies. And the new integrated approach of Sustainable Food Systems (SFS) appears in another five studies. Characteristically, this approach promotes a comprehensive approach to food issues, taking into account the economic, social, health, and general context aspects of food "systems."

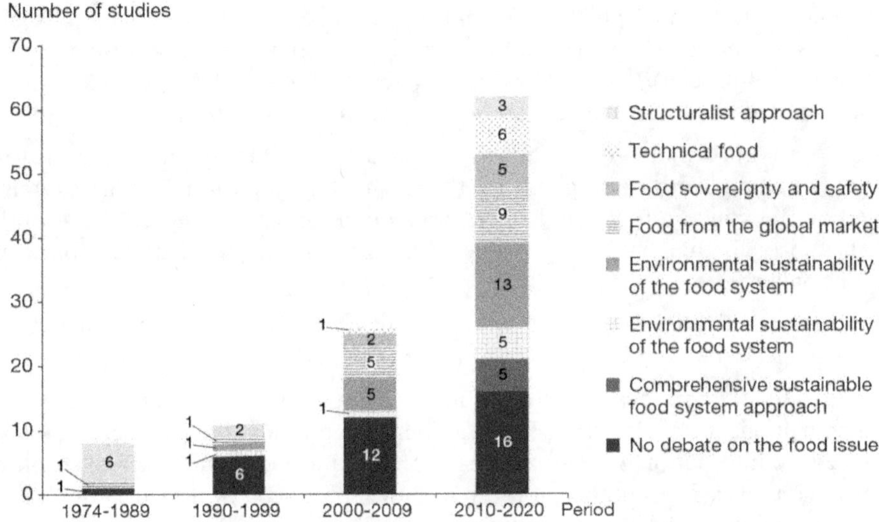

Figure 2.7. Mode of treatment of the food issue in the studies reviewed.

The analysis of the papers of this period highlights the fact that the majority of the studies that address the issue from this comprehensive approach are those that take a territorial foresight approach, i.e., they are analyzing the possible futures of food, but in a way that is situated within a particular region or territory. This is particularly true of the futures studies produced in Argentina and Colombia linked to territorial planning processes, the analysis of horticultural production systems and situated analyses of land use and innovation processes.

▸▸ Summary and conclusion

Beyond the general descriptions detailed in the different sections, there remains a synthesis and a concluding reflection on the uses of foresight and future studies to address food problems in Latin America.

In the set of works analyzed, it is possible to see the incidence of different spaces in the promotion of foresight techniques and a forward-looking approach to the analysis of agricultural and food issues. The influence of international organizations in promoting the medium- and long-term view can clearly be observed at the beginning of the period considered (the 1970s and 1980s), with a strong incidence of the structuralist approach by ECLAC, emphasizing the promotion of regional integration strategies and promoting critical visions of the future that had no place within national technical organizations.

From the start of the 1990s and during the 2000s, in particular, there was also an increase in the number of organizations analyzing the potential for expansion of Latin American agriculture, repositioning the region as a supplier of food for international markets. In the 1990s, technological foresight approaches gained prominence, an emphasis that was maintained throughout the subsequent decades, although other approaches were adopted from the 2000s onwards. In the 2000s, the number of foresight studies multiplied and so did the approaches to food-related problems. Emphasis was placed on policy analysis in order to promote food security (the food sovereignty approach), technical issues related to food and the association with environmental issues. To a lesser extent, there is an association with health issues.

The adoption of a more comprehensive approach to food problems within futures studies coincided with an emphasis on the territorial dimension. Certain exercises, based on the concern for specific territories in particular, prompted the adoption of an approach that made it possible to coordinate different dimensions, with greater balance.

It is also worth highlighting the recent reprioritization of food issues, which had been largely neglected in agricultural sectoral foresight studies, particularly in the 1990s.

The analysis of the types of publication of the studies also shows that they only appear in publications aimed at a broad audience in the early stages, when the studies financed by international organizations are co-published with a mass distribution publishing house. The number of futures studies produced by agricultural research institutes and universities (generally in postgraduate theses related to rural administration or agribusiness, with some examples of studies in the social sciences) has grown over time. There has also been an increase, albeit very gradual, in the number of studies produced by other public bodies. The mass publications produced for a general audience are, however, no longer being produced. Discussions on productive and food futures thus seem to be limited to technical arenas, and the impact on decision making, instead of being oriented towards generating debates and raising awareness among a broader public.

The survey and analysis carried out suggest that a revival of the treatment of the food issue with a long-term and comprehensive view is essential, especially in these times of economic, health and environmental crisis that we are experiencing today. It would be useful to promote these long-term analyses within different arenas, not only within technical spheres, but also as part of political debate, to a more general audience.

One aspect that remains to be comprehensively examined is the effective capacity of these studies to influence the definition of public policies, both in their formulation and in the problematization that defines public problems. Through this initial study, and by means of a quantitative approach, it has been possible to determine the focus of agrifood studies, where they are produced, to whom they are addressed, and the scale they propose for the analysis of problems. In order to further examine their power to influence, other methods of investigation need to be implemented and different sources triangulated.

In this case, the contribution has been limited to analyzing certain chapters of the history of foresight studies in the Region, the scales that have been relevant for thinking about agrifood futures, and the methods and strategies of approach that have been promoted by these studies. Much remains to be explored on the basis of this first investigation, which we hope will also be useful for other teams taking on the task of learning about the use of futures studies in the Region.

▸▸ References

Aichholzer, G. (2009). The Delphi method: eliciting experts' knowledge in technology foresight. In Bogner A., Littig B., & W. Menz (Eds.), *Interviewing experts* (pp. 252–274). Palgrave Macmillan. https://doi.org/10.1057/9780230244276_13

Bell, W. (2003). *Foundations of futures studies. Volume 1: History, purposes, and knowledge.* Transaction Publishers.

Börjeson, L., Höjer, M., Dreborg, K.-H., Ekvall, T., & Finnveden, G. (2006). Scenario types and techniques: towards a user's guide. *Futures*, *38*(7), 723–739. https://doi.org/10.1016/j.futures.2005.12.002

Cap, E. J. & González, P. (2002). *Argentina: una exploración de la frontera de posibilidades productiva del sector de granos y oleaginosas*. Instituto de Economía y Sociología, INTA. https://inta.gob.ar/sites/default/files/script-tmp-frontera.pdf

Cimoli, M. & Porcile, G. (2013). *Tecnología, heterogeneidad y crecimiento: una caja de herramientas estructuralistas*. ECLAC. http://hdl.handle.net/11362/4592

Durance, P. (2010). Reciprocal influences in future thinking between Europe and the USA. *Technological Forecasting and Social Change*, *77*(9), 1469–1475. https://doi.org/10.1016/j.techfore.2010.06.006

ECLAC & FAO. (1974). *La alimentación en América Latina dentro del contexto económico regional y mundial*. ECLAC. http://hdl.handle.net/11362/32327

Funtowicz, S. O., & Ravetz, J. R. (1993). The emergence of post-normal science. In R. Von Schomberg (Ed.), *Science, politics and morality* (pp. 85–123). Springer. https://doi.org/10.1007/978-94-015-8143-1_6

Gallopín, G. C. (2015). Desarrollo sostenible, complejidad y anticipación del futuro. *Cartografías del Sur, 1*, 146–163. https://doi.org/10.35428/cds.v0i1.7

Gallopín, G., Hammond, A., Raskin, P., & Swart, R. (1997). *Branch points: global scenarios and human choice. A resource paper of the global scenario group (PoleStar Series #7)*. Stockholm Environment Institute. https://mediamanager.sei.org/documents/Publications/Future/branch_points.pdf

Godet, M., & Durance, P. (2011). *La prospectiva estratégica para las empresas y los territorios* (K. G. Cortina, Trans.). UNESCO-DUNOD. http://www.laprospective.fr/dyn/traductions/contents/1dunod-unesco-vspan-ext-15-06-2011.pdf

Herrera, A. O., Scolnick, H. D., Chichilinsky, G., Gallopín, G. C., Hardoy, J. E, Mosovich, D., Oteiza, E., de Romero, G. L., Suárez, C. E., & Talavera, L. (1977). *Catástrofe o nueva sociedad: modelo mundial latinoamericano*. Centro Internacional de Investigaciones para el Desarrollo. https://www.idrc.ca/en/book/catastrofe-o-nueva-sociedad-modelo-mundial-latinoamericano-30-anos-despues-segunda-edicion

Herrera, A., Corona, L., Dagnino, R., Furtado, A., Gallopín, G., Gutman, P. & Vessuri, H. (1994). *Las nuevas tecnologías y el futuro de América Latina*. Siglo XXI Editores, Editorial de la Universidad de las Naciones Unidas.

Inayatullah, S. (1998). Causal layered analysis: poststructuralism as method. *Futures*, *30*(8), 815–829. https://doi.org/10.1016/S0016-3287(98)00086-X

Kozel, A., Martínez, L., Taraborrelli, D. S. & Carvalho, N. (2017). *El sistema agroalimentario del Área Metropolitana de Buenos Aires al 2030–2050. Ejercicio exploratorio de prospectiva territorial*. Ediciones INTA. http://hdl.handle.net/20.500.12123/1975

Kozel, A., & Patrouilleau, M. M. (2016). La exploración científica del futuro, antes de la última dictadura. In Biagini, H. & G. Oviedo (Eds.), *El pensamiento alternativo en la Argentina contemporánea. Tomo III: Derechos humanos, resistencia, emancipación* (pp. 103–119). Biblos.

Manuel-Navarrete, D., Gallopín, G., Blanco, M., Díaz-Zorita, M., Ferraro, D., Herzer, H., Laterra, P., Morello, J., Murmis, M. R., Pengue, W., Piñeiro, M., Podestá, G., Satorre, E. H., Torrent, M., Torres, F., Viglizzo, E., Caputo, M. G., & Celis, A. (2005). *Análisis sistémico de la agriculturización en la pampa húmeda argentina y sus consecuencias en regiones extrapampeanas: sustentabilidad, brechas de conocimiento e integración de políticas* (ECLAC, Serie Medio Ambiente y Desarrollo, n° 118). United Nations. http://hdl.handle.net/11362/5656

Manzanal, M. (2017). Territorio, poder y sojización en el Cono Sur latinoamericano: el caso argentino. *Mundo Agrario, 18*(37), e048. https://doi.org/10.24215/15155994e048

Martner, G. (Ed.). (1986). *América Latina hacia el 2000: opciones y estrategias*. Nueva Sociedad.

Masini, E. (1993). *Why futures studies?* Grey Seal Books.

Masini, E. (2005). International futures perspectives and cultural concepts of the future. In R.A. Slaughter (Ed.), *The Knowledge Base of Futures Studies* (CD-ROM). Foresight International.

Meadows, D. H., Meadows, D. L., Randers, J., & Behrens, W. (1972). *Los límites del crecimiento: informe al Club de Roma sobre el predicamento de la humanidad* (M. S. Loaeza de Graue, Trans.). Fondo de Cultura Económica.

Merello, A. (Ed.). (1969). *Prospectiva. Revista de Futuribles*. Association Internationale des Futuribles.

Miller, R. (Ed.). (2018). *Transforming the Future. Anticipation in the 21st Century*. Routledge.

ODEPA (Oficina de Estudios y Políticas Agrarias). (2005). *Agricultura chilena 2014: una perspectiva de mediano plazo*. Ministerio de Agricultura, Chile. https://www.odepa.gob.cl/odepaweb/servicios-informacion/publica/Agricultura2014.pdf

Orlansky, D. (2007). Investigación social y políticas públicas. *Sociedad, Revista de la Facultad de Ciencias Sociales, UBA, 26*, 81–98. http://www.sociales.uba.ar/wp-content/uploads/8.-Investigación-social-y-pol%C3%ADticas-públicas-N°26.pdf

Palma, C. (1991). Análisis prospectivo de las agroexportaciones. In *Cadenas agroexportadoras en Chile: transformación productiva e integración social* (pp. 35–106). ECLAC. http://hdl.handle.net/11362/29871

Patrouilleau, M. M. (2020). Los estudios del futuro y el Análisis Causal por Capas (CLA). Entrevista a Sohail Inayatullah. *Utopía y Praxis Latinoamericana, 25*(91), 266–274. https://www.produccioncientificaluz.org/index.php/utopia/article/view/34101

Quintero, R. (1993). *Prospectiva de las agrobiotecnologías* (Serie Documentos de programas, Vol. 34). IICA. https://repositorio.iica.int/handle/11324/16701

Robinson, J. (2003). Future subjunctive: backcasting as social learning. *Futures, 35*(8), 839–856. https://doi.org/10.1016/S0016-3287(03)00039-9

Subirats, J. (1989). *Análisis de políticas públicas y eficacia de la administración*. Instituto Nacional de Administración Pública, Spain.

Torres Torres, F. (1990). *La segunda fase de la modernización agrícola en México: un análisis prospectivo*. Instituto de Investigaciones Económicas, UNAM. http://ru.iiec.unam.mx/id/eprint/1342

Varsavsky, O. & Calcagno, A. E. (Eds.). (1971). *América Latina: modelos matemáticos. Ensayos de aplicación de modelos de experimentación numérica a la política económica ya las ciencias sociales*. Editorial Universitaria. https://repositorio.esocite.la/id/eprint/893

van Bers, C., Bakkes, J., & Hordijk, L. (Eds.). (2016). *Building bridges from the present to the desired future. Evaluating approaches for visioning and backcasting*. TIAS. https://www.tias-web.info/wp-content/uploads/2016/12/TIAS-Report-Series_no1_2016_Backcasting_Evaluating_Approaches.pdf

Wright, J. T. C., dos Santos, S. A., & Johnson, B. B. (1992). *Análise prospectiva da vitivinicultura brasileira: questões críticas, cenários para o ano 2000 e objetivos setoriais*. EMBRAPA-CNPUV. https://www.infoteca.cnptia.embrapa.br/bitstream/doc/535054/1/CNPUVDOC.692.pdf

Yero, L. (1993). Los estudios del futuro en América Latina. *Revista Internacional de Ciencias Sociales, 137*, 413–424.

▸▸ Appendix 2.1. Approaches to the food issue identified in Latin American futures studies

Approach	Description
Structuralist approach	Linked to the school of Latin American structuralism. Poverty, food deficit and malnutrition are linked to the structural problems of Latin American development.
Technical food	Prioritizes the issue of food safety and food security.
Food sovereignty and security	Assumes that food security is subordinated to the possibility of implementing policies at regional, national or local level to regulate aspects related to production, distribution and access to food, and even to define research, development and innovation agendas.
Food from the global market	The market and global food trade are the ways to promote and guarantee food security (exclusively through market mechanisms).
Does not problematize the food issue	–
Classical concept of food security	Emphasizes the analysis of the conditions of access and availability of food, taking the concept of food security from the initial problematization of the issue in post-war Europe (in terms of *food security*: access and availability). To a lesser extent, it addresses the problems of nutrition and safety.
Environmental sustainability of food security	Addresses food security as it is affected or affects the environmental issue, particularly in connection with climate change.
Integrated approach to Sustainable Food Systems (SFS)	Considers food and nutritional security in relation to the given context as an economic, social and environmental system, taking into account economic, sanitary and environmental aspects of production processes and marketing and consumption circuits.

▸▸ Appendix 2.2. Foresight methods mentioned in the chapter (brief description)

Method	Description
Mathematical digital simulation models	Simplified representation of reality that facilitates the correlation of quantitative variables through formulas and mathematical functions and is operated by means of a computer program capable of processing large amounts of data. There are different types: econometric, statistical, input-output, optimization, dynamic simulation or evolutionary models (Gallopín, 2015).
Scenario	The scenario method is one of the most widely used methods in foresight studies. The method proposes the elaboration of a set of scenarios to explore the possible futures of a given problem based on the analysis of multidimensional causal chains and bifurcation points. Each scenario is composed of a time horizon, a hypothetical sequence of events articulated around a narrative and a future condition or image derived from this sequence. There are different classifications and types of scenario methods. A simple classification is that which distinguishes predictive, exploratory or normative scenarios.

Method	Description
Delphi	Expert consultation method that is carried out in different rounds of inquiry, collecting anonymous feedback from the different contributors, promoting the elaboration of collective visions on possible futures.
Roadmapping, especially technology roadmaps	Method involving the strategic planning of a possible route for the development of an innovation, company or goal, using visual representations and integrating the contribution of different experts or stakeholders through participatory workshops. Widely used in technology foresight to analyze the future trajectory of a certain innovation.
M. Godet Methods	A set of four methods designed for strategic foresight processes. They are methods that support the process of constructing possible scenarios and defining a strategy. Together they are designed to conduct a "strategic foresight" process (Godet and Durance, 2011). They are based on other existing methods or techniques, to which a dynamic of consultation and interaction between experts is added, as well as certain conceptual elements from the scenario method. Each method is also associated with software for information processing and analysis. The four methods that make up this set of methods are: MIC MAC (a cross-impact matrix version), MACTOR (a stakeholder analysis version), MORPHOL (a morphological analysis version), SMIC-PROB-EXPERT (a Delphi version).
Regulatory scenario or backcasting	A normative scenario or "backcasting" is a method that sets out to reconstruct the decisions and causal chains that should be developed over the defined time horizon to achieve an image of the future, based on an image or vision of a desirable future. There are different types of *backcasting,* some are based on mathematical models (optimization, for example) and others are built on participatory workshops with social actors.
Normative methods (or approaches)	Refers to methods that aim to analyze possible trajectories to reach a desirable (not only possible) future. The Backcasting method, the Three Horizons method and the vision building method are examples of normative methods.
Causal Layered Analysis (CLA)	Method designed by Sohail Inayatullah that proposes to deepen the analysis of used and alternative futures, based on a layered analysis of the official and current discourses in which the problems are presented, scientific explanations, social discourses and worldviews, and the myths and metaphors that support the previous layers. The method is based on a logic of deconstruction and reconstruction of discourses.

Source: Based on Börjeson et al. (2006); Gallopín (2015); Van Beers et al. (2016); Godet & Durance (2011); Patrouilleau (2020).

▸▸ Annex. List of studies identified in the survey

Studies with global scale

Da Silva, A. T. B., Spers, R. G., Wright, J. T. C., & da Costa, P. R. (2013). Cenários prospectivos para o comércio internacional de etanol em 2020. *Revista de Administração, 48*(4), 727–738.

FAO. (2018). *The future of food and agriculture – Alternative pathways to 2050.* https://www.fao.org/3/I8429EN/i8429en.pdf

Herrera, A. O., Scolnik, H. D., Chichilnisky, G., Gallopín, G. C., Hardoy, J. E., Mosovich, D., Oteiza, E., de Romero Brest, G. L., Suárez, C.E., & Talavera, L. (1977). *Catástrofe o nueva sociedad: modelo mundial latinoamericano.* CIID.

Regional studies: Latin America

Ardila, J. (2010). *Extensión rural para el desarrollo de la agricultura y la seguridad alimentaria: aspectos conceptuales, situación y una visión de futuro.* IICA. http://repiica.iica.int/docs/B1898e/B1898e.pdf

Arnold, D., Fajgenbaum, J., Goel, V. K., Ortega Goodspeed, T., Kharas, H., Kohli, H. S., Kohli, H. A., Loser, C. M., Lustig, N., Puryear, J. M., Shifter, M., Sood, A., & Szyf, A. Y. (2010). *América Latina 2040. Rompiendo con las complacencias: una agenda para el resurgimiento.* The Asian Development Bank. https://scioteca.caf.com/handle/123456789/713

Bitar, S. (2014). *Las tendencias mundiales y el futuro de América Latina.* CEPAL. http://hdl.handle.net/11362/35890.1

ECLAC. (1974). *La alimentación en América Latina dentro del contexto económico regional y mundial.* http://hdl.handle.net/11362/32327

ECLAC. (1977). Tendencias y proyecciones a largo plazo del desarrollo económico de América Latina (Resolución 366(XVII)) [Long-term trends and projections of Latin American economic development (Resolution 366(XVII)]. In *Annual Report (7 May 1976–6 May 1977) Vol. 1* (pp. 221–224). United Nations. http://hdl.handle.net/11362/16047

ECLAC. (1979). *Las transformaciones rurales en América Latina: ¿desarrollo social o marginación?* http://hdl.handle.net/11362/27897

Gallopín, G. C. (Ed.). (1995). *El futuro ecológico de un continente. Una visión prospectiva de la América Latina.* Editorial de la Universidad de las Naciones Unidas.

Gazzoni, D. L. (2009). *Biocombustibles y alimentos en América Latina y el Caribe* (Serie Crisis Global y Seguridad Alimentaria, no. 7). IICA.

Gómez Oliver, L. (1994). *La política agrícola en el nuevo estilo de desarrollo latinoamericano.* FAO.

Herrera, A., Corona, L., Dagnino, R., Furtado, A., Gallopín, G., Gutman, P., & Vessuri, H. (1994). *Las nuevas tecnologías y el futuro de América Latina.* Siglo XXI Eds-Editorial de la Universidad de las Naciones Unidas.

Martner, G. (Ed). (1986). *América Latina hacia el 2000: opciones y estrategias.* Editorial Nueva Sociedad.

Pinto, A. (1982). Estilos de desenvolvimento e realidade latino-americana. *Brazilian Journal of Political Economy, 2*(1). https://centrodeeconomiapolitica.org.br/repojs/index.php/journal/article/view/1949

Quintero, R. (1993). *Prospectiva de las agrobiotecnologías* (Serie Documentos de Programas, Vol. 34). IICA. https://repositorio.iica.int/handle/11324/16701

Timmer, P. (1997). Tendencias de la agricultura en la era de la globalización: una visión prospectiva. *COMUNIICA, 2*(7), 40–50. http://repiica.iica.int/docs/B1741E/B1741E.PDF

Sub-regional studies: South America and the Southern Cone

Acioly, L., & Moraes, R. F. D. (Eds.). (2011). *Prospectiva, estratégias e cenários globais: visões de Atlântico Sul, África Lusófona, América do Sul e Amazônia.* IPEA. https://www.ipea.gov.br/portal/images/stories/PDFs/livros/livros/livro_prospectiva_cenariosglobais.pdf

Gauna, D., Oviedo, S., Campos, S. K., Gomes Pena Jr., M. A., Vial, A., & Szostak, J. (2019). *El Cono Sur ante una instancia crucial del desarrollo tecnológico global.* PROCISUR-IICA. https://repositorio.iica.int/handle/11324/13245

PROCISUR. (2010). *Rol del Cono Sur como reserva alimentaria del mundo y los posibles escenarios para la investigación, la innovación y el desarrollo.* https://www.procisur.org.uy/adjuntos/161234.pdf

UNASUR. (2018). *Estudio prospectivo: Suramérica 2025. Segunda Parte.* CEED-CDS UNASUR.

Studies on Argentina

Baum, G., & Artopoulos, A. (Eds.). (2019). *El Libro Blanco de la prospectiva TIC, Proyecto 2020.* Ministerio de Ciencia, Tecnología e Innovación Productiva, Argentina. http://cdi.mecon.gov.ar/bases/docelec/va1028.pdf

Bocchetto, R., Ghezan, G., Vitale, J., Porta, F., Grabois, M., & Tapia, C. (2014). *Trayectoria y prospectiva de la agroindustria alimentaria argentina: agenda estratégica de innovación.* Ministerio de Ciencia, Tecnología e Innovación Productiva, Argentina. https://inta.gob.ar/sites/default/files/libro_trayectoria_y_prospectiva_de_la_agroindustria_alimentaria_argentina_-_agenda_estrategica_de_innov.pdf

Cap, E. J., & González, P. (2002). *Argentina: una exploración de la frontera de posibilidades productiva del sector de granos y oleaginosas.* INTA, Instituto de Economía y Sociología. https://inta.gob.ar/sites/default/files/script-tmp-frontera.pdf

De Martinelli, G. (2017). Los escenarios sociales del desarrollo agropecuario pampeano. Un ensayo de prospectiva. In De Martinelli, G., & M. Moreno (Eds.), *Cuestión agraria y agronegocios en la región pampeana: tensiones en torno a la imposición de un modelo concentrador* (pp. 401–432). Universidad Nacional de Quilmes. http://www.iesac.unq.edu.ar/wp-content/uploads/2018/05/Cuestión-agraria-y-agronegocios-en-la-región-pampeana-Guillermo-De-Martinelli-Manuela-Moreno-digital.pdf

Falivene, G., Artusi, J. A., Arrejoria, G., & Curró, C. V. (2018). Prospectiva y desarrollo local: el caso del municipio de Santa Anita, departamento Uruguay, provincia de Entre Ríos, Argentina. In Cuervo, L. M., & F. Guerrero (Eds.), *Prospectiva en América Latina: aprendizajes a partir de la práctica* (pp. 83–100). CEPAL.

Fasciolo, G. E. (Ed.). (2011). *El futuro ambiental de Mendoza: escenarios* (Serie Estudios no. 72). EDIUNC.

Ferro Moreno, S. (2015). *Análisis estratégico de los sistemas agroalimentarios agroindustriales de carne caprina de las provincias de La Pampa y San Luis* [Doctoral thesis, Universidad Nacional de Córdoba]. Ediciones INTA. http://hdl.handle.net/20.500.12123/995

Gallopín, G. C. (2004). *La sostenibilidad ambiental del desarrollo en Argentina: tres futuros.* CEPAL. https://repositorio.cepal.org/bitstream/handle/11362/5624/1/S049721_es.pdf

Gómez Oliver, L. (1974). *Posibilidades de la Provincia de Tucumán como exportadora de limones.* Consejo Federal de Inversiones (CFI).

Kozel, A., Martínez, L., Taraborrelli, D. S., & Carvalho, N. (2017). *El sistema agroalimentario del Área Metropolitana de Buenos Aires al 2030–2050. Ejercicio exploratorio de prospectiva territorial.* INTA. http://hdl.handle.net/20.500.12123/1975

Leavy, S. (2007). *Análise prospectiva dos agronegócios no município de Pergamino, Buenos Aires, Argentina* [Master's thesis, Universidad Federal Rio Grande Do Sul]. UFRGS Digital Repository. http://hdl.handle.net/10183/8550

Loewy, T. (2008). Indicadores sociales de las unidades productivas para el desarrollo rural en Argentina. *Revista Iberoamericana de Economía Ecológica, 9,* 75–85.

Mariano, R., & Ferro, S. (2019). Escenarios prospectivos estratégicos a 2030 de las producciones agrícolas bajo riego de la cuenca del río Colorado (La Pampa, Argentina). *Revista de Estudios Políticos y Estratégicos, 7*(2), 40–68. https://revistaepe.utem.cl/?p=1051

Molina, N. A. (2017). *Información para la toma de decisiones en cadenas agroalimentarias: una propuesta metodológica para un entorno citrícola internacionalizado* [Doctoral thesis, Universidad Nacional de Rosario]. Facultad de Ciencias Económicas y Estadística, Universidad Nacional de Rosario. http://hdl.handle.net/20.500.12123/6792

Moreno, S. F., Mariano, R. C., & Balestri, L. (2015). Análisis estratégico-prospectivo de un caso: matadero frigorífico de carnes alternativas en la provincia de la pampa. *Semiárida*, *25*(2), 9–18. https://cerac.unlpam.edu.ar/index.php/semiarida/article/view/2517/2405

Patrouilleau, R. D. (Ed.). (2012a). *Escenarios del sistema agroalimentario argentino al 2030*. INTA. https://inta.gob.ar/sites/default/files/script-tmp-inta_escenarios_del_saa_al_2030_13-9-12.pdf

Patrouilleau, R. D. (Ed.). (2012b). *Prospectiva del desarrollo nacional al 2015: Las fuerzas que impulsan los futuros de la Argentina*. INTA.

Patrouilleau, R. D., Kozel, A., & Lacoste, C. (2015). *Un nudo en el foco: Vigilancia prospectiva del Sistema Agroalimentario Argentino 2015*. INTA. https://inta.gob.ar/sites/default/files/inta_-nudo_en_el_foco-completo_baja.pdf

Ruíz, A. M., & Vitale, J. A. (2011). *Prospectiva y estrategia: el caso del Plan Estratégico Vitivinícola 2020 (PEVI)* (Estudios socioeconómicos de los sistemas agroalimentarios y agroindustriales, No. 7). INTA. https://inta.gob.ar/sites/default/files/script-tmp-7_prospectiva_estrategia_pevi.pdf

Schuff, P. V., Gonzalez, L., Moltoni, L. A., Sanchez, G., Carrapizo, V. N., & Cladera, J. L. (2017). *La producción y gestión del conocimiento científico y tecnológico en el CNIA: una experiencia prospectiva*. INTA. http://hdl.handle.net/20.500.12123/1978

Subsecretaría de Agricultura. (2020). *Prospectiva Agrícola*. Ministerio de Agricultura, Ganadería y Pesca, Argentina. https://www.magyp.gob.ar/sitio/_pdf/Prospectiva-agricola-2030.pdf

Subsecretaría de Planificación Territorial de la Inversión Pública. (2013). *Los territorios del futuro. Escenarios prospectivos del territorio argentino y sus regiones hacia el año 2026*. Ministerio de Planificación Federal, Inversión Pública y Servicios, Argentina. https://www.mininterior.gov.ar/planificacion/pdf/Los-territorios-futuro-(abril-2013).pdf

Studies on Brazil

Angelotti, F., Petrere, V. G., Teixeira, A. D. C., Sá, I. B., & de Moura, M. S. B. (2010). Cenários de mudanças climáticas para o semiárido brasileiro. In Sá, I. B., & da Silva, P. C. G. (Eds.), *Semiárido brasileiro: pesquisa desenvolvimento e inovação* (pp. 159–198). EMBRAPA. https://www.embrapa.br/busca-de-publicacoes/-/publicacao/861948/cenarios-de-mudancas-climaticas-para-o-semiarido-brasileiro

Bensussan, J. A. (1996). Prospectivas da economia brasileira para 1995–2015. *Indicadores Econômicos FEE*, *24*(2), 286–320. https://revistas.planejamento.rs.gov.br/index.php/indicadores/article/download/1218/1569

Bodini, V. L. (2001). *Uso da análise estrutural prospectiva para a identificação de fatores condicionantes da competitividade na agroindústria brasileira* [Doctoral thesis, Universidad Federal de Santa Catarina]. Repositório Institucional UFSC. http://repositorio.ufsc.br/xmlui/handle/123456789/79466

Camargo, M. E., Priesnitz Filho, W., Silva, J. L. S., Malafaia, G. C., Cruz, M. R., & Motta, M. E. V. (2014). Construção de cenários prospectivos em sistema agroalimentar de Vacaria, Rio Grande do Sul, Brasil. *Agroalimentaria*, *20*(38), 137–149. http://www.saber.ula.ve/handle/123456789/38144

Da Silva, V., de Mello, N. T. C., Junqueira, J. R. C. D. M., Elizabeth, A., Verdi, N. A. R., Torquato, S. A., & Chabaribery. (2008). Diagnóstico e prospeccao para o setor agropecuário paulista: subsídios para acoes de políticas públicas. Plan Plurianual del Estado de Sao Paulo.

ECLAC & Patrimonio Natural. (2013). *Amazonia posible y sostenible*. http://hdl.handle.net/11362/1506

EMBRAPA. (2014). *Visão 2014–2034: o futuro do desenvolvimento tecnológico da agricultura brasileira*. https://ainfo.cnptia.embrapa.br/digital/bitstream/item/103222/1/Visao-2014-2034-O-Futuro-de-Desenvolvimento-Tecnologico-da-Agricultura-Brasileira-sintese.pdf

Gaya Scandiffio, M. I. (2005). *Análise prospectiva do álcool combustível no Brasil-Cenários 2004–2024* [Doctoral thesis, Universidad Estadual de Campinas]. Repositorio UNICAMP. http://www.repositorio.unicamp.br/Busca/Download?codigoArquivo=473836

Nogueira, A. C. L. (2001). Mecanização na agricultura brasileira: uma visão prospectiva. *Caderno de Pesquisas em Administração*, *8*(4), 77–87.

Núcleo de Estudos Estratégicos (NAE) e Centro de Gestão e Estudos Estratégicos (CGEE). (2004). *Projeto Brasil 3 Tempos. Documento de referência da dimensão do conhecimento*.

Porto, C. (2010). Macrocenários da Amazônia 2000–2020. *Parcerias Estratégicas, 6*(12), 185–213. http://seer.cgee.org.br/index.php/parcerias_estrategicas/article/viewFile/190/184

Quirino, T. R. (1998). Agricultura e meio ambiente: tendências. In da Silveira, M. A., & S. L. de O. Vilela (Eds.), *Globalização e a sustentabilidade da agricultura* (pp. 109–138). Embrapa-CNPMA. https://ainfo.cnptia.embrapa.br/digital/bitstream/item/151814/1/1998PL002-Quirino-Agricultura-3592.pdf

Ribeiro, E. M. (2013). *Prospectiva e sustentabilidade do ecoturismo: uso da técnica de construção de cenários no Estado do Amazonas* [Doctoral thesis, Universidade de Brasilia]. Repositorio Universidade de Brasilia. https://repositorio.unb.br/handle/10482/14319

Schmidt Cortez, A. (2007). *Métodos de cenários prospectivos como ferramenta de apoio ao planejamento relativo a substituição do atual uso do solo por florestamento: estudo de caso: a Bacia do Rio Ibicuí – RS* [Doctoral thesis, Universidade Federal de Santa Maria]. Repositorio UFSM. http://repositorio.ufsm.br/handle/1/3564

Silva, J. E. A. D. (2014). *Cenários prospectivos em redes de cooperação: o caso da Associação dos Produtores Rurais dos Campos de Cima da Serra – APROCCIMA* [Master's thesis, Universidade de Caxias do Sul]. Repositorio UCS. https://repositorio.ucs.br/handle/11338/112

Torres, D. F. U. (2015). *Análise prospectiva para o setor atacadista de flores e plantas ornamentais no Brasil e suas tecnologias da informação e comunicação* [Master's thesis, Universidad Federal Rio Grande Do Sul]. UFRGS Digital Repository. http://hdl.handle.net/10183/158929

Wright, J. T. C., dos Santos, S. A., & Johnson, B. B. (1992). *Análise prospectiva da vitivinicultura brasileira: questões críticas, cenários para o ano 2000 e objetivos setoriais*. EMBRAPA-CNPUV. https://ainfo.cnptia.embrapa.br/digital/bitstream/item/60695/1/CNPUV-DOC.-6-92.pdf

Studies on Chile

Andrade, L. I., Ríos, S. M., & Torres, G. E. (2012). Análisis prospectivo de la estructura agraria en Chile basado en un modelo de dinámica de sistemas. In *Memorias del X Congreso Latinoamericano de Dinámica de Sistemas* (pp. 403–409). https://comunidadcolombianads.com/wp-content/uploads/2017/08/2012-Memorias-X-CLADS-Argentina.pdf

Apey, A., Barrera, D., & Rivas, T. (Eds.). (2017). *Agricultura chilena. Reflexiones y desafíos al 2030*. Oficina de Planificación Agrícola, Ministerio de Agricultura, Chile. https://www.odepa.gob.cl/wp-content/uploads/2018/01/ReflexDesaf_2030-1.pdf

GFA Chile Consultores. (2015). *La fruticultura chilena al 2030: Principales desafíos tecnológicos para mejorar su competitividad*. FIA. http://bibliotecadigital.fia.cl/handle/20.500.11944/145492

Larraín Prieto, R., Melo Contreras, Ó., Díaz Rojas, L. G., & Fernández López, J. (2018). *Estudio prospectivo: industria de la carne bovina y ovina chilena al 2030: principales desafíos tecnológicos para mejorar su competitividad*. FIA. http://bibliotecadigital.fia.cl/handle/20.500.11944/146251

Martínez, E. A., Jorquera-Jaramillo, C., Veas, E., & Chia, E. (2009). El futuro de la quínoa en la región árida de Coquimbo: lecciones y escenarios a partir de una investigación sobre su biodiversidad en Chile para la acción con agricultores locales. *Revista Geográfica de Valparaíso, 42*, 95–111.

Ministerio de Agricultura de Chile, Fundación para la Innovación Agraria, & Banco Mundial. (2010). *Ejercicio 'Chile agroalimentario, forestal y rural 2030.' Implicancias de los escenarios*. http://bibliotecadigital.fia.cl/handle/20.500.11944/147908

ODEPA. (2005). *Agricultura chilena 2014: una perspectiva de mediano plazo*. Ministerio de Agricultura, Chile. https://www.odepa.gob.cl/odepaweb/servicios-informacion/publica/Agricultura2014.pdf

Palma, C. (1990). Análisis prospectivo de las agroexportaciones. CEPAL. http://hdl.handle.net/11362/29871

Ulloa, P., Coper, S., Rovira, A., Azurmendi H. (2005, October 18–21). *Estudio prospectivo para la provincia de Valdivia en el año 2020* [Conference presentation]. X Congreso del CLAD sobre la Reforma del Estado y la Administración Pública, Santiago, Chile.

Studies on Uruguay

Ackermann, M. N., & Cortelezzi, Á. (2017). *Mercado de trabajo agropecuario: situación y prospectiva hacia el Uruguay Agrointeligente de 2030*. Anuario OPYPA.

Arocena, R., & Bortagaray, I. (1996). *Competitividad: ¿Hacia dónde puede ir Uruguay? Primera etapa de un ejercicio colectivo de prospectiva" tipo Delfos."* Ediciones Trilce.

Oficina de Planeamiento y Presupuesto (OPP). (2017). *Visión 2050: Tacuarembó en la Región Norte. Memoria del proceso prospectivo.* Presidencia de la República Oriental del Uruguay. https://otu.opp.gub.uy/gestor/imagesbiblioteca/Visión_2050_Tacuarembó_Prospectiva_Fase_2.pdf

Schwoob, M.-H., Dobermann, A., & Treyer, S. (2016). *Agricultural transformation pathways initiative: 2016 report.* IDDRI, Rothamsted Research. https://www.iddri.org/sites/default/files/import/publications/atpi---report-2016.pdf

Studies on Venezuela

Giacalone, R., Hernández, M. Y., & Zerpa, S. (2013). El Estado como actor del Sistema Alimentario Venezolano (SAV): conflicto y escenarios prospectivos. In Gutiérrez, S. A., (Ed.), *El Sistema Alimentario Venezolano a comienzos del Siglo XXI. Evolución, balance y desafíos* (pp. 367–416). Facultad de Ciencias Económicas y Sociales, Consejo de Publicaciones de la ULA. http://www.saber.ula.ve/bitstream/handle/123456789/39056/capitulo9_giaca_hern_zer.pdf?sequence=1

Gutiérrez, A. (2015). El sistema alimentario venezolano (SAV): evolución reciente, balance y perspectivas. *Agroalimentaria, 21*(40), 19–60. https://www.redalyc.org/pdf/1992/199241170002.pdf

Mazzonelli, J., Jelambi, F., Alvarez, E., de la Canal, H., & Nava, B. O. (1979). Estudio prospectivo de la leptospirósis porcina en granjas organizadas de Venezuela. *Boletín de la Oficina Sanitaria Panamericana (OSP), 87*(1). https://iris.paho.org/bitstream/handle/10665.2/17351/v87n1p60.pdf

Studies on Mexico

Camacho-Sanabria, J. M., Juan-Pérez, J. I., & Pineda-Jaimes, N. B. (2015). Modeling of land use/cover changes: prospective scenarios in the State of Mexico. Case study – Amanalco de Becerra. *Revista Chapingo Serie Ciencias Forestales y del Ambiente, 21*(2), 203–220. http://doi.org/10.5154/r.rchscfa.2014.10.049

Castro, M. C. (2000). La seguridad alimentaria de México en el año 2030. *Ciencia Ergo Sum, 7*(1), 49–55. http://www.redalyc.org/articulo.oa?id=10401706

García, J. A. E., Flores, J. L. D., & López, G. M. (2014). Investigación biotecnológica pecuaria en México: situación actual, prospección y estrategias de fortalecimiento. *Revista Legislativa de Estudios Sociales y de Opinión Pública, 7*(14), 147–176. https://dialnet.unirioja.es/servlet/articulo?codigo=5656167

López, G. M., García, J. A. E., Reyes, V. C., Santillán, F. R., & Barrera, J. L. J. (2008). Estudio prospectivo al año 2020 sobre la importancia de la calidad de la leche y queso en la cadena agroalimentaria en el Estado de Hidalgo, México. *Revista Mexicana de Agronegocios, 22*, 551–569. https://www.redalyc.org/pdf/141/14102210.pdf

Martínez, D. H. F. (2013). Agenda prospectiva de investigación de la cadena productiva de la panela y su agroindustria. *Tecnura: Tecnología y Cultura Afirmando el Conocimiento, 17*(36), 72–86. https://dialnet.unirioja.es/descarga/articulo/4239477.pdf

Méndez, Á. A. C., López, G. P., & Alejandre, F. J. V. (2010). Análisis al plan nacional de desarrollo: una visión prospectiva. In *XV Congreso International de Contaduría, Administración e Informática.* ANFECA. http://congreso.investiga.fca.unam.mx/docs/xv/ponencias/179.pdf

Miklos, T., & Arroyo, M. (2016). El futuro de México a debate (prospectivas de ayer, de hoy y para mañana). *Multidisciplina, 23,* 120–147. http://revistas.unam.mx/index.php/multidisciplina/article/view/58589/51801

Patrón Ramírez, Z. (2002). *Hacia el desarrollo sustentable de México: una visión prospectiva* [Master's thesis, Instituto Technologico y de Estudios Superiores de Monterrey]. RITEC. http://hdl.handle.net/11285/567387

Pérez Haro, E. (2013). Prospectiva de la agricultura en el desarrollo de México. *El Cotidiano, 177*, 47–60. https://www.redalyc.org/articulo.oa?id=32527004006

Perló Cohen, M., & Inclán Oseguera, S. (Eds.). (2018). *El futuro de México al 2035: una visión prospectiva*. Instituto de Investigaciones Sociales, Universidad Nacional Autónoma de México. http://ru.iis.sociales.unam.mx/bitstream/IIS/5471/2/futuro_mexico_2035_final.pdf

Torres Torres, F. (1990). *La segunda fase de la modernización agrícola en México: un análisis prospectivo*. Instituto de Investigaciones Económicas, UNAM. http://ru.iiec.unam.mx/id/eprint/1342

Ysunza, M. (2001). Escenario futuro de la agricultura en México: un referente para el perfil del ingeniero agrónomo. *Sociedades Rurales, Producción y Medio Ambiente, 2*(2), 31–42. https://sociedadesruralesojs.xoc.uam.mx/index.php/srpma/article/view/25/25

Studies on Colombia

Angulo Cuentas, G. L., Charris Polo, M., & Ariza Pertúz, V. D. (2009, November 25–27). *Prospectiva tecnológica aplicada a la agroindustria en el Departamento del Magdalena* [Conference presentation]. ALTEC 2009: Innovación y creatividad para el desarrollo sostenible, Cartagena de Indias, Colombia.

Castellanos Domínguez, Ó. F., Torres Piñeros, L. M., & Flórez, D. H. (2010). *Agenda prospectiva de investigación y desarrollo tecnológico para la cadena productiva de la panela y su agroindustria en Colombia*. Ministerio de Agricultura y Desarrollo Rural, Colombia. https://repositorio.unal.edu.co/handle/unal/3408

Gómez Sabogal, P. (2018). Estudio competitivo y prospectivo 2030 a las Mipymes sector horticultor – Sabana Centro Cundinamarca – Colombia, ante los retos del futuro con la seguridad alimentaria Bogotá – Región. *Global Conference on Business and Finance Proceedings, 13*(2), 846–859. https://www.theibfr.com/wp-content/uploads/2018/09/ISSN-1941-9589-V13-N2-2018.pdf

Gualteros Sánchez, J. M. (2011). *Estudio prospectivo de la cadena productiva del biodiésel a partir de palma africana en Colombia* [Master's thesis, Universidad Nacional de Colombia]. Repositorio UNAL. https://repositorio.unal.edu.co/handle/unal/7856

Herrera, T. J. F., Schmalbach, J. C. V., & López, J. A. (2009). Construcción de cadenas productivas del sector agrícola mediante modelos de redes: caracterización, simulación de escenarios y prospectiva. *Dimensión empresarial, 7*(1), 11–21. https://dialnet.unirioja.es/descarga/articulo/3990193.pdf

López, L. F., Mejía González, D., Gómez, J. A., & Albarracín, C. (2009). *Agenda prospectiva de investigación y desarrollo tecnológico para la cadena productiva de plantas aromáticas, medicinales, condimentarias y afines con énfasis en ingredientes naturales para la industria cosmética en Colombia*. Ministerio de Agricultura y Desarrollo Rural, Colombia. http://repository.humboldt.org.co/bitstream/handle/20.500.11761/33656/Agendaprospectivainves.pdf

Martínez Torres, J. C., & Amézquita López, J. A. (2013). Análisis prospectivo del talento humano del sector agrícola en el departamento de Bolívar al año 2033. *Escenarios: Empresa y Territorio 2*(2), 297–319. http://revistas.esumer.edu.co/index.php/escenarios/article/view/105/129

Mojica, F. J., Trujillo Cabezas, R., Castellanos, D. L., & Bernal, N. (2007). *Agenda prospectiva de investigación y desarrollo tecnológico de la cadena láctea colombiana*. Ministerio de Agricultura y Desarrollo Rural, Colombia. https://sioc.minagricultura.gov.co/SICLA/Normatividad/2007-06-30%20Agenda%20Prospectiva%20Investigacion%20Cadena%20Lactea.pdf

Morales Trochez, W. M. (2009). *Estudio prospectivo sobre la producción y comercialización de la carne del curí desde el resguardo indígena de Guambia, una alternativa alimentaria para Colombia y el mundo hacia el 2014* [Undergraduate thesis, Universidad Nacional Abierta y a Distancia]. UNAD. https://repository.unad.edu.co/handle/10596/1220

Moreno-Valderrama, M. N., Flórez Martínez, D. H., Yepes Vargas, L. A., & Uribe Galvis, C. P. (2017, September 25–28). *Articulación de la oferta y la demanda en ciencia, tecnología e innovación a través de agendas dinámicas territoriales y focos prospectivos: caso de estudio sector agropecuario colombiano* [Conference presentation]. VIII Congreso GIGAPP, Madrid, Spain.

Ortiz-Guerrero C, Arango J. H., Solórzano V. G., & Barón P. R. (2017). *Trayectorias históricas y escenarios prospectivos de las reformas de tenencia colectiva de la tierra en áreas forestales comunitarias en Colombia* (Documento de trabajo 233). CIFOR. https://www.cifor.org/publications/pdf_files/WPapers/WP233Ortiz-Guerrero.pdf

Pinto Hernández, J. E., & Salazar de Cardona, M. S. (2014). Cambio climático y vulnerabilidad: prospectivas para la región nororiental de Colombia-Santanderes. *BISTUA Revista de la Facultad de Ciencias Basicas,* *12*(1), 16–45. https://revistas.unipamplona.edu.co/ojs_viceinves/index.php/BISTUA/article/view/1817/807

Sanchez, E. G., Peña, A. A. R., Torres, M. Y., & Rojas, N. A. B. (2014). Plan prospectivo para el desarrollo agrario en las Regiones colombianas a partir del posconflicto al año 2025. *El Ágora USB,* *14*(2), 397–417. https://www.redalyc.org/pdf/4077/407747670004.pdf

Uribe Galvis, C. P., Sarmiento Moreno, L. F., Bochno Hernández, E., Andrade Benitez, G., Sánchez Rojas, M. R., & Hernández Iglesias, N. (2011). *Agenda prospectiva de investigación, desarrollo tecnológico e innovación para la seguridad alimentaria colombiana, vista desde la disponibilidad de alimentos.* Ministerio de Agricultura y Desarrollo Rural, Colombia. http://hdl.handle.net/20.500.12324/12692

Studies on Peru

Noriega Araníbar, M. T. (2009). Estudio prospectivo en la cadena vid-pisco. *Ingeniería Industrial,* *27,* 141–167. https://hdl.handle.net/20.500.12724/2480

Polo Campos, Á. F. (2019). Sostenibilidad agroindustrial en el ámbito de la tercera etapa del proyecto Chavimochic: un enfoque prospectivo. *Scientia Agropecuaria,* *10*(1), 125–135. http://dx.doi.org/10.17268/sci.agropecu.2019.01.14

Studies on Costa Rica

Segura Ramírez, L. (Ed.). (2016). Análisis prospectivo sobre el impacto de la producción de café de exportación en el desarrollo futuro del cantón de Tarrazú en Costa Rica [Special section]. *Revista Centroamericana de Administración Pública,* *71,* 101–131. https://ojs.icap.ac.cr/index.php/RCAP/article/download/61/121/288

Studies on Ecuador

Ministerio de Agricultura, Ganadería, Acuacultura y Pesca, Ecuador. (2016). *La política agropecuaria ecuatoriana: hacia el desarrollo territorial rural sostenible: 2015-2025.*

Studies on Cuba

Feria, U. P., Dorado, R. M. C., & Pérez, M. M. (2018). Evaluación prospectiva de la eficiencia económica de la producción de frijol en la provincia Santiago de Cuba. *TERRA: Revista de Desarrollo Local,* *4,* 71–97. https://ojs.uv.es/index.php/TERRA/article/view/10090/11730

International organizations and the evolution of food security public policy *référentiel* in Latin America and the Caribbean

FERNANDA C. FRANÇA DE VASCONCELLOS

▸▸ Introduction

Over the last two decades, Latin America and the Caribbean have become global benchmarks in the formulation and implementation of food and nutrition security (FNS) public policies, which has led to the region being recognized for meeting the target set by the Millennium Development Goals (MDGs) in 2000 – to reduce hunger by 50% by 2015 (FAO, 2015). The region was a pioneer in creating the Community of Latin American and Caribbean States (CELAC) Plan for Food and Nutrition Security and the Eradication of Hunger 2025, with the support of the Food and Agriculture Organization of the United Nations (FAO), the Economic Commission for Latin America and the Caribbean (ECLAC), and the Latin American Integration Association (LAIA).

In the middle of the last century, with the creation of international agencies and organizations in the period after World War II, the region began to receive development and assistance aid across different political spheres, such as food. In addition to institutional aspects, we have seen the emergence of a political and cultural environment in which public policies can be transferred. In view of this influence of international models in Latin America and the Caribbean, we analyzed the literature on the transfer and dissemination of public policies as a way of understanding the relevance of multilateral organizations in the process of formulating and implementing FNS public policies in the region.

The impact of communication, globalization and information flows has allowed policymakers to gain a broader view of political systems that differ to those existing in their countries, in their quest for knowledge and ideas about institutions, programs and policies. Dolowitz and Marsh (2000) state that public policy transfer can be understood as a process by which knowledge of public policies, administrative arrangements, institutions and ideas in a given political environment (past or present) is used to develop policies, administrative arrangements, institutions and ideas in another political environment. The objects of transfer can be public policies, institutions, tools (regulatory, administrative or judicial), ideologies, justifications, attitudes and ideas, or even "negative" lessons that point out what should not be done (Stone, 2012; Dolowitz, 1997).

Various social actors are involved in public policy transfer processes, such as government officials, academia and even the private sector. In this chapter, however, we will focus particulary on international organizations, which act autonomously, creating policy networks and focusing their activities on disseminating and transferring the "best practices" of policy models and standards. The standards and objectives to be achieved in different regions of the world are defined by a host of agreements, dialogues, task forces, conferences, summits and guidelines of a global scope. It is commonly the case that this agenda is defined, incorporated and disseminated by these multilateral organizations. International consultants and specialists (visitors) provide an opportunity to gauge experiences abroad, which allows international organizations and transnational political partnerships to take policy transfer processes and transform them from a simple bilateral relationship into a more complex multilateral environment (Stone, 2012).

Stone (2004) states that the importance of these organizations in the choices and dissemination of what we call "best practices" is linked to their power to create spaces for the exchange of experience and knowledge, and to determine which countries and, thereby, which public policies will receive their funding, services and projects. International agencies have research departments and organize and take part in the largest and most relevant conferences and consultations in their fields. This gives them "scientific validation" for legitimizing their objectives and engaging governments and civil society in their agendas, and, as we will show, for forming a "large epistemic community" (Stone, 2004).

This chapter sets out to analyze the ideas, interpretations, arguments and recommendations in terms of food policies used by international organizations in the dissemination and transfer of actions for FNS in Latin America and the Caribbean. Based on contributions on the approach of Public Policy Transfer (mentioned above), and the concepts of Narrative Policy Analysis and of Public Policy *Référentiels* (Grisa, 2011; Jones & McBeth, 2010; Fouilleux, 2003; Roe, 1994; Stone, 1989), this paper focuses on the narratives used by international organizations in defending certain food policy *référentiel*.

The analysis of policy narratives is a relatively new field, with Emery Roe's 1994 book providing the starting point for studies of this approach. According to the author, policy narratives are stories (with settings and plots) that guarantee and provide stable ground for assumptions when it comes to drafting public policies in situations in which there is a great deal of uncertainty, a high degree of interdependence, and little or no agreement. These stories permeate political dialogues and have a beginning, a middle and an end; characters, victims, heroes and villains; a context or a backdrop; a plot; and more than anything else, a "moral of the story" – the solutions proposed in the form of public policies (Roe, 1994). Similarly, Jones and McBeth (2010) argue that narrative is a story featuring a temporal sequence of events that results in a plot made up of dramatic moments, symbols and characters, and which culminates in the moral of the story. The approach understands that narratives are also part and parcel of the creation of public policies, help to structure and consolidate paths, and have the symbolic power to silence other arguments (Stone, 1989).

The public policy *référentiel* approach was initially proposed by Bruno Jobert and Pierre Muller in the 1980s. It sought to analyze public policies as processes through which society comes up with representations for understanding and acting on the basis of what is real. Subsequently reworked by Fouilleux (2017, 2003, 2000), the concept of the public policy *référentiel* refers to a set of ideas – negotiated and agreed on by diverse actors – that guide the construction of public policies and are manifested in their development and their instruments. This concept is our starting point and we consider in our analysis the eight food policy *référentiel* identified in chapter 1 of this book, with their respective "secondary *référentiels*," namely: the *food-for-market référentiel*; the *urban food supply référentiel*; the *productivist référentiel*, divided into *agribusiness-based productivist* and *productivist based on family farming*; the *social food welfare référentiel*; the *technical food référentiel*, divided into *nutritionist* and *hygienist*; the *food sovereignty and food security référentiel*, divided into *food autonomy* and *heritage*; the *gastronomic référentiel*; the *environmentalized food référentiel*; and the *integrative référentiel* (Grisa et al., 2021).

The narratives analysis involved a documentary analysis of 29 reports produced by international agencies present in Latin America and the Caribbean. This research method seeks to gain an understanding of and approach social reality indirectly, through reference to various types of published documents that provide evidence of their actions, as well as the construction of their ideas, opinions and ways of acting and living. According to Cellard (2008, p. 295), documentary analysis "allows the dimension of time to be added to the understanding of the social." In other words, it facilitates the observation of the maturation or transformation of individuals, groups, concepts, knowledge, mindsets, etc.

The objective of this analysis is to identify "the stories told in these narratives," the way in which they construct arguments for the defense of a given public action. To this end, we sought to identify the characters involved, the context in which the narrative is constructed, and the "morals of the story," which appear as policy proposals. Among the various international organizations operating in the region, we identified some of those most central to the issue of FNS, chief among them FAO, the World Food Programme (WFP), the International Fund for Agricultural Development (IFAD), and the Inter-American Institute for Cooperation on Agriculture (IICA). On a smaller scale, reports from the World Bank (WB) and the Economic Commission for Latin America and the Caribbean (ECLAC), which often share reports and projects with the other agencies under analysis, were also assessed.

In terms of the timeframe, the reports we analyzed are from the period between 2009 and 2019. This period was chosen on the basis of the food price crisis, which began in 2008, after the U.S. subprime crisis in 2006. This is an important event when it comes to understanding the connection between agriculture and food in Latin America and the Caribbean, as price rises negatively impact consumers, not least the poorest, but also benefit the commodities producers and exporters in the region, which accounts for a significant proportion of world food production and is also very relevant to the aggregate production of local economies.

2009 also saw the reform of the World Committee on Food Safety (CFS), which was created in the 1970s in response to another food price crisis. Its main function on its inception was to control grain stocks and thereby guarantee price stability in the markets. The 2009 reform triggered a series of transformations in food security concepts and international debates of the issue, shifting the focus from "availability" to "access" and "use" (Duncan, 2015).

This chapter is divided into two sections, the first of which traces the data collected from the reports of each agency analyzed and seeks to identify the food policy *référentiel* actioned as a result of analyzing the policy narratives contained in these documents. The second section discusses the results, taking into account the transformations that have occurred over the time period in question in the agencies and in the region as a whole, as well as the impact of changes in the global FNS agenda. The chapter concludes with final considerations.

▸▸ International organizations and their public policy *référentiel*

FNS is crucial to the consideration of sustainable development and has appeared with increasing frequency across a wide range of international reports, guidelines and directives. The approach to this issue has essentially become more multidisciplinary in nature since the 1996 World Food Summit in Rome and the creation of the current definition of food security[1]. This approach involves various processes that encompass productive, economic, social and human health aspects. In more recent times, with the debate on climate change now front and center in political and scientific spaces, the concept of FNS has been gaining ground and occupying an increasingly important place in debates on development.

As part of this new approach, and in order for food needs to be met, it is not only necessary to have enough food stored to feed everyone; new ways of making food available and distributing it must be created as well. As well as broadening the debate and casting light on global food issues, the multidisciplinary nature of the issue involves a very large number of political actors and international organizations, many of them specializing in agriculture and food, as we will see in the following analysis, which seeks to identify the public policies espoused or proposed in the narratives of organizations.

Food and Agriculture Organization of the United Nations (FAO)

Founded in 1945, FAO leads global thinking on FNS. The organization has its regional office for Latin America and the Caribbean in Santiago, Chile, and is present in virtually every country in the region, dividing its activities into sub-regions (Caribbean, Mesoamerica and South America), and operating desks in each member country. Thanks to this extensive presence in the region, its reports use many examples of local

---

1. The definition of food security formulated at the summit addresses four main areas: availability – the maintaining of sufficient quantities of food in a constant and consistent manner; access – the ability of individuals within a population to have sufficient resources to obtain adequate food for a nutritious diet; use or absorption – the ability of individuals to use the food they access to ensure their health; stability – the strength of the structure of food systems over time.

"best practice." It also produces more publications on FNS than any other organization (global or regional), which is due to the status it has acquired as a "knowledge agency" (Vasconcellos, 2018). For this analysis, we have selected the report, The State of Food and Nutrition Security in Latin America and the Caribbean. Published annually, it assesses the state of FNS. Having given thought to public policy *référentiels*, we can affirm that one *référentiel* is not always sufficiently large to cover recommendations for every aspect of FNS, which explains why two *référentiels* sometimes appear in the same report. However, these recommendations and their respective *référentiels* are not necessarily contradictory. In many cases they complement each other, addressing problems and solutions both in terms of agriculture and food.

Between 2009 and 2014, the most commonly found narratives in FAO publications were linked to the *productivist based on family farming* and *social food welfare référentiels*. These were articulated in every report during this period. The former was present in public policy proposals designed to solve problems of food availability, and the latter was present in sections relating to access. The *technical nutritionist food référentiel* was also seen, in particular from 2014, most probably as a result of the Second International Conference on Nutrition (ICN2), promoted on the basis of a partnership between FAO and the World Health Organization (WHO), which also led to the drafting of the report in partnership with the Pan American Health Organization (PAHO).

With regard to food availability, the first report published – within the context of the 2009 crisis – stated that the solution lay not only in increasing production, as the region's agricultural production actually rose during the crisis. The document stated that predominant crops required major technological and financial resources, were mainly intended to suppress the international market (given that the increase in commodity prices benefited the region's exporting producers) and were controlled by large agrifood industries. In this respect, the public policies put forward sought to boost the domestic market, driving family farming production through credit, access to institutional markets and land, and encouraging the creation of family farmers' cooperatives (FAO, 2009).

In that first report, the issue of access to food remained in the background, but was present nevertheless. FAO built this narrative on the assertion that the concentration of resources devoted to rural areas (inputs, land, credit) increased social inequality, and that the crisis directly affected consumers in the cities, not least the poorest. The consumers most affected by the crisis were considered victims, in particular children, women and indigenous peoples, and the solution put forward was to strengthen social protection policies, including the conditional cash transfer and food assistance for "the victims" of the story (FAO, 2009).

In the year that followed, the "relevance" of the *référentiels* was reversed. The *social food welfare system* became more prominent, although the *productivist based on family farming référentiel* remained present. Poverty and inequality in the region provided the central theme of the 2010 report, with a new villain emerging, namely the *El Niño*[2] phenomenon, which highlighted inequalities and poverty by seriously

---

2. The climate event as a key character in the story is an important aspect, as the organization later points to an even more prominent character later in the period: global warming.

impacting on the most vulnerable. As in the previous year, and as we also identified in the years that followed, children, women and indigenous peoples remained the main victims, while the heroes continued to be family farmers and the government, the former providing food for the domestic market at more affordable prices, and the latter implementing public cash transfer and food assistance policies (FAO, 2010).

Between 2011 and 2014, the organization began to address food insecurity as a "two-way" problem, in that it was not just a question of hunger or food shortages but of a lack of supply and access to healthy and adequate food. The *technical nutritionist food référentiel* was more prominent in these reports, with new villains of the piece appearing, such as ultra-processed food, and food and agrifood giants. It is also worth noting a slight change in the story: as well as the disaster posed by hunger, Latin America and the Caribbean also had to contend with overweight and obesity and all the other comorbidities caused by overeating. It was during this period that the region, while recording its best results in terms of reducing hunger, also consolidated its position as the region with the highest rates of overweight and obesity in adults and children (FAO, 2011, 2012, 2013, 2014). It was a period in which children became the main victims in FAO reports, although women and indigenous peoples remained part of the story. The presence of these characters underlined the moral of the story: from 2013 reports offered up school feeding programs as one of several solutions.

In 2016, with the end of the deadline to achieve the MDGs, a new development agenda began to be promoted and implemented by United Nations agencies, giving rise to a new framework and a new *référentiel: integrative*. The Sustainable Development Goals (SDGs)[3] offer a multidisciplinary approach to development and, to some extent, we can say that FNS is at the heart of the sustainability debate. The very notion of FNS has gained new ground in FAO reports. Where *technical nutritionist food* was once a supporting *référentiel*, from 2016 it acted as a link for several *référentiels*. Nutrition has emerged as a character in production, consumption, and environmental preservation, and has become essential to social, racial and gender equality.

Where the narratives of FNS in Latin America and the Caribbean were once founded on a promising context and setting, from 2016 onwards, the story began to play out as more of a disaster, not least because of the climate issue. As a result, the policies proposed by the integration *référentiel* had the capacity to address several problems at the same time: socioeconomic, health, environmental, and production.

> The sustainability of food systems means not only the conservation of natural resources, but also the delivery of food, economic, environmental and nutritional products and services that their operation involves; that is, everything that allows the role they play in food and nutritional security, income generation and ecosystem diversity to be extended over time, among other interactions that maintain human and natural systems (FAO, 2016, p. 24).

The increase in undernutrition and hunger remained present in the storylines of FAO publications, reinforcing the idea that the problem is not food scarcity but unequal access to healthy food. The narratives highlighted the risk that this reality poses for children, and pointed school feeding and institutional purchasing policies among

---

3. Also known as the 2030 Agenda, in reference to the year in which its 17 goals and 169 targets must be achieved.

some of the solutions put forward. Family farmers continued to be important figures, playing the role of heroes, but did not have the mission of increasing the food supply.[4] In these reports, they were responsible for producing in an environmentally friendly manner, increasing the availability of healthy food, and preserving local biodiversity.

Then, in 2019, FAO published a report on healthy food, the focus of which was on combating all forms of malnutrition. The document shows that the *integrative référentiel* defines the problem of obesity and overweight as a consequence of a food system that is systemically unhealthy, from agricultural production through to final consumption. FAO and the other agencies that contributed to the publication highlighted the importance of public policies that regulate both agricultural production and the ultra-processed industry, identifying it as the main villain of the story. The proposed solution was industry regulation based on labeling, advertising control, and nutritional warnings. It was hoped that this would combat the problems of availability (expanding the range of food available), access (controlling prices and distribution channels), stability (preventing the existence of food industry monopolies and oligopolies), and, most importantly, use and absorption, which would have a direct impact on population health (FAO, 2019).

World Food Programme (WFP)

The WFP was created in 1961 as a FAO program and was established as a permanent United Nations program in 1965. Today, the two agencies are practically on a par in terms of infrastructure, technical capacity, and geographical dispersion, though their modes of operation differ considerably. The WFP has its regional office in Panama City, but it goes about its activities in a different way to FAO's regional office as it has a more decentralized structure. As we will see below, it also has a more specific focus on actions aimed at emergency food and school feeding programs. Having said that, since 2011, with the creation of the Center of Excellence against Hunger in Brasília, Brazil, the agency has grown in size and relevance, in putting forward public FNS policies for the region.

No international food organization runs more projects and has more resources available to it than the WFP. In terms of frequency, content and format, it produces materials in a different way to FAO. These materials are usually specific to a project or a theme and often contain short stories and case studies, frequently in the form of a brochure or short briefing.

The agency does not operate in every country of the region, which is why it does not publish a single report for Latin America and the Caribbean. The WFP also publishes different documents on its projects in specific countries or small groups of countries, such as the Andean region, the Caribbean and Central America.

---

4. The documents report that the region has one of the highest food availability scores in the world, exceeding daily calorie requirements, a fact that sheds light on the region's biggest problem: inequality. The World Bank (2014) states that large-scale production in Latin America and the Caribbean are essential to global supply and that inequalities are the real causes of food insecurity in the region. As a result, the region would have benefited from high food prices, which would have allowed the implementation of public social protection policies, though these were not enough in themselves to eradicate inequality and hunger (or the absence of FNS) in the long term.

The emerging nature of its operations can be seen in the presence of the *technical food and nutrition référentiel* in the documents analyzed. However, from 2016 onwards there is a change in its content, which essentially involves the issue of school feeding. Though it has always been important to the WFP, in the period under analysis a transformation in the approach to this issue can be detected, with a shift in focus from use and absorption to availability, access, and stability. This development can be seen in its reports on *home-grown school feeding*, which foresees school feeding based on natural products, produced by local family agriculture.

In addition, as is the case with IFAD and as we will see below, one of the main focuses of WFP reports is forms of international cooperation on the issue of FNS, with the emphasis on South-South[5] and triangular[6] cooperation. This can be seen not only in the content of the documents, but also in their preparation, as many of the agency's publications are produced in partnership with other organizations, such as research centers and universities, and even with ECLAC and FAO itself. These partnerships between agencies result in reports with more than one *référentiel*, some more coherent than others.

The 2009 report *Food and Nutrition Insecurity in Latin America and the Caribbean*, published jointly with ECLAC, included three *référentiels*: a *technical food référentiel* (containing both the *nutritionist* and *hygienist* secondary *référentiels*), the *productivist based on family farming*[7] *référentiel* and the *social food welfare référentiel*. Also written in the context of the food price crisis, the report tells the story of a region that produces enough food for its population, but which suffers hunger and undernutrition due to poverty and inequality (aggravated by the food price crisis) (Martinez et al., 2009).

The report begins by setting out the narrative of a disaster scenario and is then divided into two parts: the first comprising policy recommendations for social protection and food assistance and also proposing measures for food benefits, food supplementation, distribution, and the expansion of food safety control (the *technical food référentiel*); and the second proposing solutions for unemployment, boosting food markets, introducing credit policies for small producers, and reducing taxes on foodstuffs (the *productivist-based on family farming référentiel*). The central players in this publication are the poor (rural or urban), and children, the main targets of the policies proposed in the first part of the report (Martinez et al., 2009).

In 2011, the organization assisted with the publication of an issue of the journal *Border-Lines*,[8] which drew a link between FNS issues in Latin America and the Caribbean with the growing migration flows in the region. The main villain depicted in the articles was global warming, and the solutions put forward aimed to increase

---

5. According to the UN, South-South cooperation is a mode of economic, scientific, technical and technological cooperation between developing countries, and aims to address the challenges and realities they have in common.

6. According to the UN, this cooperation involves three actors – two from the Southern Hemisphere and one from the North, which may be an international organization, responsible for providing resources to allow the countries of the South to exchange technical assistance in a specific area.

7. It should be noted that the report does not use the term "family farming" but "small-scale production."

8. The publication is the result of the discussions that took place at the Food Security and International Migration: Perspectives from the Americas conference in 2010.

the resilience of the various populations assessed in the case studies, thus triggering the *food security and sovereignty* and *social food welfare référentiels* (Latino Research Center, 2011; WFP, 2011).

The narratives in the WFP reports have also been influenced by the new 2030 agenda, as evidenced by the presence of new and more multi-faceted storylines relating to FNS, and, consequently, solutions and policies designed in a more integrated way. In the 2017 report Gender-Sensitive Social Protection for Zero Hunger, the organization explained that its actions in the region's countries included the provision of technical support services for the development of public policies, with an emphasis on conditional cash transfer policies[9] (WFP, 2017).

Many of the WFP's reports are small in scale and focus on its actions and missions in countries in response to a wide range of emergency situations. For example, the agency offered a strong response to the earthquake in Haiti in January 2010, through humanitarian relief missions and food distribution. It remains present in the country today, responding to the drought it has suffered since 2014. Haiti and its people are one of the main subjects of the publications and reports on WFP missions and continue to be depicted in the agency's narratives as the victims of natural and climate disasters and the poverty that results from these events.

In a 2019 report, the agency pointed to the large percentage of people affected by environmental disasters in the region – close to 30% of the total population of Latin America and the Caribbean – which led it to tie its narrative to the issue of global warming. The document identified another major and well-known villain afflicting the most vulnerable: the double burden of malnutrition, thus highlighting the danger posed by the growth of obesity and overweight in Latin America and the Caribbean. Based on an integrative *référentiel*, the report identified school feeding programs and gender-sensitive social protection policies as the heroes of the story (WFP, 2019).

> The double burden of malnutrition imposes a significant economic and social price on every country in the region, as shown in the recent study on the cost of the double burden of malnutrition conducted in seven countries (WFP & ECLAC 2017, 2019). Responding to the growing threat of obesity and overweight while also addressing the unfinished agenda for the eradication of undernutrition requires action across many sectors. Nutrition-sensitive social platforms can be used along with school feeding programs to achieve better outcomes for the most vulnerable groups, particularly young children, mothers, and adolescent girls (WFP, 2019, p. 6).

Throughout its history, we can see a significant change in the nature of the school feeding programs proposed by the WFP. Whereas the agency initially sought to distribute food, very often acquired from a large international aid agency, to developing countries (Maluf, 2007), in more recent times it has begun to propose programs with a multidisciplinary approach, moving towards an *integrative référentiel*, the result of a new international agenda adopted by United Nations organizations.

Another project and narrative that stands out among the most recent WFP actions is the Beyond Cotton project, conducted with the cooperation of the Center of

---

9. According to the report, conditional cash transfer policies have the function of compensating the "care economy," marked by the unequal sexual division of labor, which leads to time poverty for women.

Excellence against Hunger, the Brazilian Cooperation Agency and the Brazilian Cotton Institute. Although this project is not being implemented in Latin America or the Caribbean, it stands out owing to its similarity with the More Cotton project, the result of cooperation between FAO and Brazil, specifically the Brazilian Association of Technical Assistance and Rural Extension Companies (ASBRAER), the Brazilian Association of Cotton Producers (ABRAPA), and the Brazilian Agricultural Research Corporation (EMBRAPA).[10] Unlike FAO, which sees agriculture as one of its main issues and focal points, the WFP does not view the transformation of agriculture as a central objective. The reasoning behind the Beyond Cotton project is, therefore, the possibility of integrating its production with the agroecological production of food that will be publicly purchased for school feeding. The project represents the integrative *référentiel* adopted more recently in the agency's narratives, linking food to agriculture, but also embracing social agents and family farmers as possible heroes in the quest for sustainable alternatives leading to the eradication of hunger and all forms of food insecurity.

Inter-American Institute for Cooperation on Agriculture (IICA)

IICA was created in 1942 and had its head office in Turrialba, Costa Rica. In 1948, following the creation of the Organization of American States (OAS), it became the specialist agricultural agency of the inter-American system. The main focus of its activity, as its name indicates, is agricultural cooperation projects between the nations of the Americas. IICA is currently headquartered in São José, Costa Rica, and is present in 34 countries in the Americas. Its regions of activity are as follows: Northern Region, Caribbean Region, Central Region, Andean Region, and Southern Region. As a technical cooperation body, its projects and publications focus mainly on agricultural production, technological innovation, and productive inclusion. In addition to agriculture, the institute also supports and monitors trade agreements and the international food market.

IICA differs from the other agencies in several respects. It is the only one of them that operates exclusively in the Americas; the other three are global in scope. Furthermore, the main focus of its actions is agricultural production and not food, which is directly reflected in its concept of FNS and the policy solutions it proposes. Another aspect that sets IICA apart from the other organizations under analysis here is that, while it adheres to the discourses and narratives of sustainability, it does not cite the Sustainable Development Goals (SDGs) or even the 2030 Agenda until the 2018 report. This is despite the fact that during the course of the period under analysis it has often acted in partnership with agencies in the UN system.

Like the FAO, IICA produces a vast number of reports and informative material. Our attention is focused on the agency's annual reports, which allow us to identify its stance on food and agriculture issues and on public policies that inspire its projects. Aside from rolling out projects, IICA is often called upon by member countries to assist with the creation of public policies on agricultural trade, rural development, animal and plant health and safety, sustainable agriculture, and family farming.

---

10. Information obtained from the following sources: https://centrodeexcelencia.org.br/alem-do-algodao/ and http://www.fao.org/in-action/programa-brasil-fao/proyectos/setor-algodoeiro/pt/

IICA has thus created narratives that defend its choice and implementation of projects, which often inspire policies. From these narratives we were able to identify the policy *référentiels* implemented.

Given its status as an agricultural cooperation body, it stands to reason that *productivist* is the most frequently mentioned *référentiel* across the period, with both the *agribusiness-based* and *based on family farming* secondary *référentiels* appearing in its narratives. The introduction of innovations, sustainability and adaptation to climate change were themes (and heroes) in every story, both as characters that increase competitiveness and expand international markets, and as characters committed to increasing the FNS of the region's population and to bringing about greater well-being for family farmers (especially women and young people) in rural areas. In this regard, since 2009 IICA has attached great importance to addressing the development of rural communities through land valorization. It also encourages the introduction of technological innovations and promotes agriculture that guarantees food safety[11] and more competitive trade for the region's agribusiness (IICA, 2010).

Although one of the institute's strategic objectives is the expansion of international trade and the competitiveness of the region's agribusiness, we were unable to identify the *food for market référentiel* in the narratives. Although "the market" appears as a character in the narratives and plays the part of the hero, its objective in these documents is not to feed the population but to aid the region's economic development. In the 2016 annual report, IICA highlighted the certification of family farmers' cooperatives produce as one of the main policy tools used to achieve this objective, to ensure that they could introduce their products in more demanding importing countries, mainly in North America and the European Union (IICA, 2017). The report also pointed to the importance of investments in more sustainable value chains capable of breaking into new international markets.

In the 2017 annual report, we again note the coexistence of two *référentiels*, which are again coherent with each other: *productivist* (from agribusiness and from family farming) and *technical nutritionist and hygienist*. Many of the projects presented and policies put forward pointed to investment in technological innovation as a solution (and in this report the focus on family farming and peasants is fairly extensive), especially investment in seeds and techniques that provide for greater security in agricultural and livestock production (IICA, 2018).

Family farming is a character that has grown in stature in the organization's narratives. Although it has always been present, in 2017 it was mentioned alongside the UN Decade of Family Farming, which was just about to begin. The increase in credit and technology emerged as IICA's main recommendation, though there was also growing concern about land management. Meanwhile, family farmers (a wide-ranging group that includes indigenous and traditional peoples, rural women, and rural youth) have gone from being the victims of narratives to heroes, capable of making better use of natural resources, promoting sustainable techniques and mitigating global warming, as well as integrating different aspects of the agency's territorial development proposals.

---

11. It should be pointed out here that "food *safety*" does not equate to "food *security*," as it deals with the nutritional composition and hygiene-related aspects of food products.

The 2018 report began with a message from IICA's new director general, the Argentinian Manuel Otero, who took office for the 2018–2022 period. This message saw a small change in the agency's narratives and public policy *référentiels*. Otero stressed the importance of investing in projects that promote sustainable agriculture through the intelligent use of resources and the bioeconomy and stated that these projects should be aimed at rural women and young people. Though present in previous years, in the 2018 and 2019 reports these characters became central to consideration of land development and family farming policies (IICA, 2019, 2020).

Finally, the 2019 report also proposes policies promoting the bioeconomy and land development, and although FNS was not the main objective of these proposals, we note a higher level of integration in the strategies compared to the beginning of the period under analysis. The *référentiel* was still *productivist*, and though family farming became a more relevant figure after 2017, the two secondary *référentiels* were noted in the institute's reports (IICA, 2020).

International Fund for Agricultural Development (IFAD)

IFAD was created at the 1974 World Food Security Conference[12] as one of several actions put forward to alleviate the damage caused by the food crisis of the 1970s. The predominant narrative at the conference was that of a disaster scenario, in which global food production was insufficient to meet the basic needs of the world's population, which triggered actions designed to increase productivity, mainly through the use of what became known as the "Green Revolution technological/agricultural package."[13] In this respect, IFAD's role was to provide international funding for projects that sought to boost food production in developing countries (IFAD, 2018; Vasconcellos, 2018).

The organization's participation in Latin America and the Caribbean is relatively low-key compared to other regions. Before the 1980s, it was present in only a few countries in the region, concentrating its activities in the Andean countries and Central America. It stepped up its operations from the 2000s onwards, supporting the activities of the Southern Common Market (MERCOSUL) and promoting dialogue and knowledge transfer on family farming at the Specialized Meeting on Family Farming (REAF) (IFAD, 2018). Its regional Latin America and the Caribbean desks are currently located in Brasilia,[14] Panama City and Lima. Of all the regions covered by the organization worldwide, Latin America has the lowest number of funded projects (and, as a result, values).

IFAD does not publish a periodical on its activities in Latin America and the Caribbean, although it does produce a great deal of material on the region. It should

---

12. The first World Food Security Conference, promoted by the FAO and proposed by the United States, in 1974, as a response to the reduction of world food stocks (due to the collapse of the oil industry in several countries) and the consequent increase in prices (which were also impacted by the first oil crisis). (Vasconcellos, 2018).

13. This technological package corresponds to the instruments developed in an attempt to increase agricultural productivity, which aimed to respond to the demands of the growing industrial and urban society. For the growth of capitalist agriculture, the increase in the use of agricultural machinery, the use of fertilizers and the improvement of seeds were widely disseminated (Silva, 1987).

14. It is also the South-South and triangular cooperation and knowledge center.

also be noted that the organization's annual report contains data and updates on its projects and actions in the region. In analyzing the organization's narratives, therefore, we used the material created for Latin America and the Caribbean, in addition to the organization's annual reports.[15]

Some characters are present in IFAD publications throughout the period under analysis: rural women and young people, indigenous peoples, rural peasants, and rural poverty. One of IFAD's objectives in the region (2010) was to empower rural populations. Since 2009, the organisation has been promoting projects aimed at supporting actions proposed by rural women and indigenous peoples with a view both to expanding the FNS of these groups and to making them more resilient by encouraging locally designed innovations. The existence of these characters in the narratives, along with proposals for the development of the knowledge and technologies present in the territories, allows us to conclude that the main *référentiel* actioned by the agency throughout the decade under analysis was *food sovereignty and security*, either by means of the *food autonomy* or the *heritage référentiels*.

> Indigenous peoples in rural areas face economic, social, political and cultural marginalization, which results in extreme poverty and vulnerability for a great many of them. Although indigenous peoples account for an estimated 5 per cent of the world's population, they make up 15 per cent of people living in poverty. In order to reach this important target group, IFAD uses tailored approaches that respect their values and build upon their strengths (IFAD, 2010, p. 15).

The creation of the IFAD Markets Program has also resulted in the inclusion of new issues and public policies. The promotion of knowledge transfer for the inclusion of vulnerable groups in food markets, for example, is encouraged in the 2011 annual report. Given that environmental guidelines were already present in the organization's reports, even before the confirmation of the 2030 Agenda, we note that the narrative was in line with the *integrative référentiel*, highlighting a scenario in which climate change, poverty, inequality and the health of the world's population are central topics when considering agricultural development. However, the characters in the narrative were already known (rural women and youth and indigenous peoples), and the proposed policies and projects did not differ either from those found in previous accounts, focusing investment on local initiatives with the objective of empowering minority groups and placing value on the area's existing knowledge. The *référentiel* activated, therefore, continues to be that of food sovereignty and security (IFAD, 2012).

In addition to the *food sovereignty and security référentiel*, the 2013 annual report also includes the *productivist based on family farming référentiel*, even though the same characters continue to play a part. According to IFAD (2014, p. 21):

> Enabling poor rural women and men to increase their productivity is vital to reducing rural poverty in Latin America. However, this goal is dependent upon strengthening rural people's access to a broad range of public goods and services such as education and training, infrastructure, and financing sources and services. Small producers

---

15. Like the WFP, IFAD devotes its actions and publications mainly to cooperative projects. Many of the organization's reports, which focus specifically on Latin America and the Caribbean, are written in conjunction with other institutions, including the other organizations studied in this chapter.

require financial products designed particularly to meet their needs. Most countries in the region are in a position to achieve this through national development banks and microfinance institutions.

The United Nations declared 2014 the International Year of Family Farming, which explains another IFAD report, this one regional and which also focused on the *productivist based on family farming référentiel*. The report compared the policies and programs of different countries in the region aimed at this group, with the ultimate aim of providing more general guidelines in terms of public policies. It is worth noting that in addition to public policies designed to increase productivity, such as increasing access to credit and creating specific markets and rural extension, the report proposed specific policies for the pluriactivity of farming families, in other words for farmers whose income is not entirely generated by agricultural activity and who, therefore, need more and better access to services, such as transportation, education and information technology (Schneider, 2014).

A 2015 report on investments in a "new" rural population cast light some of the projects developed by the organization in the region. The theme of the productive inclusion of family farming, and increased access to resources and markets for minority groups (rural women and youth), began to gain ground in the narratives of the *integrative référentiel*. Projects focusing on the creation of rural cooperatives, adaptation to and mitigation of climate change, and the increased availability of healthy food were central topics in this report (IFAD, 2015).

More recent reports have continued to fuse *sovereignty* with the *food security* and *integrative référentiels*. This is because the organization has continued to pursue similar policies and proposals – increasing resilience and striving to empower and strengthen minority groups through policies that value their knowledge and traditions – that are targeted at known characters (rural women and youth and indigenous peoples), but which are put forward as integrated solutions to the problems of sustainable development. The result of the fusion of two *référentiels* can be seen in Chakrabarti (2019, p. 35):

> A passionate defender of the Amazon rainforest, he had seen for years how açaí palms were felled for the *yagua* (sheath) and heart of palm, commonly known as *palmito* in Bolivia. But he also knew that the palms with tiny fruit were important. When he decided to return from the city with his wife, he also realized that the key to a better future for his family and the forest lay in the collection and processing of açaí. So, he joined forces with other young families to create the Trinchera Association of Harvesters, Producers and Processors of Amazonian Fruits. With the support of ACCESOS and third parties, from their community in particular, they opened a small factory to process açaí pulp and send it to markets in the city of Cobija and further afield in Bolivia. They also set up an açaí bar in Cobija and attended food fairs across Bolivia, spreading the message about the benefits of the fruit and how it could secure the future of their forests and their children.

In short, the narratives created by IFAD have not undergone substantial change in the period in question. The characters and solutions are virtually always the same. However, the organization reorganizes contexts and characters, turning people who were seen as victims (rural women and youth and indigenous people) into the heroes needed to mitigate global warming, food and nutrition insecurity, poverty and inequality across the world.

▸▸ The rise of the *integrative référentiel* in international organizations

Before looking at the changes that have taken place in the time period in question, we need to highlight certain events that have influenced and continue to influence the creation of FNS policy narratives in the international arena. The 2009 crisis, chosen here as the initial timeframe for analysis, triggered a series of changes in the actions, concepts and operations of international organizations. The restructuring of the CFS increased the participation of and cooperation between other international agencies responsible for contemplating global food and agriculture. The increase in hunger caused by the crisis was not directly related to the control of grain stocks, a function performed by the committee since its creation, but to difficulty of access. The upshot was that the issue of poverty could no longer be ignored if hunger were to be eradicated (Duncan, 2015).

This new approach to food insecurity, as a consequence of poverty, has gained ground in debates, as shown by reports on the issues of cash transfer and rural employment. Latin America is the most urbanized region in the world, as a result of which it has suffered an increase in inequality, a factor that was already prevalent even before the urban and industrial development projects embarked on in the middle of the last century (ECLAC, 2016). With more than 80% of its population concentrated in urban areas, mainly on the outskirts of major cities, the region's rural areas are becoming increasingly empty.[16] International organizations focused on agriculture and food are taking part in this discussion, proposing and giving thought to policies that keep rural populations in the countryside, as a way of preventing poverty from concentrating in urban peripheries, with the emphasis on those with the greatest motivation to leave: women, young people, and indigenous peoples.

For these people (victims), agencies propose policies that increase access to food, especially conditional cash transfer policies. The reduction of poverty in rural areas aims to avoid rural exodus and the overcrowding of urban peripheries. While such phenomena are a consequence of the evils of poverty and hunger, they also result in their increase. We can safely say that this vision, which links the reality of the countryside to the consequences lived out in the city, is the first step towards an integrated vision of public food policies as they relate to urban and rural areas. Up until recently, rural-related FNS policies were limited to efforts to increase productivity through the dissemination of technologies designed to increase the availability of food, while policies targeted at urban areas involved emergency aid and the distribution of additional food supplies for children and populations at risk.

The period in question also saw greater debate on climate change and global warming, with the word "sustainability" adopting an increasingly prominent position in policy proposals. In 2016, this debate led to the formulation of the 2030 Agenda, with the Sustainable Development Goals (SDGs) being created and several other actions instigated or absorbed by these new guidelines. Launched by the UN and FAO in

---

16. This has not impacted on production, however. While the "countryside has emptied," there has been an increase in the expanse of cultivated areas, which extend beyond and often disrespect environmental protection and biodiversity boundaries.

2012, the Zero Hunger Challenge frames the sustainability agenda advocated and embraced by international agencies and has led to the FNS issue taking up a central role in the most recent narratives.

Generally speaking, the narratives of international organizations have been fairly well aligned throughout the decade under analysis. However, only FAO and IICA have specific public policies *référentiels* and publications for Latin America and the Caribbean. For their part, the WFP and IFAD have global *référentiels* that are used and implemented in the projects and public policies espoused in the region. It is no surprise to learn that the actions proposed by FAO and IICA are more consistent in nature. Even so, it is worth noting FAO's greater capacity to integrate food and agriculture issues, a fact explained by the nature and objective of each organization.

The most noticeable of the narrative transformations across the time period occurs in the WFP. To begin with, the organization operates in such a way as to prevent or mitigate crises and emergencies. One of the most specific ways in which it does so is by implementing school feeding policies. It goes on to propose the adoption of the Home-Grown School Feeding model, which encompasses food and environment, through local purchasing, and gradually phasing in food purchasing from family farming. Ultimately, the organization develops and implements rural extension projects, such *Beyond Cotton*, which increases opportunities for organic and sustainable cotton production and the production of agroecological food that can be sold to the public authorities for school feeding.

The 2030 Agenda has had a significant impact on the changing role of FNS in the narratives of every organization. Initially positioned as a goal to be achieved, FNS has since 2016 become a hero of sustainable development. FNS policies, especially school feeding policies, have become examples of practices that can respond to the SDGs, combining public health and the environment, while encouraging the creation of markets and the use of clean technologies preferably created and designed by actors in rural areas.[17]

In more recent times, the appearance of an integrative *référentiel* in the FNS narratives of international agencies has revealed a capacity to propose policies and programs that provide solutions for the various problems posed by other *référentiels*. Production and consumption are taken into consideration in the creation of these proposals, as well as the socioeconomic and environmental impact. Threats to public health and to planet health are seen as problems emanating from the same source, with the result that policies need to be targeted at solving these issues simultaneously. It should also be pointed out that, in terms of this integrative *référentiel*, the preservation of the environment and biodiversity is directly related to respect for and the protection of indigenous or traditional peoples.

It is not surprising, therefore, that, in the last two years in particular, the issue of food quality has been expanded on in debates on FNS. The term "healthy eating" is now

---

17. Many of these narratives on sustainability, however, had already been adopted by international agencies after the 2012 United Nations Conference on Sustainable Development, also known as Rio+20, in which countries made a global commitment to adjust their socioeconomic development models in an effort to slow global warming. What the SDGs provided was the capacity for an integrated agenda of sustainable development policies (with an emphasis on FNS policies).

beginning to replace FNS, as the food crisis takes on other aspects besides hunger, such as overweight, obesity and other diseases related to malnutrition. Although the narrative of population growth is present in many of the international agency reports analyzed, the need to increase food production has been replaced by the urgent need for production that is less harmful to the environment and that provides consumers with healthy and safe food that is rich in nutrients and respectful of the identities and values of the territory where it is produced and consumed.

Finally, although the issue of migration is not directly connected to the *integrative référentiel*, it is one that cuts across other problems present in the narratives. We cannot fail, therefore, to highlight its growing importance in the reports compiled by international agencies. Migration, be it international or internal, played a prominent role in debates on FNS and poverty in the period between 2016 and 2018 in the main. On many occasions, the objectives of the public policies proposed for the most vulnerable groups, such as rural women and young people, and indigenous peoples, include the reduction of rural poverty as a means of curbing rural exodus and mass migration. It should also be noted that the issue of migration is frequently associated with climate change and the impossibility of remaining in the countryside, thus driving policies aimed at increasing the resilience of rural populations.

▸▸ Final considerations

The aim of this chapter is to analyze and understand how the policy narratives and the public policies *référentiels* adopted by international agencies were used in FNS policy recommendations in Latin America and the Caribbean. The results suggest that FAO, WFP, IICA and IFAD use public policy *référentiels* that are strongly interlinked throughout the period under analysis and that even when different *référentiels* coexist in the narrative, this does not lay bare proposals that reveal any expressive conflict between organizations.

In the period in question and of the agencies studied, we can state that FAO has the most complex narratives, the most detailed proposals, and is most consistent in terms of integrating food and agriculture issues. For their part, the narratives of the WFP underwent the greatest changes during the time period, while the proposals of IICA and IFAD saw less in the way of change.

Across the decade, narratives became increasingly catastrophic in nature, with more people going hungry in the region, at the same time as there was an increase in the occurrence of obesity, overweight and diseases caused by poor nutrition. It proved to be an ideal context for proposing policies based on the *integrative référentiel*, in line with the premises of a wider international agenda, made up of by the SDGs.

Scrutiny of the FAO reports shows that elements that were once crucial to the creation of narratives became less important in the agency's most recent publications, particularly the character of family farming and the concern with raising productivity. In contrast, the WFP included the character of family farming in its narratives, even proposing projects and policies aimed at this group.

Finally, FNS ceases to be a final objective and becomes a hero in the narratives put forward by international agencies, responsible for ensuring that the new objective is achieved: the slowdown of global warming and the promotion of healthy eating.

▸▸ References

Chakrabarti, S. (2019). *La ventaja de América Latina y el Caribe – La agricultura familiar: un factor decisivo para lograr la resiliencia de la seguridad alimentaria y la nutrición* [The Latin America and Caribbean advantage: family farming – a critical success factor for resilient food security and nutrition]. FIDA https://www.ifad.org/es/web/knowledge/-/the-latin-america-and-caribbean-advantage

Dolowitz, D. (1997). British employment policy in the 1980s: learning from the American experience. *Governance*, 10(1), 23–42. https://doi.org/10.1111/0952-1895.271996027

Dolowitz, D. & Marsh, D. (2000). Learning from abroad: the role of policy transfer in contemporary policy-making. *Governance*, 13(1), 5–23. https://doi.org/10.1111/0952-1895.00121

Duncan, J. (2015). *Global food security governance: civil society engagement in the reformed Committee on World Food Security*. Routledge. https://www.routledge.com/Global-Food-Security-Governance-Civil-society-engagement-in-the-reformed/Duncan/p/book/9781138574861

ECLAC, UN. (2016). *Horizontes 2030: a igualdade no centro do desenvolvimento sustentável* [Horizons 2030: Equality at the centre of sustainable development]. https://www.cepal.org/pt-br/publicaciones/40161-horizontes-2030-igualdade-centro-desenvolvimento-sustentavel

ECLAC, UN, FAO, & ALADI (2014). *Food and nutrition security and the eradication of hunger CELAC 2025: Further discussion and regional cooperation*. http://hdl.handle.net/11362/40355

FAO, PAHO. (2017). *América Latina y el Caribe: Panorama de la seguridad alimentaria y nutricional: Sistemas alimentarios sostenibles para poner fin al hambre y la malnutrición, 2016*. https://iris.paho.org/handle/10665.2/33680

FAO, PAHO. (2017). *2017. Panorama de la seguridad alimentaria y nutricional en América Latina y el Caribe* [2017. Panorama of food and nutrition security in Latin America and the Caribbean]. https://doi.org/10.37774/9789275319727

FAO, PAHO, WFP and UNICEF. (2018). *Panorama de la seguridad alimentaria y nutricional en América Latina y el Caribe 2018*. https://doi.org/10.37774/9789251310595

FAO, PAHO, WFP and UNICEF. (2019). *Panorama de la seguridad alimentaria y nutricional en América Latina y el Caribe 2019*. https://doi.org/10.37774/9789251319581

FAO. (2009). *Panorama de la seguridad alimentaria y nutricional en América Latina y el Caribe 2009*. http://www.fao.org/3/a-as580s.pdf

FAO. (2010). *Panorama de la seguridad alimentaria y nutricional en América Latina y el Caribe 2010*. http://bibliotecadigital.ciren.cl/handle/123456789/12752

FAO. (2011). *Panorama de la seguridad alimentaria y nutricional en América Latina y el Caribe 2011*. http://www.asocam.org/sites/default/files/publicaciones/files/99bb85a90d84c32d-6ed99498ab9048b9.pdf

FAO. (2013). *Panorama de la seguridad alimentaria y nutricional en América Latina y el Caribe 2012*. https://www.fao.org/documents/card/en/c/fd965f41-6192-55e9-97ba-7da947139596/

FAO. (2013). *Panorama de la seguridad alimentaria y nutricional en América Latina y el Caribe 2013*. https://www.fao.org/americas/publicaciones-audio-video/panorama/2013/es/

FAO. (2014). *Panorama de la seguridad alimentaria y nutricional en América Latina y el Caribe 2014* [Panorama of Food and Nutritional Security in Latin America and the Caribbean 2014]. https://www.fao.org/publications/card/en/c/78f2e88c-8da7-494b-9e53-ff45db7944b9/

FAO. (2015). *2015 Panorama de la inseguridad alimentaria y nutricional en América Latina y el Caribe: La región alcanza las metas internacionales del hambre* [2015 Regional overview of food insecurity latin america and the caribbean: the region has reached the international hunger targets]. https://www.fao.org/publications/card/en/c/a61ce773-fefc-4481-8396-e4641f62b4f1/

Fouilleux, E. (2003). *La politique agricole commune et ses réformes: une politique européenne à l'épreuve de la globalisation*. Editions L'Harmattan. https://www.editions-harmattan.fr/livre-la_politique_agricole_commune_et_ses_reformes_une_politique_a_l_epreuve_de_la_globalisation_eve_fouilleux-9782747546805-15026.html

Grisa, Catia. (2011). As ideias na produção de políticas públicas: contribuições da abordagem cognitiva. In: Bonnal, P. & Leite, S.P. (Eds.), *Análise comparada de políticas agrícolas* (pp. 93–110). Mauad X. https://mauad.com.br/analise-comparada-de-politicas-agricolas

IFAD. (2010). *Annual Report 2009.* https://www.ifad.org/en/web/knowledge/-/publication/annual-report-2009-full-version?p_l_back_url=%2Fen%2Fweb%2Fknowledge%2Fannual-reports

IFAD. (2011). *Annual Report 2010.* https://www.ifad.org/en/web/knowledge/-/publication/annual-report-2010-full-version

IFAD. (2012). *Annual Report 2011.* https://www.ifad.org/en/web/knowledge/-/publication/annual-report-2011-full-version?p_l_back_url=%2Fen%2Fweb%2Fknowledge%2Fannual-reports

IFAD. (2013). *Annual Report 2012.* https://www.ifad.org/en/web/knowledge/-/publication/annual-report-2012-full-version?p_l_back_url=%2Fen%2Fweb%2Fknowledge%2Fannual-reports

IFAD. (2014). *Annual Report 2013.* https://www.ifad.org/en/web/knowledge/-/publication/ifad-annual-report-2013?p_l_back_url=%2Fen%2Fweb%2Fknowledge%2Fannual-reports

IFAD. (2015). *Annual Report 2014.* https://www.ifad.org/en/web/knowledge/-/publication/ifad-annual-report-2014?p_l_back_url=%2Fen%2Fweb%2Fknowledge%2Fannual-reports

IFAD. (2015). *Una nueva generación para la transformación rural: El FIDA en América Latina y el Caribe* [A new generation of rural transformation: IFAD in Latin America and the Caribbean]. https://www.ifad.org/es/web/knowledge/-/a-new-generation-of-rural-transformation-ifad-in-latin-america-and-the-caribbean-1

IFAD. (2018). *Agents of rural change: The IFAD story.* https://www.ifad.org/es/web/knowledge/-/publication/agents-of-rural-change-the-ifad-story

IICA. (2010). *2009 Annual Report: IICA's Contribution to the Development of Agriculture and Rural Communities in the Americas.* http://repositorio.iica.int/handle/11324/5485

IICA. (2017). 2016 *Annual Report: Agriculture, opportunity for development in the Americas.* http://repositorio.iica.int/handle/11324/2788

IICA. (2018). 2017 *Annual Report: Agriculture, opportunity for development in the Americas.* http://repositorio.iica.int/handle/11324/6855

IICA. (2019). 2018 *Annual Report. Agriculture, opportunity for development in the Americas.* http://informeanual.iica.int/

IICA. (2020). 2019 *Annual Report: Agriculture, opportunity for development in the Americas.* https://repositorio.iica.int/handle/11324/15258

Jones, M. & McBeth, M. (2010). A narrative policy framework: clear enough to be wrong? *Policy Studies Journal*, 38(2), 329–353. https://doi.org/10.1111/j.1541-0072.2010.00364.x

Latino Research Center (2011). Conference food security and international migration: perspective from the Americas. *Border-Lines*, 5. https://nevada.app.box.com/v/Border-LinesMarketplace

Martínez, R. et al. (2009). Inseguridad alimentaria y nutricional en América Latina y el Caribe. CEPAL; WFP. https://www.cepal.org/es/publicaciones/38913-inseguridad-alimentaria-nutricion-al-america-latina-caribe

Roe, E. (1994). *Narrative policy analysis: theory and practice.* Duke University Press. https://doi.org/10.1215/9780822381891

Schneider, S. (2014). La agricultura familiar en América Latina. Un nuevo análisis comparativo. IFAD, RIMISP, 2014. https://www.ifad.org/documents/38714170/39135645/Family+farming+in+Latin+America+-+A+new+comparative+analysis_s.pdf/9330a6c4-c897-4e1c-9c05-1144ebec0457

Stone, D. (1989). Causal stories and the formation of policy agendas. *Political Science Quarterly*, 104(2), 281–300. https://doi.org/10.2307/2151585

Stone, D. (2004). Transfer agents and global networks in the 'transnationalization' of policy. *Journal of European Public Policy*, 11(3), 545–566, 2004. https://doi.org/10.1080/13501760410001694291

Stone, D. (2012). Transfer and translation of policy. *Policy Studies*, 33(6), 483–499. https://doi.org/10.1080/01442872.2012.695933

Vasconcellos, F.C.F. (2018). *As narrativas da FAO sobre segurança alimentar: uma análise sobre a convivência de paradigmas políticos conflitantes*. [Master's thesis, Universidade Federal do Rio Grande do Sul]. UFRGS LUME digital depository. https://lume.ufrgs.br/handle/10183/210773

WFP. (2019). *Estrategia de protección social del WFP en América Latina y el Caribe*.

WFP. (2017). *Protección social sensible al género para el Hambre Cero. El papel del PMA en América Latina y el Caribe*. https://reliefweb.int/report/world/protecci-n-social-sensible-al-g-nero-para-el-hambre-cero-el-papel-del-pma-en-am-rica

Part II

A historical overview of national food policies

Chapter 4

Social history and institutional change in Nicaragua's agricultural and food policies[1]

SANDRINE FREGUIN-GRESH, GENEVIÈVE CORTES

▸▸ Introduction

Food security has been a driving force in public policy for over thirty years, particularly since the World Food Summit in 1996, which was one of the most important events of the millennium and brought together more than 120 heads of state and government. However, food policies face major challenges: food-related issues are highly complex, as they involve the coexistence of social, economic and political dimensions that must be coordinated within a context of uncertainty and multifaceted constraints.

This chapter examines changes in Nicaraguan food policy over the past century. It is based on an approach that combines the socio-history of public action with historical institutionalism. Firstly, however, it is important to establish the context of the study. Since Nicaraguan agriculture plays a central role in the economy and in food availability, agricultural policy and food policy are closely interconnected. While agricultural policies have long promoted agro-exports at the expense of food production, in a context of large-scale food importation, food policy has for many years been limited to regulating the health aspects of production, marketing and consumption. Although the focus on food self-sufficiency began in the 1980s, it was only in the mid-2000s that food sovereignty was placed on the political agenda, bringing about a gradual institutional change and a new bifurcation of the trajectory of agricultural and food policies.

This context raises some pertinent questions. How do food policies interconnect with agricultural and rural development policies in order to tackle the food challenges of the Nicaraguan population? Why have political and social actors in Nicaragua converged towards the development of policy instruments aimed at food self-sufficiency and sovereignty? How have they been able to place this paradigm at the heart of the agenda that actually assimilates agricultural and food policy?

---

1. This chapter is a translated and revised version of an article accepted for publication in the *Economie Rurale* journal, no. 377, 2021.

This chapter mobilizes several sources of information: Nicaraguan agricultural[2] and food policy documents (strategies and action plans, laws, regulations, programs, etc.); academic literature on the evolution of agricultural, rural and food policies in Nicaragua; and the results of field work carried out in the northern department of Chinandega.

After outlining the context in relation to the agricultural sector and the food security situation in Nicaragua, the section that follows introduces the approaches mobilized, and the materials and method used in the analyses. These are presented chronologically in four stages, focusing on the trajectory of agricultural and food policy instruments. The chapter then draws to a close with a conclusion and some reflections on the way forward.

➼ Contextualization

Although Nicaragua's population has been predominantly urban since the mid-1980s, rural communities still represented 40% of the total population in 2020. The majority of this sector of the population are employed in agriculture (80% of the rural population), an important sector of the economy that, according to the World Bank (2015), contributes 15% of the GDP,[3] ahead of mining (14%) and trade (11%).

The central role played by agriculture in food availability

National agricultural production is even more central in Nicaragua, providing 80% of the basic foods consumed by the population: maize, beans, sorghum, rice, meat and dairy products. Agricultural production takes place in a wide variety of biophysical conditions that allow various types of farms to produce a wide range of crops and livestock (Maldidier and Marchetti, 1996). On the one hand, the large estates, legacy of the Conquest, produce mainly export crops (sugar cane, bananas, peanuts, sesame, coffee, cacao) and cattle on the fertile plains of the Pacific and in the mountains of the Central North, or practice timber extraction and mining in enclaves on the Atlantic Coast. And on the other, in the areas left by these large landowners, smallholdings and family farms produce food, sometimes combined with diversified animal husbandry, coffee or cocoa (Merlet,1990).

Nicaragua is a net exporter of agricultural products for which food imports represent only 10% of total imports (Bornemann et al., 2012). The fact that food is imported, although in small volumes, is an indicator of the vulnerability of the Nicaraguan agrifood system. At present, food cultivation for the local market involves many families in a limited amount of space, as livestock farming and plantations for export occupy the lion's share of the available land. However, domestic production does not cover the country's needs and, as such, it imports cereals (rice, wheat), oil and other food products (raw and processed), while also benefiting from food aid.

---

2. This literature review, started in 2012, has resulted in the publication of a chapter on policies targeting family farming (Pérez and Fréguin-Gresh, 2015) and another on public policies in support of agroecology (Fréguin-Gresh et al., 2016).
3. GDP 2018 = US$13.1 million (current dollars).

According to FAOStat, the trend points towards a growing food deficit. Providing the basis for a poorly diversified diet, Nicaraguan agriculture is characterized by low productivity and is subject to climatic challenges and natural disasters. These factors affect the variability of the quantities of food produced and the market prices (Solornazo, 2016; Bornemann et al., 2012) which, moreover, have increased in the last 15 years (ECLAC, 2017). Finally, gender and generational inequalities, the degradation of natural resources, isolation and limited accessibility to services also have a negative impact on agricultural production.

Food insecurity continues to prevail, despite some improvement

In this context, 17% of the Nicaraguan population suffers from hunger (ECLAC, 2017). As in other countries, food insecurity mainly affects poor populations, and as 94% of the rural population is in a situation of multidimensional poverty (INIDE, 2016; FAO, 2018), the Nicaraguan rural communities are those most affected by food insecurity. While the Global Hunger Index (GHI) has improved over the last 30 years, Nicaragua remains one of the Central American countries most affected by hunger (FAO and PAHO, 2017). Despite a sharp decline in the number of under-nourished people over the last 20 years, the food insecurity trend has remained level (see figure 4.1). In addition, while maintaining low levels of acute malnutrition (>4%) and global malnutrition (6%), and reducing chronic malnutrition (Solornazo 2016), there is a high percentage of overweight and obese adults, estimated at 10% of the population (FAO and PAHO, 2017). Nicaragua thus faces the double burden of malnutrition (FAO, 2019; FAO and PAHO, 2017).

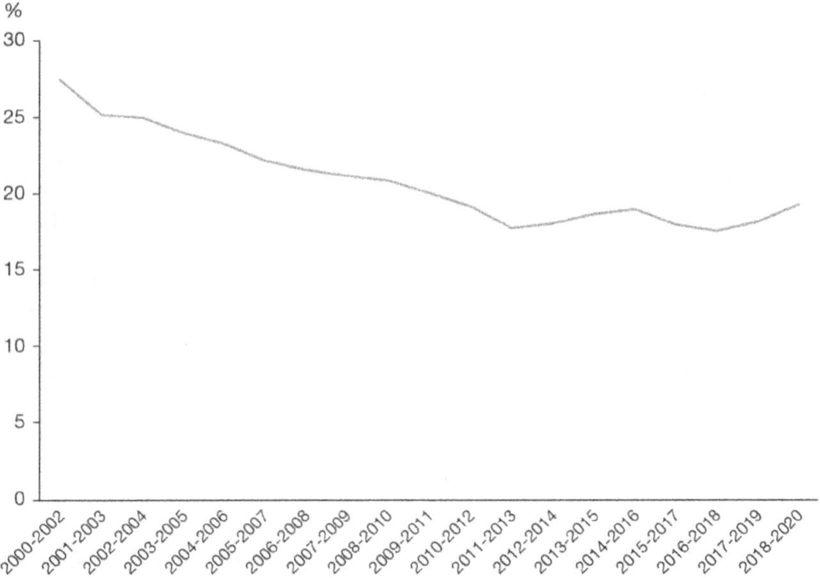

Figure 4.1. Evolution of the number of undernourished people (3-year average) (source: FAOStat).

▸▸ A sociohistorical approach to public action and institutional change based on the analysis of the trajectories of policy instruments

Theoretical approaches

This section analyzes the evolution of public policies in the field of agriculture and food in Nicaragua, mobilizing various approaches.

On the one hand, this chapter is aligned with other research conducted in the sociohistorical field of public action (Payre & Pollet, 2005; Dubois, 2003; Cossart & Hayat, 2015). These investigations do not form a homogeneous corpus nor do they claim to provide a specific theoretical framework. However, they offer keys to the analysis of the complexity of public action based on the past in order to evaluate the lessons learned (Baudot, 2014). The sociohistory of public action makes it possible to question the power relations at play in the elaboration of public policies and forms of social governance, highlighting continuities and ruptures beyond the classic temporalities linked to the alternation of governments and regime changes, including the most radical, such as revolutions (Lascoumes & Le Galès, 2007). In the sociohistorical approach, time is not mobilized as a variable external to social processes, but rather as a component of the context within which actors must act and that positions them as part of social processes situated in space and time. This chapter uses the practice of social history, which is defined by a focus on temporalities and, empirically, on the social production of categories of knowledge and action. Thus, we consider, in particular, the administrative categories (Baudot, 2014) mobilized in policy documents (e.g., the poor in a policy to fight poverty, small-/medium-/large-scale producers in an agricultural policy, etc.), especially those instruments that are "effective markers of change" (Lascoumes, 2007; Lascoumes & Le Galès, 2007). Charting the instruments and their evolution over time, the practice of social history, in our opinion, facilitates the analysis of institutional changes in agricultural and food policies in Nicaragua. In social history, the term instrument refers to the, "technical device with a generic vocation that entails a concrete conception of the political/society relationship and is based on a conception of regulation," (Lascoumes, 2007, pp. 776–777) that embodies the relationships between social actors. The hypothesis is then based on the idea that, "instruments embody one or more convergent political rationality and that they are supported, in the implementation of a program, by a social group that finds its potentialities in line with its interests. The instrument produces, in fact, a specific representation of the issue it addresses" (Ibid., p. 77). Unlike the sociology of public action, social history analyzes instruments as a transmission belt of policy intentions and content up to their implementation (Baudot, 2014).

On the other hand, the chapter mobilizes historical institutionalism approaches, drawing on research by Hall et al. (Hall, 1993, 1997; Hall & Taylor, 1997), which show that the choice of instruments is indicative of institutional change, accumulations of institutions and ruptures, which may reflect a path of dependence or critical junctures that reshape the national trajectory (Mahoney, 2001; Collier & Collier, 2003; Capoccia, 2015; Capoccia & Kelemen, 2007). These works further show that the analysis of policy instruments allows us to qualify levels of change ranging from policy mix adjustment, as defined by Flanagan et al. (2011), reflecting a first order of change to the introduction of new instruments designed as part of a paradigm shift (a deeper level of change).

Material and Method

In addition to a literature review on the evolution of agricultural, rural and food policies in Nicaragua, the research is based on an analysis of "gray literature" (documents on strategies and action plans, laws, regulations, programs, etc.). It is complemented by empirical data collected from fieldwork conducted in 2016 in the department of Chinandega, which is located in the Mesoamerican dry corridor, where poverty and food insecurity persist. Qualitative surveys were conducted in eight villages illustrating the disparities of this rural region (in terms of biophysical conditions, access, agricultural production). Data collection was based on six focus groups, 11 semi-directive interviews with local leaders and seven interviews with administrative actors involved in the implementation of flagship programs of the current agricultural and food policy (the Food Production Program (PPA), known as "Zero Hunger" and the Comprehensive School Nutrition Program (PINE), known as "School Meals"). The focus groups, which brought together between six and 15 women and men representing different age groups, addressed a range of topics: changes in the production system, the institutional environment related to agricultural production and food, the population's diet, its evolution and the strategies deployed to overcome crises, and gender and intergenerational relations. Interviews with local leaders and administrative actors focused on the interpretation of the content of public programs, their operation and effects. These stakeholders were selected on the basis of their role in the organizations involved in the implementation of the programs (table 4.1). A number of stakeholders allowed us to consult their work files.

Table 4.1. Selection of surveyed administrative stakeholders

Affiliation of administrative stakeholders surveyed	Level of participation in the implementation of agricultural and food policies
Family cabinets (at municipal and local level)	Pre-selection of PINE and PPA beneficiaries
Nicaraguan Agricultural Technology Institute (INTA)	Agricultural technical assistance provided under the PPA
Ministry of Family, Community, Cooperative and Associative Economy (MEFCCA) at municipal and departmental level	Application of the PPA
Ministry of Education (MINED)	Application of PINE
World Food Programme (WFP)	
Public schools (local)	Distribution of food to schoolchildren

▸▸ Socio-historical analysis of agricultural and food policy instruments in Nicaragua: evidence of institutional change

A combination of instruments that promote large landholdings and agro-exports and regulate trade and food production in a context of high market dependence

Until the end of the 1970s, agriculture was the main sector of the Nicaraguan economy and had to meet the objectives of economic growth and macroeconomic equilibrium: it played the role of supplier of items for export (Grigsby Vado & Pérez, 2007). This situation is the result of power relations, established during colonization, skewed in favor of the ruling classes composed of an agrarian bourgeoisie of landowners and a commercial oligarchy (Merlet, 1990).

In the 19th century, the ruling classes formed a political and economic elite (Paige, 1985), as in other Latin American countries (Hurtado, 2000), and promoted the opening of the economy and a model based on integration into the world market, the development of export crop plantations (indigo, cotton, sugar, bananas) and cattle ranches, on farms supported by different public policy instruments. These estates were also potentially involved in timber and mineral extraction on the Atlantic Coast, financed by foreign investment. In this context, a series of laws allowed the privatization of land and its concentration in the hands of the elite, such as the agrarian laws of 1858, 1862 and 1877, which promoted the development of haciendas and large estates. They also promoted the individual appropriation of land (IDERU, 2001), while the labor laws (1841) guaranteed the availability of labor for large landowners (Merlet, 1990). Thus, large landholdings benefited from support provided by public policies that facilitated land concentration and capital accumulation (Gould, 2008).

This policy orientation was accentuated with the coming to power of the Somoza family that controlled Nicaragua under a military dictatorship for several decades (IRAM & AEDES, 2000). The Somozas benefitted from the support of the United States and foreign investment focused on the development of large plantations for agro-export and the extraction of natural resources, enabling Nicaragua to, at that time, profit from the most dynamic and prosperous economy in Central America (Wiggins, 2007). This dynamic was made possible as a result of the coercion and repression of workers and peasant farmers that represented both a threat and an opportunity, as it sparked the emergence of cohesive social movements against the dictatorship, mixing social struggle and guerrilla activity (Sánchez González et al., 2016).

The most illustrative example of the dynamics of intense modernization of export agriculture (Hurtado, 2000) was the expansion of large cotton plantations benefiting from significant public support: access to land, technical advice, credit conditional on the adoption of technical packages (Fréguin-Gresh, 2017). The creation of a Ministry of Agriculture and Livestock in 1952 to promote and diversify export products was accompanied by funding and technical assistance from the United States: the creation of a National Agricultural Technical Service (STAN), closely linked to the United States Department of Agriculture (USDA), stimulated the expansion of cotton, coffee and livestock breeding and, to a lesser extent, sugarcane and tobacco. Specialist institutes, such as the Nicaraguan Coffee Institute (INCAFE), were set up that, among other things, were in charge of relations with exporting companies, often close to the government and/or linked to US interests, in order to enforce international price-fixing agreements (Craipeau, 1992). Other public incentives indirectly promoted agro-exports with various development programs aimed at establishing economic and communication infrastructures (railroads, ports, roads, telegraphs, etc.), facilitating the financing of production (system of usurious loans to establish plantations for export) and promoting large-scale ownership and circulation of their products for export, as well as facilitating the expansion of the sector and the advance of the agricultural frontier. The corollary was the forced displacement and poverty of the peasantry (Maldidier & Marchetti, 1996). Food production was marginalized and imports of commodities (textiles, food) increased in the context of the free market (Wiggins, 2007).

Thus, the only policy instruments related to food at that time referred to trade (regulation of the sale of milk of 1936) and food safety (regulation of milk pasteurization of 1949) or sanitary regulations of production (Animal Health Law of 1954 and Plant Health Law of 1958, Law of Production, Marketing and Use of Improved Seed Varieties for Planting of 1967) (Perez Martinez, 2019). Despite a lack of reliable figures showing a rise in poverty and hunger, the INCAP (Institute of Nutrition of Central America and Panama), created in 1946 as a specialist center for the study of food and nutrition within the framework of the Central American Integration System (SICA), flagged up concerns about the situation of malnutrition, particularly in Nicaragua. Faced with this situation, the government's response was limited to creating a Nicaraguan Agrarian Institute (IAN), which has been organizing agricultural colonization and road construction since 1963, the year in which an agrarian reform law was passed in name only, since far from giving rural communities access to land, it only ensured the availability of labor for low-cost work in the plantations (IRAM & AEDES, 2000).

The 1972 earthquake destroyed the capital and caused the collapse of the economy and public institutions. This marked a critical juncture (Stuart Olson and Gawronski, 2003), opening a window of opportunity in the fight against poverty, while popular discontent peaked, mobilized around Sandino's project seen as an alternative to the economic and social model controlled by the ruling classes (Sánchez González et al., 2016). While socioeconomic structures deteriorated and aid was diverted (Rueda Estrada, 2013), the agro-export model was in crisis (Maldidier & Marchetti, 1996). The majority of the population was suffering from hunger (Barroso Peña, 2011). It is in this context that a popular uprising led by the Sandinista National Liberation Front (FSLN) triumphed and led to a regime change (Figueroa Ibarra, 2005; Austin, Fox and Kruger, 1985).

The reform of instruments in support of food self-sufficiency: new options under pressure?

The Sandinista Popular Revolution marked a turning point in public policy and prompted profound social and economic transformations (Núñez Soto, 1984, 1987) and institutional changes (Mahoney, 2001). The focus of the new political agenda was on defense, literacy, health, the transformation of the food system, particularly the land issue and the social conditions of production (IRAM & AEDES, 2000). While agro-exports remained crucial for generating foreign exchange, the policy instruments guiding public interventions in the food system were directed at the peasantry: agrarian reform (agrarian reform laws of 1981 and 1988), credit programs, technical assistance, price guarantees, marketing services (Zalkin, 1987). Agricultural policy prioritized food self-sufficiency (Austin, Fox & Kruger, 1985) through the reactivation of food production and the strengthening of rural employment, which responded, in part, to the demands of the population. Indeed, social movements, particularly workers' movements, which represent the collective expression of the interests of traditionally marginalized groups in Nicaraguan society, were gaining significant ground (Terán & Quezada, 2005) and weighing in on the political agenda. However, the country was also experiencing hostile action from both internal and external forces opposed to the Revolution that influenced the direction of policy: food self-sufficiency is called for in a context of blockades.

The Ministry of Agricultural Development and Agrarian Reform was created to manage the production of state farms and cooperatives (Grigsby Vado & Pérez, 2007), prioritizing associative forms of activity that were almost nonexistent prior to 1979, and that benefit from significant public support: credit, technical assistance and agricultural services. The Government established a food marketing and distribution network with the Basic Food Company (ENABAS), which purchased food production at controlled prices and subsidized consumption (Rueda Estrada, 2013). These instruments formed part of a National Food Plan focused on local production and consumption of "indigenous" products such as corn and its derivatives, which had previously been devalued in favor of imported foods, while the Ministry of Culture organized fairs to promote peasant production (Berth, 2014). In 1985, an incentive program for self-sufficiency was introduced to promote food production, including in the city (for example, 800 hectares were cultivated in the outskirts of Managua).

While changes in agriculture were mitigated (Grigsby Vado & Perez, 2007), particularly due to the agrarian counter-reform carried out (Roux, 2010), the objectives of reducing inequality, poverty and malnutrition were achieved (Redclift, 1986). However, after the mid-1980s, conflicts hindered production (Ortega, 1986). Economic imbalances, the gap between prices and wages, the use of subsidies, complex solutions (work for food) and the concentration of defense expenditures were difficult to manage. The end of the 1980s was characterized by a low-intensity conflict, which, combined with years of drought, minimized the effects of public interventions.

Elaboration of a draft SSAN policy in a post-crisis context marked by the affirmation of the role of external actors in the fight against poverty

Following the 1990 elections, there was another change of direction, with the return to power of the liberal governments in a context of national peacemaking and reconciliation, in which a strategy that once again favored free markets and agricultural exports was rethought. The lifting of the US blockade against Nicaragua enabled the country to re-enter the world market. Macroeconomic stabilization policies initiated in the late 1980s continued and programs were initiated to liberalize trade, privatize national enterprises, rebuild a network of traders and distribution chains for private goods and services, and reduce the budget deficit and inflation. The economy and infrastructure were in ruins and society remained polarized along both partisan and socioeconomic lines. Although growth had picked up since the mid-1990s, poverty and hunger continue to plague the population, particularly in rural areas.

In the mid-1990s, agricultural production recovered, with the return of producers to their farms following displacement as a result of fighting and new farm installations on the agricultural frontier, which generated large-scale deforestation (Maldidier & Marchetti, 1996). However, the recovery of losses suffered due to the conflicts did not succeed in reducing tensions that were further aggravated by austerity and agrarian counter-reform affecting the peasantry (Roux, 2010). With the reduction of the role of the State in the agricultural sector (cancellation of input subsidies, reduction of rural credit and technical assistance), the Ministry of Agriculture and Livestock, which had replaced the Ministry of Agricultural Development and Agrarian Reform, reoriented its strategy towards the regulation of production and trade, while transferring sector support to the newly created agencies: the Nicaraguan Agricultural Technology Institute (INTA) for agricultural expansion, the Rural Development Institute (IDR) and the National Forestry Institute (INAFOR).

The majority of the population were in a fragile state caused by the years of conflict and restrictions of the mid-1980s, and by the austerity of the 1990s. In the late 1990s, 44% of the population were surviving on less than $1 a day and 75% on less than $2, making Nicaragua the poorest country in the Americas after Haiti (OXFAM, 1998). Poverty was a stark reality when Hurricane Mitch, one of the country's worst natural disasters, hit the region in 1998. According to ECLAC, it affected almost 20% of the population (causing 6,000 deaths), mostly among the poorest communities living in precarious housing (ECLAC, 1999). According to Bradshaw and Linneker (2003), these disasters tend to, "reveal existing power structures and relations (…) provoking profound changes (…) and opportunities for transformation" (p. 148). However, Mitch also offered hope for reconstruction and an opportunity for civil society, government and international agencies to work together with the common goal of improving the lives of the population (Bradshaw & Linneker, 2003).

As the international community made pledges to reduce poverty (Copenhagen Declaration of 1995, G7 Summit of 1999), the Government of Nicaragua submitted an intermediary Poverty Reduction Strategy Paper (PRSP) in 2000 (McIlwaine, 2002) to apply for the Heavily Indebted Poor Countries (HICP) initiative operated by the World Bank and the International Monetary Fund. These international actors, as well as external cooperation in the broad sense, asserted themselves as central actors of development, investing funds in the revival of the agricultural sector and establishing themselves as guarantors of public programs under conditionality (Le Coq et al., 2013). The role of international actors represents a singular element in the orientation of public policies in Nicaragua and, more generally, in the elaboration of public policies in developing countries (Darbon, 2004). A Poverty Reduction Strategy (PRS, 2000), an Enhanced Economic Growth and Poverty Reduction Strategy (ERCERP, 2001), and then a National Development Plan (2003, 2004) were drawn up with the same objective: economic growth and poverty reduction. However, none of these strategies met expectations, minimizing commitments to education and health (McBain-Haas & Wolpold-Bosien, 2008; Hazell, 2004). Nevertheless, a change seemed to be underway, with the orientation shifting towards poverty reduction. Thus, new instruments for integrated rural development were being implemented, such as the PRORURAL sectoral plan, which promoted the growth of the agricultural sector by increasing productivity and implementing a territorial approach. However, public interventions failed to solve the problems, neglecting environmental and social aspects, as well as food production.

It was also during this period that food sovereignty emerged in the debates as a reaction to the conclusion of the Uruguay Round of the World Trade Organization that advocated food security through the market (Godek, 2014). Civil society and

Sandinista parliamentarians took up the concept of food sovereignty promoted by the *Via Campesina* movement and introduced it into the debates corresponding to a new bill aimed at the formulation of a National Food and Nutrition Security Policy (PNSAN). The bill never made it to the assembly vote, however, due to the political forces in play at that time (Godek, 2014), but an Action Plan (2001–2006) was proposed including new food policy instruments, such as the Comprehensive School Nutrition Program (PINE) for the provision of milk and snacks to children in state schools, following on from the National Action Plan for Nutrition (1990–1995) that provided for the monitoring of children's growth, and an Education for All program that included school nutrition as one of its priority areas of action. However, the level of school attendance in the country was low, which limited the scope of these instruments. The Ministry of Health and UNICEF provided support to several NGOs for maternal education and improved child nutrition (MINED & Nicaragua, 1995). A Social Safety Net (SSN), funded through a loan from the Inter-American Development Bank, provided cash transfers for poor families (Moore, 2009). The World Food Programme played a central role in coordinating and distributing food aid to implement these programs. With limited impact and implementation problems (concentration in certain areas, lack of coordination between implementing institutions, duplication of efforts), these instruments made it possible to improve food diversity (Hoddinott & Wiesmann, 2010) without improving the nutritional level of children (Gajate Garrido & Inurritegui Maúrtua, 2002).

The affirmation of institutional change with the enactment of the SSAN Law promoting food production and support for vulnerable populations

The election of Ortega as President of the Republic of Nicaragua in 2007 marked the start of a turning point, which continued with his reelection in 2011 and 2016. His government ensured a continuity of policies that promoted macroeconomic stability and investment. However, his National Human Development Plan (PNDH) strategy enabled the reorientation of policies (Le Coq et al., 2013) in the fight against poverty and hunger. The role of the State was strengthened with a new form of management for external cooperation (GRUN, 2012) and its prioritization of marginalized populations and the family economy, as evidenced by the creation of the Ministry of Family, Cooperative, Community and Associative Economy in 2012, which marked a change in agricultural policies in support of family farming (Perez & Fréguin Gresh, 2014) and historically marginalized populations.

This turning point in agricultural policy represented a third order institutional change: the frameworks for interpreting action changed during a period of alternation and normative uncertainty (Hall, 1997). The leaders relied on a set of existing diagnoses and instruments, including the Food Production Program (PPA) inspired by the work of a Sandinista ideologue (Núñez Soto, Cardenal & Morales, 1995; Núñez Soto, 1984), who became a presidential advisor on social issues in 2008. The latter put forward a document (CIPRES, 2007) that resulted in the development of a Food Production subprogram (MAGFOR, 2008) definitively oriented towards food production, referring to Hurricane Mitch as an event that allowed "highlighting the rural panorama of the tragedy (…) of Nicaragua" (p. 14).

Food security and food sovereignty became central to the poverty reduction strategy and its production and social programs, as well as education and health (McBain-Haas & Wolpold-Bosien, 2008). The PPA instrument prioritized food sovereignty, which formed the basis of Act No. 693, Nicaragua being one of the few countries to incorporate this concept within its policy (Godek, 2014). The change was consolidated with the development of another instrument, *PRORURAL Inclusivo*, which focused on family farming, poverty reduction, adaptation to climate change and strengthening food security (GRUN, 2009). In 2014, an intersectoral coordination system was established: the National System of Production, Consumption and Commerce (SNPCC) that integrated the Ministry of Family, Community, Cooperative and Associative Economy, Ministry of Agriculture and Livestock, the Ministry of Environment and Natural Resources, the Institute of Agricultural Protection and Health (IPSA), INTA and INAFOR. The SNPCC was piloted by *PRORURAL Inclusivo*. In fact, the positive evaluation of *PRORURAL Inclusivo* (GRUN, 2015; Kester, 2009) highlighted the need to consolidate intersectoral and interinstitutional coordination and organization. The SNPCC is responsible for developing annual plans,[4] the latest of which (2017–2018) is aligned with PRORURAL Inclusivo, with the objective of "guaranteeing unrestricted access of Nicaraguan products to internal and external markets, promoting research and market promotion, in accordance with foreign trade treaties and agreements, and signing new trade agreements to diversify goods and trade partners" (GRUN, 2018).

Act No. 693 of 2009 on Food and Nutritional Sovereignty and Security (SSAN) sought to guarantee the population's right to food (enshrined in the constitution since the 1980s) and position national food production, the promotion of environmental and economic sustainability of the food system and inclusion, with emphasis on women, children and youth, as central issues. Its objective was to provide services along value chains, giving priority to food chains (rice, beans, maize, sorghum, meat, milk and derivatives), as well as to increase food productivity, conditions of access to employment and productive resources, food education, food sanitary controls, coordination of public institutions and private organizations. It adopted a territorial approach to development and inclusion to address risks (Asamblea Nacional de la República de Nicaragua, 2010). This policy focused on the availability and stability of food production in terms of both quantity and quality. The priority beneficiaries were marginalized populations in poor regions (micro-, small and medium-sized agricultural producers, indigenous populations), while other types of producers benefited from incentives for innovation and technology transfer. Finally, the law stipulated that non-agricultural populations may benefit from food aid and other types of support to help them enter the labor market (aid to promote handicraft activities or MSMEs).

The SSAN law made it possible to establish the instruments of the, now sole, agricultural and food policy: i) the National Food Program (PNA), an integration of the Food Production Program (PPA) also known as "Zero Hunger" and the "Healthy Kitchen Gardens" (*Huertas Sanas*) Program, aimed at the production of basic grains, improving access to and consumption of healthy food; ii) the National Rural Agro-Industry Plan

---

4. The SNCPP also operates at the subnational level with the creation of spaces for dialogue (working groups, for a time called Territorial Research and Innovation Hubs) that aim to strengthen the participation of stakeholders in certain sectors in sectoral discussions.

(PNAIR) aimed at increasing the added value of production by promoting post-harvest and processing activities; iii) the National Forestry Program (PNF), promoting the rational exploitation of forests. At the same time, other public institutions developed instruments to improve the road network in milk and coffee production areas and access to financial services, education, health, tourism development and security. After 2014, while the main policy instruments remained in place, others were also implemented, such as programs focused on agricultural production in general (rice, sorghum, coffee, cattle), and social policy instruments based on earlier versions: the Comprehensive School Nutrition Program (PINE or School Meals) continued as part of the revised Strategic Education Plan. The innovative nature of most of these instruments is the result of the emphasis on the active participation of women and parents as key figures in the development of their families and villages.

The content of the SSAN Law is, in some respects, reminiscent of the policy of the 1980s, but applied in a completely different context, in the absence of internal and external tensions and pressures. The role of external cooperation agreements signed by the Nicaraguan government that facilitated the implementation of the instruments should be highlighted, as is the case of the Treaty signed within the framework of the Bolivarian Alliance for the Peoples of Our America (ALBA), ratified in 2008, which aimed to jointly formulate and implement SSAN instruments. Within this framework, Venezuela purchased Nicaraguan products in exchange for financial support for the agricultural sector (McBain-Haas & Wolpold-Bosien, 2008), which was funded by the PPA until 2016, along with other funds allocated from the government budget and other donors (Kester, 2009). The funds to finance the PINE came from the State budget, the European Union, the WFP, and other bilateral cooperation and international organizations.

An institutional change with mitigated effects: the limitations of the implementation of the PPA and the PINE in Chinandega

Field surveys reveal, on the one hand, that the content of the programs studied is consistent with the content set out in the SSAN law. The beneficiaries of the PPA are women, which is a positive aspect of the program (Carrión Fonseca, 2015). It is based on donations of productive assets to identified poor women to strengthen food production and capitalization of their families. The donation of inputs (cows and sows, chickens and roosters, seeds, tools, construction materials, etc.) follows criteria defined by the law and interpreted by officials operating in the region, and is combined with technical assistance, training and funding. The beneficiaries are required to manage a savings account to repay 20% of the value received through a rural credit cooperative for the development of their communities. The Ministry of Family, Community, Cooperative and Associative Economy is responsible for operational planning and its officials for local implementation. As for PINE, its impact is reflected in the daily provision of balanced meals in state-run preschool, primary and secondary schools, with the hope of increasing school attendance.

The surveys also, however, indicate the existence of very specific conditions for program implementation. In fact, they reveal the central role played by certain local organizations, as in the 1980s, when many revolutionary leaders lacked technical and managerial skills (Austin, Fox & Kruger, 1985). The Family, Community and Livelihood Cabinets (*Gabinetes de Familia, Comunidad y Vida*), inspired by the Sandinista Defense Committees (CDSs) of the 1980s, play an important role in the selection of beneficiaries for public programs. Set up in 2008, as National Citizen Power Councils (*Consejos del Poder Ciudadano* or CPCs) and renamed in 2013, the *Gabinetes* are closely linked to the Sandinista Leadership Councils (CLS) (they exist in rural communities and urban and peri-urban neighborhoods) that form the structure of the FSLN (Ortega's governing party) at local level. Since their inception, the role and legitimacy of these actors has been debated (Stuart Almendárez, 2009). All interviewees cited them as key and influential actors in local decision-making and program implementation: while their mandate is to strengthen links between the state and communities, and to stimulate social participation in decision-making on the ground, the *Gabinetes* carry out technical inspections and influence the choice of beneficiaries, in theory (but not always in practice) in coordination with administrative actors.

According to the surveys, the *Gabinetes* are made up of individuals that are (self-) appointed as community leaders, which makes them, for some, legitimate decision-makers in view of their knowledge of the local people. However, their legitimacy depends on local political forces and the personal ethics of their representatives, making their role in the implementation of the instruments a sensitive element (Finnegan, 2011; Kester, 2009). Surveys confirm a selection bias between women who are part of the governing political party, and the others. The archival account of Ministry of Family, Community, Cooperative and Associative Economy officials does not explicitly reveal this criterion, but it is mentioned by the administrative actors interviewed. If that indeed has been the case, it calls the universality and non-discrimination promoted in the SSAN law into question. This is even more so because the surveys reveal that other non-explicit criteria exclude some of the potential beneficiaries: this is the case of families with relatives that migrate that the officials do not classify as being poor and are, therefore, excluded from the programs. This situation illustrate the role of front-line officials or "field bureaucrats" (Dubois, 2012), who interact directly with the population. Their role is not limited to a strict application of the instruments, these being non-univocal and offering the possibility of reinterpretations and adaptations for singular cases. For example, many respondents reported a clientelistic system that has existed for decades (Pérez Márquez, 2007; Envío, 2015), which is an indicator of path dependency.

Finally, the analyses demonstrate the limited effectiveness of the instruments. According to the surveys, Zero Hunger is implemented in all municipalities of Chinandega. But, the records of the officials in charge of implementation in 2016 indicate the low number of beneficiaries: only 537 beneficiary families in the municipality of Cinco Pinos at the date of the survey, for a population of about 6,800 people, a percentage below the level of poverty and involvement in the agricultural sector at municipal level. In addition, men may have also benefited from the program in some cases because, given that women rarely own plots of land, the gender criteria for implementing the program could not be met. Certain limitations of the PINE program have also been highlighted. According to MINED, it is implemented in all state-run schools and only applies to school-age children. However, the capacity of the educational infrastructure is limited, although we could not obtain the enrollment rate in the study region. But it can be assumed that it follows the trends in the country: although the numbers are increasing, only five out of ten children complete elementary school (Näslund-Hadley et al., 2012). According to the surveys, many children, especially girls, do not attend school, particularly at preschool level. Some parents revealed that they did not wish to send their daughters to school because they help with household chores and, as schools are often far away, they are not always able to walk them there.

Another limitation in the implementation of these instruments is the nature of the products delivered. The food donated by PINE comes from ENABAS and processors linked to the state. According to the WFP interviews, the products are largely sourced from national family producers, who receive quality controls. They are rarely imports, but in some cases, the donations do not correspond to the preferences of the children (case of dates donated by Saudi Arabia as school snacks for indigenous children, who were unfamiliar with the product). In the case of Zero Hunger, while the composition of the donation is fiercely debated, the limitations correspond to the quality of the donations (Finnegan, 2011; Kester, 2009). Concerns about supplies from potentially sick animals, led to a requirement for the screening of suppliers in IPSA health records. After donations, respondents mentioned the low level of follow-up by officials. In the case of PINE, MINED conducts regular meetings with school directors and makes three annual visits to monitor food quality and storage, and the Ministry of Health provides them with health and nutrition training. In the case of the PPA, the Ministry of Family, Community, Cooperative and Associative Economy provides training to beneficiaries to increase productivity and promote women's association and empowerment. However, according to the surveys, these training sessions fail to make provision for power relations that exist within households, or address the gender division of labor, which remains a source of women's vulnerability in the study region. Although the Ministry of Family, Community, Cooperative and Associative Economy organizes visits so that the beneficiaries can become promoters of the knowledge acquired, these visits are irregular and insufficient due to the short supply of officials to serve families living in dispersed and isolated communities. As such, the instability (sometimes deficiency) of the support is mentioned.

▸▸ Conclusion and outlook

This chapter highlights the existence of a critical juncture in the late 2000s in Nicaragua that led to a gradual institutional change in the area of agriculture and food, further consolidated in 2007 following Ortega's election. It is observed that the antecedents to change, particularly those of the 1980s, a crucial period in national history, were remobilized in part to reorient the trajectory and reposition the fight against poverty and hunger at the center of policy. While Nicaragua has historically followed a model based on agrarian capitalism and agro-export to the detriment of food and peasant production, the turning point that occurred during the 1998–2001 period, which was subsequently affirmed following the 2008 election, again prioritized family agriculture, fundamentally food-producing. According to the last agricultural census of 2011, family farming is the predominant form of production in Nicaragua, in operation on more than 85% of farms (Perez & Freguin-Gresh, 2014).

With this new trajectory, the instruments that intervened in the food system reoriented social and production programs aimed at the poorest and historically excluded populations, with the objective of promoting food sovereignty and security.

At the same time, we show that the effects of institutional change are ambivalent, a conclusion underscored by other studies (Finnegan, 2011; Kester, 2009; Solornazo, 2016; Tschirley, Flores & Mather, 2010; Carrión Fonseca, 2015). The implementation of the two flagship instruments of the agricultural and food strategy, the "Zero Hunger" and "School Meal" programs, demonstrate significant limitations in terms of governance, particularly in relation to the beneficiary selection process in which surveys reveal the existence of partisan clientelism. This calls the universality and inclusiveness of these programs into question, including the PPA. There is also the question of the sustainability of these instruments that have faced significant difficulties since the support from ALBA has been withdrawn.

At a time when the country is once again going through a period of turbulence and uncertainty since the 2018 demonstrations, worsening the economic situation for a large proportion of the, still vulnerable, population, the question arises as to whether the progress facilitated by this latest institutional change in the country's history has managed to effectively combat poverty and hunger in a sustainable manner. According to official figures, poverty in the country has been significantly reduced in the past ten years (INIDE, 2016). However, FAO and PAHO (2017) reveal that the number of undernourished people in Nicaragua has remained the same since 2013, which would suggest that the policies implemented have not be as effective as advertised. If we draw a parallel with the situation in another country, Brazil, that implemented social and production policies for more than a decade (some of which inspired the PPA and PINE and even share the same names), with significant results that were thought to be sustainable until 2016, there is an urgent need to understand whether or not these institutional changes have provided effective responses in the fight against hunger in Nicaragua.

▶▶ References

Asamblea Nacional de la República de Nicaragua. (2010). *Ley no. 693. Ley de soberania y seguridad alimentaria y nutricional* (Publicada en La Gaceta, Diario Oficial No. 133 del 16 de Julio de 2009). https://base.socioeco.org/docs/ley-ssan.pdf

Austin, J., Fox, J., & Kruger, W. (1985). The role of the revolutionary state in the Nicaraguan food system. *World Development, 13*(1), 15–40.

Barroso Peña, G. (2011). Reforma agraria en Nicaragua bajo el Sandinismo. *Historia Digital, 11*(18), 1–29.

Baudot, P.-Y. (2014). Le temps des instruments. Pour une socio-histoire des instruments d'action publique. In Halpern, C., Lascoumes, P., & P. Le Galès (Eds.), *L'instrumentation de l'action publique* (pp. 193–236). Presses de Sciences Po. http://www.doi.org/10.3917/scpo.halpe.2014.01.0193

Berth, C. (2014). Las politicas de consumo y de alimentacion en Nicaragua: 1965–1995. *Diálogos, 15*(1), 79–108. http://www.doi.org/10.15517/DRE.V15I1.10427

Bornemann, G., Neira Cuadra, O., Narváez Silva, C., & Solorzano, J. L. (2012). *Desafios desde la seguridad alimentaria y nutricional en Nicaragua*. OXFAM.

Bradshaw, S., & Linneker, B. (2003). Civil society responses to poverty reduction strategies in Nicaragua. *Progress in Development Studies, 3*(2), 147–158. https://doi.org/10.1191/1464993403ps058ra

Capoccia, G. (2015). Critical junctures and institutional change. In Mahoney, J., & K. Thelen (Eds.), *Advances in comparative historical analysis* (pp. 147–179). Cambridge University Press. http://www. doi.org/10.1017/CBO9781316273104.007

Capoccia, G, & Kelemen, R. D. (2007). The study of critical junctures. Theory, narrative, and counterfactuals in historical institutionalism. *World Politics, 59*(3), 341–369. http://www.doi.org/10.1017/ S0043887100020852

Carrión Fonseca, G. (2015). *Análisis del Programa Productivo Alimentario en Nicaragua desde un enfoque de género.* CIAT.

CEPAL. (1999). *Nicaragua: evaluacion de los daños ocasionados por el Huracan Mitch, 1998. Sus implicaciones para el desarrollo económico y social y el medio ambiente.* http://hdl.handle.net/11362/25356

CEPAL. (2017). *Seguridad alimentaria y nutricional en Centroamérica y la República Dominicana: explorando los retos con una perspectiva sistémica.* http://hdl.handle.net/11362/42588

CIPRES (Centro para la Promocion, la Investigacion y el Desarrollo Rural y Social). (2007). *Programa Productivo Alimentario* (3rd Ed.)

Collier, R.B., & Collier, D. (2002). *Shaping the political arena: critical junctures, the labor movement, and regime dynamics in Latin America.* University of Notre Dame Press.

Cossart, P. & Hayat, S. (2015). Étendre le domaine de la sociologie historique du politique. *Revue française de science politique, 65*(5–6), 833–834. https://doi.org/10.3917/rfsp.655.0833

Craipeau, C. (1992). El café en Nicaragua. *Anuarios de Estudios Centroamericano, 18*(2), 41–69. https://revistas.ucr.ac.cr/index.php/anuario/article/view/2275

Darbon, D. (2004). Comparer les administrations du nord et du sud: orientations méthodologiques à partir des administrations des Afriques. In Thiriot, C., Marty, M., & E. Nadal (Eds.), *Penser la politique comparée: un état des savoirs théoriques et méthodologiques* (pp. 95–118). Karthala.

Dubois, V. (2003). La sociologie de l'action publique, de la socio-histoire à l'observation des pratiques (et vice-versa). In Laborier, P., & D. Trom (Eds.), *Historicités de l'action publique* (pp. 347–364). PUF. https://halshs.archives-ouvertes.fr/halshs-00464322

Dubois, V. (2012). Le rôle des street-level bureaucrats dans la conduite de l'action publique en France. In Eymeri-Douzans, J.-M., & G. Bouckaert (Eds.), *La France et ses administrations. Un état des savoirs.* Bruylant-De Boeck. https://halshs.archives-ouvertes.fr/halshs-00660673

FAO. (2018). *Panorama de la pobreza rural en América Latina y El Caribe. Soluciones del Siglo XXI para acabar con la pobreza en el campo [Overview of rural poverty in Latin America and the Caribbean. Solutions for eliminating rural poverty in the 21st century].* https://www.fao.org/publications/card/en/c/ CA2275EN

FAO, IFAD, UNICEF, WFP, & WHO. (2019). *L'Etat de la sécurité alimentaire et de la nutrition dans le monde. Se prémunir contre les ralentissements et les fléchissements économiques [The state of food security and nutrition in the world 2019. Safeguarding against economic slowdowns and downturns].* https://doi.org/10.4060/CA5162EN

FAO & PAHO. (2017). *Panorama de la seguridad alimentaria y nutricional en América Latina y el Caribe [Panorama of Food and Nutrition Security in Latin America and the Caribbean].* https://iris. paho.org/handle/10665.2/34500

Figueroa Ibarra, C. (2005). La revolucion sandinista y los contratiempos de la utopia centroaméricana. *Bajo el Volcán, 5*(9), 67–85. http://www.redalyc.org/articulo.oa?id=28650904

Finnegan, K. (2011). *Sostenibilidad y autogestión en programas de seguridad alimentaria en Nicaragua* [Independent study project, School for International Training]. ISP Collection 1042. https://digital-collections.sit.edu/isp_collection/1042

Flanagan, K., Uyarra, E., & Laranja, M. (2011). Reconceptualising the 'policy mix' for innovation. *Research Policy, 40*(5), 702–713. https://doi.org/10.1016/j.respol.2011.02.005

Fréguin-Gresh, S. (2017.) Agroecología y agricultura orgánica en Nicaragua. Génesis, institucionalización y desafíos. In Sabourin, E., Patrouilleau, M. M., Le Coq, J. F., Vásquez, L., & P. Niederle (Eds.), *Políticas públicas a favor de la agroecología en América Latina y el Caribe* (pp. 311–350). Red PP-AL, FAO. https://inta.gob.ar/sites/default/files/e._sabourin_et_al_2017_politicas_publicas_a_ favor_de_la_agroecologia_en_alyec.pdf

Gajate Garrido, G., & Inurritegui Maúrtua, M. (2002). *El impacto de los programas alimentarios sobre el nivel de nutrición infantil: una aproximación a partir de la metodología del "Propensity Score Matching."* GRADE. https://www.mef.gob.pe/contenidos/pol_econ/documentos/VaspLeche_GajateInurritegui.pdf

Godek, W. (2014). *The institutionalization of food sovereignty: the case of Nicaragua's law of food and nutritional sovereignty and security* [Doctoral dissertation, Rutgers University]. Rutgers University Community Repository. https://doi.org/doi:10.7282/T3TD9VMH

Gould, J. L. (2008). *Aquí todos mandamos igual: lucha campesina y conciencia política en Chinandega, Nicaragua, 1950–1979.* IHNCA-UCA.

Grigsby Vado, A. H., & Pérez, F. J. (2007). *RuralStruc Program. Structural implications of economic liberalization on agriculture and rural development. First phase: national synthesis.* Universidad Centroamericana, Nitlapán.

GRUN (Gobierno de Reconciliación y Unidad Nacional). (2009). *Plan sectorial PRORURAL Incluyente. 2010–2014.* Gobierno de Nicaragua.

GRUN. (2012). *Plan nacional de desarollo humano 2012–2016.* Gobierno de Nicaragua.

GRUN. (2015). *Informe de evaluación final del PRORURAL Incluyente 2010–2014.* Gobierno de Nicaragua.

GRUN. (2018). *Plan anual de produccion, consumo y comercio del ciclo 2017–18.* Gobierno de Nicaragua.

Grupo Venancia. (2015). El Programa Hambre Cero: como se les va a las mujeres. *Envío, 396.* https://www.envio.org.ni/articulo/4972

Hall, P. A. (1993). Policy paradigms, social learning, and the state: the case of economic policy-making in Britain. *Comparative Politics, 25*(3), 275–296. https://doi.org/10.2307/422246

Hall, P. A. (1997). The role of interests, institutions, and ideas in the comparative political economy of the industrialized nations. In Lichbach, M., & A. Zuckerman (Eds.), *Comparative politics* (pp. 174–207). Cambridge University Press.

Hall, P. A., & Taylor, R. C. R. (1997). La science politique et les trois néo-institutionnalismes. *Revue française de science politique, 47*(3–4), 469–496. https://www.persee.fr/doc/rfsp_0035-2950_1997_num_47_3_395192

Hazell, P. (2004). *Small-holders and pro-poor agricultural growth.* OECD.

Hoddinott, J. F. F., & Wiesmann, D. (2010). The impact of conditional cash transfer programs on food consumption in Honduras, Mexico, and Nicaragua. In Hoddinott, J. F. F., & M. Adato (Eds.), *Conditional cash transfers in Latin America.* International Food Policy Research Institute, Johns Hopkins University Press.

Hurtado, R. V. (2000). Desarrollo rural y pobreza en Centroamérica en la década de 1990. Las políticas y algunos límites del modelo "neoliberal." *Anuario de Estudios Centroamericanos, 25*(2), 139–157. https://dialnet.unirioja.es/descarga/articulo/5075962.pdf

IDERU (Iniciativa por el Desarrollo Rural de Nicaragua). (2001). *Bases para un plan de desarrollo rural de Nicaragua. Una propuesta para la discusión y la acción.* Universidad Centroamericana, MAGFOR, AECI.

INIDE (Instituto Nacional de Información de Desarrollo). 2016. *Reporte de pobreza y desigualdad EMNV 2016.* Gobierno de Reconciliación y Unidad Nacional, Nicaragua. https://www.inide.gob.ni/docs/Emnv/Emnv17/Reporte%20de%20Pobreza%20y%20Desigualdad%20-%20EMNV%202016%20-%20Final.pdf

IRAM (Institut de Recherches et d'Applications des Méthodes de Développement). (2000). *Estudios sobre la tenencia de la tierra. Parte I. Marco legal institucional* (Contrato de consultoría Nº OPA-001-2000).

IRAM & AEDES (Agence Européenne pour le Développement Economique et la Santé). (2000). *Sécurité alimentaire et politiques agricoles des pays en développement: problématiques nationales et enjeux des négociations internationales.* https://www.iram-fr.org/ouverturepdf.php?file=193.pdf

Kester, P. (2009). *Informe evaluativo (2007 – 2008) Programa Productivo Alimentario (PPA) "HAMBRE CERO."* Embajada del Reino de los Países Bajos.

Lascoumes, P. 2007. Les instruments d'action publique, traceurs de changement. L'exemple des transformations de la politique française de lutte contre la pollution atmosphérique (1961–2006). *Politique et Sociétés*, 26(2–3), 73–89. https://doi.org/10.7202/017664ar

Lascoumes, P., & Le Galès, P. (2007). *Sociologie de l'action politique*. Armand Colin.

Le Coq, J.-F., Fréguin-Gresh, S., Saénz Segura, F., & Pérez, F. J. (2013, March 20–23). *Transfert de la notion de développement durable dans les politiques publiques centroaméricaines: lecture croisée des évolutions de référentiels et des trajectoires de politiques rurales au Costa Rica et au Nicaragua* [Conference presentation]. CANAL 2013 – Circulations et appropriations des normes et des modèles de l'action locale, Montpellier, France. https://agritrop.cirad.fr/569982/1/document_569982.pdf

MAGFOR (Ministerio Agropecuario y Forestal). (2008). *Subprograma Productivo Alimentario (PPA)*. Gobierno de Reconciliación y Unidad Nacional, Nicaragua. http://extwprlegs1.fao.org/docs/pdf/nic138399anx.pdf

Mahoney, J. 2001. Path-dependent explanations of regime change: Central America in comparative perspective. *Studies in Comparative International Development*, *36*, 111–141. https://doi.org/10.1007/BF02687587

Maldidier, C., & Marchetti, P. (1996). *El campesino-finquero y el potencial economico del campesinado nicaraguense*. Nitlapan, Universidad Centroamericana. https://www.nitlapan.org.ni/1996/02/el-campesino-finquero-y-el-potencial-economico-del-campesinado-nicaraguense-tomo-i/

McBain-Haas, B. & Wolpold-Bosien, M. (2008). *El derecho a la alimentación y la lucha contra el hambre en Nicaragua. Un año del Programa Hambre Cero*. FIAN Internacional. http://www.oda-alc.org/documentos/1340857537.pdf

McIlwaine, C. (2002). Perspectives on poverty, vulnerability and exclusion. In McIlwaine, C., & K. Willis (Eds.), *Challenges and change in Middle America: perspectives on development in Mexico, Central America and the Caribbean* (pp. 82–109). Longman.

Merlet, M. (1990, July). *El siglo diecinueve en Nicaragua. Auge y derrota de la vía campesina. (1821–1934)* [Conference presentation]. Simposio "Las sociedades agrarias centroamericanas," Costa Rica. https://www.agter.org/bdf/_docs/merlet_1990_nicaragua_sigloxix_final1.pdf

MINED (Ministerio de Educación). (1995). *Avances logrados por el ministerio de educación de Nicaragua en el marco del programa de "educación para todos" periodo 1990–1995. Examen a mediados de decenio*. Gobierno de Nicaragua.

Moore, C. (2009). *Nicaragua's Red de Protección Social: an exemplary but short-lived conditional cash transfer programme*. International Policy Centre for Inclusive Growth, United Nations Development Programme. https://www.ipc-undp.org/pub/IPCCountryStudy17.pdf

Näslund-Hadley, E., Meza, D., Arcia, G., Rápalo, R., & Rondón, C. (2012). *Educación en Nicaragua: retos y oportunidades*. Banco Interamericano de Desarrollo. https://publications.iadb.org/publications/spanish/document/Educación-en-Nicaragua-Retos-y-oportunidades.pdf

Núñez Soto, O. (1984). *Revolucion y Desarrollo: Revista Trimestral de MIDINRA*.

Núñez Soto, O. (1987). *Transición y lucha de clases en Nicaragua (1979–1986)*. Coordinadora Regional de Investigaciones Economicas y Sociales (CRIES).

Núñez Soto, O., Cardenal, G., & Morales, J. M. (1995). *Desarrollo agroecológico y asociatividad campesina: el caso de Nicaragua*. Centro para la Investigación, la Promoción y el Desarrollo Rural Social.

Ortega, M. (1986). La reforma agraria sandinista. *Nueva Sociedad*, *83*, 17–23. https://static.nuso.org/media/articles/downloads/1387_1.pdf

Oxfam International. (1998). *Debt relief for Nicaragua: breaking out of the poverty trap*. https://oxfamilibrary.openrepository.com/bitstream/handle/10546/115033/bp-debt-relief-nicaragua-011098-en.pdf;jsessionid=4C029CDB04EB8AB4ADA2FAF1C832CB0F?sequence=1

Paige, J. M. (1985, March 28). *Coffee and politics in Central America* [Conference presentation]. Ninth Annual Conference of the Political Economy of the World System, New Orleans, Louisiana. https://deepblue.lib.umich.edu/bitstream/handle/2027.42/51098/330.pdf

Payre, R., & Pollet, G. (2005). Analyse des politiques publiques et sciences historiques: quel(s) tournant(s) socio-historique(s)? *Revue française de science politique*, *55*, 133–154. http://www.doi.org/10.3917/rfsp.551.0133

Perez, F. J., & Freguin-Gresh, S. (2014, September 25–26). *Caracterizar y clasificar el agro nicaragüense: una propuesta de tipologia de explotaciones agropecuaria en base al IV CENAGRO* [Conference presentation]. CONICYT Simposio-Taller: Articulando ciencia, educación y desarrollo rural en Nicaragua, Managua, Nicaragua. https://agritrop.cirad.fr/574143/

Pérez, F. J., & Fréguin-Gresh, S. (2015). Nicaragua: evoluciones y perspectivas de las políticas agrarias y la agricultura familiar. In Sabourin, E., Samper, M., & O. Sotomayor (Eds.), *Políticas públicas y agriculturas familiares en América Latina y el Caribe: nuevas perspectivas* (pp. 261–291). IICA. https://hal.archives-ouvertes.fr/hal-02842158

Pérez Márquez, R. (2007). Relaciones clientelares y desarrollo económico: algunas hipótesis sobre la persistencia de "enclaves" de Antiguo Régimen en las democracias de América Latina. El caso de Nicaragua (1990–2005). *Nuevo Mundo Mundos Nuevos.* https://doi.org/10.4000/nuevomundo.3225

Pérez Martínez, B. J. (2019). *Aportes del Programa Hambre Cero a la seguridad alimentaria en Nicaragua 2014–2017* [Dissertation, UNAN]. Repositorio UNAN. http://repositorio.unan.edu.ni/10966/

Redclift, M. (1986). Radical food for thought: food policy in Cuba and Nicaragua. *Food Policy, 11*(3), 202–204. https://doi.org/10.1016/0306-9192(86)90004-7

Roux, H. (2010). *Contre-réforme agraire au Nicaragua, instrument de reconquête du pouvoir 1990–2010* [Doctoral dissertation, Université Paris 1 (IEDES)].

Rueda Estrada, V. (2013). El campesinado migrante. Políticas agrarias, colonizaciones internas y movimientos de frontera agrícola en Nicaragua, 1960–2012. *Tzintzun, Revista de Estudios Historicos, 57*, 155–198. http://tzintzun.umich.mx/TZN/article/view/148

Sánchez González, M., Castro Quezada, D., Rodríguez Ramírez, R., & Guerra Vanegas, J. (2016). Movimientos sociales y acción colectiva en Nicaragua: entre la identidad, autonomía y subordinación. *Amnis, 15.* https://doi.org/10.4000/amnis.2813

Solórzano, J. L. (Ed.). (2016). *Perspectivas sobre la seguridad alimentaria en Nicaragua en el contexto del cambio climático. Reflexiones y propuestas.* UCA Publicaciones.

Stuart Almendárez, R. (2009). *Consejos del Poder Ciudadano y gestion publica en Nicaragua.* Centro de Estudios y Análisis Político. http://www.oas.org/juridico/spanish/mesicic3_nic_consejos.pdf

Stuart Olson, R., & Gawronski, V. T. (2003). Disasters as critical junctures? Managua, Nicaragua 1972 and Mexico City 1985. *International Journal of Mass Emergencies and Disasters, 21*(1), 5–35. http://ijmed.org/articles/87/

Terán, S., & Quezada, F. (2005). Partidos políticos y movimientos sociales en la Nicaragua de hoy. In A. Serrano Caldera (Ed.), *La democracia y sus desafíos en Nicaragua* (pp. 83–124). Fundación Friedrich Ebert, CIELAC. http://biblioteca.clacso.edu.ar/Nicaragua/cielac-upoli/20120813014431/caldera.pdf

Tschirley, D., Flores, L., & Mather, D. (2010). *Análisis de políticas agrícolas y de seguridad alimentaria en Centroamérica: evaluación de la capacidad institucional local, la disponibilidad de datos y la demanda efectiva para datos e información [Agricultural and Food Security Policy Analysis in Central America: Assessing Local Institutional Capacity, Data Availability, and Outcomes]* (MSU International Development Working Paper No. 105S). Michigan State University. https://www.canr.msu.edu/fsg/publications/idwp-documents/idwp105_English.pdf

Wiggins, S. (2007). *Poverty reduction strategy review. Country case: Nicaragua.* Chronic Poverty Research Centre. https://assets.publishing.service.gov.uk/media/57a08c0fe5274a31e0000f86/prs-wiggins-nicaragua.pdf

World Bank. (2015). *Agriculture in Nicaragua: performance, challenges, and options.* http://documents.worldbank.org/curated/en/532131485440242670/Agriculture-in-Nicaragua-performance-challenges-y-options

Zalkin, M. (1987). Food policy and class transformation in revolutionary Nicaragua 1979–1986. *World Development, 15*(7), 961–984. https://doi.org/10.1016/0305-750X(87)90045-3

Chapter 5

Food security and sovereignty in Paraguay: who are public policies aimed at?

DANIEL CAMPOS, MARÍA C. BENAVIDEZ

▸▸ Introduction

While other countries in the Latin American region, such as Brazil, Bolivia, Argentina, Uruguay and Ecuador, already have a Law on Public Policies in Pursuit of Food Security and Sovereignty, Paraguay has made little progress in this area, even though the right to food promoted by international organizations is undisputed. To understand this situation, it is necessary to analyze the contradictions and inequalities that arise within the country with reference to public policies, which are the reflection of a state vision that has changed little over the years, despite its liberal enunciations in programs and projects.

Paraguay continues to be a country characterized as a producer of raw materials and food production, albeit with a concentration of wealth and land that generates enormous economic and social inequalities. In 2018, the incidence of extreme poverty registered a slight increase of 0.4 percentage points (4.4% in 2017), mainly in rural areas, where it increased from 9% to 10%. This means that 29,000 of the 34,000 thousand new poor are located in rural areas (EPH, 2019).

Anti-poverty policies, with the *Tekoporã* ("live well") and *Tenondera* ("moving forward") programs since 2005, have served as important palliatives, especially in terms of the health and education of women and children in rural areas (Campos & Benavidez, 2017a and 2017b), as well as the protection of the elderly that has recently been introduced. During the pandemic in 2020, Paraguay developed programs to help the unemployed with the *Pytyvo* ("help") *and Ñangareko* ("protection") programs.

In this context, the general objective of this study is to examine the public policies developed in Paraguay to ensure food security and food sovereignty for the population. The specific objectives are: i) To review the concepts of food security and sovereignty that shed some light on public policies in Latin America; ii) To briefly outline the historical trajectory of public policies for food security and sovereignty; and iii) To identify the social and structural actors that intervene, block or support the management and fulfillment of public policies for food security and sovereignty.

This paper is conceived within the framework of a methodological triangulation, using quantitative and qualitative methodology and the research, reflection, participatory action methodology. Quantitatively, data from different national and international sources was analyzed. The agricultural statistical data used is taken from the last National Agricultural Census (CAN) conducted by the Ministry of Agriculture and Livestock (MAG) in 2008 and, as such, is *référentiel* as it is not able to fully reflect the current reality[1]. Qualitatively, a number of in-depth interviews were carried out with key figures. In terms of the participation-action methodology, participatory workshops were held on the right to food and land with members of peasant organizations and rural family farming community leaders.

▸▸ Conceptual review

The conceptual construction of the Right to Food is essentially based on human rights. This point is highlighted in the Universal Declaration of Human Rights adopted by the United Nations in 1948[2] and included as a socio-economic right in the International Covenant on Economic, Social and Cultural Rights – ICESCR (1966). At the World Food Summit in 1996, the United Nations Food and Agriculture Organization (FAO) defined the term *food security* as the availability, access in quantity and quality and the biological use of food by people, both individually and collectively.

At the same Summit this definition was questioned by the *Via Campesina* peasant movements on the basis of its productivist vision, proposing rather the concept of *food sovereignty,* that is, the peoples' right to define their public and food policies, prioritizing local agricultural food production, with access to land for peasants through Agrarian Reform, with seed support, credit and fair markets. They have also defended the right of consumers to decide what they eat, as well as the right of nations to protect themselves from detrimental imports and to control the market (Via Campesina, Declaration of Managua, 1992).

This vision was supported by Latin American scholars (Yuryevic, 1996; Altieri & Toledo, 2011; Rosset & Altieri, 2017), committed to safeguarding the ancestral wisdom of indigenous and peasant communities disregarded by capitalism in Latin America and the Caribbean. One of the major questions is whether technology per se is able to eradicate hunger and ensure food security and sovereignty, which leads to a questioning of the concepts underlying its creation, which should address the root causes of hunger and food insecurity based on the philosophy and spirituality of peasant and indigenous wisdom.

It is possible to salvage a new agricultural science that complements historical materialism (Yuryevic, 1996; Altieri & Toledo, 2011; Rosset & Altieri, 2017), with access to technology, knowledge and agroecological vision based on its participatory action with rural communities. This needs to be complemented by the vision of the solidarity economy, heritage of the peasant community mobilized (Campos, 2016) within the framework of an agroecology and social, political, economic, cultural, spiritual, anthropological economy (Meliá, 1977).

---

1. The government plans to conduct the next agricultural census in 2021.
2. Universal Declaration of Human Rights, Article 25, Point 1.

In recent years, the complementary concept of *food autonomy* that highlights the right of peasant communities and indigenous peoples to preserve and protect their food production and consumption process (UN Declaration, session 73, 2018), such as the Declaration of the Rights of Peasants, Indigenous communities, Pastoralists, Small-scale Fishermen, and Poor Rural Sectors, was developed as a tool to strengthen human rights for this sector.[3] This approach was reinforced with the targets set out in the Sustainable Development Goals (SDG 2030) and the Declaration of the Decade of Family Farming 2019–2028.[4]

▶▶ Historical trajectory of public policies on food security and food sovereignty

In the pre-colonial and colonial period, indigenous peoples cultivated crops for family and community consumption, developing a division of labor system in which all their wisdom regarding the land and its management, water, the sun and the moon was held by women, while men were in charge of hunting, fishing, and wild and domesticated livestock. The arrival of the Spanish disturbed this balance, to the detriment of women, who continued as producers and reproducers, but under a patriarchal system of slavery and servitude (Meliá, 1973).

During the period of independence (1811-1864), Paraguay benefitted from a time of outstanding endogenous development, both economically and socially. Throughout the period, the Paraguayan State was the owner of the land, which was managed by the peasant community for internal and external production. Although the country was relatively shut off to defend itself against the annexation of Argentina and Brazil, it exported yerba mate, cattle and fruits, particularly to Argentina, in a controlled manner. It produced abundantly since it was obliged to produce food for domestic consumption.

This autonomy was shattered by the War of the Triple Alliance (1865-1870), ending the economy of abundance that the country had enjoyed up to that point. Public lands being farmed by the peasantry were requisitioned and sold off to multinationals, giving rise to large estates. The lands were concentrated under the management of a few families, while the majority of the peasant and indigenous populations settled in marginal lands around the Paraguay River in the eastern region. In the western region, or Chaco, the local people lived together in factories owned by Anglo-Argentine entities.

In the period of British colonialism throughout Brazil and then Argentina (1871–1954), the economy was based on the extraction and export of natural resources and raw materials, such as cotton and beef. In this period, the reconstruction of the devastated country took place through the work of peasant women who cultivated the land; who also again took over responsibility for producing the food that was sent to the battle front during the period of the Chaco War (1932–1935).

---

3. See: http://agriculturafamiliar.co/adoptada-la-declaracion-de-la-onu-sobre-de-los-derechos-de-las-campesinas-y-campesinos/
4. See: http://www.fao.org/family-farming-decade/home/es/

The period of the military dictatorship (1954–1988) saw the beginning of the aggressive penetration of capital within rural areas, facilitated by the repeal of the 100 km border protection security law. This allowed the massive inflow of foreign capital to occupy the prime land in the regions of Alto Paraná and Itapúa, later extending to Misiones, Caaguazú, San Pedro and Concepción, for the development of extensive crops and food industries, especially transgenic grains.[5] This was a period marked by peasant resistance to the land requisitions, which was fought and subjugated.

In this period of authoritarian regime, the agrarian reform took place, which was reduced to a distribution of land, most of which went to the dictatorship's officials and a minimal part to the peasants. This situation exacerbated the land concentration in Paraguay. According to the 2008 Agricultural Census, 86% of the land is owned by 2% of the population, while less than 20% of the total land mass belongs to the remaining 98% of the population, with a Gini index of 0.97%, which makes the country one of the most unequal (Campos, 2013).

The development of public policies in the country only took place with the fall of the dictatorship and the beginning of democratization in the country (1989–2020), although in the progressive National Constitution of 1992, the treatment of the right to food is very ambiguous and refers only tangentially to the elderly and children, failing, therefore, to guarantee this right for the entire population or make any reference to the necessary public policies. Even so, it remains the first reference for any law related to the subject.

The presentation of a draft Framework Law on Sovereignty, Food and Nutritional Security and the Right to Food to Congress in October 2013 was an important milestone, following a year of work on the proposal initially drafted with the technical support of the FAO. This law sought to establish a state policy on the human right to food, organizing the different provisions, plans, programs and projects already in place in the country under a National System for Food and Nutrition Sovereignty and Security (SNSSAN), through the articulation of several public institutions and the monitoring of civil society.

The law was finally passed in Congress on 17 September 2018, with the characteristic of having been drafted in a participatory manner with organized society (Martín, 2011). However, on 22 May 2019, the Executive Branch, vetoed the Law, arguing that several public policies on the subject are already being implemented and additional resources would be needed if this law were to be enacted.

Faced with this lack of a comprehensive vision of the right to food in Paraguay, the management of the issue is reduced to laws, provisions and plans that are disconnected and disjointed from each other. These include, among others:
– Law 2051/2003 *on Public Procurement and Decree 3000/2015,* which regulates purchases from family farming, through a simplified process and enables individual producers and organizations to be suppliers of the State. Pilot projects have been carried out for its implementation, which have been successful for as long as the

---

5. In 2019 alone, 13 licenses for transgenic soybean, corn and cotton production were registered for the BASF SYNGENTA, BAYER, MONSANTO AND INDEAR companies in the 2019 Statistical Yearbook of the Ministry Of Agriculture and Livestock.

support of international organizations lasted. Some of the factors affecting its efficiency are: i) the legal informality of the organizations; ii) a lack of prior productive preparation on the part of the peasant organizations to comply with the law; iii) deficient and insufficient technical assistance from the State; iv) an ineffectual and irresolute implementation system; v) the absence of any credit provision for this purpose.

– Law 3481/2008 *on Promotion and Control of Organic Production and its Decree 4577/2010*, which establishes the procedures for the promotion and control of organic agriculture, for its contribution to food security. It is implemented with a Technical Committee for the Promotion of Organic Production composed of public agencies and civil society, under the tutelage of the MAG, but without guaranteed funding to meet its objective.

– Law 5210/2014 *on School Feeding and Sanitary Control*, which establishes the provision of food for primary school students, implemented by the Ministry of Education and Science with the operational support of the governorates and municipalities. Articles 2 and 10 of this law have subsequently been modified (6277/2019) to include other food sources.

– Law 5446/2016 *on Public Policies for Rural Women*, which highlights the productive role of women and the obligation of the institutions to make institutional services, such as technical assistance, credit, technology and others, available to them. It contemplates the representation of women's organizations for the monitoring of the Law. However, to date, it has not been regulated by the Executive Power, which makes its implementation difficult.

– Law 6286/2019 *on the Restoration, Promotion and Strengthening of Peasant Family Farming*, was enacted under pressure from peasant movements that occupied the streets of the capital Asunción for forty days. However, as of now, it has not been given the regulatory decree by the Executive for its full implementation.

Other programs and projects exist alongside these laws, such as the National Plan for Food and Nutrition Sovereignty and Security (PLANAL) (2009–2012), the National Program for the Reduction of Extreme Poverty (2013–2018), the Comprehensive Nutritional Food Program (PANI), providing nutritional support for pregnant women and children (2005 to date), Food Pension for the Elderly in Poverty (2010 to date). Most of these programs and others have not been maintained over time and, with the exception of some, they last as long as international financial support lasts and disappear once their continuity is reliant on the State.

Regardless of these government measures, Family Farming continues to contribute to food production in the country with a diverse range of products (table 5.1).

The importance of family farming also lies in the fact that it is present in all the country's departments of the Eastern Region, with diversified and healthy food, as shown in table 5.2.

Within family farming, women play a preponderant role in agricultural production as they are responsible for an important part of food production, as well as in the preservation of biodiversity and the guarantee of food security and sovereignty, based on the production of healthy food (Nobre et al., 2017). They act as food processors and preservers, even when they have limited access and control in relation to land services, credit, technical assistance and training (Ballara et al., 2012).

Rural women in Paraguay are no exception, as they are subject to the same circumstances seen in other countries. They contribute to family and community nutrition on a daily basis. They also generate income by selling their products at local fairs, which corresponds to a significant flow of money that sustains families and cities. They must, therefore, be included in public policies on food security and sovereignty (Benavidez, 2008, 2013; Campos, 2016).

Table 5.1. Participation of Family Farming (FF) in the total production of food items in the country

Products	Production (thousands of tons)		%FF of the total
	National	**FF**	
Corn, normal harvest	990.6	206.6	20.9
Corn, between harvests	1384.3	119.3	8.6
Chipa corn	85.8	79.1	92.2
Unshelled beans	44.6	41.9	93.9
Cassava	2,218.5	2,075.6	93.6
Stevia	0.973	0.911	93.6
Unshelled peanuts	30.0	11.3	37.7
Sesame	50.0	44.5	89.0
Banana	59.9	55.8	93.2
Pineapple	54.3	52.7	97.1
Sweet potato	43.4	38.8	89.6

Source: National Agricultural Census 2008. Ministry of Agriculture and Livestock.

Table 5.2. Participation of Family Farming (FF) in terms of items, by department

Items	Departments of origin of production
Sesame	San Pedro, Concepción, Boquerón, Caazapá, Amambay, Canindeyú
Tomato	Caaguazú, Central, Paraguarí, San Pedro, Cordillera, Concepción, Alto Paraná, Itapúa
Carrot	Itapúa, Caaguazú
Banana	Caaguazú, San Pedro, Cordillera, Concepción, Alto Paraná, Itapúa, Canindeyú, Amambay
Lemon	Alto Paraná, Itapúa, Concepción, Cordillera
Mandarin	Itapúa, Alto Paraná, Guairá, Caaguazú
Orange	Itapúa, Alto Paraná, Caazapá, Cordillera, San Pedro, Caaguazú, Canindeyú
Pineapple	San Pedro, Concepción, Cordillera, Canindeyú, Caaguazú, Alto Paraná, Paraguarí

Source: Based on the Statistical Synthesis of Agricultural Production, MAG/DCEA, 2008.

Despite the contribution of family farming, the country's food autonomy is constantly at risk, due to the absence of sound public policies to strengthen the productive system of this sector. This dependence is reflected in the dependence on imports from Brazil, Argentina and Uruguay for some strategic food items and the fluctuating behavior of domestic production (table 5.3).

Table 5.3. Imports of basic foodstuffs versus production in Paraguay (%)

Items	Source	2013	2014	2015	2016
Onion	Argentina	43%	83%	20%	33%
	Brazil	34%	7%	73%	52%
	Paraguay	23%	10%	7%	15%
Potato	Argentina	96%	96%	98%	98%
	Brazil	–	3%	2%	2%
	Paraguay	4%	1%	–	0%
Bell pepper	Paraguay	94%	65%	46%	27%
	Brazil	6%	35%	54%	73%
Tomato	Argentina	24%	26%	30%	34%
	Brazil	14%	2%	7%	14%
	Paraguay	62%	72%	63%	52%
Orange	Argentina	84%	91%	95%	32%
	Brazil	–	1%	1%	55%
	Paraguay	16%	8%	4%	13%

Source: Extracted from Imas (2019).

Moreover, Paraguay is considered a food producing country, ranking among the top five soybean producers in the world, alongside Brazil, the United States, Argentina and China, and also among the top exporters of corn, meat and wheat. However, an analysis of import data shows a growing dependence on basic foodstuffs (table 5.4).

Table 5.4. Food imports, five-year period 2013–2017 (in tons)

Items	2013	2014	2015	2016	2017	Five-year increase
Citrus	29.9	29.4	32.2	14.5	32.5	109%
Vegetables	43.2	45.7	63.2	85.1	92.1	213%
Legumes	111	125	300	205	212	191%
Fresh fruit	12.9	14.3	20.6	25.4	27.7	214%

Source: Extracted from Franceschelli and Lovera (2018).

In addition, Paraguay, despite its potential to produce more food through family farming, is among the seven countries with more than 10% prevalence for undernourishment along with Honduras, Guatemala, Nicaragua, Bolivia, Venezuela and Haiti, as well as the increase in overweight children and adults, indicating malnutrition (FAO, 2020).

In 2019, the Ministry of Agriculture promoted the cultivation of seven crops (tomato, sesame, sugarcane, onion, cotton and banana) for family farming, adding hemp in 2020. Most of these crops, in addition to not representing the basic traditional food products of family farming, are focused on income generation rather than food security and sovereignty (MAG, Anuario 2019).

▸▸ Identification of institutional and social actors

There are several public institutions that have the role of addressing public policies for food security and sovereignty. Among the main ones are:
– *The Ministry of Agriculture and Livestock (MAG),* the managing entity of public agricultural policy and sustainable development. It includes an Agricultural Advisory Council (CAA), made up of representatives of private sector production associations, such as the Paraguayan Agricultural Coordinator (CAP), which represents the soybean sector, the Paraguayan Rural Association (ARP), which represents medium- and large-scale livestock farmers, the Paraguayan Industrial Union (UIP), which represents industrialists, the Paraguayan Federation of Industrialists (FIP), the Federation of Production, Industry and Commerce (FEPRINCO), representing these sectors, the Federation of Production Cooperatives (FECOPROD), representing medium- and large-scale cooperatives, mostly Mennonite, and others. Neither women nor the peasant sector of family agriculture are represented in this national body.
– *The National Institute for Rural and Land Development (INDERT),* created in 1963 to develop land policy, has undergone several transformations in nomenclature and new roles throughout its half century of existence, but without substantive changes to its vision. Moreover, there have been no changes in the staff, reproducing the same archaic paradigms of the militarist past. It maintains an Advisory Council, with a balance of power negatively tipped against the peasant sector, with only one representative, while the same business associations that make up the CAA hold the rest of the seats. There is also no female representation.
– *The Paraguayan Institute of Agricultural Technology (IPTA),* an autonomous entity corresponding to the MAG, created in 2010, with the objective of leading the generation, safeguarding, adaptation, validation, dissemination and transfer of agricultural technology and the management of agricultural and forestry genetic resources to increase the productivity and competitiveness of crops. It is assisted by an Advisory Council composed of eight people, of whom two represent public institutions, five represent sectors related to business agriculture and one represents the National University, with no representation from the family farming food sector or women.
– *The Crédito Agrícola de Habilitación (CAH) agricultural loans facility,* an autarkic entity corresponding to the MAG established 76 years ago, responsible for the awarding of exclusive credit to small low-income rural producers. It has a Board of Directors with representatives from the public sector, including women. There are no participatory bodies representing the peasant sector, despite being central to the policy. It manages several products, among them the financial product for the items prioritized by the MAG.

– *The National Science and Technology Council (CONACYT),* created in 2003, at the request of public institutions and the private business sector to promote research and innovation with biochemical laboratories of pharmaceutical, livestock and soybean industries (Goulet et al., 2019). Its mission is, "formulating and proposing to the National Government the national policies and strategies on science, technology and innovation and quality for the country, in accordance with the State's economic and social development policy" (Law 2.279/03 "which modifies and expands articles of the general law 1028/97 on science and technology"). It is made up of 14 councilors, of which 36% represent public institutions, while 21% of the remaining 64% of representatives are from business associations, academia and civil society; among which only four are women. These councilors occupy their positions by presidential decree, and the selection mechanism is unclear. The majority represent public institutions, with a negligible representation of academia. The same business associations mentioned above are, once again, part of this institution. Moreover, as in the case of the previous administration, it is headed up by an industrialist.

– *Ministry of Social Action:* the former Secretariat of Social Action was elevated to a Ministry in 2018 (Law No. 6.137 of 10 August 2018) with the objective of designing public policies to reduce inequalities and improve the quality of life of the vulnerable and poor population. One of its most important functions is to implement the non-contributory policy through the two star programs, *Tekoporã* and *Tenodera,* through which vulnerable families receive a conditional (monetary) transfer for food and education. As a social assistance policy, both programs have had an impact, but they are far from contributing to food security, with the exception of the promotion of kitchen gardens. In this regard, the family guides possess little agricultural training, and are, therefore, unable to properly support the diversified food production of the beneficiary families. This shortcoming has been addressed through agreements with the Ministry of Agriculture and Livestock, but no concrete results are yet known.

– *Ministry of Health and Social Welfare,* in charge of the health and welfare policies for the population. Among its attributions, it has the function of regency and standardization to ensure the safety of industrialized or semi-industrialized food, through the National Institute of Food and Nutrition (INAN). In the National Health Policy 2015–2030, the topic of food is barely mentioned, while the issue of food security does not form part of public health policy.

Organized civil society should be included among the relevant social actors for the implementation of public policies, particularly in the form of social movements and non-governmental organizations. This sector, however, does not have a visible presence because of the absence of participatory bodies or mechanisms, either within institutional structures or the mechanisms for the implementation of laws. Coincidentally, the bills presented in the past five years that promote the participation of civil society are those that have been subject to rejection or bureaucratic stagnation.

For almost 30 years, the National Peasant Federation (FNC), part of the *Via Campesina* movement,[6] has been holding marches in Asunción to protest against

---

6. In addition to the FNC, the Paraguayan branch of the *Via Campesina* Latin American Coordination of Rural Organizations Body (CLOC) is made up of the Rural and Indigenous Women's Workers' Organization (CONAMURI), the Struggle for Land Organization (*Organización de Lucha por la Tierra,* OLT), the Northern Rural Women's Workers' Organization (OCN) and the *Cultiva* Organization.

the neoliberal policies of successive governments. The issue of food is always included among the demands. Other peasant unity platforms, such as the National Intersectional Coordinator (CNI), which brings together more than 12 national organizations, together with the Peasant, Indigenous and Popular Unity Committee (ACIP) and the Paraguayan Peasants' Movement (MCP) and other departmental and regional coordinators have also included the right to food and the need for support and assistance for food production among their demands.

For its part, the National Committee for Family and Indigenous Agriculture (CNAFCI),[7] made up of national and regional organizations, has focused its efforts on establishing public policies for the sector. This body is a member of the World Rural Forum, which advocated for the United Nations to establish, not only the International Year of Family Farming in 2014, but also the Decade of Family Farming 2019–2028.

All the peasant protests have succeeded in raising awareness in relation to peasant issues and have forced governments to establish agreements with the organizations, but their real impact is still limited. As for urban consumers, they are still fragmented and lack organic capacity, so they have little impact on public policies. This is seen with the *slow food* trend which, although it has been introduced in the country, has not found support in a representative way, although the number of young urban couples interested in healthy eating is increasing.

The advisory councils corresponding to ministries and governmental entities are important participatory bodies and highly influential in terms of public decision-making. However, there is a prevalent lack of representation of civil society and the peasantry in these spaces, which are mostly made up of businessmen and politicians. From the composition of these bodies, it can be inferred that their ultimate goal is the safeguarding of their trade union interests, capitalizing on the strategic position they hold.

For example, the Minister of Agriculture and Livestock recently convened the Agrarian Advisory Council (CAA)[8] for a consultation meeting on institutional reform and ministerial restructuring, to date, failing to consult other important sectors, such as agricultural professionals, peasant organizations and academia, among others.

Thus, on the one hand, there is the peasantry vying to be a social subject, battling to be a representative actor in the production of food from family farming, although still unable to unite under a single objective, setting aside exclusionary leaderships and personal and petty interests. And, on the other, there is the hegemonic bloc made up of the agricultural and industrial business associations that accumulate power and control public policies, forming a solid body that defends their economic interests.

For this group, peasant family farming is not a productive system as such, but rather a cultural system, and therefore does not merit the development of public policies to reinforce it. However, it does not fail to recognize that it generates a dual system of self-sufficiency and income from food products (cassava, corn, beans, sesame, sugarcane, fruit trees) (UGP, 2015).

---

7. This group led the negotiations for the approval and enactment of Law No. 6286 on Peasant Family Farming, which addresses food security and sovereignty.
8. Created by Presidential Decree No. 10,353 in 2007 and comprising the MAG, the Production Guilds Union (UGP) and the Paraguayan Cotton Producers' Chamber of Commerce (CADELPA).

The indisputable fact is that family farming has been losing power over its production and food; its native seeds have been replaced by imported seeds, agrochemicals and foods, and its system has been almost absorbed by agribusiness due to the pressure on its lands. Its vulnerability is not, therefore, due to its own incapacity, but to the incapacity of the State that has failed to provide for its protection and revitalization. The same applies to rural women, whose daily work contributes productively and economically to the country, and despite the formulation of a specific law to support them, there is an absence of any regulatory structure to facilitate its implementation.

The context of economic crisis, with the stagnation of growth in 2019, has been exacerbated as a result of the Covid-19 health crisis. According a report produced by the Central Bank of Paraguay – BCP (2020), the national economy is expected to contract by 3.2%. Under these circumstances, peasant family agriculture has emerged as a key activity through the work of thousands of people who have supplied their products for domestic consumption, although without substantive changes in public policies to support the sector.

The reports published by the United Nations Commission on Human Rights in 2001 and 2006 state that poverty reduction and the right to food in Paraguay are affected by three factors: i) the lack of access to land for rural and indigenous communities; ii) the increase in migratory flows and the uprooting of rural communities; iii) the concentration of land under the control of multinationals that monopolize food production (Draft Bill on the Right to Food in Paraguay).

More recently, the 2017 Report of the Special Rapporteur on the right to food urges Paraguay, among other initiatives, to: i) ratify the Optional Protocol to the International Covenant on Economic, Social and Cultural Rights and the Optional Protocol to the Convention on the Rights of the Child; ii) review the impacts of the mechanized agriculture model to ensure that it does not impede on the right to food of small producers; iii) develop and adopt a national right to food framework law with implementation plans for each region and with budgetary measures, long-term sustainability, implementation mechanisms and full and active participation of all stakeholders, including those most vulnerable to hunger; iv) enact pending legislation, such as the Act on Native and Creole Seeds and fast track the Bill on the Right to Food Sovereignty and Nutrition with budgetary and human resources for its effective implementation; v) enact a law introducing tariffs on the export of grain, including soya; vi) protect and promote family agriculture as a productive model; vii) develop comprehensive nutrition policies to eradicate malnutrition; viii) create seed banks; ix) promote organic farming and agroecology including financial mechanisms for the introduction of training programs.

▸▸ Future prospects

Food security and sovereignty policies are necessary in any country context, but even more so in the context of poor countries. The trend of worsening hunger in the world, the food production crisis and the national debt are challenges that must be addressed with public policies that guarantee food for the population. In Paraguay, unlike other countries, peasant and indigenous family farming is being undermined, which is precisely the sector that produces the most food for domestic consumption.

The reduction of the agricultural lands allocated to small farmers and indigenous people due to the pressure exerted by large-scale agriculture is forcing these social groups to abandon their communities and relocate to the periphery of the cities, forming urban poverty belts.

To remedy this situation, there is firstly an urgent need to reverse the environmental changes attributed to human activity, such as the indiscriminate clearance of forests, the uncontrolled expansion of extensive production, the pollution of watercourses, and the use of toxic inputs in agriculture. These destructive factors have high environmental costs that must be covered by high tax levies that can be reinvested to mitigate the damage caused. This also implies a revolutionary revision of the agrarian reform in the country, which has not yet been carried out, despite the provisions in the Constitution of Paraguay and the democratic reforms. It is not possible to seriously address any issue if the conditions are not in place to fairly tackle the equitable distribution of land due to the strong resistance it causes from the power groups that represent, or protect, the large estates.

Throughout this process, the clear definition of a nationwide vision is fundamental to direct efforts towards the achievement of goals in line with a sustainable planet. To that end, the discussion at all levels, not only academic, of the type of development that is sought involves breaking away from the paradigm of the green revolution, mechanized technology and exported seeds as a panacea for agricultural economic development. An openness to another type of more humane development, without renouncing economic and technological development, must be adopted in all sectors, particularly within those with political-financial power.

However, this whole process must be accompanied by an organized civil society that is united, aware and determined to demand its own rights. In this sense, the maturity of this sector calls for common objectives and short-, medium- and long-term strategies to be sought that lead to a clear and unwavering position on the fulfillment of the population's human rights.

Future prospects do not look promising for Paraguay in the current context. On the one hand, no substantive progress has been demonstrated on the part of the State and the institutions. Recently, the draft bill that sought to increase the tax on tobacco, soybeans and sugared beverages from the current minimum of 18% to 30%, has suffered a new setback having been rejected for the third time in recent years.[9] On the other hand, civil society has not developed viable mechanisms to contribute to the efforts for unity and joint endeavors between the social movements and organizations of the commercial, industrial and rural sectors.

▸▸ Conclusions

The economy of abundance that Paraguay maintained until 1870 has remained irretrievably in the historical anecdotes. Its recovery, which had the potential to take place in the subsequent 100 years, was also destroyed by the military dictatorship, which has exacerbated the social and economic inequalities that still persist.

---

9. See: https://www.ultimahora.com/senado-rechaza-proyecto-que-aumenta-impuesto-al-tabaco-n2819708. html; https://www.abc.com.py/nacionales/2020/05/05/senado-rechazo-la-creacion-del-impuesto-a-la-soja/

Although public policies have been initiated and developed across all areas in the past 30 years of democratization, the issue of the right to food, food security and food sovereignty are still absent.

The construction of a food and nutrition security system that ensures the full realization of the right to food is still a distant reality in the country. What has been done so far has been palliative, without detracting from some social protection efforts. However, no profound changes have been initiated that could promote strategies for a state policy on food security and sovereignty, coordinating the participation of all social, private and public sectors, including vulnerable groups that are mostly from the rural and peri-urban sector.

Business groups with economic power are the representatives that advise the State on public policies. These groups occupy strategic sectors in which public policies are designed to protect their sectoral economic interests rather than social ones. They preclude the inclusion of other groups to participate and debate on the crucial points that should be addressed by public policies, as in the case of family agriculture and food, which affects the majority of the population. Nor does the State fulfill its role as a mediatory agent between sectors, nor does it encourage intersectoral cooperation to promote and strengthen the different levels of cooperation and promote food security and sovereignty.

The Paraguayan State has failed to develop any policies that promote the creation of mechanisms to eliminate obstacles to food production. The little that does exist is focused on trade, thereby reinforcing the idea that income generation leads to welfare and therefore to food security, but the reality is more complex. The Law regulating purchases from family farming serves as an example of this complexity.

Food, like the science of agroecology and solidarity economy, is the heritage of peasant and indigenous communities and forms the basis of human rights. It was the thousands of peasant family farmers who saved the national society from a famine crisis because they continued to produce. The right to food as a human right, whether individual or collective, also demands the right to land and to the production of one's own food, just as the right to ecological justice is a human right, for which the strengthening and consolidation of peasant family agriculture is essential.

Within this framework, the State must guarantee the application of laws and norms, with sufficient budget to eradicate hunger and strengthen human rights. With exhortations lodged by international organizations and the reality of the country, it is up to the Paraguayan State to fulfill its social function.

▸▸ References

Ballara, M, Ninoska, D., & Rodrigo V. (2012). *Mujer, agricultura y seguridad alimentaria: una mirada para el fortalecimiento de las políticas públicas en América Latina.* United Nations. http://www.marcelaballara.cl/genydes/2012%20Mujer,%20agricultura%20y%20seguridad%20alimentaria%20Ballara%20Damianovic%20Valenzuel.pdf

Benavidez, M. C. (2008). *Situación de las mujeres rurales en Paraguay.* FAO. https://www.fao.org/3/a1591s/a1591s.pdf

Benavidez, M. C. (2013). *Estudio acceso de las mujeres de la agricultura familiar a la asistencia técnica y al crédito. Paraguay.* Reunión Especializada sobre Agricultura Familia (REAF) del Mercosur.

Biedermann, G., Blanco, C., Diz, S., Legal, D., Molinas, L., & Ortiz, G. (2020). *Boletín sobre el COVID19 y su impacto*. Banco Central del Paraguay. https://repositorio.bcp.gov.py/bitstream/handle/123456789/135/bm_nro9.pdf?sequence=6&isAllowed=y

Campos, C. (2016). *Políticas y experiencias territoriales relevantes para el empoderamiento de las mujeres rurales en Paraguay: un análisis desde el enfoque territorial.* ONU Mujeres/RIMISP. https://lac.unwomen.org/es/digiteca/publicaciones/2017/05/politicas-y-experiencias-territoriales-para-el-empoderamiento-de-las-mujeres-rurales-en-paraguay

Campos Ruiz Díaz, D. (2013). *Reforma agraria: Una causa nacional pendiente*. Editorial Arandura.

Campos Ruiz Díaz, D. (2018). *Empoderamiento de la agricultura familiar campesina.* SER.

Campos Ruiz Díaz, D., & Benavidez, M. C. (2017a). *Estudio sobre gastos de bolsillo para el cumplimiento de corresponsabilidades de parte de las familias participantes del programa TEKOPORA.* SAS/SER.

Campos Ruiz Díaz, D., & Benavidez M. C. (2017b). Diseño de gestión del programa TENONDERA. SER/CPES/SAS.

FAO, OPS, WFP, & UNICEF. (2018). *Panorama de la Seguridad Alimentaria y Nutricional en América Latina y el Caribe.* FAO. http://www.fao.org/3/CA2127ES/CA2127ES.pdf

Franceschelli, I., & Lovera, M. (2018). *¿Derecho a la alimentación? No. Paraguay camina en sentido contrario. Derecho a la alimentación.* HENOI. https://henoi.org.py/2018/12/18/derecho-a-la-alimentacion-no-paraguay-camina-en-sentido-contrario-derecho-a-la-alimentacion/

Freire, P. (1970). *Educación como práctica de la libertad.* Edit. Tierra Nueva.

Goulet, F., Le Coq, J.-F., & Sotomayor, O. (Eds.). 2019. *Sistemas y políticas de innovación para el sector agropecuario en América Latina.* Editora E-papers.

Imas, V. (Ed.). (2019). *Seguridad alimentaria y soberanía alimentaria en Paraguay. Sistema de indicadores y línea de base.* CADEP/CONACYT/PROCIENCIA. https://www.conacyt.gov.py/sites/default/files/upload_editores/u294/seguridad_soberania_alimentaria_cadep.pdf

Meliá, B. (1973). El guaraní dominante y dominado. *CEADUC Suplemento Antropológico, 8*(1–2).

Meliá, B. & Temple, D. (2003). *El don, la venganza y otras formas de economía guaraní.* Centro Paraguay de Estudios Paraguayos.

Ministerio de Agricultura y Ganadería. 2008. Dirección de Censos y Estadísticas Agropecuarias (DCEA). Censo Agropecuario

Nobre, M., Hora, K., Brito, C., & Parada, S. (2017). *Atlas de las mujeres rurales en América Latina y el Caribe.* FAO. http://www.fao.org/3/a-i7916s.pdf

Paraguay, Dirección General de Estadística, Censos y Encuestas. (2017). *Encuesta Permanente de Hogares.* Ministerio de Hacienda

Paraguay, Dirección General de Estadística, Censos y Encuestas. (2019). *Encuesta Permanente de Hogares.* Ministerio de Hacienda.

United Nations. (1948). *Declaración universal de los derechos humanos* [Universal declaration of human rights]. United Nations. https://www.un.org/es/about-us/universal-declaration-of-human-rights

United Nations. (1966). *Pacto Internacional de los Derechos Económicos, Sociales y Culturales* [*International Covenant on Economic, Social and Cultural Rights*] (Resolution 2200ª (XXI) of 16 Dec. 1966). United Nations. https://www.ohchr.org/sp/professionalinterest/pages/cescr.aspx

United Nations. (2015). *Transformar nuestro mundo: la Agenda 2030 para el desarrollo sostenible* [*Transforming our world: the 2030 agenda for sustainable development*] (Resolution A/RES/70/1). United Nations.

United Nations. (2017). *Informe de la relatora especial sobre el derecho a la alimentación acerca de su misión al Paraguay* [*Report of the special rapporteur on the right to food, addendum – mission to Paraguay*] (Report A/HRC/34/48/Add.2). United Nations. https://acnudh.org/load/2018/03/G1701976.pdf

United Nations. (2018a). *Decenio de las Naciones Unidas de la agricultura familiar* (2019–2028) [*United Nations decade of family farming (2019–2028)*] (Resolution A/RES/72/239). United Nations. https://documents-dds-ny.un.org/doc/UNDOC/GEN/N17/468/08/PDF/N1746808.pdf?OpenElement

United Nations. (2018b). *Declaración de las Naciones Unidas sobre los derechos de los campesinos y de otras personas que trabajan en las zonas rurales* [*United Nations declaration on the rights of peasants and other people working in rural areas*] (Resolution A/HRC/RES/39/12). United Nations. https://digitallibrary.un.org/record/1650694/files/A_HRC_RES_39_12-ES.pdf

Unión de Gremios de la Producción. (2015). *Agricultura y Desarrollo en Paraguay*. UGP. https://www.mre.gov.py/v2/novenoconcurso/docs/materias/agricultura%20y%20desarrollo.pdf

Vía Campesina. (2003, 15 Jan). *¿Qué es la soberanía campesina?* https://viacampesina.org/es/que-es-la-soberania-alimentaria/

Yurjevic, A. (1996). *Un desarrollo rural humano y agroecológico*. CLADES/FIAD. https://www.mapa.gob.es/ministerio/pags/Biblioteca/fondo/pdf/569_10.pdf

Chapter 6

Historical evolution of institutions and public policies for sustainable food security in Mexico: continuities and ruptures

HECTOR ÁVILA-SÁNCHEZ

▸▸ Introduction

Government focus on the food requirements of the Mexican population dates back to the beginning of the 20th century, in the context of the first governments carrying out post-revolutionary national reconstruction. Food plans and programs in Mexico arose at specific junctures (natural, social, economic and political crises) with effects on food security and nutrition (Barquera, Rivera & Gasca, 2001), with the aim of meeting specific requirements (social assistance, school breakfasts, promotion of the production of staple foods, milk and meat products). This was mainly due to the rapid urbanization and gradual population growth experienced by the country's main cities. This is one of the central facts in the definition of food policies in Mexico: the greater the number of inhabitants concentrated in cities, the greater the requirements and strategies for meeting supply needs; the greater the marginal population and the greater the nutritional lags in both rural and urban areas, the greater and broader the requirements for social coverage policies and programs.

A lag exists in Mexico with regard to ensuring access to sufficient and healthy food for the marginalized rural population. Since the year 2000, obesity and other related diseases have worsened and have permeated the lower-income strata, affecting more than 70% of the country's adult population. The situation is critical in rural areas as a result of the lack of quality health services for the prevention, diagnosis and timely treatment of these conditions (Ávila, Flores & Rangel, 2011).

This chapter aims to present the historical sequence under which the Mexican agrifood system has been instituted. It will address the transcendental sociopolitical and economic processes in the national history and the specific conjunctures where diverse actors, governmental and social organizations, established agreements, dynamics, negotiations, conflicts and disputes, present in the conformation of the institutions responsible for managing and constructing the diverse processes for feeding the population and, at the same time, creating the conditions for the insertion of small producers (family and peasant economies) within this system. A socio-political system that, even though the food issue has been considered central to its agenda, has incorporated nutritional and public health issues on a gradual basis.

The development of the agrifood system in Mexico has always been characterized by the constant presence of the state intitutions, as the guiding institution of economic and social development, from which policies and/or programs have been created to meet the needs of the population. Governmental interventions have, from the outset, always been characterized by their welfare-based approach, given the widespread poverty in most post-revolutionary urban and rural societies. In the post-war period, with the Mexican economy undergoing significant growth, the Mexican State became a benefactor and, at the same time, adopted a clientelistic approach characterized by political control in relation to its interaction with Mexican rural communities. Their demands were met, especially at election time, and their requests for land and participation in the rural dynamic, which was biased in favor of agricultural companies, were temporarily resolved. After the crisis of capitalism in the 1970s, poverty became a constant among small producers and peasant communities, and Mexican food programs were integrated into broader social welfare programs to address serious problems of malnutrition and lack of school coverage for rural children. Since then, rural development programs have been used as a palliative measure to address the marginality of peasant groups.

As of 2019, Mexico has undergone a change of government, shifting from those constituted from the corporatist model to a different one, which has initially identified poverty as its central priority and therefore key to the orientation of its policies, including rural policies. The challenge is even greater, as there is a need to act within corporatized institutions with structural vices that have served to maintain political control over peasant organizations. At this point, therefore, the analysis of this change of government includes more uncertainties than certainties. The new government proposes a different model where peasant societies revitalize traditional farming systems (the *milpa*) in the face of forms and models that prioritize productivity. What is required from the institutions that will promote a more egalitarian and sustainable agrifood system in terms of production and trade? How do peasant organizations position themselves here? How can a sustainable agrifood system, linked to addressing nutritional lag, be constructed? Which roles and strategies will the various social actors adopt in this new scenario? Which mechanisms strengthen the constitution of an authentic territorial governance in an agrifood system that differs from the traditional, historic ones? What roles should peasant organizations play in the resolution of conflicts, disputes and negotiations with governmental actors and institutions?

These are central questions that need to be clarified for the reorientation of a food production system that aims to be inclusive and democratic, while having a positive impact on the economy and welfare of small- and medium-scale rural producers – eternally marginalized within the countryside context – incorporating a focus on public health problems (obesity, diabetes and other diseases), which has permeated into the lower income strata and already affects more than 70% of the nation's adult population, as part of the national agenda (Ávila et al., 2011).

▸▸ The New Historical Institutionalism as a framework of analysis in the constitution of food policies

The creation of a system to meet food requirements is a fundamental process within rural, social and nutritional-health public policies. Food production and supply needs have led to the emergence of historically constituted institutions, the product of vicissitudes, negotiations and exchanges between political actors. In Mexico, it has been a matter of welfare needs being covered by a national State that, through subsidies, has provided access to basic foodstuffs to large sectors of the impoverished population in a context of the national reconstruction of a country that experienced a violent revolution in the second decade of the 20th century. An institutional order therefore began to be developed that, based on major events and other social processes of transcendent magnitude (a second agrarian reform under Cardenismo, 1934–1940), positioned food production at the center of its priorities, as an element that would contribute to the reduction of economic inequality within the country.

Through government interventions at different historical moments, the foundations were laid for the emergence of institutions dedicated to the construction of public policies. The existence of a food system was sustained, which, through strategies and programs at specific junctures, has tried to ensure the national demand for food. According to Farfán (2007), crucial historical phenomena are decisive for the success of a project that aspires to become an institutionalized policy. The different historical moments that have marked the construction of the national food system has involved the participation of different actors and social groups, formal and informal associations of individuals, agricultural entrepreneurs and landowners, organizations with party affiliation, pressure groups, trade unions, and even international organizations (for example, scientific groups and capitalized farmers, promoters of the green revolution), which influenced the decision making of governmental bodies (Ibid.).

New Institutionalism (NI) is a current of thought that considers rules, norms and identities as instruments of stability and change scenarios. It regards institutions as independent and autonomous actors, capable of pursuing their own objectives and, in doing so, affecting society based on the view of collective action as the aggregation of calculated individual decisions (March & Olsen, 1984, cited by Torres, 2015). NI illustrates how political struggles are mediated by inherent institutional settings that contribute to defining institutions as formal organizations, informal rules and procedures that structure behavior (Steinmo, Thelen & Longstreth, 1992).

There are different trends and/or variants of New Institutionalism; however, the most recurrent are those identified by Hall and Taylor (1996): historical NI, rational choice NI and sociological NI. For the purposes of this chapter, the approach of historical New Institutionalism is considered in order to characterize the temporal evolution and succession of processes and phenomenologies from which a food system has been constituted in Mexico. This current supports the processes and mechanisms through which rural policies, and, specifically, those related to food production, emerge and become institutionalized. A central role is given to the processes of gestation and adoption of political decisions in their historical evolution and sequence (Farfán, 2007).

The new historical institutionalism views institutions as the legacy of specific historical processes. They are generally considered to maintain long periods of stability and institutional arrangements, which can be interrupted by brief, intense and profound changes (Thelen, n.d.).

Two concepts are central to the corpus of the new historical institutionalism, namely, that of path dependence, which considers that new institutions should emerge on the basis of already existing institutional models, with defined interests of the actors, where differentiated spaces and positions of power are manifested, which are reflected both in the shaping of public policies and in institutional change. The second, related to critical junctures, considers that institutional creation and/or emergence comes from changes based on discontinuities or interruptions within social processes, in established institutional spaces that result in new institutions and political structures (Mahoney & Thelen, 2010, cited by Torres, 2015).

Under these guidelines, this chapter considers the conformation of a food system in Mexico, as social actions have been constituted on the basis of a permanent historical dynamic in terms of intervention mechanisms, participation and bargaining power of the actors. The process has peculiarities in Mexico, since most of the social organizations in rural areas, those of small producers and peasant economies, have been permanently subject to political and clientelistic control by the governments in office. Thus, rural public policies have always been biased in favor of large agricultural enterprises. This does not necessarily mean the paralysis of actors and struggles or social processes, although the actions of peasant and workers' social movements during different historical eras has had a limited effect on the formation of institutions dedicated to food production and supply.

▸▸ Small-scale producers and farmers, marginal actors in food policies

In the case of Mexico, state institutions have played a fundamental role in the design and implementation of public food policies. Perhaps uniquely in Latin America, the political system functioned for 70 years on the basis of solid corporate mechanisms, through which it exercised strict control and political clientelism over peasant movements and organizations. This was the case of the National Peasant Confederation (CNC), established during the *Cardenista* government in the 1930's, which was co-opted by subsequent governments and the agreements reached in the meetings of the peasant associations did not transcend to the grassroots. Despite this situation, there were some independent movements and struggles that did affect local transcendence and limited action in the construction of rural and food public policies. Other independent peasant organizations, created in response to the intense agrarian dynamics of *Cardenismo*, the General Union of Workers and Peasants of Mexico (UGOCEM), as well as the Independent Peasant Center (CCI), brought together sectors of the peasantry that were on the margins of official control. Due to their political positioning and their ongoing involvement in land disputes and struggles, they had restricted access to credit and material support for agricultural production. Their territorial impact was minor, as they were mainly based in the areas of large agricultural enterprises, where economic and political power exercised effective control.[1] The post-war governments, which promoted the "modernization" of the Mexican economy based on intense technified agricultural activity, assigned a marginal place to peasant family producers (*ejidatarios* and agrarian communal farmers), because "large agricultural capitalists figured prominently among the political support bases of the incumbent president" (Hewitt, 2007). *Caciques*, oligarchic figures who exercised an archaic system of economic and political power, played a fundamental role in these political control mechanisms, where usury, extortion and land dispossession were their main forms of enrichment and political power. They frequently influenced the election of associated local authorities (Ibid.).

▶▶ Historical evolution of food policies

Throughout the different stages of economic and social development, the food issue has been at the center of the agenda of government agencies concerned with rurality. Table 6.1 details the particularities of the policies and programs instituted.

The official incidence dates back to the 1920s, when actions assumed a charitable and/or welfare approach (Barquera et al., 2001) with the provision of breakfasts and food supplies to impoverished urban sectors. Subsidies for agricultural production were also granted in order to improve livestock production. To channel such aid, popular cooperatives, systems and savings banks were set up to promote agricultural and livestock activities (Ibid.).

These initiatives continued during the 1930s, through programs of indirect access to food, by means of subsidies on the price of basic foodstuffs (initially corn and beans) for workers in the cities. The governmental period corresponding to President Lázaro Cárdenas (1934–1940) established or provided guidelines for the active participation of workers' and peasants' organizations in the construction of public policies, in a country that was still being rebuilt in the wake of a revolutionary movement (1910–1918), which was highly detrimental for the population, the economy and the territory. To that end, the *Cardenista* government implemented core actions to continue the national reconstruction process: i) a new and comprehensive agrarian reform, where the land and the assistance to work it had greater scope and benefits for the peasant *ejido* societies; ii) the strengthening of the national infrastructure (road system, irrigation and electrification), which would enable isolated rural regions to access the circuits of the national economy; iii) the intervention of social organizations in the commercial management of basic products and subsistence for the popular classes; iv) the promotion of the mechanization of national agriculture, as part of the national development planning (*Six-Year Plans*).[2]

---

1. In fact, the experiments for the introduction of the Green Revolution and technological development in Mexican agriculture were carried out in agricultural production regions dominated by private, highly technified properties with irrigation in the north and northwest of Mexico, as well as in the center and west of the country (Bajío). Agricultural production in these regions was largely tailored to the market of the Southern United States.

2. This was such an extensive program that by the 1950s, Mexico had the highest levels of agricultural mechanization in Latin America (Esteva, 1980).

Table 6.1. Historical evolution of food policies in Mexico

Period	Program and/or policy	Objectives
1920–1930 **Post-revolutionary national reconstruction**	Various support programs; Production subsidy	Consumer support; Increase and improvement of agricultural and livestock production
1930–1970 **Cardenism** **Post-war Industrialization;** **Growth Economy;** **Benefactor State;** **Economic miracle;** **Economic and social crisis**	Food policy; National Storage Facilities; School Breakfast Program; Popular supply; Food production subsidy; National Food Program; Consumption subsidy for agricultural products	Control of grain market prices (guaranteed prices); Access to basic necessities; Lowering the cost of food items; Focus on tackling child malnutrition and price control of basic food basket; Regulation and importation of foodstuffs
1976–2019 **Petroleum economy;** **Market opening;** **Dismantling of public policies;** **Privatization of the economy;** **Market economy**	Mexican Food System (SAM); Program to Support Ejido Commerce (PACE); National Food Program (PNA); Social Food Welfare Program; National Program for Integral Rural Development; Family Food and Nutrition Program (PANF); Law for Sustainable Rural Development; Special Program for Food Security (PESA); Food Support Program; Special Combat Program for Rural Development National Crusade against Hunger, Mexico without Hunger Program; Food Sovereignty Program	Revival of food self-sufficiency; Address hunger, health and malnutrition problems in marginalized areas; Expansion of the commercial network; Ensure Food Sovereignty and strengthen food security; Access to basic food basket; Fulfilment of education, health, food and housing needs; Eradication of hunger and malnutrition; Increased agricultural production and income for small farmers
2019–2024 **New political and social regime?**	National Development Plan 2019–2024; Rural Development Public Policy; Mexican Food Security Agency (SEGALMEX)	Promote and ensure the right to nutritious and quality food; Strengthening food securityAccess to marketing for small producers; Reduce economic and social inequality in rural areas; Strengthening agroecological production and biodiversity

Source: Ávila et al. (2011), CONEVAL (2010), López and Gallardo (2015), SADER (2019).

Social and economic actions and impacts	Linked social programs	Actors involved
School breakfasts for children; Popular cooperatives; Agricultural and livestock producers		
Basic food subsidies; Inexpensive food; Storage and price regulation; Import products; Price control; Creation of the National Company of Popular Subsistence (CONASUPO); DICONSA rural stores and LICONSA milk; Objective: subsidies for the impoverished population	Regulatory Committee of Popular Subsistence Markets	Workers' centers; Farmers' organizations; Independent traders
Production, collection, transformation, supply and consumption of food for the poor and extremely poor population; Actions in the health field; Technological support for the rural economy; Price liberalization of the basic food basket; Subsidies in rural stores and tortilla subsidies; School breakfasts; Monetary subsidies for poor families and products; support for family spending; Community kitchens; Promoting food projects for the population with nutritional deficiencies	COPLAMARSocial Food Welfare Strategy; Rural Food Supply Program, for Families (DIF); the National Solidarity Program (PRONASOL); Children in Solidarity; Progresa and Prospera; *Oportunidades* Human Development Program; Social Milk Supply Program (LICONSA); Rural Supply Program (DICONSA); 100 x 100 and Living Better Strategies	Ejido companies; Private farmers; Farmers' organizations; Citizen organizations
Promotion of programs to improve food sufficiency; Strengthening of the peasant economy; Anchoring the rural migrant population; Economic benefits for communities with high poverty rates	Sowing Life; Institute of Health for Welfare (INSABI)	Small producers; Indigenous peasant societies; Peri-urban agricultural producers; Private producers

The most far-reaching policies regarding agricultural production and the strengthening of the food system were the creation of the National Storage Warehouses (ANDSA), the Regulatory Committee of Wheat Markets (CRMT) and the Regulatory Committee of Popular Subsistence Markets, forerunner of the National Company of Popular Subsistence (CONASUPO), of great importance in later years for the national food supply (Barquera et al., 2001). In terms of food supply, the Regulatory Committee of Subsistence Markets was established in 1938, in order to protect rural producers and consumers. Actions were carried out through workers' and peasants' organizations, which operated cooperative stores and collection centers with government support (Ibid., 85); in the 1950s, official guaranteed prices were established for basic products (corn, beans, rice, wheat), which ensured profitability in production and regulated market prices against speculation. Subsidy programs for the low-income population were strengthened, such as the school breakfast program (Esteva, 1980; Barquera et al., 2001).

Beginning in the 1940s, the operation of programs and policies corresponding to popular food supply was inconsistent, particularly in terms of the sales companies and cooperative stores, as a result of pressure from private commerce, which alleged unfair competition. The workers' and popular citizen organization movements eventually succeeded in maintaining these modalities, despite significant changes in their geographic distribution and number. In the context of post-war economic growth (also called "agrarian developmentalism," 1945–1960), peasant movements played a marginal role in the construction of the national economy. Official efforts and policies for rural development prioritized mechanized and irrigated capitalist agriculture, which had already found its the main source of income in the export market. By 1947, the Mexican State carried out constitutional reforms, which obstructed peasant demands for land, with the intention of creating a climate of "confidence and security" in land tenure and stimulating private investment in rural areas (Esteva, 1980).

In the 1950s and 1960s, government policies were strengthened through fundamental measures and actions that had an impact on basic food production. The National Seed Production Agency (PRONASE) was set up with the purpose of making basic crop seeds available at low prices to small producers and *ejidatarios* in a market dominated by transnational companies. The same function was carried out with the nationalization of the fertilizer producing industry (*Guanos y Fertilizantes de México*), which became a government monopoly, although most of its production was distributed to commercial irrigated agricultural areas. Finally, the National Company of Popular Subsistence (CONASUPO) was created, which intervened in fundamental food system processes: price control of basic foodstuffs and collection, storage and distribution as part of different agricultural programs for food supply (Hewitt, 1991; Esteva, 1980, pp. 88-89). Two key initiatives that served to ensure access to and the distribution of food for the majority of the population were the DICONSA (CONASUPO Distribution Agency) network of stores, mainly in rural areas, and the CONASUPO Industrialized Milk Supply Program (LICONSA), which mainly served the population living in poor districts of the cities and their periphery (Ávila et al., 2011). Peasant organizations participated in this context, establishing alliances with the government for the establishment of marketing support programs (subsidies for the transport and packaging of agricultural products), as well as a network of warehouses for grain storage controlled by the communities. Popular consumer cooperatives, which promoted regional supply committees throughout the country, played an important role in the management of the DICONSA rural stores network (Hewitt, 1991).

By the 1970s, the Mexican "economic miracle" (sustained economic growth, consolidation of an emerging middle class, management and prioritization of poverty issues) was over. The authoritarian corporatist model (the government and its almost single-party system) began to break down, resulting in a major social and economic crisis that heavily impacted rural areas and, fundamentally, the *ejido* and peasant communities (Esteva, 1980).

Rural policies for food production were subject to several phenomena: firstly, the guaranteed pricing system for basic products (corn, beans, wheat, rice) was removed; secondly, the increased importation of some of these products, to the detriment of local rural producers, mainly *ejidatarios*, which accelerated the migration of impoverished peasants to the cities. Both policies were decisive in the loss of food self-sufficiency[3], which was identified in the literature as the so-called "cattle ranching" of agriculture, caused by the increased amount of land allocated to the cultivation of fodder crops to the detriment of basic food grains (Barkin & Suárez, 1983, cited by Zepeda, 1988). Thus, the loss of food self-sufficiency was ensured due to the role assigned to agriculture by the growth model within the global trends of the Mexican economy and the gradual profusion of the transnationalization of agriculture and its growing impact on the food model and on the consumption patterns of the urban population (Zepeda, 1988).

Three fundamental programs were established between 1979 and 1980, to address marginality and the problems of malnutrition and health in rurale areas: i) the National Commission for Deprived Zones and Marginalized Groups (COPLAMAR); ii) the Public Investment for Rural Development Program (PIDER) and iii) the Mexican Food System (SAM). All three programs were directly linked to productive development and social protection. COPLAMAR operated in areas where marginalized groups lived, fundamentally in indigenous regions; PIDER was mainly oriented to the construction and reinforcement of rural infrastructure – roads, bridges, electricity; SAM, focused on the production and commercialization of basic crops, as well as the nutritional requirements of the population (Gordillo, 2017).

The objective of the SAM was to restore food self-sufficiency through the production of basic crops and strengthen the participatory role of peasant and *ejido* economies; it also sought to make food distribution systems more efficient, mainly among the low-income population. The production of grains for consumption was one of the strategic areas to which the government assigned significant resources. The funds for the subsidies awarded through the SAM (guaranteed prices, soft loans, low-cost fertilizers and seeds, low-cost insurance, shared risks, technological improvements), would come from the high oil revenues being banked by the government at the time (Spalding, 1985).

The SAM prioritized the production of basic grains for human consumption over fodder crops. Participating actors were given an important role, as the thinking

---

3. Due to the direction taken by agricultural public policies, in 1973 Mexico imported 25% of the corn consumed (Barquera et al., 2001), while, at the same time, the deficit in the production of other grains increased. Imports of basic grains increased at an exponential rate (9% of the sectoral total in 1965, 67% in 1975 and 80% in 1980) (Luiselli, 1980).

was that the creation of the SAM would be a response to the dynamics of an independent peasant movement during the 1970s (Esteva, 1980). Its organizations would be the strategy's stakeholders, with the aim of increasing their participation in the income generated across the entire food chain[4]. They would also be responsible for controlling the production process and would participate in the generation and appropriation of value with other agents in the food chain (Luiselli, 1980). The program mainly benefited private commercial agriculture, although it had a positive impact on employment, reduced rural migration and mitigated the inequality of income distribution in rural areas (Esteva, 1980).

In general terms, it was felt that the SAM was constrained by virulent bureaucracy and by the political interests and priorities of large producers. It has been pointed out that most rural measures and programs have traditionally benefited the country's major agricultural production areas (Spalding, 1985; Appendini & Salles, 1975). However, in terms of this program, some of the states with large regions based on the peasant-farming economy did increase their corn yields (Spalding, 1985). The SAM, established and sustained with resources garnered through oil revenue, was finally scrapped in 1981, due to fiscal problems resulting from the fall in oil prices.

▶▶ The dismantling of food policies

At the beginning of the 1980s, Mexico was already operating an open economy model, with the progressive withdrawal of State intervention in the management of the economy. The change of government in 1982 marked the start of a gradual dismantling of historically instituted food policies.

The National Food Program (PRONAL) was implemented in 1983, operated by the National Food Commission, whose objective was to raise basic food consumption for low-income population sectors by increasing the production of agricultural and fishery foodstuffs, in order to improve the population's nutrition. It was implemented through subprograms related to production, supply, consumption, health and nutrition, with little coordination among them. Subsidies granted by previous programs were drastically reduced, so the issue of food self-sufficiency was relegated to a lower ranking priority (López & Gallardo, 2015; Spalding, 1985).

Since the 1990s, food strategies have been linked to social programs with greater scope and coverage, with the intention of reducing marginalization and poverty.[5]

---

4. The Mexican Food System consisted of "a production and income strategy in the interest of feeding the people. In response to the trend for denationalization in agriculture and the brutal attack on the peasant economy; (...) it seeks to reactivate agricultural and fishing production and to act on the transformation and distribution devices of public food in order to strengthen the peasant economy (...) it also proposes the renewed development of agriculture, as it aims to reorient it to the domestic market, so that it can substitute its own imports and free up foreign currency for productive investments with a broad and rapid impact on rural employment, production and positive links to industry and distribution" (Luiselli, 1980).

5. Poverty and food insecurity are two of the most serious problems affecting a large part of the Mexican population. According to the National Council for the Evaluation of Social Development Policy (CONEVAL) in 2014, 46.2% of the national population (55.3 million people), was in poverty, and 23.4% (27 million people), suffered from food poverty. The situation was even more serious in rural areas, with 61.1% of the population living in poverty and 30.9% lacking access to food (Coneval, 2015).

As is the norm in Mexico, each presidential administration tries to make its own mark, maintaining, with some variations, the prioritization of food and nutrition programs.[6]

Around the same time, the importation of basic grains was reinforced to ensure the operation of food programs. Foreign food purchases multiplied, to the detriment of small producers and peasant-farming economies. In government plans and programs, the reference to food self-sufficiency was definitively eliminated, officially replaced by the term, food sovereignty, which almost immediately became food security, in response to guidelines promoted by international organizations (FAO, World Bank, International Monetary Fund). The continuity of the National Food Program was maintained, with numerous subprograms, integrated and operationalized on the basis of the core social policy for the 1988–1994 six-year term, the National Solidarity Program (PRONASOL). The gradual withdrawal of subsidies for production and consumption continued, except for staple products, such as milk and tortillas, in addition to maintaining food support programs, such as the DICONSA rural stores network, school breakfasts and LICONSA dairies (López & Gallardo, 2015; Ávila et al., 2011).

During the 1994–2000 government administration, the food security approach continued, based on the Family Food and Nutrition Program (PANF), with three core areas: i) school breakfasts, ii) basic food basket for the poorest families in rural areas, and iii) food basket and support for families in marginalized urban areas (López & Gallardo, 2015). The distribution of family food parcels was increased, mainly in peripheral urban and rural areas, with emphasis on areas with indigenous populations in order to address food and nutritional deficits. The measure was part of a broader social strategy, the Education, Health and Food Program (Progresa), focused on combating extreme poverty in rural areas and providing direct subsidies to strengthen food security through the provision of food of higher nutritional quality and school scholarships. They operated as targeted policies, with comprehensive health, nutrition, food support and education interventions run in 250 rural micro-regions in poverty and extreme poverty (Ibid.).

The year 2000 marked the beginning of a different era in Mexico's political life. The Institutional Revolutionary Party (PRI), which built the corporatist system for 70 years, left power. The successor party, the National Action Party (PAN), with a right-wing ideological orientation, did not make any fundamental changes in the economic and social model of government programs and marked an era of continuity in neoliberal policies. It tackled the problems of poverty, food and nutrition from a welfarist perspective.

The prioritization of food and nutritional requirements was implemented through Rural Development programs linked to numerous subprograms within the main government project 2000–2006, the *Oportunidades* Human Development Program (CONEVAL, 2010). Government aid was provided through in-kind or economic transfers, as well as the granting of food supplements, aimed at addressing lags in the areas of malnutrition, malnutrition and the growth of obesity and overweight.

---

6. In the 1988–1994 period, the *Solidaridad* program was established; in 1994–2000, *Progresa*; in 2000–2006, *Oportunidades*, with continuity in the 2006–2012 period; for the 2012–2018 period, the *Prospera* program was established.

In 2001, the Law for Sustainable Rural Development was enacted, under which the legal instruments that regulate the dynamics and activities inherent to rural territories in Mexico are contemplated, with particular focus on food production, emphasizing security, sovereignty and self-sufficiency. It identified the basic products required for an adequate diet, the domestic demands for the consumption of basic and strategic products, the needs for food supply and imports, working in conjunction with environmental protection programs (Chamber of Deputies, 2001). It is worth noting that none of the laws or regulations referring to food made any mention of involving the producers' organizations in the discussion or design of the food programs and policies. Similarly, said law exhibits a dangerous bias that, as a matter of national policy, could face institutional limitations at subnational and/or local territorial levels (Valencia et al., 2019).

The food security framework continued to be implemented through the Special Program for Food Security (PESA), a model recommended by the FAO to increase the productivity of small farmers in low-income food-deficit countries in order to reduce the incidence of hunger and malnutrition. With the necessary adjustments, the PESA was adopted in Mexico in 2002, operating as a pilot project in six Mexican states to adapt its methodology to national conditions. As of 2005, its operation was expanded to 16 states in the central, southern and southeasterly regions of the country with high poverty rates, representing one of Mexico's most important government rural programs (Gordillo, 2017). As central objectives, it proposed the strengthening of food security, as well as the impact on increasing the income of families in highly marginalized communities. This would be achieved by increasing agricultural production, promoting food self-sufficiency and expanding local markets. At the same time, it would implement technological innovations and improvements, in addition to generating interactions that would impact local and community organization (Cherrett, 2017). A central role was played by the rural development agencies (ADRs), a novel form of extensionism carried out by a multidisciplinary team that advised and guided productive and marketing processes (Gavotti, 2017). In 2008, a change of name saw it become the Strategic Food Security Project.

During the 2006–2012 governmental period, the *Oportunidades* program continued to be run with some minor adjustments. The main actions concerning food issues were channeled to the population through food subsidies[7]. Although some external evaluations (CONEVAL, 2010) reported benefits in terms of improvements in child nutrition as a result of being supplied with healthy food and regular milk rations, the painful problems associated with social inequality were also exposed: in terms of access to food, 18.2% of the Mexican population was in food poverty in 2008, that is, they did not have enough income to acquire the goods for a basic food basket, even if they used their entire income. Moreover, according to the multidimensional measurement of poverty, more than 10% of the nationwide population were lacking access to adequate food in 2008 (Ibid.).

In the 2012–2018 period, the issue of food was addressed through an intergovernmental program, the *Prospera* Human Development Program, with the participation of the

---

7. Rural Food Supply Program; Social Milk Supply Program; Food Support Program (monetary or in-kind transfer); Food Supplements; Nutritional Status Monitoring. Other programs related to food issues were: School Breakfasts; Care for at risk children under the age of 5; Food Aid for Vulnerable Families.

Secretariat of Social Development (SEDESOL) and the Secretariat of Agriculture, Livestock, Rural Development, Fisheries and Food. One of the main components, the *National Crusade against Hunger*, was established in 2013, and adopted the PESA methodology, as a comprehensive and participatory national social policy strategy aimed at addressing the problem of hunger in Mexico. It focused its strategy on territorial units in multidimensional poverty based on the provision of adequate food and nutrition to strengthen food security. Among its central objectives, it aimed to reduce acute child malnutrition and hunger, as well as to improve the weight and height indicators for children. In terms of production, it sought to increase food production volumes to raise the income of farmers and small producers, as well as the production of grains, meat and fish products, which are fundamental to the basic rural food basket. It sought to improve the eating habits and dietary diversity of rural families, and to strengthen mechanisms to facilitate access to markets. It also sought to minimize post-harvest losses of food during storage, transportation, distribution and commercialization, as well as to promote community participation (Martínez, 2017).

Unfortunately, the programs aimed at addressing these social problems failed to fully meet their objectives for many reasons, including inadequate methodological strategies, excessive bureaucracy and lack of continuity in fundamental mechanisms, such as the ADRs. Other vices and phenomenologies rooted in the Mexican rural milieu also exist, such as the duplicity of functions of governmental production support agencies, corruption in the management of material and financial resources; intangible structural situations, such as despotism and local and regional political control, which have historically acted as major obstacles to the development of local initiatives. Other issues mentioned include, a lack of understanding of the historical and cultural environment associated with the communities targeted for intervention (Cherrett, 2017), a lack of comprehensive understanding in relation to poverty and marginality, particularly within indigenous communities, in order to act, as well as the absence of any strategy that identified the territorial scope of the communities, their cultural, environmental, economic and social dynamics, and the ways in which they interact with institutions (Ibid.). Some evaluative analyses of the PESA program indicate marginal benefits in terms of the contribution of food to food baskets.[8]

As of 2017, the PESA changed its name once again to the Food Security Project for Rural Areas, as a component part of the Small Producers' Support Program, which was run until the 2012–2018 government left office. The component's purpose was to contribute to ensuring food security in highly and very highly marginalized areas with communities living in poverty and extreme poverty.

▶▶ A new national project: a shift towards food and social policy?

Rural development policies in Mexico have, historically, assigned a marginal role to family and small producers in terms of their participation in the national agrifood system. The established rural development model has prioritized corporate

---

8. In some cases, "the PESA contributes less than 10% of the value of the beneficiaries' per capita consumption, which makes its contribution to the basic food basket insignificant, thus beneficiaries are not incentivized to stay or improve the projects" (Gimate & Muñoz, 2017).

agribusiness, particularly in agricultural regions linked to the international markets of the United States and Canada. They also provide large volumes of food produced under conventional methods to supply the large urban food supply markets, in which small and medium-sized producers (*ejidatarios*, communal and private landowners) also participate to a lesser extent (figure 6.1).

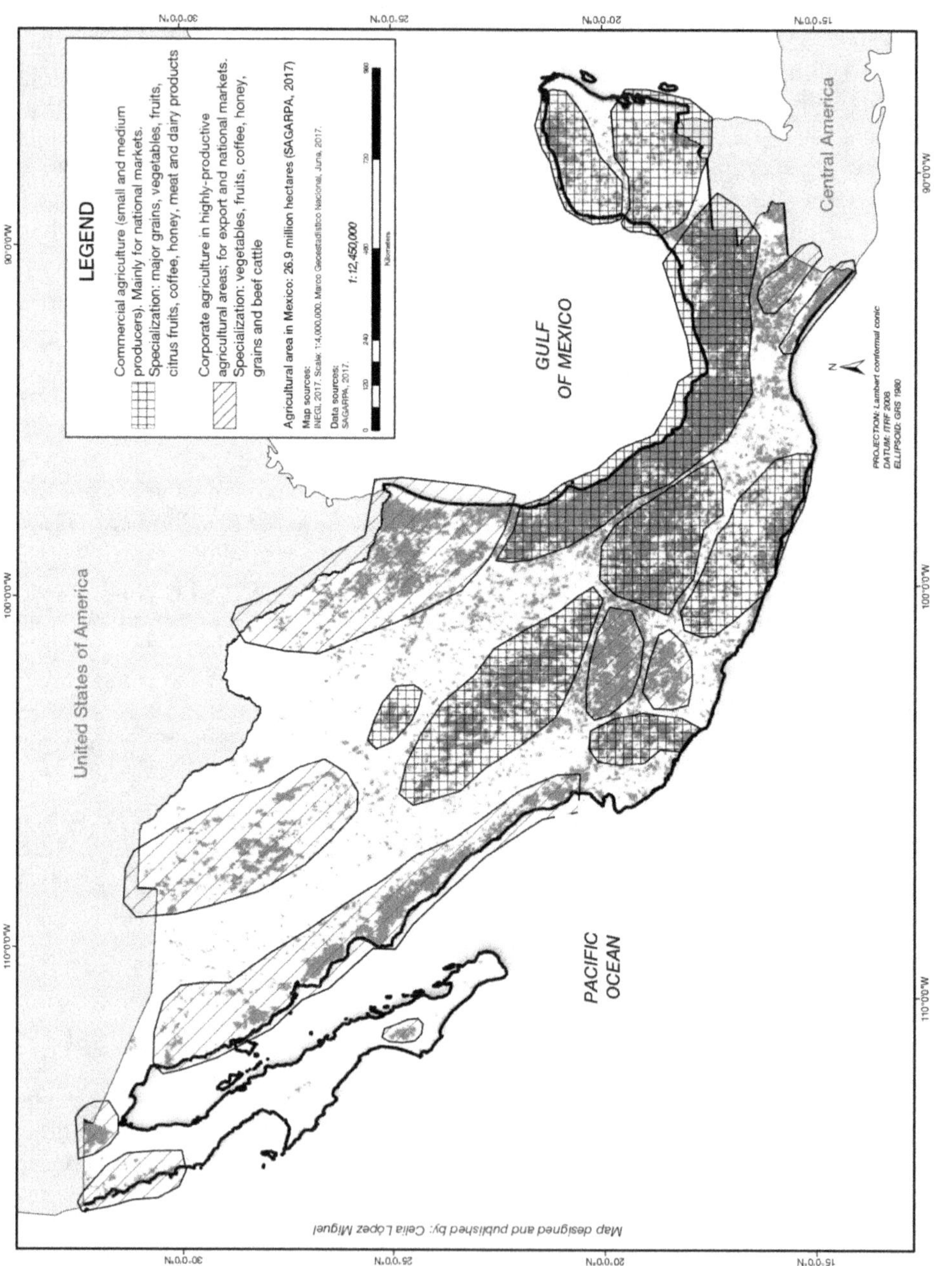

Figure 6.1. Mexico's agricultural frontier (source: SIAP, 2017, Agricultural Atlas of Mexico).

The year 2019 marked the beginning of a new stage in Mexican social history: the rise to power of a political formation that differed from the corporativism party, in power for seventy years and the continuity of subsequent governments, identified with market economies (2000–2012), and a predominantly center-left political formation that set out, as its main goal, to reduce the economic and social inequality existing in the country. Its government program emphasizes the development of a Welfare Economy, where the market does not replace the State, but rather the State places the focus on marginalized sectors as a priority for public policies (Mexico, Secretaría de Gobernación, 2019).

One of the fundamental policies for inclusive rural development is related to food production. It tries, as far as possible, to reverse the conventional food model and influence healthy nutrition practices, which will have a positive impact on the decrease of malnutrition and malnutrition rates. Similarly, it seeks to address the serious problems of obesity, diabetes and other diseases suffered by the Mexican population. It brings back the essence of the self-sufficiency and food sovereignty approach in order to guarantee a higher standard of living for peasant groups.

The government's guiding framework, the National Development Plan 2019–2024, states that one of its objectives is to promote and ensure the right to nutritious, sufficient and quality food, through the National Agreement for Food Self-Sufficiency. Thus, the State once again plays a central role in the management of the process, in addition to vindicating small and medium-sized producers as productive subjects with rights, to whom support will be granted directly without intermediaries, with technical support and access to productive assets. Such measures are proclaimed as acts of social justice (Mexico, Secretaría de Agricultura y Desarrollo Rural, 2019).

Food self-sufficiency is established as a concept for the revitalization of domestic food production. Similarly, there is a focus on strengthening agriculture as a profitable activity, with the aim of preventing rural decline. An important objective is to achieve sustainable food production as a means of combating malnutrition. The relative production chains for basic grains, meat, dairy, fisheries and aquaculture will be promoted, with the joint participation of the public and private sectors in public policies. Gradually, such processes will have an impact on the reconstruction of social and community cohesion and solidarity in rural societies, especially in peasant economies and marginal rural and urban-rural groups (Ibid.). The aim is to promote programs that will strengthen the country's food sovereignty (figure 6.2).

There are two main programs through which the strategy will be operationalized. Firstly, the Mexican Food Security Agency (SEGALMEX), established to take charge of food production and distribution and backed by a central subprogram, Guaranteed Prices, which consists of granting minimum per-ton payments to small producers of corn, bean, rice and wheat to ensure their profitability in the process, which is affected by increasingly unprofitable market prices. In addition to promoting the income of small and medium-sized agricultural producers, it aims to encourage food self-sufficiency by promoting the cultivation and production of staple foods (corn, beans, rice, bread wheat and milk) in order to limit imports as much as possible. In a complementary manner, it will seek to improve the system of collection, distribution and rural supply and promote the production and sale of basic food basket products. The operation of this program prioritizes its actions

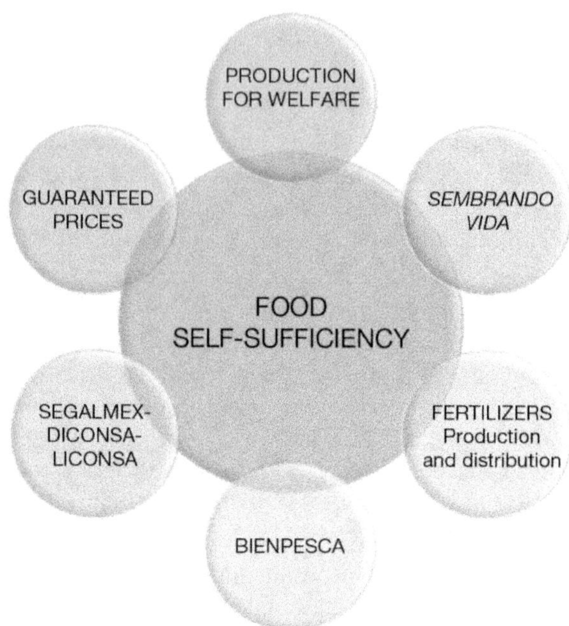

Figure 6.2. Government programs in Mexico's Agrifood System, 2019–2024 (source: Suarez, 2020).

in areas characterized by high and very high marginalization and will benefit indigenous producers. It is linked to other parallel subprograms run by the Secretariat for Rural Development (SADER), such as Production for Welfare and Fertilizers for Welfare, which will grant credit and in-kind support, in addition to providing low-cost fertilizers to small and medium-scale producers (Mexico, Secretaría de Agricultura y Desarrollo Rural, 2019) (table 6.2).

Other programs included in SEGALMEX's strategy (already in place during previous governments) are the Rural Supply Program (DICONSA), which will distribute basic grains and other processed foods in rural areas at below-market prices for the benefit of the local population. The role played by the Social Milk Supply Program (LICONSA) is crucial in that, in addition to providing subsidized food, it addresses nutritional deficiencies throughout the country[9] (SADER, 2019).

The strategy is complemented by the *Bienpesca* program, which seeks to provide small-scale producers with access to fishery products that provide nutritional variety in their diet. Monetary support and credits will be granted for investment in fishing infrastructure in priority coastal areas with low technical development in the activity. Table 2 shows the programs linked to SEGALMEX.

A second core area of work linked to the food issue is established around the Sowing Life (*Sembrando Vida*) Sustainable Communities Program. The central subprogram,

---

9. According to the National Health and Nutrition Survey (Ensanut, 2012), 13.6% of children under five years of age nationwide suffer from chronic malnutrition. In rural areas of the country, that percentage rises to 27.4%.

Agrarian Subjects and Rural Areas of Mexico, will be implemented in the areas with the greatest wealth of biodiversity (south and southeast of the country), where a large proportion of the native peoples also live. This territorial area has the highest rates of poverty and backwardness, due to the ineffectiveness of previous institutional programs implemented to meet basic food needs.

The *Sembrando Vida* program seeks to increase agricultural productivity by positioning sustainable family (peasant) agriculture at the top of the agenda. The aim is to gradually strengthen the strategy, with a long-term regional development vision that will help reduce the vulnerability of its population. Support will be provided to rural producers farming tracts land of less than 2.5 hectares to encourage the establishment of agroforestry production systems in rural areas with living standards below the social welfare line (poverty and extreme poverty).

The agroforestry project will combine traditional crops with fruit and timber trees. As the program will be developed with family economies at its core, it will be based on the *Milpa Intercalada* mixed agriculture system (MIAF System), that is, the cultivation of grains and vegetables in areas of fruit trees (rubber, pepper, cocoa, coffee, cinnamon, etc.) cultivation. The program's goal is to recover the forest cover of one million hectares in the country, while strengthening family and peasant agriculture. Government support will be provided in kind (plants, inputs and tools), in addition to technical assistance. Financial support will also be provided for minor expenses and the promotion of savings (establishment of self-managed savings banks) (Mexico, Secretaría de Bienestar, 2019).

Parallel to the operation of the above programs, a body called the Intersectoral Group for Health, Food, Environment and Competitiveness (GISAMAC) has been created, "whose objective is to build a fair, healthy, sustainable and competitive agrifood system that improves the nutritional status of the Mexican population through food production that preserves and restores agroecological systems, using an intersectoral cross-cutting approach" (Suárez, 2020). The challenges and modalities of operation in this shift towards food policy are assumed by this entity, broadly linking policies to health issues such as child malnutrition, malnutrition, obesity and weight problems.

GISAMAC is made up of different ministries and institutions involved in the operation and management of agriculture, health, economy, use of and access to water, care for indigenous groups and everything related to a different, more inclusive vision of rural development in Mexico. It includes national research centers and international organizations, such as the FAO, as consultants. The group has developed as a result of a need to establish an agrifood system with solid institutional links to address nutritional problems in the country. Its function would be to coordinate policies and actions related to food in the context of the Covid-19 pandemic, with the aim of influencing the comorbidities associated with poor nutrition. The program is still under construction and, as such, no precise results currently exist.

In this sense, the direction that food policy in Mexico will take is still uncertain, as the conditions for implementing change are not clearly set out. It will not be an easy task to dismantle a corporatized framework, deeply embedded in the agrarian structure, with deep-rooted mechanisms of power and control.

Table 6.2. Food security programs, 2019–2024

Program	General	Target	Objectives
Production for welfare	Food self-sufficiency and rural recovery	Beans, coffee and sugarcane producers	Financing for producers; Liquidity to invest in inputs and services for production
Guaranteed Prices	Incentivize the income of small producers; Food self-sufficiency; Reduction of imports	Producers of corn for human consumption, rice, beans, wheat for breadmaking and milk	Improve rural collection, commercialization and supply; Food sales and production; Guaranteed prices in highly marginalized areas
Agrifood Safety and Health Program 2020	Improve the country's phytosanitary heritage and food, aquaculture and fishery safety	Epidemiological Surveillance of Pests and Diseases; Food, Agriculture, Aquaculture and Fisheries Safety	Prevention and combat of pests and diseases affecting agriculture, livestock, aquaculture and fishing
Rural Supply Program (DICONSA)	Inclusive Economic Development	Rural communities characterized by high or very high marginalization with no local supply service	Facilitate economic access to food products; Food security in highly marginalized areas
Social Milk Supply Program (LICONSA)	Actions to supply milk to marginal and low-income sectors with nutritional deficits	Inhabitants of mostly indigenous areas, high marginalization and high rates of violence	Improved access to food for the members of beneficiary households
Fertilizers for Welfare Program	Increased productivity	Small-scale corn, bean and rice producers	150 kg of Urea/ha 150 kg of DAP/ha
Program for the Promotion of Agriculture, Livestock, Fishing and Aquaculture. (BIENPESCA)	Promoting nutritious food production and consumption	i) Welfare for fishermen and fish farmers. ii) Varied use of aquaculture resources	Promoting small and medium-sized fishery and aquaculture producers for food self-sufficiency

Source: SADER (2019); Agriculture Programs. SADER (2020), Agricultural Programs.

Scope of operation	Actors involved	Link	Impact on Nutrition and Health
Payment of labor, machinery rental, purchase of seeds and fertilizers; social effects	Small producers under 5 ha and up to 20 ha. Indigenous population	Various SADER programs	Improve nutrition
National; Particularly in highly marginalized areas; Farmer economy regions	Small producers; Medium-sized producers	State (SEGALMEX), producers	Improve the safe and pricing of commodities
National coverage; Phytosanitary and food, aquaculture and fisheries safety measures	Regional controls; State governments; Poor, vulnerable and elderly producers	Strengthening of the Technical; Operational and Administrative Capacity of SENASICA	Risk coverage for contamination in agrifood, aquaculture and fishery production units
Rural areas characterized by high and very high marginalization; Peripheral urban spaces	Impoverished, disadvantaged and vulnerable communities characterized by marginalization. The elderly	Interinstitutional Coordination; Secretariat for Rural Development and Environment	Strategy to combat poverty and hunger; Promoting food safety and nutritional improvement
Rural areas and urban peripheries throughout the country	Child and youth population in poor and vulnerable situations; Disadvantaged, marginalized elderly people	Liconsa, direct beneficiary; Liconsa, community supply; SEGALMEX	Malnutrition; Food Security
Edo. Guerrero (81 marginalized municipalities); 95 SEGALMEX distribution centers	Small-scale producers up to 2 ha	Three levels of government; Educational and research centers	Environmentally sustainable practices
Coasts and lakes in rural areas throughout the country	Small fishery and aquaculture producers	CONAPESCA; INAPESCA; Fishery and Aquaculture Producers' Register	Increased production and consumption of fishery and aquaculture products for a healthy and nutritious diet

The flagship programs for food security (SEGALMEX and *Sembrando Vida*), revive strategies and/or mechanisms to achieve the goals, i.e., the strengthening of the country's peasant agriculture, and greater participation in the food process. To that end, the figure of the ADR, implemented as one of the pillars of the PESA program, will be reoriented as a means for direct delivery of benefits and advice on the execution of the various modalities in the program.

Other structural vices that have developed in Mexican institutions relate to the existence of corporatized areas that promote rural development and where political control still prevails. The inefficiency and bureaucracy of the public institutions, as well as a high degree of corruption, prevail within the agencies in charge of rural development. There is also a frequent overlapping of functions and programs between national and state governments. The main peasant organizations continue to engage in clientelistic dynamics, particularly when it comes to electoral processes. In Mexico, as in many countries, the eradication of corruption and clientelism has become the principal hurdle for rural development programs. As a result, innumerable difficulties and ruptures are encountered with regard to the institutionalization of innovations and territorial development.

The shift or transformation in the organization of the Agrifood System in Mexico, aims to become an intergovernmental State policy, aligned with the fulfillment of the 2030 Sustainable Development Goals, as well as with the Paris Agreements on Climate Change (Mexico, Secretaría de Agricultura y Desarrollo Rural, 2019). This constitutes a territorial transformation project that requires the mobilization of resources and capacities corresponding to various public and private actors; however, to be effective, it must have the capacity to develop mechanisms in which effective links are established between the state, rural society and the market to promote the food requirements of the national population and raise the welfare levels of small producers in processes that foster social and environmental cohesion. According to Berdegué and Favareto (2019), it is essentially a matter of substantiating the food issue in terms of inclusive and sustainable rural territorial development.

▸▸ Conclusion: outlook for the agrifood system in the current context of health and economic crisis

The shift in Mexico's agricultural public policy is an important opportunity to strengthen the role of family and peasant agriculture, especially in the current circumstances of the pandemic. The purpose of rural policies, the recovery of food security, places marginal rural dwellers, small-scale producers, at the center of the agenda as actors to whom policies are directed and as participants in their management.

In more recent times, interaction between social actors has occurred at different levels; it is the result of social movements that they developed during the 1990's, when the greatest dismantling of rural public policies was in progress, with the transfer of state functions to private companies and an increase in the importation of basic foodstuffs. A number of social movements emerged during that period, one of the most important of which was the No Corn, No Country (*Sin Maíz no hay País*) campaign, a social agreement between numerous national and local peasant organizations, which reclaimed important negotiation spaces regarding their participation in food production and distribution. However, these were only temporary gains in an economic and social environment where decisions and public policies followed neoliberal guidelines, in which state involvement was largely semi-symbolic. This trend continued until well into the 21st century.

The current government's shift towards the implementation of rural development policies returns to these antecedents of struggles and social movements in which the role of small and medium-sized producers is vindicated, with the intention of linking rural programs and policies to food sovereignty.

The great challenge that lies ahead relates to the actual operation of these policies; with how they might transcend and achieve their remit. How government action might promote greater participation for small rural producers in strengthening and consolidating the food system. This is the unfinished business of the Mexican Food Security program (SEGALMEX), which, in the first instance, needs to generate a nationwide impact on the rural sector, while, in a differentiated manner, with greater emphasis on highly marginalized regions, work towards anchoring the local population and raising welfare levels.

The aim is also to reinforce the role of the state as a promoter of agricultural development and of small and medium-scale rural producers as productive subjects, a role that was curtailed during previous governments.

Experiences and programs developed in the past could be recovered, which, when adapted to current modalities, would reactivate the original purpose for which they were created. For example, the various programs of the PESA strategy contain useful guidelines that could have an impact on strengthening community-based peasant agriculture (urban kitchen garden programs, among others). It would be necessary to reactivate efforts, such as those of the ADRs, once considered new forms of extensionism, where the relationship between farmers and extensionists would generate broad and diverse channels of participation, resulting in the development of local development processes and, therefore, higher welfare levels.

And here it will be necessary to establish solid mechanisms of interaction between different actors. Some major conflicts exist that need to be resolved: first, the monopolistic expansion of transnational agribusinesses and its impact on the food system.[10] In parallel, greater access to markets needs to be facilitated for small producers. In order to move towards a sustainable agrifood regime, agreements need to be negotiated with large capitalist farmers for the transition towards more sustainable forms of production and the promotion of agroecological agriculture. An example of this is the dispute over the reduction and gradual prohibition of the use of glyphosates, promoted by the Mexican government for the year 2024.

---

10. In 2016, the Mexican government introduced legislation to limit the sale of sweet pastries and drinks with high sugar content in schools. The initiative came under strong pressure as a result of lobbying from ultra-processed food manufacturing companies, some of them transnational (Bimbo, Unilever, Coca Cola), which neutralized the initiatives. In June 2020, in some Mexican provinces (Oaxaca, Tabasco, Mexico City), laws were passed prohibiting the sale of soda, sweets and fried foods (junk food) to minors; an insufficient measure, with no adequate controls to reduce childhood obesity currently in existence. In addition, traditional Mexican cuisine contains high amounts of fats and carbohydrates. Front-of-pack warning labeling for processed and ultra-processed foods, as well as for sugar-sweetened beverages was also approved in 2020. There is currently no evidence or data available to evaluate this measure.

Although food production for supplying Mexico's urban centers is largely carried out under the conventional productivist model, the current health crisis scenario highlights the importance of alternative models and forms of agriculture carried out in rural areas close to cities, in peri-urban areas, mainly for the cultivation of vegetables and perishable products, as a contemporary trend at international level. In the present circumstances, online food supply is gaining ground as one of the ways in which agricultural producers and restaurant businesses are facing the economic crisis resulting from the pandemic. In the local peri-urban agricultural areas there are family producers, as well as small private producers, involved in this process, forming short commercialization and consumption circuits, whereby the production is delivered directly or almost directly to the consumer through the supply of baskets and/or boxes of various foods (usually organic, healthy).

The high environmental and social costs of the prevailing agrifood model have been discussed in both academic and scientific circles. The importance of building "(...) social participation initiatives aimed at promoting agroecological and supply production for segments of metropolitan populations, from markets and street markets that promote short circuits and solidarity economies (...) strengthening processes, bonds of solidarity and trust between producers and consumers, with significant participation of local peri-urban agricultural production areas," (CONACYT, 2020) has been debated.

Major obstacles and difficulties exist in Mexico in relation to building public policies based on conditions of social equality and full democratic processes. Firstly, due to weak or non-existent governance, as civil society organizations have little influence on decision-making, which generates uncertainties regarding the generation of economic benefits for small producers. In the broad sense of territorial development, it is imperative to enhance technical capacities in order to introduce new content in public action, strengthened with political capacities that ensure the dissemination of innovative processes. The absence or partial presence of these elements would limit their scope and the interactions between actors in a territory (Berdegué and Favareto, 2019).

Hence the initial question posed in the title of the article. Will food policy in Mexico follow a process of continuity, or will it be capable of establishing modalities that lead to the rupture of trends and practices that hinder the development of rural economies (particularly those of peasants and small producers), in order to reduce economic and social inequalities? That is the question.

▸▸ References

Appendini de A., K., & Almeida Salles, V. (1975). *Agricultura capitalista y agricultura campesina en México (diferencias regionales en base al análisis de datos censales)*. Volume 10, Cuadernos del CES (Centro de Estudios Sociológicos). El Colegio de México.

Ávila Curiel, A., Flores Sánchez, J., & Rangel Faz, G. (2011). *La política alimentaria en México*. Centro de Estudios del Desarrollo Rural Sustentable y la Seguridad Alimentaria (CEDRSSA) H. Cámara de Diputados LXI Legislatura / Congreso de la Unión.

Barquera, S., Rivera-Dommarco, J. & Gasca-García A. (2001). Políticas y programas de alimentación y nutrición en México [Food and nutrition policies and programs in Mexico]. *Salud Pública de México*, 43(5), 464–477.

Barkin, D. & Suárez, B. (1983). *El fin del principio: las semillas y la seguridad alimentaria*. Centro de Ecodesarrollo: Editorial Océano.

Barradas, P. & Baca J. (2017). *El PESA en México*. Universidad de Quintana Roo. https://www.researchgate.net/profile/Julio_Moral/publication/331825060_EL_PESA_en_MexicoISBN/links/5c8ea18245851564fae478e9/EL-PESA-en-MexicoISBN.pdf

Berdegué, J. A. & Favareto, A. (Eds.) (2019). *Desarrollo Territorial Rural en América Latina y el Caribe. 2030 – Alimentación, agricultura y desarrollo rural en América Latina y el Caribe*. No. 32. FAO. Licence: CC BY-NC-SA 3.0 IGO. http://www.fao.org/3/ca5059es/ca5059es.pdf

Cherrett, I. M. (2017). El Pesa en México: reflexiones de un participante. In: P. Barradas & J. Baca (Eds.) *El PESA en México* (pp. 35–59). Universidad de Quintana Roo.

CONACYT. (2020). *Programa de soberanía alimentaria de pronta respuesta en agriculturas de pequeña y mediana escala*. Dirección Regional Centro. Internal document.

Consejo Nacional de Evaluación de la Política de Desarrollo Social (CONEVAL). (2010). *Dimensiones de la seguridad alimentaria: Evaluación Estratégica de Nutrición y Abasto*. https://www.coneval.org.mx/rw/resource/coneval/info_public/pdf_publicaciones/dimensiones_seguridad_alimentaria_final_web.pdf

Consejo Nacional de Evaluación de la Política de Desarrollo Social (CONEVAL) (2015) *Informe de Evaluación de la Política de Desarrollo Social en México 2014*.

ENSANUT. (2012). *Encuesta Nacional de Salud y Nutrición. Resultados nacionales*. Instituto Nacional de Salud Pública.

Esteva, G. (1980). *La batalla en el México rural*. Siglo XXI editores.

Farfán, G. (2007). El nuevo institucionalismo histórico y las políticas sociales. *Polis,* 3(1): 87–124.

Gavotti, S. (2017). Originalidad del Programa de Seguridad Alimentaria en México. In: P. Barradas & J. Baca (Eds.) *El PESA en México* (pp. 31–34). Universidad de Quintana Roo.

Gimate-Baños, S. & Muñoz, M. (2017). Evidencias del Proyecto Estratégico para la Seguridad Alimentaria en México. In: P. Barradas & J. Baca (Eds.) *El PESA en México* (pp. 237–260). Universidad de Quintana Roo.

Gordillo, G. (2017). El PESA en el contexto de la teoría del cambio en el ámbito rural. In: P. Barradas & J. Baca (Eds.) *El PESA en México* (pp. 19–29). Universidad de Quintana Roo.

H. Cámara de Diputados. (2001). *Ley de Desarrollo Rural Sustentable*. LVIII Legislatura/Congreso de la Unión.

Hall, P. A., & Taylor, R. C. R. (1996). "Political Science y the Three New Institutionalisms. *Political Studies*, XLIV, (936–957).

Hewitt, C. (1991). La economía política del maíz en México. *Comercio Exterior*, 41(10): 955–970.

Hewitt, C. (2007, September-December). "Ensayo sobre los obstáculos al desarrollo rural en México. Retrospectiva y prospectiva" [An essay about the obstacles to rural development in Mexico: a retrospective and prospective study]. In *Desacatos*, 25:79-110.

López Salazar, R., & Gallardo García, E. D. (2015). Las políticas alimentarias de México: un análisis de su marco regulatorio [Food policy in Mexico: an analysis of the legal framework]. *Estudios Socio-Jurídicos*, 17(1): 11–39. https://doi.org/10.12804/esj17.01.2014.01

Luisselli, C. (1980). ¿Por qué el SAM? *Nexos*. https://www.nexos.com.mx/?p=3679

Mahoney, J. & Thelen, K. (2010). A Theory of Gradual Institutional Change. In: J. Mahoney & K. Thelen (Eds.), *Explaining Institutional Change. Ambiguity, Agency y Power* (pp. 1–37) Cambridge University Press. https://doi.org/10.1017/CBO9780511806414.003

March, J. G. & Olsen, J. P. (1984), The new institutionalism: organizational factors in political life. *American Political Science Review*, 78(3): 734–749. https://doi.org/10.2307/1961840

Martínez, J. (2017). El Pesa en la política de la Sagarpa. Evolución de una metodología para territorios rurales marginados. In: P. Barradas & J. Baca (Eds.) *El PESA en México*. Universidad de Quintana Roo, pp. 117-145.

Mexico. (n. d.) *Programa Fertilizantes para el Bienestar*. https://www.gob.mx/cms/uploads/attachment/file/554053/triptico_uso_Fertilizantes__1__compressed.pdf

Mexico. Diconsa S.A. de C.V. (2020). *Programa de Abasto Rural Diconsa*. https://www.gob.mx/diconsa/documentos/reglas-de-operacion-del-programa-de-abasto-rural-a-cargo-de-diconsa-s-a-de-c-v

Mexico. LICONSA S.A. de C.V. (2016). *Programa de Abasto Social de Leche Liconsa S.A. de C.V.* https://www.gob.mx/liconsa/acciones-y-programas/programa-de-abasto-social-de-leche

Mexico. Secretaría de Agricultura y Desarrollo Rural. (2020). *Programa de Fomento a la Agricultura, Ganadería, Pesca y Acuicultura de la Secretaría de Agricultura*. https://www.gob.mx/agricultura/documentos/programa-de-fomento-a-la-agricultura-ganaderia-pesca-y-acuicultura-de-la-secretaria-de-agricultura

Mexico. Secretaría de Agricultura y Desarrollo Rural. (2020). *Programa Producción para el Bienestar*. https://www.gob.mx/agricultura/es/videos/produccion-para-el-bienestar-243457

Mexico. Secretaría de Bienestar. (2020). *Programa Sembrando Vida*. https://www.gob.mx/bienestar/acciones-y-programas/programa-sembrando-vida#acciones

Mexico. Servicio Nacional de Sanidad, Inocuidad y Calidad Agroalimentaria (2019). *Programa de Sanidad e Inocuidad*. https://www.gob.mx/senasica/documentos/lineamientos-tecnicos-especificos-para-la-ejecucion-y-operacion-del-programa-de-sanidad-e-inocuidad-agroalimentaria-2019

Mexico. Secretaría de Agricultura y Desarrollo Rural. (2019). *Programas Agricultura 2020*. https://www.gob.mx/agricultura/acciones-y-programas/programas-agricultura-2020-258752

Mexico. Servicio de Información Agroalimentaria y Pesquera [SIAP]. (2017). *Atlas agroalimentario de México*. https://www.gob.mx/siap/prensa/atlas-agroalimentario-2017

Mexico. Secretaría de Bienestar. (2019). *Sembrando Vida. Programa de Comunidades Sustentables*. https://www.gob.mx/bienestar.

Mexico. Secretaría de Gobernación. (2019). *Plan Nacional de Desarrollo 2019-2024*. https://www.dof.gob.mx/nota_detalle.php?codigo=5565599&fecha=12/07/2019

Spalding, R. J. (1985). El Sistema Alimentario Mexicano: ascenso y decadencia. *Estudios Sociológicos* (pp. 317–320). Vol. III, No. 8. El Colegio de México,

Steinmo, S., Thelen, K. and Longstreth, F. (1992). *Structuring Politics*. Cambridge UniversityPress.

Suárez, V. (2020). Hacia un nuevo sistema agroalimentario en el México de la 4T. *XL Seminario de Economía Agrícola* [Conference presentation]. Instituto de Investigaciones Económicas-UNAM, Mexico.

Thelen, K. (n.d.). *Historical Institutionalism in Comparative Politics*, Centro de Investigación y Docencia Económica (CIDE), División de Estudios Políticos. Working document no. 91.

Torres, E. (2015). El Nuevo Institucionalismo: ¿hacia un nuevo paradigma? [The new intitutionalism: toward a new paradigm?]. *Estudios Políticos*, *(January–April)*34: 117-137. http://www.scielo.org.mx/scielo.php?script=sci_arttext&pid=S0185-16162015000100006

Valencia, M., Le Coq, J.F., Favareto, A., Samper, M., Sáenz, F., Sabourin, E. (2019). Hacia una nueva generación de políticas públicas para el Desarrollo Territorial Rural en América Latina [Towards a new generation of public policies for rural territorial development in Latin America]. *Info Note DTR*. Red Políticas Públicas y Desarrollo Rural en América Latina; Red Brasileña de Investigación y Gestión en Desarrollo Territorial. https://agritrop.cirad.fr/595511/

Zepeda Patterson, J. (1988). *Las sociedades rurales hoy*. El Colegio de Michoacán, A.C.

Chapter 7

Public policies and agriculture: historical analysis of their effects on the structure of the Bolivian food system

JORGE ALBARRACIN

▸▸ Introduction

In order to understand the structure of the current Bolivian agrarian and agrifood system and its situation, a historical review is required that shows how the co-optation and management of institutions and the formulation of public policies by one group of actors, namely businessmen, oligarchs, landowners and large landowners, versus another neglected and disregarded group, known as natives, peasants and, more recently, family production units, have been in conflict for centuries, leading to the configuration and constitution of the current system.

A central aspect of our history is that the State has always lacked the ability to recognize the social and cultural bases on which it was created. Thus, it has systematically formulated laws and policies that fail to consider the original indigenous communities and communal land ownership, which has acted as a barrier to the objective of modernizing the productive apparatus and, as such, the constitution of a new agrifood system.

Within the framework of these processes, the access to power that certain actors have been afforded has undeniably enabled them to create an institutional structure that has borne fruit in the agrarian sector, influencing the formulation of policies, looking to see who would benefit and who would be excluded. These aspects have become latent in the current scenario in Bolivia, with a transitional government that is on its way out and a new administration under the MAS party, in relation to issues of interest to the business sector and within the context of demands and protests being staged by peasant organizations.

Underlying all of this is the protagonism and the role played by each of the actors in food production, food security and income generation. The paradox is that, despite the existence of a set of regulations that promote food security and sovereignty, the construction of the food system is actually showing a tendency towards fragility and structural collapse.

⇥ The indigenous tax and the precolonial food system

Before the constitution of the Bolivian Republic, the agrarian structure and production regime did not change substantially. Although the Crown had disarranged the native food system, it ratified and respected the survival of the indigenous communities in the highlands and valleys. It thus recognized the origin and identity of the indigenous communities, their individual and collective property and their political and organizational system (Antezana, 2006). One of the central elements for this decision was the indigenous tax (*tributo indígena*), which kept the Spanish authorities happy, as it was the main source of economic income. This system was complimented by the payment of labor in money, private property and the exchange of money instead of bartering. A mixed form of payment for labor was created.

It is important to note that the *tributo indígena* was paid both by the indigenous people who lived in the communities (known as *originarios*) and by those who worked on the haciendas (known as *yanaconas*). The *tributo indígena* and servitude were both abolished in 1825, with the constitution of the republic. But the tribute was reinstated in 1827, as it represented one of the main sources of income for the new state. This led to the development of new measures, among which was a single direct tax on property (rural and urban) and on personal income. Later, in 1831, the name was changed to "*contribución indigenal*," which remained in force until the end of the 19th century.

Moreover, and for the purposes of this chapter, it is important to ask ourselves about what the food system was like in pre-colonial times and how it operated. The answer to this question will help us to understand the social and cultural factors or barriers that have been considered an impediment to the creation of a modern productive system that overcomes the so-called primitive and anachronistic system of the indigenous peoples.

In this first section, we seek to characterize the precolonial food system as a starting point. That is to say, the objective is to understand how it was structured, the functional rationale behind its conception and its mode of operation. Within this framework, we will be able to pinpoint the changes that have been introduced through certain modernization policies imposed and developed by the governments of the time, which responded to particular interests.

Before describing the food system, however, we need to gain an understanding of the relative common components that can be identified with regard to the definition of what constitutes a food system. Taking into account the definitive components outlined by two organizations, the FAO and CIAT: a food system is composed of four components; the elements (the environment, population, resources, processes, institutions, and infrastructure), the activities (related to pre-production, production, processing, distribution, preparation, and consumption of food), the actors (research centers, producers, food processors, companies, service providers, and consumers) and the results of these activities (food security and nutrition, health status, socioeconomic growth, equity, and environmental sustainability).

Within this order of components, and considering that it dates back to pre-colonial times, this food system can be defined as a "vertically integrated" system. Before we go on, it is important to note that several of the characteristics of this system can still be found, particularly in the inter-Andean areas of the country.

So, what is meant or implied by the concept of a vertically integrated food system? This concept implies that the food system was, first and foremost, contingent upon the ecological and natural conditions of the environment where the productive activities were carried out (which are the elements). This means that agriculture was managed on the basis of the productive potential and ecological conditions of each of the ecosystems. These conditions also implied the development of technologies to carry out productive activities, ranging from tools for planting, to the construction of terraces, platforms (*takanas*) and irrigation systems, among others. The organization was based on *ayllus* and communities, where production was carried out communally on collective plots, called *aynokas,* that were rotated and rested from one year to another, and where each family had individual access to plots, granted by the Inca.

In terms of results, this system responded to several criteria, such as, minimizing the risk of losses, management of biodiversity based on ecological tiers, a stable flow and supply of food, and varied diets. In summary, studies have verified that food production and supply based on "vertical integration" and "archipelago" management (Murra, 1975, quoted by FIDA-CEDLA, 1985), ensured a stable food supply for the population. This management system also explains why, with no knowledge of the wheel and its use in transportation and as a concept for shortening distances, the food transportation system was not based around roads, but rather around trails on which indigenous people, and their llamas, transported their loads of food. The vertical management of ecological tiers was aimed at facilitating production in areas where, within short distances of less than 20 km, products from the puna and subtropical zones could be accessed. The need to obtain additional products from other regions further away implied the management of the concept of "archipelagos," i.e., territories that were not necessarily continuous, but over which they had access and management of production.

The handling and trading of products with the consumer was done through a system of exchange (bartering), for which a meeting point, known as a center or *taypi*, was required. This was the place where producers from the highlands and valleys came together. Thus, as a state, the Inca Empire managed a land distribution[1] system, and a production storage system in *pirwas*, as well as the administration of transformed products, such as chuño, tunta, caya, etc., for their conservation and consumption in times of drought, aspects that illustrate the organization and structure of the food system before colonization. It is also important to consider the significant structural changes that were introduced to the food system during colonization. The Spanish introduced livestock (cattle and sheep), agricultural crops (wheat, beans, peas, etc.), and a range of agricultural tools, the most important of which was the Egyptian plow. The latter, together with the oxen, represented a transcendental change in the way of preparing the land for crop sowing.

There was also a change in the power dynamic during this period, with the Inca rulers and their court no longer in charge of the direction, management and organization of the productive and food system, and the Spanish, representatives of the King of Spain, taking over the administration and management of the conquered lands.

---

1. Land distribution corresponded to three categories: the lands of the Sun, the lands of the Inca and the lands of the people or *ayllus*.

Within this framework, a food system was developed during the colonial period that can be qualified as "dual." On the one hand, there was the demand of a large number of workers in the mines, urban populations and cities, who began to be supplied by the production of the Spanish-owned haciendas, as well as significant quantities of food imported from the north of Chile and later from Argentina. On the other hand, the rest of the population, mostly rural indigenous communities, were reliant on traditional food systems that provided food on the basis of self-sufficiency with an essentially local character (FIDA-CEDLA, 1985).

In summary, at these two junctures, a food system existed that was the product of an authoritarian system, the "Inca empire," as some call it, but which provided crops with a high nutritional value for a population that, according to the chroniclers, was well fed and that developed a system according to their needs and according to that point of historical experience. That is not to say it was a perfect system, what we are trying to show is that it was a food system, also under construction, that was disrupted, and on which the Republic and, subsequently, the Plurinational State, has been constituted and is trying to change and adapt under a new national outlook and vision. It is, however, the pre-colonial and colonial system that directly conflicts with the new vision of the State and the production system that the Mestizos and Creoles who took over the reins of the new country wanted to build. It is worth stressing this point once again, as it forms the crux of the struggles, uprisings and revolutions that have taken place and explains the tensions generated in relation to the agricultural policies enacted by the State and the indigenous communities; and between the new actors in the agricultural sector and the indigenous, peasant and native peoples.

▸▸ Agricultural policy in the 19th century and the development of the haciendas

With the constitution of the Republic, several historians and political scientists, including Mesa (2003), are united in the belief that the founders of the country were unaware of or did not understand the social, organizational, cultural and population context on which the foundations of the new republic were based. For the Creoles and Mestizos, who assumed power and control over the country's destiny, the indigenous population represented an obstacle to their class interests and modernization objectives and, as such, did not recognize them as citizens.[2] This is one of the reasons behind the fierce indigenous opposition to a proposal that wanted to put an end to feudal exploitation and create a modern, fairer system that would promote the elevation of productive forces, but was a project that had failed to consider them.

The new nation that was to be built implied a change in the productive and food structure that had been inherited. The new actor (political and food system), formed part of the 5% of the population with the power and right to elect, be elected, and,

___

2. A Bolivian was a person born in the country but, in order to become a citizen, certain requirements had to be met, such as being male, over 21 years of age or married, able to read and write, have a job or trade and have a minimum annual income. Only around 50,000 of Bolivia's one million inhabitants were citizens.

therefore, had the option of accessing power to enact public and agrarian policies. It was the manager of these proposals for change that allowed these citizens to enact policies that shaped a new institutionality and structure aimed at favoring certain actors, who would ultimately become the direct beneficiaries themselves.

The non-recognition of more than 90% of the population as citizens led, throughout the 19th century, to the fact that, under policies that threatened communal land ownership, the indigenous peoples assumed their strategies of opposition and defense against these governmental measures and the ongoing lobbying of the State in search of recognition of their communal land rights and their forms of production.

In this new scenario, and referring back to one of the previously mentioned components of the food system, the environmental element, in addition to the Altiplano and valleys areas contemplated within the "vertical integration" system, the lowland region of Santa Cruz was now also included as a new area with a high potential for food production. This is very important detail, as it constituted the basis of the lands where the new food system would be established, which would emerge at the end of the 20th century and be consolidated at the start of the 21st century.

This historical backdrop highlights the fact that agrarian policies throughout the 19th century were focused on the elimination of collective property, communal grazing lands (*Ayjaderos*), and the entire system of production and management of tracts of land on various ecological tiers. The objective of moving towards a modern capitalist production system, characterized by individual ownership and the concentration of land in favor of the haciendas, has been contentious, involving dispossession secured by military force and the legal dispositions of governments that protected the interests of the large landowners. This explains the principal land dispossession policies enforced, which ranged from granting a 60-day term to process their land titles and forcing them to pay, to laws that prohibit the use of the name, community or *ayllu* (Valenzuela, 2008; Soux, 1999).

In fact, these measures allowed the purchase of land by private parties, which led to the consolidation and expansion of the haciendas. The claim that the lands in the hands of the indigenous people were not productive and were not being fully exploited was the focus of the debate at that time, in a country that was seeking modernization. The orientation sought by means of these policies is clear, as land expropriation laws, dictated by the conservative governments, were subsequently continued under the liberal governments. The "review[3] of native lands and cadastral provisions and legal norms on vacant lands" laws serve as a good example of this continuity (Demeure, 1999, p. 270).

Another element that forms part of the change, in contrast to the situation in the Andean region, is the emergence and scope of food production in the Bolivian lowlands. In this area, the so-called forest communities lived isolated from society and without a production system linked to markets. Under the vacant lands (*tierras baldías*) law, land was given to people from the Andean region. These actions of expansion and subjugation were interpreted by the State as a new territorial conquest.

---

3. The purpose of the review was to confirm the rights of ownership on an individual basis through the inspection of documents and land.

The pressure of the ranchers on the indigenous lands, manifested through violence, sparked an indigenous uprising that was repressed by the army, ending in what is known as the *Kuruyuki* massacre, in which thousands of indigenous people died.

Essentially, it is possible to state that, as a result of the policies developed and the influence exerted by the relevant power groups, by the end of the 19th century, the expansion and consolidation of the haciendas had modified the colonial food system. The Creole landowners took control of the commercialization to the population centers. The traditional local markets disappeared almost completely, and food security became extremely precarious for the rural population during the hacienda period (FIDA-CEDLA, 1985).

▸▸ Haciendas, agrarian and agricultural policy in the 20th century

The 20th century, in historical terms of agricultural and rural development, can be divided into two eras. The first, running from 1880 to the agrarian revolution of 1953, is characterized by the continuity of measures aimed at consolidating the hacienda regime, which included colonization and forced labor under the *pongueaje* system (Demeure, 1999). The second, characterized by the impacts of the agrarian reform and the expansion of the agricultural frontier to develop modern commercial agriculture in the lowlands, known as the "march to the east."

Until the first half of the 20th century, only a few haciendas could be considered agricultural enterprises. Commercial production came from smallholdings, where more than 20% of the food was imported to cover the food requirements of the cities and mining centers. Meanwhile, community production was almost entirely destined for self-sufficiency and bartering. Data from the 1950 agricultural census paints a picture of a very precarious, disjointed and low-productivity production system (Demeure, 1999).

The agrarian reform of 1953 represented a significant historical milestone, as it marked a turning point in land and food production. Under the slogan "the land belongs to those who work it," the Reform Law Decree established a new land ownership regime, abolished the system of unpaid peasant labor services, recognized communal lands and returned the lands to the indigenous people, which led to the disappearance of the haciendas in the western part of the country. While it is true that the demands of the indigenous peoples were recognized, this resolution did away with everything that had been developing for 125 years in terms of establishing the haciendas as the engine of the production model and the basis of the food system.

The dissolution of the large estates meant that production ceased or was only produced to supply the peasant communities. This led to food shortages in the cities (FIDA-CEDLA, 1985). Breaking this structure, in fact, generated a new system of food production, since peasants went from being semi-slaves to being owners of their land and, as they were no longer tied and obliged to hand over their produce to the haciendas, they became potential food producers for the domestic market. In other words, they were given the role of substituting the haciendas, of supplying food to the domestic market.

The national economic and inflationary crisis of 1956 led to the implementation of policies, such as the removal of price controls and subsidies for the production of consumer goods. It was in this context that food aid and donations started to be supplied, with the support of the American Point Four and PL-480 programs. This aid restricted the prospects and objectives of the Revolution, making it dependent on the American model and the financiers, which, in the long run, resulted in insufficient focus being given to how to solve economic and development problems. For its part, Bolivia entered into this type of aid, which corresponds to what McMichael (2016) calls the second food regime model,[4] based on the donation of food from countries that have achieved an excess of production, and therefore supposedly can feed the world, within the framework of a new model of trade and cooperation.

Within the framework of "the march to the east" slogan, a program called "Plan Bohan" was developed in 1941, one of the mainstays of which was the support and promotion of intensive agricultural development. Among its goals was to achieve self-sufficiency, increase the population's consumption levels, promote import substitution and the development of tropical products for export. Significant resources generated by mining and hydrocarbons were used to finance these objectives (Demeure, 1999; Albarracin, 2015). Between 1960 and 1970, the vast majority of public and private investments were made to stimulate the establishment of large commercial farmers and an agro-industrial complex in the Santa Cruz region. The government strategy sought to substitute agricultural imports and promote exports. This was to the detriment of support for staple food production in the Altiplano and Valle regions.

Returning to idea of food system components, one of the factors that was ultimately crucial to finally affecting radical change in the pre-colonial food system, of vertical integration, was the road connection and the construction of the Santa Cruz-Cochabamba-La Paz highway. This connection enabled intermediaries to reach the communities and village fairs. In this way, bartering and the sale of food changed the previous rationale drastically. Producers began to trade their products with middlemen and peddlers, not only for food (sugar, rice, noodles), but also for other products, such as batteries, radios, school supplies, canned food and others. This new system generated the change from a rationale of vertical exchange to one of "horizontal exchange" that, in fact, represented the incorporation of rural peasants into the commercial market circuits of products and, in short, the beginning of a change in their eating habits and the substitution of traditional foodstuffs.

It is important to consider, once again, that the western zone, which appeared to benefit most from the agrarian reform measures, ended up being the one that lost out in this whole process. The ineffectiveness of the policies to implement the necessary components to support an agrarian reform generated agricultural regression and stagnation. As can be seen in figure 7.1, the cultivated area in the eastern lowlands, where the modernization model was developed, expanded, with the western region left out of this model. Paradoxically, despite this situation, the highlands and valleys are the country's main areas of food supply for domestic consumption and, as such, are, ultimately, ensuring food security for the nation.

---

4. McMichael (2016) talks about three food regimes: i) that of the colonies, as food providers; ii) that of food donation, marked mainly by the presence of the United States, and iii) that of corporations or transnationals.

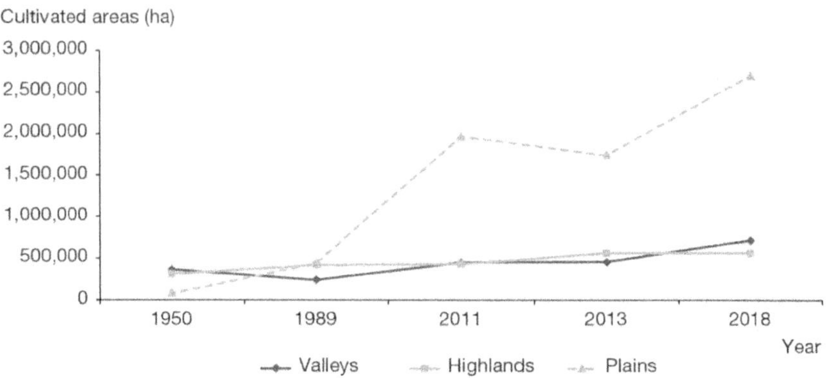

Figure 7.1. Increase in cultivated area by region (in ha) in Bolivia. Source: Data updated based on Albarracin (2015).

This situation demonstrates a contradiction in the decisions taken, reflecting the positioning of food security as a secondary factor within this modernization scheme. The orientation and prioritization of support for productive development is focused on reaching external markets. The implementation of import substitution policies and policies to promote foreign trade were those that directed the strategies towards the development of a strong, modern agricultural system, capable of supplying the required food (Zeballos, n.d.).

One of the consequences of the agrarian reform generated over the years with regard to the structure of the food system was the establishment of smallholdings. The absence of mechanisms to determine and establish a rational exploitation, with sufficient and/or collective size, as demanded by the communities, made it impossible to increase levels of productivity and adapt to the Aymara and Quechua collective work system (Mesa, 2003). This situation that represents production systems with low yields and high costs, makes peasant economies inefficient, with high production costs and uncompetitive in relation to imported products. In recent years, this situation has caused the negative trade balance to grow and, as a result, has led to the displacement of farmers to the cities in search of alternative livelihoods, with government policies failing to generate outcomes to reverse the situation.

In relation to the point above, it should be noted that IFAD's concept of food system, outlined in its 1985 study on peasant economies in Bolivia, states: "It is that which has the capacity to generate a sufficient domestic food supply (through national production and imports) to meet increased demand and basic food needs" (IFAD, 1985, p. 21), thereby, emphasizing the fact that food supply can be both domestic and imported, which leads us to reflect once again on the debate on dependence and vulnerability that may be generated. Dependence, in the sense that the country depends on an external factor, the availability and prices of food products from the exporting country; and vulnerability or food sovereignty, in the sense that, in the event of climatic phenomena, social or political turmoil or bad relations, the country loses all domestic capacity to confront the situation. As such, and from a nationalist point of view, the constitution of a food system with these characteristics only compromises the country's food security and sovereignty.

Similarly, in operational terms, the following five criteria are set out for the situational analysis of the food system: i) Sufficiency; ii) Dependable/Reliability; iii) Autonomy and self-determination; iv) Long-term stability (environmental); and v) Equity. IFAD's analysis of these criteria indicates that they form the basis of the problems identified and are a priority for the development of policies and actions to be implemented by governments through agricultural policies in their different spheres (IFAD, 1985). Be that as it may, these criteria have not been used as a decision-making tool in plans, strategies or programs.

Finally, although the agricultural and agrarian policies of the 20th century manifest processes of progress, setbacks and contradictions in relation to the agrarian structure and modernization of rural areas, it is the agro-industrial and agribusiness model based on the expansion of the agricultural frontier, particularly in the lowlands of Bolivia, which ends up imposing itself as the development imaginary in Bolivia, displacing, in terms of both importance and contribution to the gross production value, the peasant economies established in the lands of the highlands and valleys (see figure 7.2). The decrease in the contribution of peasant production to production value should be duly noted, as it not only reflects the preponderance of modern commercial agriculture, which is highly mechanized and oriented towards foreign trade, but also shows that peasant agriculture is not growing, thereby implying a process of de-peasantization towards urban centers or a migration towards rural labor markets in the lowlands, which is also one of the factors leading to fires and deforestation.

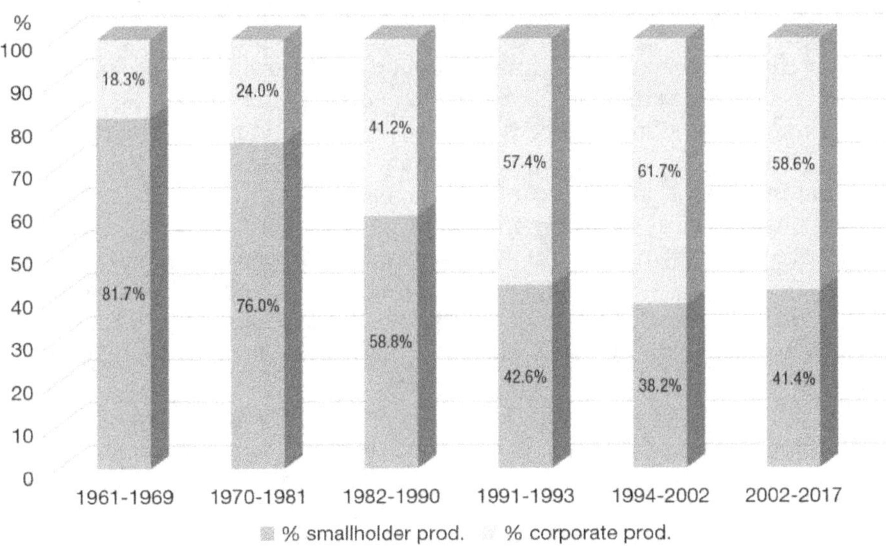

Figure 7.2. Contribution of peasant and entrepreneurial production to Gross Production Value in Bolivia. Source: Based on Zeballos (2006) and BDP Complexities Map (2020).

▶▶ Agricultural policy in the 21st century

The first 20 years of the 21st century have represented a period of important change in terms of development policies. In 2006, the neoliberal policies that had governed the country since 1984 were profoundly questioned and changed under the government of Evo Morales. State intervention with regard to the economy is the clearest policy example highlighting the orientation of the new government.

With respect to the agricultural sector, the government enacted Law 3545 "Redirection of the agrarian reform," seeking to achieve three objectives: i) transformation of the unequal agrarian structure and eradication of unproductive large estates; ii) providing communities and new human settlements with fiscal lands; and iii) titling of Native Community Lands (TCOs). According to Fundación Tierra (2020)[5], this reform started very well but ended very badly. Between 2006 and 2009, more than 7.5 million hectares were titled in favor of 130 TCOs. Between 2010 and 2011, due to the conflict over the TIPNIS, the government clashed with the indigenous peoples, due to its policy of expansion of the agricultural frontier and the non-recognition of the rights of indigenous peoples to consultation, suspending all the titling that was in progress. During the "Sowing Bolivia" agricultural summit in 2015, the MAS government aligned itself more closely with the agribusinesses by agreeing to extend the agricultural frontier in favor of large entrepreneurs and to the detriment of the forests. Small farmers ended up definitively omitted from the productive model scheme supported by the government and part of the supposed political alliance union with the government (Fundación Tierra, 2020). Behind these decisions is the imposition of large economic interests that form part of the agribusiness production model that, ultimately, prevail in the eyes of a State in need of fresh sources of economic income, due to the fall in prices of raw materials in the world market.

A characteristic of the new generation of public and agricultural policies in the first seven years of the government is the development of a series of policies very similar to those implemented in the process of import substitution industrialization (ISI), which occurred in the second half of the twentieth century, with the State seeking to resume the role of entrepreneur in the management and control of markets. In line with this, a strategy of intervention in the product market and food system was implemented with the creation of several public food companies (EMAPA, EBA, LACTEOS BOL and others), which carried out the tasks of purchase, sale, storage and marketing of food products and inputs. One of the largest companies, EMAPA, had the function of price control and food supply and, as such, the company focused on managing the market for wheat, rice, corn and other products of importance to the food basket and family diet. The company worked under a strategy of direct subsidies for producers, with a 15% higher payment on the purchase of each ton of their production, as well as the provision of inputs, credit at a zero interest rate and technical assistance, while the consumer also received a subsidy, since they could buy the products at a price of less than 10% of the market price.

One of the issues that led to the closure and liquidation of state-owned companies in the neoliberal era, and which is again part of the debate, was the discretionary, clientelistic management and sustainability of these businesses. Let us return to EMAPA as an example. This company suffered from serious operational problems and the loss of its initial incorporation capital, due to the significant subsidies

---

5. Fundación Tierra's video for Agrarian Reform Day (July 2020).

awarded. According to Prudencio (2017), between 2008 and 2015, the company generated a revenue 4.2 times higher than the initial capital, but spent 4.8 times more. Moreover, its mode of operation was also called into question, as, ultimately, its subsidy policy did not reach the supposed beneficiary, the small producer, with payments rather going to the entity carrying out the commercialization. Another indirect effect was that it led to the disbanding of the few producers' organizations that it was supposed to strengthen, as the company negotiated directly with the producers and not through their organizations, substituting them in several operations, such as the provision of inputs. In short, EMAPA never became an efficient operational instrument for intervention policies, such as: temporary regulations, application of subsidies, export bans, imports of staple foods to supply the population and direct food sales, in order to avoid price hikes (regulatory agent) (Prudencio, 2017; Baudoin & Albarracin, 2014).

Government intervention in trade policies, such as the control of exports through quotas for certain products (soybeans, corn and sugar), established in 2007, was unsuccessful. This control, which initially helped to confront the world food crisis of 2008/2009, had a negative effect in subsequent years, as it discouraged planting in subsequent agricultural campaigns, leading to a shortage of food, and forcing the government to import them.

Moreover, the results of these experiences throw up an important question. As Baudoin and Albarracin (2014, p. 68) put it: "the delimitation of the roles and functions of each of the actors in the production chain is unclear. In other words, how far are state-owned companies expected to intervene, and how should they be supported? As a strategic partner, as a promoter and strengthener of producers' associations? Moreover, the question still remains: have the transformation and commercialization inequalities been reduced?

The specific policy proposals that seek to address the problems of food security are presented in table 7.1. We can see that, out of a total of 13 development plans and strategies reviewed, only 6 of them deal with the issue, and these have been developed at the end of the 20th century and are concentrated in the 21st century, with the incorporation of the issue of food sovereignty. Also in this period, despite not being named in the policies, great emphasis has been placed on recognizing the Right to Food, a concept introduced and acknowledged in the Political Constitution of the State (CPE) approved in 2009 (Article 16, Paragraph I).

Notable among the programs implemented under the umbrella of food security that also played an important role is PASA (Food Security Support Program), which allocated a large part of its resources to the construction of infrastructure, mainly bridges and roads, leaving technical support, training and infrastructure, such as silos, irrigation, etc. as a secondary priority. In the mid-1990s, the SINSAAT (National Food Security Monitoring and Early Warning System) was designed and implemented, which subordinated food security activities to early warnings and the effects of natural disasters. In 1998, producers' organizations presented a proposal for a "Food Security Policy in Bolivia," which was rejected by ministries and prefectures simply because the proposal had been tabled by peasant organizations. In 2001, the "Nutritional Improvement and Promotion of Local Production" Bill was presented to National Congress.

Table 7.1. Policies related to food security and food sovereignty in Bolivia's strategies and pl[

Course of action	PIPEGRN 1955-62	NDP 1962-1071	ESEDN 1971–1991	PQA 1976–1980	PNRD 1984–1987	EDES 1989–2000	END 1992–2000
Productive infrastructure							1
Supply and storage							
Food security		1				1	1
Food sovereignty							
Environmental sovereignty with integral development							
Productive sovereignty with diversification and integrated development							
Food Aid							1
Food production							
Response to food emergencies							
Assurance of food supply to the population							
Assurance of adequate food and nutritional status							
Agricultural health and food safety services							
Transformation of Production and Food Patterns							
Total policies	**0**	**1**	**0**	**0**	**0**	**1**	**3**

Programs of the laws enacted by MAS. Source: Albarracin (2015).

As of 2008, there was a radical change across all the programs that had been implemented, with the approval of the PRRAyF Plan (2008–2012) and within the Rural Revolution framework, specifically within the policy for the transformation of production and food patterns, as a result of which a number of programs were implemented, which are shown in figure 7.3.

The final milestone that we encounter in this policy path is the incorporation of the concept of malnutrition, which not only refers to the classic concept of malnutrition, but also incorporates the issues of obesity and malnutrition. In 2007, a proposal was made to implement a food security program from a multisectoral perspective.

	ETPA 1996–2000	NDP 1999–2004	ENDAR 2002–2007	PND 2006–2011	PRRAyF 2007–2012	LRPCA N 144 of 2011	LMadre Tierra 2012	LAPARB n 337 OF 2013	13 PA Pillars 2013–20225	Total no. of times
		1	1							3
						1				1
		1		1	1		−	−		6
						1	−	−	1	2
					1					1
									1	1
								1		1
						1				1
						1	−			1
						1	−			1
						1				1
					1		−			1
	0	**2**	**1**	**1**	**3**	**6**	**0**	**1**	**2**	**18**

The Multisectoral Zero Malnutrition Program (PMDC)[6] sought to implement policies to eradicate malnutrition with a multisectoral approach. The program was initially created to solve nutrition problems in the poorest municipalities. It later incorporated the problems of obesity and malnutrition into its activities, as these

---

6. The management of resources was entrusted to the National Council for Food and Nutrition (CONAN/PMDC), a body created with the main objective of coordinating actions of the PMDC/Health at the national level. The National Council for Food and Nutrition (CONAN) is chaired by the President of the Plurinational State and is made up of 10 ministries (Health, Education, Development Planning, Environment and Water, Economy and Public Finance, Justice, Productive Development and Plural Economy, Rural Development and Land) and representatives of Social Organizations and Civil Society.

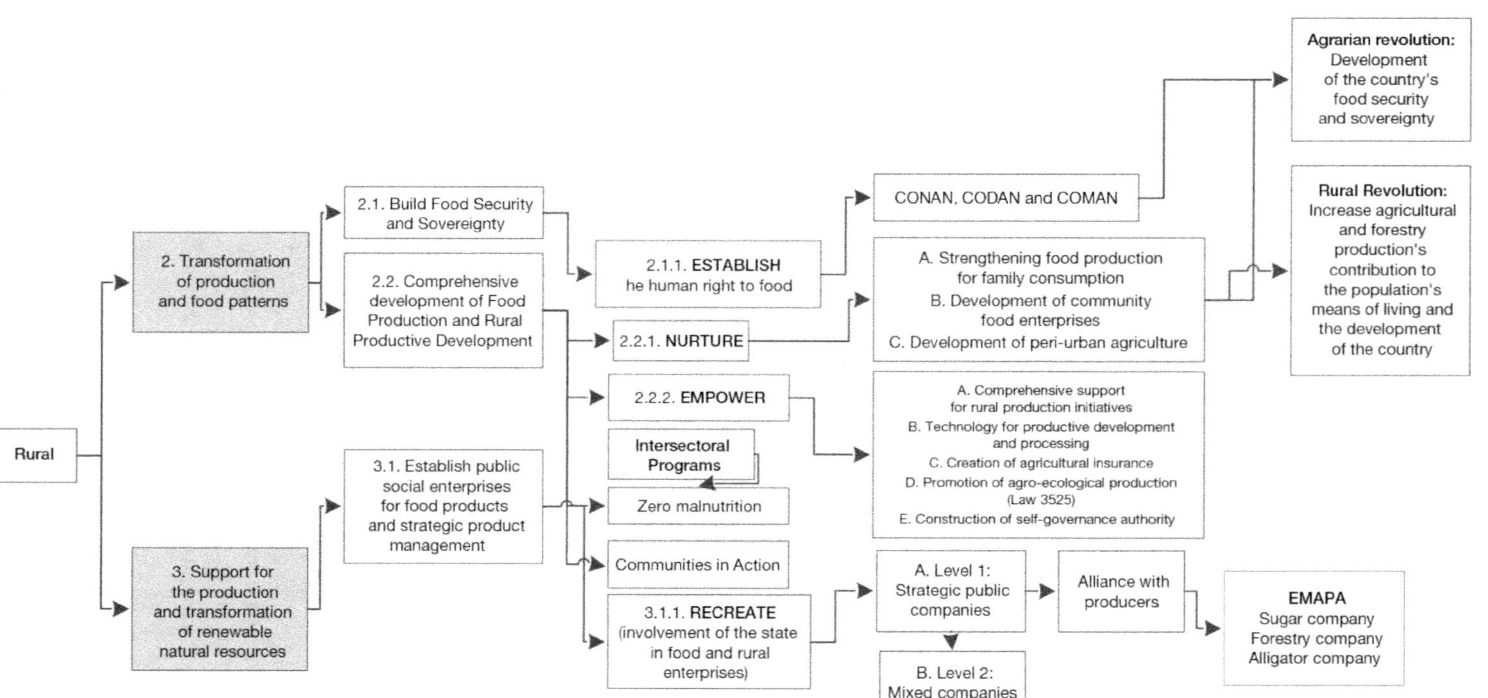

Figure 7.3. Plan for the Rural, Agrarian and Forestry Revolution 2008–2012. Source: Albarracin (2015).

have become a serious national problem, particular in the main cities. Among the problems it faced, was that the program was installed in the Ministry of Health, which automatically generated a sectorial vision, with other ministries assuming a remote position, making it impossible or very difficult to develop an integral project. Although theoretically and conceptually, we start by indicating that food security is multisectoral, this experience once again demonstrates that in practice and in the field, particularly within the institutions in charge of operating them, administrative biases and sectoral jealousies still prevail, preventing their implementation.

In this overview of policies and programs, as well as in the way food security is dealt with, we have identified the implementation of five definitive approaches. The first, the second and the fifth are seen as operational changes that occurred sequentially over time (figure 7.4).

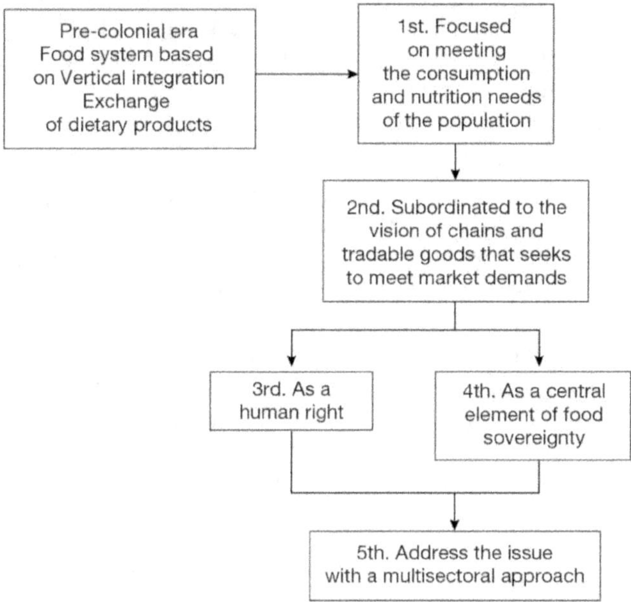

Figure 7.4. Shift in approach to food security.

▸▸ Institutional actors and cooptation

In support of the thesis being undertaken, this section briefly outlines the fact that the structure and changes in the Bolivian food system are the result of the intervention of a group of actors who have had access to power that has been used to develop a set of policies that have directly favored their own interests and sectarian vision, over and against the majority of the native indigenous population. In fact, generically speaking, two very clear groups of actors have historically been constituted according to the orientation of their demands and positions. To achieve this objective, we have used the social network approach in the field of public policies (*policy network*), with which we analyze and present some of the facts supplied in the institutions and sectorial policies.

The persistent pressure from the business actors we have been observing responded to two motives: on the one hand, the accelerated process of complexification and diversification of the public agenda; and, on the other, the growing mobilization and influence of private interests in the definition of policies (Porras, 2001). The importance of the national revolution of 1952 is that:

> (...) this process represented the displacement of classes at decision-making level within the government and society as a whole. The tiny ruling class that ran the country was replaced by a middle class (...) that also had a significant impact on the interests of the elites through the expropriation of large mines and large estates. The emergence of peasant farming in agriculture (...) radically modified the power structures (Mesa, 2003, p. 651).

As a result, the economic structure that had been built up over time was modified. The transition from a semi-feudal economy to an economy controlled by the State was not an obstacle to the interests of the elite groups. These groups advocated for the change in power relations between the social classes, which was generated in agriculture, should be limited solely to the highlands and valleys. Leaving the lowlands to constitute a new bourgeoisie that would serve as the engine of the country's modernization model.

With this objective, the military governments, which would be in power for 18 years (1964–1978), gave continuity to the measures associated with the "march to the east" and the revolution under the concept of nationalism, although, in practice, they supported the developmentalist model by supporting the most powerful sectors. During these dictatorships, the largest land concessions were carried out constituting large[7] and medium-sized enterprises (see table 7.2). This is a reflection of what Urioste (1988, p. 16) indicates: "There is, in fact, a direct relationship between the dictatorships and the emergence of neo-latifundism."

Table 7.2. Land concessions during successive agrarian reform governments in Bolivia

Government	Period	Hectares	%
MNR	1952–64	6,000,000	19.6
Barrientos	1065–69	4,000,000	12.5
Banzer	1971–78	17,800,000	57.0

Source: Urioste (1988).

For its part, the Eastern Agricultural Chamber (CAO), which had little significance until the seventies, through its alliance with the civil-military government, began to have a growing role and a high degree of influence and direct control in the centers of state decision making, becoming, as Porras points out:

> (...) a determining factor in the process of formation and implementation of agricultural policy, displacing the hitherto hegemonic peasant organizations and unions that represented the interests of western agriculture and were closely linked organically to the Revolutionary Nationalist Movement (MNR). The influence of the CAO and its affiliated entities in the Ministry of Peasant and Agricultural Affairs

---

7. Although only 7% of the land on the large estate was cultivated (Urioste, 1988).

(MACA) was accentuated by the wide degree of autonomy with which this institution exercised its competencies and employed its important resources in the sector. (Porras, 2001, p. 726).

The intra-organizational network built by the CAO with its affiliated entities (Fegasacruz, ANAPO and ADEPA)[8] allowed it to maintain its decisive role in policy development and execution. Referring to the power of the network built by the CAO, Porras states:

> (…) the degree of control exercised by the CAO over the ministry continues to be extensive. Proof of this is the appointment of the former manager of the Santa Cruz union for eight years as Deputy Minister of Agriculture and Livestock (VAG), the executive position with the greatest resources and competencies in the ministry (Porras, 2001, p. 741).

In the 21st century, especially during the early years of the MAS government, the influence exercised by the CAO over the Ministry of Agriculture was overturned and was even threatened by the emergence of new groups and organizations representing indigenous communities (Porras, 2001), which called for resources and policies to be oriented towards their sector.

In order to see the influential power of these groups, we need to review the positions held by the MAS government during its 14 years in power, which can be divided into three periods. The first, extending from 2006 to 2009, was characterized by the State participation and representation of small peasant, indigenous and middle-class producers. With regard to the land issue, the change in the tenure structure was mainly the result of direct confrontation with landowners and businessmen in eastern Bolivia, in favor of large TCO titling (see figure 7.5). For some analysts, these titles did not represent a substantial change reflected in an increased volume of production on the part of the beneficiary actors (peasant, indigenous and inter-cultural communities); on the contrary, they have highlighted the emergence of a new type of unproductive latifundia, this time, with a peasant-indigenous base. The graph gives the impression that the lands managed by the agribusinessmen were affected and reduced, which was not, in fact, the case and demonstrates, in essence, the pact made between the government and the businessmen.

The second period, which runs from the approval of the CPE (Political Constitution of the State) in 2009 to 2015, the year of the agricultural summit. During this period, we can see the most severe governmental attrition as it attempts to impose its proposals that were steadily being undermined by a national and international system that was gradually demonstrating to the MAS the existence of agreements, international standards and an economic system that operate on the basis of certain rules that must be complied with.

The third period, running from the agricultural summit in 2015 to date, is characterized by a clear, evident, and patent double discourse on the part of the government. On the one hand, the protection of mother earth and small peasant and indigenous producers, and, on the other, comprehensive support for the business sector regarding the expansion of the agricultural frontier, the use of transgenics and the production of biofuels.

---

8. Fegasacruz, Santa Cruz Livestock Farming Federation, ANAPO: Association of Oilseed and Wheat Producers, ADEPA: National Cotton Producer Association.

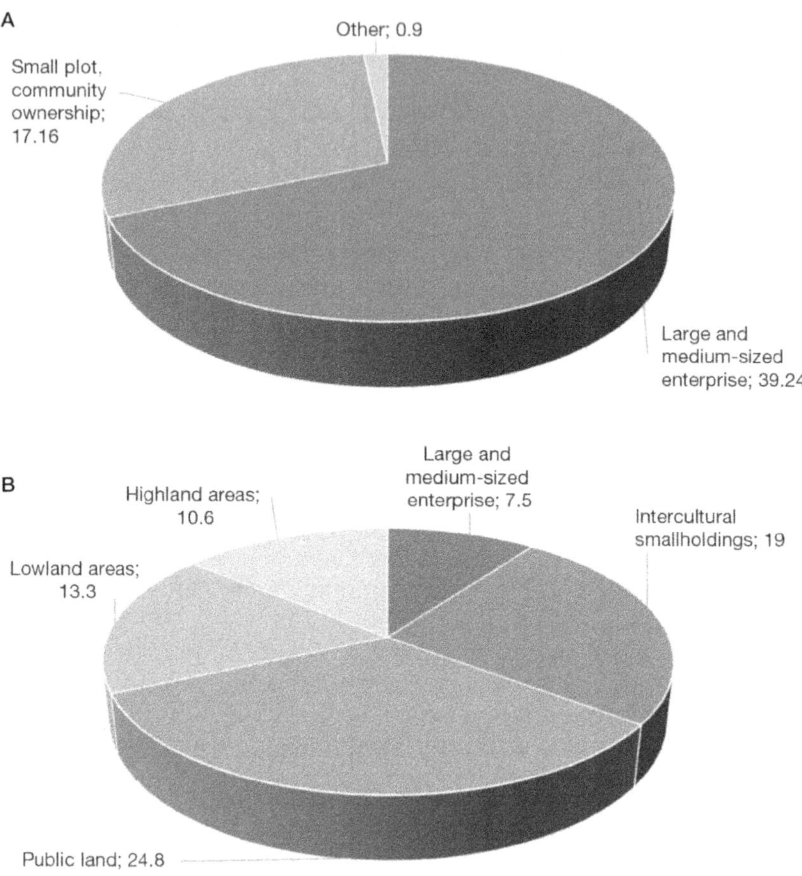

Figure 7.5. Changes in the structure of land in Bolivia. Source: Based on data from INRA (2015).

(A) Land tenure position 1953-1992 (in millions of hectares). (B) Land tenure position 1996-2015 (in millions of hectares).

The total abandonment of any efforts to strengthen the domestic market and peasant agriculture is also evident, reflected in the increased importation of horticultural crops. Figures show that between 2005 and 2015, food imports increased from US$ 242.3 million to US$ 797 million. This situation is compounded by smuggling, which increased not only in quantity and value, but also in variety and origin (Chile, Argentina, Peru, Paraguay and Brazil).

In 2020, under the transition government, the Santa Cruz business sector again took control of the public institutions in charge of managing the country's economy, land, forests and agricultural funds[9]. During this short period, they have managed, even with the pandemic, to extend more than 800 land titles; they are pushing for new

9. The businessman and former president of the Pro Santa Committee appointed as Minister of Development Planning on 5 August 2020, was managing director of Industrias Oleaginosas A.S. (OIL) and also a shareholder and director of the Banco Económico until 2010.

soybean, corn, sugar and transgenic cotton events to be authorized, fast-tracking the authorization processes, even though prohibited by the constitution. This is nothing new, as, under the previous government, they successfully persuaded them to issue regulations that promoted and allowed agricultural expansion, burning and deforestation. In fact, these regulations were responsible for the fires that destroyed 1.7 million hectares of the Bolivian Chiquitano forest. In this case, the policies and demands respond more to the economic interests of individual agribusinessmen than to those intent on the promotion of actions and investments to benefit food production and ensure food security and sovereignty.

▶▶ Conclusions

Throughout history, governments and the actors in charge of them have played a decisive role in the efforts to modernize the agricultural sector. The most striking fact, however, is that these efforts have been carried out without any recognition of the structure and social base on which the country was created, and therefore, the modernization proposals have conflicted with the rural reality of the country. The quest to implement a private property system of production, by force, gave rise to the non-recognition and, therefore, constant subjugation of the communal property of the indigenous peoples. Even the agrarian reform, which recognized their traditional property, was insufficient, as it lacked a comprehensive understanding of production systems and shared and collective land ownership, especially in the Andean area.

The agreements reached at the agricultural summit held in 2015 between the government and the agribusiness sector of Santa Cruz indicate an important shift in the vision of the government and its agricultural policies. This new public-private relationship is shaping the construction of a food system focused on the production of products and foods that follow the structure of the agro-industrial and agribusiness model. Under these agreements, Bolivia automatically signs up to the role assigned to South America in the global context, of being the suppliers of food raw materials for the world.

Agricultural policies aimed at the development of a model that promotes monoproduction are leading to a displacement of staple crops, abandoning productive diversification, healthy food, and traditional crops rich in nutrients. The mistake potentially being made by prioritizing investments and the development of public policies directed towards some actors to the detriment of other actors and regions, shows that Bolivia is ending up destroying its production systems that supply the basic foods for the family basket. Food security not only corresponds to having a guaranteed abundance of a few basic products, but the consumption of a diversity of foods that provide the nutrients and essential elements that every organism needs for the normal development of its activities. The fact that Bolivia is opting to produce an agrifood matrix of products that are largely part of a food industry that is generating issues associated with malnutrition, through the consumption of junk food and ultra-processed foods, is a huge mistake.

It has not been possible to structure a stable agrifood system given the constant struggle between the actors, peasants and entrepreneurs, where the latter have sought to remove the former and structure a production system according to their own interests. This has led to a system of agro-industrial crop production focused on supplying foreign demand. Moreover, there has also been an increase in the importation of food for the domestic market due to the abandonment and ongoing displacement of peasant family agriculture. It still remains to be seen whether this model has managed to generate a social benefit and a contribution to development, or whether the relevant costs outweigh the individual benefits.

The debate on production strategies focused on moving from traditional production systems, characterized as inefficient and low productivity, to modern, efficient, and highly competitive production systems linked to international markets. Despite all the support given to the business sector, the reflection of the national agricultural crisis shows that agriculture in the east is facing serious problems in terms of yields, productivity and the destruction of the country's forested areas and making very little contribution to national food security.

The analysis shows that agriculture in the Andean zones and valleys continues to be neglected by the State. Food production capacities are decreasing all the time, thereby increasing the reliance on imported basic foodstuffs for the family food basket, which is aggravating the highly fragile food security dynamic already in place.

The creation and diversification of markets is a fundamental aspect of any modernization effort. The Bolivian market was very small until roughly the second half of the 20th century, with low purchasing power and diets based on the consumption of basic products with little added value. This meant that the agricultural development strategies focused on international markets, primary products, with low value added and no product differentiation, which, in turn, makes products extremely vulnerable with regard to commercialization policies and prices, marketing techniques and changes in consumer tastes, aspects that can have a negative and even devastating effect on the production model the country has opted for and that is failing to account for the major food changes that are taking place worldwide.

The experience of the state-owned companies shows that an evaluation is needed that, firstly, allows us to see their relevance, and then, to think about delimiting their areas of intervention, their modes of operation, other roles, which jointly link the State with the private sector, the economy and the market.

One issue that has not been addressed and that remains pending is the quantification of the contribution of urban or peri-urban agriculture to national and family food security, to the development of employment, as a livelihood strategy and its impact in terms of the national accounts.

▸▸ References

Albarracin, J. (2015). *Estrategias y planes de desarrollo agropecuario en Bolivia: la construcción de la ruta de desarrollo sectorial (1942–2013)*. CIDES-UMA, Plural Editores. http://www.cides.edu.bo/index.php/component/content/article/24-publicaciones/coleccion-30-anos/106-estrategias-y-planes-de-desarrollo-agropecuario-en-bolivia

Antezana, L. (2006). *La política agraria en la primera etapa nacional*. Plural Editores.

Baudoin, A. & Albarracin, J. (Eds.) (2014). *Las empresas públicas de alimentos: avances, retrocesos y desafíos* (Publicaciones del proyecto Mercados Campesinos, no. 2). Agrónomos y Veterinarios Sin Fronteras. https://www.ssl.avsf.org/public/posts/1705/las_empresas_publicas_de_alimentos_bolivia_avsf_2014.pdf

Demeures, J. (1999). Agricultura. In F. Campero Prudencio (Ed.), *Bolivia en el siglo XX: La formación de la Bolivia contemporánea* (pp. 269–290). Editorial Offset Boliviana.

FIDA-CEDLA (Fondo Internacional de Desarrollo Agrícola, Centro de Estudios para el Desarrollo, Laboral y Agrario). (1985). *Propuesta para una estrategia de desarrollo rural de base campesina: informe de la misión especial de programación para la República de Bolivia. Volumen I.* CEDLA.

Fundacion Tierra. (2020, July 31). *2 de agosto declive de una fecha histórica* [Video]. YouTube. https://www.youtube.com/watch?reload=9&v=cQCc8xTMj6s

INRA (Instituto Nacional de Reforma Agraria). (2015). *La estructura de tenencia de tierras está cambiando, la revolución agraria avanza.* INRA. http://www.inra.gob.bo/InraPb/upload/SEPARATA%20 2015.pdf;jsessionid=A34D9CB0F74704EC9284C238D9E81A7B

McMichael, P. (2016). *Regímenes alimentarios y cuestiones agrarias* (G. Colque, Trans.). Fundación Tierra.

Mesa, J., Gisbert, T., & Mesa, C. (2003). *Historia de Bolivia* (5th ed.). GISBERT.

Porras, J. (2001). Policy network o red de políticas públicas: Una introducción a su metodología de investigación. *Estudios Sociológicos, 19*(3), 721–745. https://www.jstor.org/stable/40420687

Prudencio J. (2017). *El sistema agroalimentario en Bolivia y su impacto en la alimentación y nutrición (Análisis de situación 2005–2015).* https://www.derechoalimentacion.org/sites/default/files/pdf-documentos/El%20sistema%20agroalimentario%20en%20Bolivia%202005%202015.pdf

Soux, M. L. (1999). Del latifundio a la reforma agraria. *Historia y Cultura. Sociedad Boliviana de historia, 25.*

Urioste, Miguel. (1988). *Segunda reforma agraria: campesinos, tierra y educación popular* (Talleres CEDLA, no. 1). CEDLA. https://cedla.org/publicaciones/prya/segunda-reforma-agraria-campesinos-tierra-y-educacion-popular/

Valenzuela, C. (2008). *Tierra y territorio en Bolivia.* CEDIB. https://www.cedib.org/wp-content/uploads/2009/06/Tierra-Territorio.pdf

Zeballos, H. (2006). *Agricultura y desarrollo sostenible.* Plural Editores.

Zeballos, H., & Quiroga, E. (2010). *Bolivia: avances en la economía campesina.* Mimeo.

Chapter 8

Public policies and the food system in Chile

MICHEL LEPORATI NÉRON, PABLO VILLALOBOS MATELUNA

▸▸ Introduction

This chapter presents an institutional, political and economic analysis of the public policies that have shaped the Chilean food system from its beginnings to the present day. The first section discusses the ways social protection policies focused on food and nutrition have evolved, with two sub-sections that go into further detail on sectoral policies for social protection and healthy eating, and policies intended to strengthen and modernize the food safety and quality management system. The second section analyzes changes in agricultural development and production promotion policies and the development of the food industry. The third section discusses the performance of the Chilean food system. This chapter concludes by discussing various present and future challenges.

Chile has a strong tradition of food production. The country's current food system is the result of a long history of public policies and solid institutional development. Over time, Chile's efforts in the areas of public health, animal and plant health, social protection, agriculture and trade have contributed to ensuring a high level of food security to the country's people. These efforts have also boosted the development and economic growth of the country, helping it to rank among the world's leaders in the international food trade.

With regards to public health, Chile has been able to eradicate malnutrition, under-nutrition and hunger (FAO, 2013[1]); lower the incidence of foodborne diseases (WHO, 2015); and create a regulation at the international forefront in terms of nutritional quality, to address the problems of overweight and obesity that afflict the national population today. As for the agricultural sector, Chile boasts a strong phytosanitary and zoosanitary status, which has been key to preventing production, zoonotic and food diseases as well as opened the doors of global markets to the country's food exports.

From an economic and productive point of view, Chile's food industry is the second largest sector of the national economy, accounting for 23% of the total value of

---

1. According to the FAO (2013), Chile has the lowest rate of chronic child malnutrition (2%) and the lowest level of overall malnutrition in children of all countries in the region (0.5% prevalence).

exports, 20% of the value of retail sales, 31% of the number of companies, 23% of national employment and 10% to 12% of GDP. Chilean food exports reach more than 150 countries and about 63% of the global population and 86% of the world's GDP (ProChile, 2019). The export basket is made up of more than 1,500 products; Chile ranks in the top ten world exporters for around 50 of those products. Similarly, the value of exports, to date around US$ 18 billion, places Chile as one of the 15 leading food exporters in the world.

Around the world, the demand for food will only continue to grow and people increasingly want healthy, natural, sophisticated and accessible food. Nearly 900 million people in the world suffer from malnutrition and hunger, yet paradoxically, overweight and obesity have become the major public health challenge of the 21st century. Accordingly, Chile is committed to a food system approach that seeks to diversify and modernize its food supply. A focus on safe and healthy food is a virtuous mechanism that can stimulate the competitive development of the industry, while addressing the food security needs of Chile's own population as well as those of the markets to which it exports.

Undoubtedly, these approaches and strategies may need to be reconsidered in light of the causes and effects of the serious crises over the last two years on food security: first, the social upheaval in October 2019 (Waissbluth, 2020), followed by the Covid-19 pandemic that began in March 2020. Although the food system has proven to be resilient and robust in both situations – evident in the operational continuity of supply chains and the adequacy and adjustments of supplementary meal programs, school meals and emergency food distribution programs – according to the State of Food Security and Nutrition in the World report (UN, 2020), an estimated 11.8% of the Chilean population suffers from moderate food insecurity and 3.8% from severe food insecurity. Meanwhile, the study indicates that the prevalence of overweight and obesity remains above the regional average, with no change with respect to the previous report.

These social and health crises, when viewed within the context of medium-term sustainability and beyond consumer preferences or market trends, show that the Chilean food system faces major challenges. These challenges include food production and trade, public health and welfare, environmental sustainability of the industry (circular economy and green technologies), territorial equity and social inclusion in the processes of production, processing, distribution and consumption, national food culture, food sovereignty, and more. All these issues will undoubtedly be part of the discussion that will take place in the forthcoming constituent process, and will be discussed to some extent in this chapter.

▸▸ Evolution of social protection, health and food safety policies

Public policies on food, nutrition and health have been around for more than a century in Chile, beginning in the early 20th century. The aim was to address the socioeconomic and health problems of malnutrition and high infant mortality, food distribution, and nutritional health of different groups, based on how vulnerable they were. The main focus has been on obesity and noncommunicable diseases, as well as helping groups with special nutritional needs. The following is a synthesis of the historical evolution of these public policies and their implementation.

The state's participation in guaranteeing or protecting the nutritional health of the neediest population began under the government of President Arturo Alessandri Palma (1920–1925), with the creation of the Ministry of Hygiene, Social Assistance and Welfare and the enactment of mandatory workers' insurance (Act No. 4.054/1924). This initiative created a paradigm shift in the approach to solutions to social problems, from one centered on private charity[2] and emphasis on individual freedom to one that advocated science-based, planned actions with greater State participation (Ahumada, 2018). For the first time, this act enshrined in public policy medical and social care for women during pregnancy, childbirth and the postpartum period and medical care for children up to eight months, including the delivery of milk (MINSAL, 2010).

Subsequently, in 1937, the free distribution of milk to all children under age two began. This initiative was further strengthened with the enactment of the Mother-Child Act (Act No. 6.236/1938), linking the distribution of food to health control, and with the creation in 1952 of the National Health Service (Act No. 10.383/1952), which merged several units dedicated to public health into a single health authority. The new National Health Service was responsible for health protection for the entire population and the promotion and recovery of the health of workers, wives and children up to age 15 (Ahumada, 2018).

In 1954, the National Supplementary Feeding Program (PNAC) was created. The PNAC ran continuously for more than half a century with the purpose of helping maintain and improve the nutritional status of the population, evolve and adapt to demographic and epidemiological changes in the population, and generate new knowledge on food and nutrition. The universal program included various nutritional support activities for prevention and recovery, through which food was distributed to vulnerable or at-risk populations. When it began, it was the main strategic tool to control malnutrition and infant mortality, extending the benefits of supplementary food to pregnant women, wet nurses and preschool children up to age six. In 1956, under the PNAC, breast-feeding promotional campaigns began, and under the government of Salvador Allende Gossens (1970–1973), the half-liter of milk" campaign was implemented, providing each child with half a liter of milk daily through the primary healthcare service up to eight months of age, which was later extended to 14 years of age.

In 1974, during the military dictatorship (1973–1990), the National Food and Nutrition Council was created (Decree Law no. 354), a collegial body chaired by the Ministry of Health and composed of the Ministers of Economy, Education, Agriculture and Labor, the Vice President of the Chilean economic development agency (CORFO), and the Directors General of Health and of the National Commission for Scientific and Technological Research (CONICYT), and the President of the Council of Rectors.

Four years later, in 1978, the PNAC was divided into two programs, according to the target populations and their different nutritional and food needs. The "basic" PNAC provided food to children with a normal nutritional status, while

---

2. The first precursors of food programs in Chile were private initiatives run by charities. In 1912, the *La Gota de Leche* ("drop of milk") program was created to distribute milk to the most deprived and malnourished children and mothers.

the "reinforcement" PNAC provided food to undernourished children, pregnant women, wet nurses and their families. The ideas that guided this program formed the basis of the Complementary Feeding Program for Older Adults (*Programa de Alimentación Complementaria del Adulto Mayor* – PACAM), created in 1999, which distributes food fortified with micronutrients to the elderly in primary care facilities to help prevent and treat nutritional deficiencies, maintain or improve physical and mental capacities, and detect and control risk factors for their loss.

In 1996, the Ministry for Health established a food sanitary regulation (Decree no. 977). This instrument sets out the sanitary conditions for the production, importation, processing, packaging, storage, distribution and sale of food for human use to ensure healthy, nutritional and safe food. The regulation applies to all natural or legal persons related to or involved in the aforementioned activities, as well as to establishments, means of transport and distribution for such purposes. In other words, the food sanitary regulation determines specific requirements that facilities and foodstuffs must meet to guarantee safe products for human consumption.

In 1997, with a view to promoting healthy eating among the population, the Ministry of Health asked the Institute of Nutrition and Food Technology (*Instituto de Tecnología de los Alimentos* – INTA) at the University of Chile to prepare food guidelines, with the food pyramid as the official benchmark (Olivares et al., 2013). Health policies on food and nutrition at this time began to shift their focus from undernutrition to overnutrition, due to the rise of overweight and obesity problems in the population, especially among young people and children. New comprehensive policies were developed to promote healthy lifestyles and combat overweight, obesity and chronic or communicable diseases.

In 2003, PNAC coverage was extended to population groups with special nutritional needs, such as extreme premature infants and children under 18 with diseases due to inborn metabolic issues. In 2008, the program's benefits were extended to pregnant women with phenylketonuria to ensure normal fetal development, and the Program on Healthy Eating and Physical Activity (*Programa de Alimentación Saludable y Actividad Física* – PASAF) was created. This primary care program, for the obese and overweight aimed to help reduce the prevalence of obesity-related chronic diseases and improve lifestyles and the nutritional status of the overnourished population. Beneficiaries received aid through primary health care network and other establishments approved by the Ministry of Health, where they could also have access to other health prevention and treatment activities.

In 2006, under the administration of President Michele Bachelet Jeria (2006–2010), the Ministry of Health implemented the intersectoral Global Strategy against Obesity, (*Estrategia Global Contra la Obesidad* – EGO-Chile). This national strategy (based on the recommendations of the WHO's Global Strategy on Diet, Physical Activity and Health, adopted in 2004 by the 57th World Health Assembly), was an attempt to address rising obesity rates in the Chilean population, as evidenced in the 2003 National Health Survey. The EGO-Chile initiative recognized the multidimensional nature of the sociocultural factors involved in obesity, and sought to address them through public policies, plans, and multisectoral and intersectoral programs. The Ministry of Health's dietary guidelines were also revised around this time and updated in the "Eating for a healthier life" guide.

In 2007, EGO-Chile was expanded with the EGO School Strategy (*Estrategia EGO Escuela*), with a focus on nutrition and physical activity in schools. This strategy was promoted as an intervention aimed at structurally transforming school environments to help students adopt healthy behaviors. In addition to EGO-Chile, the Ministry of Health worked with the primary healthcare network to implement the Nutritional Intervention Strategy through the life cycle to prevent obesity and other noncommunicable diseases (*Estrategia de Intervención Nutricional a través del Ciclo Vital para la Prevención de Obesidad y otras Enfermedades no Transmisibles* – EINCV. Both the EGO School Strategy and EINCV took an integrated approach to nutrition, food and physical activity in the various health programs. They are part of the country's official health objectives and efforts to prevent chronic non-communicable diseases due to their impact on population morbidity and mortality.

These processes have evolved over time, giving rise to the following strategies: Healthy Cities, Towns and Communities (*Municipios, comunas y comunidades saludables*), Health-Promoting Workplaces (*Lugares de trabajo promotores de salud*), and Health-Promoting Higher Institutions (*Instituciones superiores promotoras de salud*). These strategies were initially focused on communication, and have subsequently informed public policies in local governments. Similarly, in 2011, the Ministry of Health again worked through the primary healthcare system to implement the Healthy Life Program (*Programa Vida Sana*). By promoting healthy eating and physical activity, this program aims to reduce the incidence of diabetes and hypertension, and it provides nutritional, physical and psychological support to users.

In 2012, the Ministry of Health's Act No. 20.606 on food nutrition and advertising was approved. This act represented a significant advance in terms of improving the regulatory framework for food, updating it to take into account public health challenges and issues reflected in the alarming levels of overweight and obesity in the population.[3] The main achievements of the act included the introduction of "high in" labeling for sugar, fat, sodium and calories (for which the act is probably best known) as well as requiring more exhaustive and precise labeling on the use of additives, breast milk substitutes and allergenic ingredients (soy, milk, peanuts, eggs, shellfish, fish, gluten and nuts).

The act also established restrictions on the use of food marketing strategies and advertising such as promotions and loss leaders aimed at children under 14, prohibited the sale of unhealthy foods in schools, and implemented educational and physical activities, for all ages and in all learning conditions, to help children develop healthy eating habits and warn them about the harmful effects of an unhealthy diet. Finally, a mandatory nutritional monitoring system for schoolchildren at all levels was established.

The spirit of this law is to address public health problems associated with overweight and obesity. In particular, it seeks to protect those most at risk in the population by better informing consumers about the true nature, composition and quality of food. It also aims to restrict consumers' access to unhealthy foods, prohibit advertising

---

3. In 2015, Act No. 20.606 was complemented by Act No. 20.869 on food advertising, which regulated the advertising of breast milk substitutes and established a tax levy on sugary soft drinks in the tax reform under discussion.

unrelated to the actual promotion of the product or which is misleading or uses means that take advantage of the gullibility of consumers. The law also attempts to discourage fraud and encourage people to develop healthier eating habits along with a food supply that can meet those needs.[4]

In 2017, the Ministry of Health published its National Nutrition and Food Policy, an inclusive and intersectoral policy that presents a modern focus on human rights, social determinants of health, the mainstreaming of health in all policies, gender and interculturality. This policy establishes two guiding principles: i) the right to food as an ethical principle to be included when developing programs, initiatives and standards, and ii) the social determination of food and nutrition, with people's consumption decisions being shaped by their living conditions. The purpose of this policy is to help improve the state of health and quality of life of Chileans in terms of food and nutrition.

Policies for social protection and healthy eating

Together with health policies, other institutions, through their policies, plans and social protection programs, have played a decisive role in meeting the objectives of reducing malnutrition and, more recently, in combating overnutrition (overweight and obesity) in Chile. Thus, in addition to implementing the PNAC as a pillar of nutrition and food assistance policies, under the government of President Eduardo Frei Montalva (1964–1970), the National Board for Student Aid and Scholarships (*Junta Nacional de Auxilio Escolar y Becas* – JUNAEB) was created under the Ministry of Public Education.

The JUNAEB is still active today, and is in charge of implementing the School Feeding Program (*Programa de Alimentación Escolar* – PAE). The School Feeding Program delivers daily food services (breakfasts, lunches, snacks and dinners as appropriate) during the school year to students from preschool age to adults in vulnerable conditions in municipal and subsidized private educational establishments. The aim is to improve their attendance and help prevent school dropouts.

In 2013, the Ministry of Social Development implemented the Choose to Live Healthily program (*Elije Vivir Sano*, Act No. 20.670) to encourage healthy habits and lifestyles and prevent and reduce risk factors and behaviors associated with non-communicable diseases. The policies, plans and programs that are part of the program are prepared and executed by the different bodies of the State Administration on a single or multi-sector basis, according to their respective competences, and are coordinated by the *Elige Vivir Sano* Executive Secretariat.

In 2016, high childhood obesity rates led the JUNAEB to implement the *Plan Contrapeso*, a set of 50 measures to modify the School Feeding Program policies and guidelines and provide targeted support to vulnerable students with greater exposure to the risk of overweight and obesity. To this end, two commissions of expert advisors were formed: one on obesity and another to strengthen the School

---

4. National food guidelines were also reviewed, and the Eating for a healthier life guide was replaced by the current food guides for the population. Instead of the food pyramid, the new guides use a new image that promotes variety and portion control.

Feeding Program. The commission on obesity included representatives from governmental and non-governmental sectors, such as the ministries of education, health, and sport; the FAO; the PAHO; CORFO; INTA; and civil society actors, among others. Its focus was on proposing measures in education, nutrition, physical activity and healthy living. The second commission was made up of representatives from governmental and non-governmental sectors, such as the ministries of education, economy and general secretariat of the presidency; CORFO; former directors of the JUNAEB, Institute of Agricultural Development (INDAP); National Board of Children's Gardens (JUNJI); and the INTEGRA foundation, with expertise in public policy in education and food services. This commission aimed to put forward proposals based on the School Feeding model, relating to components such as fines, technology, management, supervision, financial, business, supplier development and decentralization. Professionals from the fields of education, social sciences, agronomy and economics were involved.

Policies for strengthening and modernizing the food safety and quality management system

In 2005, under the administration of President Ricardo Lagos Escobar (2000–2006), a process was initiated to strengthen the institutional framework for food safety and quality assurance. The initiative was based on the principles of the new international food safety and quality paradigms in place since the end of the 20th century. These principles are centered on the concept of food chains, integrated action and institutions, a conceptual and methodological framework based on risk analysis and explicit recognition of the role and responsibilities of public and private actors and citizens in this area.

Against this backdrop, the Chilean Food Safety and Quality Agency (ACHIPIA, Supreme Decree 83/2005) was created, along with a presidential advisory commission to advise the Chilean president on all matters related to the identification, formulation and execution of policies, plans, programs, measures and other activities related to food quality and safety. A national food safety and quality system was also established to coordinate the various agencies involved in these matters. Specifically, this commission was entrusted with formulating a food safety policy and developing a proposal to modernize the relevant institutional framework.

To satisfy this demand, in 2009, Chile enacted its first national food safety policy. One of the policy's main characteristics is its interdisciplinary and intersectoral approach to issues on production, marketing and public health in relation to food consumption. The ministries of health, general secretariat of the presidency, agriculture, and foreign affairs; public services specialized in animal, plant and aquaculture health, public health and foodborne diseases, and international trade; and actors from academia and civil society participated in preparing the policy through a public consultation process.

In 2011, the Chilean Food Safety and Quality Agency was transferred from the Ministry General Secretariat of the Presidency to the Ministry of Agriculture (Supreme Decree 162/2011). This move provided the agency with infrastructure, equipment and permanent staff, along with new responsibilities, including the area of food quality. The agency was also tasked with coordinating the Codex Alimentarius at the national level, bringing the secretariat into the agency and establishing it as the point of contact for communication and coordination between Chile and the Codex Alimentarius Commission.

In 2018, the national food safety and quality policy update was published, with a 2030 timeline (ACHIPIA, 2018). This policy, currently in force, was designed to ensure the safety and quality of food produced, processed and marketed in Chile and intended for human consumption. The aim is to protect people's health and consumer rights and develop the competitiveness and exports of the food industry, within the framework of the national food safety and quality program.

Currently, the Chilean Food Safety and Quality Agency is a collegial body that reports to the undersecretaries of public health, agriculture, fisheries and the general secretariat of the presidency. These entities meet through a council to which the director of the agriculture and livestock department, the director of the aquaculture and fisheries department, and a representative from the undersecretariat of economic relations of the Ministry of Foreign Affairs are also permanently invited. From a technical point of view, the agency operates as a public agency that coordinates, liaises with and advises intersectoral players. Administratively, it reports to the Undersecretariat of Agriculture.

Among its many activities, the agency strives to improve the availability and access of the population, especially the most vulnerable and disadvantaged, to safe and quality food. It also promotes the competitiveness and sustainability of the food industry and trade, advances a national culture of food safety and quality, supports the establishment of adequate food environments, and stimulates citizen participation in these activities. In addition, the agency encourages the use of raw materials and Chile's biodiversity by promoting them in healthy foods, supports the development of human capital, capabilities and skills, and works to strengthen the country's image as a supplier of safe and healthy foods for domestic and foreign consumption.

▸▸ Evolution of agricultural development and production promotion policies, and the configuration of the food industry

Unlike the historical evolution of institutional policies in social protection and health, the development of the agrifood sector over the last 150 years in Chile has gone through several phases. These phases have often been contradictory. When the country first became an independent nation, it adopted free mercantilism approaches promoting international trade, which led the country to develop its agrifood sector. At other times, state intervention was strong, such as the periods following the saltpeter crisis and the Great Depression in the 1930s. During these periods, Chile implemented protectionist policies and import substitution as a strategy for industrial development. The state also intervened during the agrarian reform process that transformed the productive and social structure of the rural world, as well as when it implemented an extreme neoliberal social economic model during the civil-military dictatorship.

After these many turbulent periods in the political and institutional course of events, marked by deep underlying ideological conflicts, a sufficient level of consensus has been reached over the last thirty years to allow for a new strategic approach. Chile's leaders have taken advantage of the country's international position, with Chile as a relevant player in international food markets, to consolidate the food sector as driver of development and economic growth on the same level as mining was previously.

After gaining its independence, Chile began a process of growing involvement in the world economy. However, it remained fundamentally a rural society with an agrarian economy for much of the 19th century. The earliest available statistics show a rural population of 71% of the national total. Agricultural labor force participation was 42%, by far the highest level in the country, and there were only six major urban centers, each with more than 10,000 inhabitants. The agricultural sector was made up of large estates (known as latifundias), characterized by low land and labor productivity and, associated with them, a myriad of smallholdings. Around 86% of the rural population did not own land, which created a labor force of tenants working in semi-slavery regimes for the owners of the large estates. This was a socioeconomic and political power system, similar to the European manorial system, with economic relations of mutual dependence between landowners and peasants and a hierarchical and coercive structure. This structure remained relatively unchanged until the mid-20th century, when it began undergoing a transformation to a capitalist-type agricultural system (Llorca et al., 2017).

With regard to agrarian policy, in 1831, under the government of President José Joaquín Prieto (1831–1836), a sectoral tax was levied on the agricultural land registry. This tax was part of a major fiscal reform and replaced the old colonial forms of tax collection. This important fiscal modernization, beyond contributing to the direct collection of taxes, marked the beginning of agricultural censuses, which in time would become fundamental tools for agrarian policy management. It is because of this and subsequent censuses that we now have early information on the evolution and economic, technological and productive development of the agricultural sector.

The information reported by the land registries revealed the sector's productive inefficiency, which was caused by the excessive concentration of land ownership, insufficient financing and public infrastructure for roads and irrigation, technological backwardness, a lack of economic and commercial incentives to intensify production, underqualified workers, and the conditions of misery and neglect in which farm workers lived in rural Chile, and inadequate regulations to stimulate the sector's development. By 1840, this was already a cause for concern among the farmers themselves, as reported in documents of the Society of Agriculture, as well as by foreign experts who visited the country at the time.

Despite the generalized backwardness of the agricultural sector, towards the middle of the century, during the so-called "first great expansionary cycle of the national economy," the increase in international demand for primary products had an impact on part of the sector. This increased demand, which followed the Napoleonic Wars and the gold rushes in California and Australia, drove the development of a fledgling food industry linked to the production and export of wheat and flour (Riveros & Ferraro, 1985). To protect this development, import duties on wheat and flour were established in 1834 and, in 1840, export taxes on these products were abolished. This activity, together with silver and copper mining, was the main source of economic growth between 1840 and 1860.

Beginning in the 1850s, under the government of President Manuel Montt Torres (1851–1861), the institutional transformation of Chile was consolidated. The country transitioned from a colonial system to a republican and capitalist one. The export boom of the period led to economic growth and modernized the country, reflected in the development of railroads, telegraphs and steamships, in the expansion of cities and ports, the emergence of a new entrepreneurial class, and the arrival of immigrants, which significantly boosted the economic and technological development of the agrifood export sector. From 1861 to 1871, under the government of President José Joaquín Pérez, a liberal policy began that facilitated scientific and technological progress in agriculture, bringing with it imported machinery, inputs and new production techniques. These activities impacted the development of the food processing industry. According to census data around 1870, of the total number of registered industrial businesses (85), 48.2% (a total of 41) were food factories. In 1884, there were 485 manufacturing facilities, of which 33.1% (152) were food processing plants (Sofofa, 1983).

The war conflict between Chile and the Peruvian-Bolivian Confederation between 1879 and 1884 represented a strong industrial stimulus for the economy in general and the food sector in particular. At first, the growing demand for food stemmed from the need to feed an army of 25,000 men in combat, and was later due to the annexation of the saltpeter regions of Tarapacá and Antofagasta as well as urban expansion.

Although the export of nitrate stimulated the economy by creating considerable domestic demand that, to a large extent, was satisfied by national production, during this second period of economic growth the national currency suffered a severe devaluation due to the constant fluctuations of international markets. As a result, the final decades of the century were characterized by a shift towards a more protectionist economy, promoted by the demands of the private sector through its two major representative organizations, the National Society of Agriculture (*Sociedad Nacional de Agricultura* – SNA) and the Society of Industrial Development (*Sociedad de Fomento Fabril* – SOFOFA). These organizations advocated for direct state intervention to protect and stimulate industrial development given the instabilities of external markets. These actions were the first indication of the subsequent import substitution policy that would be adopted in the mid-20th century.

In 1897, under the administration of President Federico Errázuriz Echaurren (1896–1901), a new customs law was enacted, setting high import duties for products manufactured in Chile. This measure had a decisive impact on the development of fruit and horticulture players in the agrifood industry in the central valley. In 1895, census data show a predominance of the food industry over other industrial sectors. Of the 1,089 manufacturing facilities in the census, 45% belonged to the food industry, accounting for 10% of Chile's GDP (Schneider, 1904).

During the first decades of the 20th century, thanks to the continuing saltpeter revenues, the Chilean state developed major institutional modernization programs and invested heavily in public infrastructure, roads, bridges, ports, railroads, irrigation, and irrigated agricultural land expansion. In turn, these activities supported the intensification and specialization of agriculture in the central valley between Aconcagua and Cachapoal, where production quickly exceeded domestic demand. This provided a new stimulus to the development of the canning and processing agroindustry and exports. During this period, Chilean canned peaches were recognized in practically all the countries of Latin America's Pacific coast. At the same time, wheat and wine production expanded (Bengoa, 1988).

Another aspect that boosted the development of Chile's food industry was the need to replace imports during the First World War. By the end of the war, the agricultural productive structure consisted of nearly 100,000 farms. Irrigated agricultural land had expanded to 1.1 million hectares, while orchards totaled 22,300 hectares and vineyards 56,200 hectares. Foodstuffs accounted for 40% of industrial production. During this period, the production of canned fruits, vegetables and legumes, dehydrated fruits and grape juice continued to be further expanded, more specialized and diversified. The first modern industrial poultry farms were established and the use of cold chambers for livestock products was developed. In 1923 there were 124 mills, 43 noodle factories, 20 fruit and vegetable canneries, 11 sugar refineries, 53 beef jerky factories, 19 chocolate and cookie factories, four flour factories, two condensed milk factories, six salt refineries and 12 factories producing fats. Meanwhile, the first food canneries were installed, and in the southern part of the country, the production of industrial canned meat and seafood began.

Towards the end of the 1920s, under President Carlos Ibañez del Campo (1927–1931), the economy entered into a severe recession due to the fall in saltpeter exports, which was replaced by the production of synthetic nitrate developed in Germany, and the global effects of the Great Depression that followed the 1929 stock market crash. During this period, the country experienced the decline of exports, the growing influence of urban groups and mining towns in political life, the global rise of Keynesian economic thought and the application of counter-cyclical policies, which reinforced the "inward development" paradigm founded on import substitution and strong protectionism (Rojas, 2007). In this context, import tariffs were raised to 35% for products produced in Chile and the Agricultural Credit Fund and the Industrial Credit Institute were created as investment promotion mechanisms.

In 1939, under President Pedro Aguirre Cerda (1939–1941), CORFO, the Chilean economic development agency, was created to promote a plan to foster national production by directly executing the actions to industrialize the country (Act No. 6.334/1939). CORFO's actions were instrumental in developing strategic sectors and creating large companies that were indispensable to Chile's development. These companies included the National Electricity Company (ENDESA), the National Petroleum Company (ENAP), the Pacific Steel Company (CAP) and the National Sugar Industry (IANSA).

At the same time, special development plans supported the agricultural sector through mechanized equipment exports, the expansion of irrigation works, and the development of new crops, commerce and transportation. These initiatives particularly favored the development and modernization of the canning and dehydrated products industry, which by the end of the 1960s numbered nearly 50 companies, especially throughout the central valley. Support for the industrial development of meat and dairy products also had a significant impact with the establishment of a national cold storage network.

Although the import substitution model had some success in terms of the domestic market, and CORFO's actions provided essential support for the development of strategic sectors, there were complex consequences for the competitive development of the overall productive sector. Modernizing national industry required the import of machinery, inputs and technology, but doing so did not generate the necessary foreign exchange, which came mainly from the export of traditional commodities. Thus, instead of adding value to national production, industrial development ended up dependent on the export of primary products. Meanwhile, protectionist measures served to sustain increasingly inefficient companies, with lower quality production and higher prices compared to international players. As a result, capital accumulated in just a few groups supported by these policies (Meller & Blomstrom, 1990).

At the agricultural level, plans were initiated for the development of beef and dairy cattle and the diversification of meat production and consumption to stimulate growth in the poultry and pork industry. However, the domestic development strategy encouraged the development of deficit crops, such as oilseeds, corn and rice. New price regulations, associated with the inflationary process, caused the agricultural sector to stagnate. There was a substantial transfer of capital from the agricultural sector to the urban industrial sector.

This serious crisis also highlighted once again the inefficiency and inequity of the latifundia system, which, according to studies at the time, showed one of the most unequal distributions of land and capital resources in the world. According to the 1955 census, 78% of the country's agricultural and irrigated land was concentrated in 7% of farms, while smallholdings accounted for less than 10% of the country's agricultural land (Gobierno de Chile, 1965).

As a result of all these problems, the production and productivity of the sector collapsed. In 10 years, the sector's contribution to GDP fell from 15% to 11%, causing a drop in agricultural employment and a growing wave of migration towards urban centers, with serious consequences of unemployment and poverty. The imbalance between population growth and declining food production became increasingly evident. While for more than two decades the population grew at an average annual rate of 2.1%, agricultural production barely reached a growth rate of 1.8% and was unable to meet domestic demand. As a result, the annual per capita growth rate of agricultural production was negative –0.4% for the period. To address this very real food crisis, food imports had to increase steadily, using about 30% of the scarce foreign exchange that the country generated, which in turn resulted in a trade balance deficit (Ministry of Agriculture, 1968). Even so, serious food deficiencies persisted in a significant proportion of the population (Bellisario, 2013).

The progressive deterioration of the agricultural sector was a topic of deep political and ideological debate during the 1960s. Although it is now admitted that before the agrarian reform, the process of transition from the latifundia system to a capitalist one had already begun the previous decade (Schejtman, 1971), it was far from being concluded. Moreover, it was even less a solution to the political, social and economic crisis represented by the national agricultural situation. Not only was the sector stagnating due to an unbalanced industrialization strategy, which, through price regulation, made agricultural activity less profitable and discouraged investment, many believed the latifundia/minifundia system was a main cause of backwardness, poverty, injustice and marginality of the rural world, not only in Chile but across Latin America. This view was associated with the demands of mass, political and social movements calling for the end of this system of subjugation and exploitation of workers, and which led to the agrarian reform processes that would soon emerge throughout the region.[5]

At the beginning of the 1960s, popular political parties inspired by Christian and Marxist principles reached significant levels of consensus to promote the transformation of the agrarian sector. Although the first Agrarian Reform Law (Law 15.020) was enacted in 1962 under Jorge Alessandri (1960–1964), it was limited to the possibility of distributing state lands and had a very minor effect on expectations and needs, which did not satisfy popular demands. However, it was this law that established the creation of the Agrarian Reform Corporation (*Corporación de Reforma Agraria* – CORA) and the Agricultural Development Institute (*Instituto de Desarrollo Agropecuario* – INDAP). Both institutions would play a relevant role in the expropriation and distribution of land and in the technical and credit assistance provided to the peasants who benefited from the reform processes of successive governments.

In 1964, under President Eduardo Frei Montalva (1964–1970), a process of structural transformation of political, economic, labor and social relations in the sector began. During this period, laws were passed on agrarian reform (Act No. 16.640/1967) and peasant unionization (Act No. 6.625/1967), which would definitively end the latifundia/minifundia system and lay the foundations for modern agriculture. The agrarian reform process was undoubtedly one of the main claims of the struggle for social justice and heavily influenced the Frei government's political agenda, as well as that of his successor, Salvador Allende Gossens (1970–1973). Allende had to contend with the increasingly radical demands of leftist groups and the reactions of the landed oligarchies.

Under Frei and Allende, over nine million hectares (5,800 estates) were expropriated, accounting for 59% of the agricultural land accumulated under the entire latifundia system. In 1973, the reformed sector was comprised of about 76,500 rural workers, who together with their families totaled 342,000 people.

In 1973, after the military coup that kept dictator Augusto Pinochet Ugarte in power (1973–1990), a counter-reform process was set in motion that lasted until 1980. The agrarian reform laws were repealed and a process of returning land to former owners was initiated; around 32% of the expropriated land was given back. By the end of the counter-reform, 47,000 peasant families were able to keep two-fifths of the expropriated lands, in the form of family farms (Rojas, 2007). Approximately 30,000 peasant families did not have access to land, nor to any other means of subsistence. In addition to state intervention in this process, many peasants, harassed by former owners or speculators, were forced to get rid of their land, fearing reprisals and receiving insignificant payments in exchange. Finally, much of the land expropriated from the peasantry was auctioned off at public sale, and some was transferred to public or non-profit institutions. Today, about 40% of the families that received land still own it (Belisario, 2013).

---

5. This movement was supported by the signing, in 1961, of the Alliance for Progress cooperation agreement between the United States and around twenty Latin American countries that were members of the Organization of American States. The agreement provided financial support for agrarian reform processes on the condition that they were liberal and democratic.

In economic matters, the import substitution strategy was replaced by the adoption, between 1974 and 1976, of neoliberal shock policies recommended by Milton Friedman and later, in 1979, reforms known as "seven modernizations." The economic and social model imposed during the civil-military dictatorship was characterized by the orthodoxy in the application of neoliberal postulates of the Chicago school. This approach translated into unilateral tariff liberalization as the hallmark of the anti-inflationary policy, financial system reforms, trade liberalization, reduced state intervention, and the minimization of state participation in the economy, limiting it to only a subsidiary and supervisory role. In this context, public agricultural services were particularly hard hit by these economic policies; they suffered severe budget and personnel cuts, the reduction of technical and financial assistance programs for peasant settlements and cooperatives, the privatization of production infrastructure and transformation of first- and second-degree cooperative administration.

However, the agrarian reform process, despite the dictatorship's efforts to dismantle it, radically restructured the territorial organization and agricultural property of the latifundia system. As a result of the subdivision of the former expropriated estates, tens of thousands of peasant farms were created, boosting development of the land market. The labor market was also liberalized with the end of the tenancy system. Together with policies to liberalize trade and attract investment, these initiatives stimulated the emergence of numerous agricultural enterprises and laid the foundations for a modern agrifood system. Although not without its difficulties and inequities, this new system was able to overcome structural deficiencies and become one of the key drivers of development and growth in the last three decades.

In 1980 and 1982, in the midst of a deep economic recession caused by an abrupt devaluation of the real exchange rate and the fall in exports, Chile experienced one of the most serious economic crises in its history. The agricultural sector suffered a contraction in production of around 3%. To address the crisis, the economic model was adjusted, initiating a more flexible approach to the hitherto ironclad neoliberal economic policy. At the sectoral level, adjustments included the establishment of price bands and renewed support for the sugar beet and oilseed industries, which had been previously been left to their fates. This attracted external and urban capital, allowing for a rapid recovery. As a result, between 1985 and 1990, the sector grew at an annual rate of 5.6%. By the end of the decade, sectoral exports reached US$2 billion with sectoral employment accounting for 18.7% of the national labor force.[6] This marked the beginning of a period of development and growth of the Chilean agricultural sector of non-traditional products, especially fruit and vegetable exports. Chile would quickly become a paradigmatic model of growth for developing countries in the context of globalization and trade integration (Ríos-Núñez, 2013).

At the same time, however, these actions led to a growing socioeconomic and productive sectoral duality. On the one hand, capitalist agriculture was concentrated in non-traditional crops and located in the most agriculturally valuable lands. The sector was made up of a small number of medium to large companies. They boasted intensive capital and technology capabilities and a capacity for differentiation and added value, with a strong focus on export markets. On the other hand, the sector also had a large number

---

6. The Chilean economic transformation between 1973–2003. Chilean Memory. National Library of Chile. http://www.memoriachilena.gob.cl/602/w3-article-719.html

of traditional, small-scale, peasant and indigenous family farms, with low capacity for added value. Technology use was rare on these labor-intensive farms, which had no access to financial capital. These subsistence farms were distributed across the country, located on land of lower agricultural quality, and focused on supplying local markets on the margins of agro-export development. The lack of public policies to support this segment caused a progressive deterioration and impoverishment of the rural sector, accentuating the duality in the sector. At the end of the civil-military dictatorship, the country had a rural poverty level of 39.5% (Barril Garcia, n.d.).

It was not until the early 1990s, with the return of democracy, that policies to include peasant and indigenous family farms in the agrifood development strategy were resumed. The INDAP then played a stronger role of providing technical and financial support and worked more closely with public institutions for productive development and technological innovation, human capital formation, financing, commercial insertion, and irrigation promotion, among others.

During the 1990s, under presidents Patricio Aylwin (1990–1994) and Eduardo Frei (1994–2000), thanks to the country's international reintegration, free trade agreements were signed with the leading world economies and strongly promoted the development of agricultural and food industry exports. One of the results of these agreements is the sustained increase in public spending, which in real terms more than tripled in the last ten years of the period. Approximately half of this expenditure was directed towards public goods aimed at correcting market imperfections and asymmetries and protecting the environment and the country's natural resources. The other half went towards direct or indirect subsidies aimed at improving technological progress and the commercial competitiveness of the poorest farmers.

The trade agreements allowed the agrifood sector to take full advantage of its comparative and competitive advantages. Among the country's comparative advantages are the geographical isolation that offers privileged phytosanitary and zoosanitary conditions, the Mediterranean climate that covers more than a thousand kilometers of the central valley, and Chile's southern hemispheric location that supports climacteric fruit production during the off-season of northern hemisphere markets. Competitive advantages include institutional strength, a high standard of phytosanitary and zoosanitary protection, the business community's export experience, adequate public and private infrastructure for production, logistics and transportation, and a strategic public-private alliance (Villalobos et al., 2006).

With the arrival of the new century, the agro-industrial export development strategy was consolidated as one of the pillars of economic growth. It is in this context that efforts were focused on making Chile a "food power,[7]" which impacted the sectoral policy agenda for several years. A foundational milestone of this period was the signing, under President Ricardo Lagos (2000–2006), of the EU-Chile Association Agreement, an innovative agreement political and economic association between the European Union and Chile that set an international precedent. The agrifood sector deeply benefited from the agreement, which helped consolidate the country as a relevant player in the international food market.

---

7. A concept coined by Chilean businessman Mario Montanari in the early 2000s and institutionally adopted by the Ministry of Agriculture by successive governments from 2006 onwards.

The main result of this agreement was the momentum given to modernizing agricultural activity. The sector shifted towards a systemic view of food chains under the "farm-to-table" concept, rather than the classic sectoral approach of public policies, emphasizing the achievement of international standards of quality and safety. The Ministry of Agriculture took the lead in the food sector and promoted various initiatives. Among them was the creation of the public-private Commission on Good Agricultural Practices (GAPs), which developed free and voluntary technical quality specifications for different products, setting a fundamental precedent for the development of quality assurance in the food chain in the following years; the modernization of the Agriculture and Livestock Department, which was given powers to ensure health and safety in export processes; and the implementation of the first health traceability system for cattle and the system of officially certified farms for the export of livestock products to the EU.

However, breakdowns in food safety in the developed world in the late 1990s resulted in new food safety control policies and instruments with a more prevention-focused approach based on risk assessment, management and communication. The Ministry of Agriculture took the lead in strengthening the national food safety and quality system, promoting the creation of the Presidential Advisory Commission on Food Safety. The aim was to coordinate and integrate the national food safety system, improving its performance by correcting duplications, overlaps, information asymmetries, gray areas and regulatory gaps between competent institutions, in a system strained by the different approaches of the ministries of health, agriculture and economy. Among the relevant milestones of this process was the delegation of official beef inspection for the national market from the Ministry of Health to the Agriculture and Livestock Department (Leporati et al., 2010).

The concept of Chile food power strategy, developed by Mario Montanari,[8] and Chilealimentos,[9] was institutionalized by the government of President Michelle Bachelet (2006–2010). The initiatives were included in her government program as a natural long-term strategy to shift the country's economy from one based on the exploitation of non-renewable resources to one based on renewable, diversified and high value-added natural resources. The focus moved from agro-export development to the exploitation of competitive advantages based on quality differentiation (Odepa, 2011).

During this period, a national food power council was created as a public-private body to coordinate policy actions. Its aim was to promote a strategic agenda that would address the institutional, regulatory and operational needs and gaps make Chile one of the top ten food exporting countries in the world. Chile's exports of agricultural products and food rose from US$8 billion annually at that time to US$17 billion by 2014. In addition, the need to take into account sustainable development (understood as economically efficient, environmentally friendly and socially equitable production, beyond commercial profitability), along with a long series of

---

8. Mario Montanari is a sociologist, businessman, vice-president of Invertec Agrofood SA, former director of INDAP IV Region, and former Undersecretary of Agriculture.

9. Chilealimentos is a private entity acting similar to a trade union, representing companies that bring together processed food companies and machinery, equipment and services companies involved in food processing.

impacts and externalities that have accumulated over time, will likely make this this strategy unviable if public policies are not implemented to avoid, mitigate and/or compensate them adequately (Odepa, 2011).

In line with this approach, a period began where policies increasingly took on an intersectoral dimension via interinstitutional coordination, including with regards to food and its development strategy, through a set of policies, plans and programs of different ministries and departments. In the area of productive development, Chile's Institute of Agricultural Development has promoted production links between peasant and indigenous farms and the agrifood industry and retail players; created the Flavors of the Countryside (*Sabores del Campo*) program and Rural Trade Fair (*Expo Mundo Rural*) to help rural agricultural food producers position themselves in the gourmet market segments in cities; implemented funds to support best agricultural practices and quality certification from CORFO, the AFC internationalization program promoted by INDAP and ProChile, and SERCOTEC's programs to support the competitive and commercial development of food SMEs.

In the area of R&D, programs that focus on healthy foods and promoting food heritage, along with support to develop marketing strategies for agrifood micro and SMEs, should be emphasized. In 2008, the aquaculture and food sectors were included when a policy promoted by CORFO was adopted to cluster strategic sectors to support their competitive development. A major achievement of this policy for the food sector was the creation of the International Centre of Excellence for the Food Industry at Wageningen UR Chile in association with local universities, marking the first large public R&D fund aimed at the sector.

In 2015, the Ministry of Economy, as part of the pro-growth agenda, worked through CORFO to create strategic smart specialization programs. These joint public-private actions aim to enhance the competitiveness of specific sectors and regions. Among these is the strategic food program called *Transforma Alimentos*, which offers a collaborative development model for the public and private sectors, academia, and scientific and technological players to promote innovation as a way to make the food industry more competitive. The program's objective is to consolidate Chile's international leadership as a food supplier. To this end, a prioritized roadmap was agreed upon and is being implemented. The aim is to relax the main regulatory, scientific, technological, capacity-limiting and other restrictions to create a better environment for the diversification and sophistication of Chile's food supply. Through the coordinating action of *Transforma Alimentos*, both the public supply in terms of productive promotion and science and technology has been organized, and greater coordination has been achieved between the sectoral regulatory authority and the productive promotion agencies, establishing coherence and synergies in public policies.

During this period, a range of intersectoral initiatives were implemented, such as the program of short supply chains and public food procurement between the Agricultural Development Institute and National Board of Student Aid and Scholarships. Other initiatives included healthy production agreements to reduce unhealthy foods produced by micro and SMEs in light of the entry into force of Act No. 20.606, promoted by the Chilean Food Safety and Quality Agency and the Chilean Agency of Sustainability and Climate Change, with financing from a strategic innovation fund. A public-private strategic plan, called From the Sea to My Table (*Del Mar a mi Mesa*), was promoted by the Undersecretariat for Fisheries and Aquaculture to encourage Chileans to eat more seafood products to support the development and sustainability of the national fisheries sector. Smart specialization programs for regional centers were promoted by the Foundation for Agrarian Innovation (*Fundación para la Innovación Agraria* – FIA). The Scientific and Technological Development Support Fund (FONDEF) issued a special call in conjunction with the Chilean Food Safety and Quality Agency for R&D projects on food safety and quality. Finally, the sector has been increasingly involved in instruments such as CORFO's Programs and Technology Centers and the regional competitive funds, among others.

A fundamental aspect of state modernization efforts was the strengthening of the institutional framework to address the growing food safety challenges and trends imposed by new global paradigms. These challenges included not only normative and regulatory issues but also aspects related to production promotion, social protection, and scientific and technological development. In this regard, in 2009, the first project to transform the Ministry of Agriculture into the Ministry of Agriculture and Food was sent to parliamentary proceedings. The project suggested leaving under the same institutional framework all the links of the food chains, from primary production to industrial processing and including all food items. However, in the end, the project did not move forward. That same year, the Presidential Advisory Commission on Food Safety became an agency and developed the first national food safety policy with a modern approach to food chains, an intersectoral approach to food safety issues and a methodological and conceptual framework of risk analysis to manage those issues. Subsequently, on two more occasions, in 2011 and 2019, the executive branch attempted, unsuccessfully, to transform the Ministry of Agriculture. A major sticking point was the lack of consensus regarding the incorporation of the fisheries and aquaculture sector, currently under the Ministry of Economy, to this new institutional framework.

▸▸ The performance of the Chilean food system

The following is a review, based on official indicators and data, of the performance of the Chilean food system. This review considers the contributions (or lack of) of the food system to the achievement of objectives and commitments in the sanitary, productive, technological and commercial areas that Chile has assumed as part of its public policies.

In the area of health and social protection, by 1920 the infant and preschool mortality rate in Chile was 300/1000 live births. At the beginning of 1960, this figure was 120/1000, the highest in Latin America. The figure is now around 8/1000, the lowest in the region. This significant decrease is closely related to the reduction of child malnutrition, from a prevalence of 37% in the 1960s to 0.1% today. This reduction has been possible thanks to the sustained efforts over time of various public policies to support families, such as the PNAC National Supplemental Feeding Program. PNAC distributes about 16 million kilograms of food per year and is linked to various programs that have objectives and/or target populations. These programs include the child, adolescent and women's health programs, as well as the program for the control of chronic non-communicable diseases, the PACAM Complementary Feeding Program for Older Adults, the PASAF Program on Healthy Eating and Physical Activity, and the EINCV Nutritional intervention strategy through the life cycle to prevent obesity and other noncommunicable diseases run by Ministry of Health. Other programs include the Chile Grows with You (*Chile Crece Contigo*) program of the Ministry of Social Development and the PAE School Feeding Program of the National Board of School Aid and Scholarships, which provides more than 3.5 million daily rations to elementary and middle school students. All these programs incorporate different actions that include nutrition, food and physical activity.

Regarding obesity and overweight indicators, according to the latest OECD report, 74% of the Chilean adult population suffers from obesity or overweight, more than any other country and ahead of both Mexico and the United States. These data have been recently confirmed by the National Board of School Aid and Scholarships, which noted an increase of 0.3% of severe obesity in children from kindergarten through secondary school, with respect to the 2019 measurement. This information confirms what has been repeatedly pointed out in recent years by various reports from specialized agencies on the subject. In addition, the 2017 OECD Obesity Update report indicates that, between 2016 and 2017, Chile recorded a 9.3% increase in the proportion of obese adults over 15 years of age.[10] The 2018 Nutritional Map by the National Board of School Aid and Scholarships indicates that 51.7% of school children suffer from overweight and obesity, a prevalence that is increasing in young children compared to the 2017 measurement.

ECLAC estimates that around 10% of Chile's population lives in poverty, the second lowest after Uruguay. According to data from the Ministry of Social Development, both poverty and extreme poverty have fallen considerably since 1990 to date. In 2017, 8.6% of the population was in income poverty (1,528,284 people) and 2.3% in extreme poverty (412,839 people). Poverty has a higher incidence among women than among men; 9% of women and 8.2% of men face income poverty. The difference between the two groups in relation to extreme poverty is not statistically significant. Between 1990 and 2017, the incidence of poverty and extreme poverty has consistently decreased for both men and women.

Income poverty in 2017 totaled 7.4% of the population living in urban areas, and 16.5% in rural areas. Extreme poverty also has a higher incidence in rural areas, with 4.4% versus 2% for those living in urban areas. For people living in both urban areas and rural areas, poverty and extreme poverty rates by income have decreased over the 1990 to 2017 period. Rural areas continue to have the highest income poverty rates, more than double the rates in urban areas.

It is important to note that the indigenous population has a higher incidence of poverty and extreme poverty than the non-indigenous population. A total of 14.5% of the indigenous population lives in income poverty, while this figure is at 8% for the non-indigenous population. Extreme poverty levels show the same trend. The incidence of poverty, extreme poverty and income poverty have among the indigenous population has decreased, with income poverty falling significantly between

---

10. The FAO's State of Food Security and Nutrition in the World 2019 report indicates that Chile has one of the highest rates of child overweight in Latin America and the Caribbean, with 9.3% of children under 5 years of age, a figure that remains unchanged from the same report in 2018.

1990 and 2017. The incidence of income poverty is higher among the population born outside Chile than among those born in Chile: 10.8% of those born outside Chile are in income poverty, while among those born in Chile this figure is 8.5%.

The dimensions that contribute most to multidimensional poverty are work and social security, housing and environment, and education. Together, these issues account for nearly 85% of multidimensional poverty. Poor nutrition problems contribute 3.4% within the health dimension, which in turn accounts for 10.4%. No other dimensions or sub-dimensions directly associated with food are identified among the determinants of multidimensional poverty.

According to the WHO report on the global burden of foodborne diseases, Chile is located in the "Americas low-adult, low-child mortality-stratum B" (AmrB, according to Ezzatti methodology) category. In the AmrB region, the average rate of disability-adjusted life years (DALYs) caused by foodborne diseases is 140/100,000 inhabitants. This region ranks as the sixth least-affected region out of 14, which brings it closer to the AmrA region (Canada, Cuba, United States), with an average 35/100,000 DALYs, than to the AmrD region (Bolivia, Guatemala, Haiti, Nicaragua, Peru), with 315/100,000 DALYs on average.

Between 2011 and 2018 the number of foodborne disease outbreaks rose from 962 to 1,134; however, the number of people affected decreased from 6,736 to 6,050. The severity of illness also fell, with a drop in the hospitalization rate from 1.44 (248) to 1.03 (191) and in the rate of deaths from 0.05 (8) to 0.00 (0). The four main causative agents, in order of importance, are *Salmonella*, scombroid, *Staphylococcus aureus* and *E. coli*. Around 40% to 50% of foodborne disease outbreaks occur in the home and 30% to 40% in facilities where food is prepared and eaten, with fewer occurrences earlier in the supply chain.

The Chilean Food Information and Alert Network (*Red de Información y Alertas Alimentarias* – RIAL) noted in its 2017 annual report a total of 223 national alerts. Of these, 221 applied to food, one to animal feed and one to materials in contact with food. With regard to the alerts for food, 183 corresponded to products destined for the domestic market, of which 97.3% were food produced in the country and 2.7% were imported products. As for the type of hazard involved in the domestic product alerts, 93.4% were chemical hazards, 5.5% were biological hazards and the rest were physical or other hazards. Fresh fruits and vegetables accounted for 90.2% of the alerts for food for human consumption, 4.9% for livestock products, 3.3% for aquaculture and 1.1% for industrially processed foods. The types of hazards detected were chemical hazards in fruits and vegetables in 100% of the cases. Alerts for food of livestock origin were associated with biological hazards in 88.9% of the cases.

According to the Global Food Security Index prepared annually by The Economist, which measures the fundamental determinants of food security in 113 countries, until 2019 Chile ranked 24th in the world and was first among 18 countries in Latin America for the areas of access, availability, and safety and quality. The main gaps with respect to the world average were related to the volatility of agricultural production, risk infrastructure and public spending on R&D.

In 2020, with the inclusion of the natural resources and resilience dimension and the increase to 59 indicators, Chile dropped to third place in the Latin American ranking and to 34th place in the world ranking. The main gaps with respect to the world average in this new scenario are related to the: availability dimension (resources for safety programs, –58.4%); access dimension (agricultural infrastructure, –7.7%; crop storage capacity, –85%; risk infrastructure, –5.8%; food safety agency, –27.4%); safety and quality dimension (national food guidelines, –38.9%); natural resources and resilience dimension (average annual losses from severe storms, –41.3%; agricultural water availability risk, –14.6%; eutrophication, –16.8%; marine biodiversity, –18.3%; food import dependence, –15.5%; natural resource dependence, –31.3%).

The gaps in availability and access can be explained by the change in methodology, and will surely be positively corrected in future indexes. Meanwhile, those related to natural resources and resilience are mostly existing gaps that are being progressively addressed through tools such as clean food production agreements. These complex issues will be difficult to resolve, so it is likely that for Chile to return to the prominent position it once held, the national food system will have to take on the challenges that these new dimensions impose from the perspective of food security and sustainability.

▸▸ Social upheaval, the pandemic, the constituent process and food style outlook

Like the 1997 Asian financial crisis, the 2007 subprime crisis and the M_W 8.8 megathrust earthquake of February 27, 2010 that devastated large areas of the country, the serious crises that have hit Chile in the last two years (i.e., the social upheaval of October 2019 and the ongoing Covid-19 pandemic) have been tough tests for the food system and the thus far successful trajectory of more than thirty years of development of the national food industry.

With the current social and health crisis, the system has proven to be resilient and robust in the operational continuity of supply chains, adapting complementary feeding programs, school meals, and emergency food distribution programs, ensuring universal access to food for the population in sufficient quantity and quality. However, according to the latest United Nations report on world food security and nutrition (UN, 2020), 11.8% of the Chilean population suffers from moderate food insecurity and 3.8% from severe food insecurity, while overweight and obesity remain unchanged, indicating a worsening of the nutritional situation compared to previous reports. It remains to be seen whether this is the result of the current situation or, as some analysts see it, a structural deterioration of the system's capacity to meet the needs of the population under the new paradigms of healthy eating and food security.

Regardless, the social and health crises are a clear warning. In terms of sustainability in the medium term, beyond the ability to competitively respond to consumer preferences or market trends – as has been done so far – the crises have made major challenges for Chilean food system obvious: the production and trade of food, the effects on people's health and well-being, climate change, the environmental sustainability of the industry (circular economy and green technologies), livestock animal welfare, regional equality and social inclusion in the processes of production, processing, distribution and consumption, food culture, food sovereignty, and the right to food, among others. All these issues will undoubtedly be part of the debate that will take place in the constituent process soon to begin (BCN, 2020).

The complex social, economic and production scenario due to the Covid-19 pandemic has created an uncertain short- and medium-term outlook for Latin America and the Caribbean. Not only does the sector face issues regarding the adaptive capacity and sustainability of food systems and possible long-term consequences, but the overall performance of the region's economy, which is strongly based on food production and trade also faces challenges. The most alarming aspect in the immediate term has been the abrupt slowdown of supply chains as a result of the massive worker infections, as well as sanitary measures requiring isolation and movement restrictions implemented by the authorities. Beyond the obvious economic damage caused, the slowdown could have potential consequences in terms of the health and welfare of animals and the deterioration of agricultural products; the spread of pests and diseases; environmental contamination; loss of food safety and quality; loss of food security (market shortages); disruption of social peace; and situations that could in turn worsen already fragile public health, environmental, productive and socioeconomic conditions.

In view of the above and according to WHO recommendations, the governments of most countries in the region have declared the food industry as an essential sector. They have taken measures to safeguard production processes and distribution logistics and implemented prevention and contingency management protocols that are compatible with social distancing and movement restrictions imposed by health authorities. These actions have allowed the reasonable operational continuity of food supply chains, with no serious shortages to date, aside from pre-existing situations that have undoubtedly been aggravated by the crisis.

The economic damage caused to date and the economic projections for the coming years offer a bleak regional outlook. Governments will need to take significant economic measures to support industry players, especially small and medium-sized enterprises. There is also an urgent need to create new balances between the appropriate levels of protection of citizens against hazards that affect public health, food safety, industry competitiveness and food system sustainability. First, lessons must be drawn from the current pandemic experience, such as the high degree of uncertainty that, despite technological advances, continues to govern risk management; the increasingly complex multidimensionality of hazards and their impacts requires a conceptually more sophisticated (one health paradigm) and technologically more complex (digital transformation) approach to the management of sanitation, safety, sustainability, and public health risks, and the need to develop a culture of resilience in organizations that allows them to face crises and emerge stronger.

It is ironic to note that the Covid-19 crisis has acted as a catalyst for ongoing phenomena, accelerating the need to respond to growing challenges in the areas of productive and commercial development, scientific and technological development and human capital, public health, and social and environmental sustainability. These challenges can be grouped into the following areas:

– *Environmental*: the effects of climate change, the most evident manifestation of which is the intensification of climate phenomena, with consequences that include the displacement of global agroecological maps and the modification of epidemiological patterns of pests and diseases. These effects call for responses that advance mitigation, adaptation and resilience to climate change.

– *Production*: the transition from physical and analog production systems to computerized and digital systems, through the adoption of software and hardware technologies for sensorization, automation and robotization. There is also a need to ensure interoperability and analysis of large volumes of information and its use, among other things, in better risk and uncertainty management, the development of new products, as well as support in logistics, administrative and related service management processes.

– *Socioeconomic*: mitigation of the impact on employment and income of short-term phenomena such as Covid-19 and structural phenomena such as the digital transformation of industry at the labor level, through reconversion of workers' skills and capacities, the adaptation of school curricula, technical and professional education, and the promotion of innovation and entrepreneurship, as well as the implementation of plans to create new opportunities and integration in territories and social groups that are lagging behind.

– *Science and technology*: the promotion of R&D and the training of advanced human capital for the development of applied technological solutions and innovation in response to the needs of society, with a special focus on health, sustainability and digitalization.

– *Commercial*: the need to adapt the food supply to new consumption patterns and trends, taking into consideration the demand of increasingly informed, connected, conscious, critical, and challenging consumers who resort to non-traditional forms of information and distrust authority, whose consumption decisions are increasingly complex and include ethical, environmental, social, health, indulgence and economic aspects, among others. More specifically, the post-pandemic effects of increased food consumption at home must be accounted for, as well as the use of digital retail platforms, home delivery, the desire for local products and stronger trade protectionism.

– *Sanitary*: the global trend to raise the adequate level of protection in terms of health, safety and quality, with increasing demands on pathogen control through stricter biosafety, sanitation and hygiene standards, restrictions on the use of agrochemicals, veterinary drugs, synthetic additives in production processes and stronger requirements. The intensification of regulations on the nutritional composition of processed foods in response to the pandemic of overweight and obesity and its association with the incidence of chronic non-communicable diseases in the population, especially young people, focusing on the reduction of critical nutrients such as sodium, sugars and saturated fats.

– *Consumers*: in societies in transition, such as many in Latin American countries, in recent times we have witnessed the collapse of many conventions and customs in various areas, which have been, or are being, replaced by new approaches that are ever more demanding in terms of ethical, political, technical, social, organizational and other standards. The food sector has been no exception. In the last twenty years, we have witnessed a growing sensitivity of the population to the effects of food on health and well-being. It is a fact that after the current crisis, citizens will demand ever higher standards from the food supply, which will pressure industry and governments to move towards providing safety and quality guarantees that exceed current regulatory requirements.

In short, we must prepare ourselves for a post-Covid-19 world, where the vulnerability or sustainability of food systems in general and of companies in particular will depend, among other things, on the political, institutional, economic, technological, productive and cultural adaptation process that countries undertake to face this new reality of risk and uncertainty. To this end, the one planet, one health framework of the United Nations seems to be more relevant than ever in the search for integrated multivariate risk management models. Finally, the crisis will pass and the future is uncertain, but it is here now, so we must develop resilient strategies to tackle new emergencies as an urgent and unavoidable task to which we must all – public, private and civil society stakeholders – contribute.

▸▸ References

ACHIPIA (Agencia Chilena para la Inocuidad y Calidad Alimentaria). (2018). *Política nacional de inocuidad y calidad alimentara, 2018–2030*. Ministerio de Agricultura, Chile. https://www.achipia.gob.cl/wp-content/uploads/2018/03/POLITICA-DE-LA-INOCUIDAD-2018-2030-1.pdf

Ahumada Benítez, D. (2018). El proceso de formulación de la Ley de la Caja del Seguro Obrero Obligatorio de 1924. *Notas históricas y geográficas*, *21*, 89-121.

Araneda, J., Pinheiro, A. C., Rodriguez Osiac, L., & Rodriguez, A. (2016). Consumo aparente de frutas, hortalizas y alimentos ultraprocesados en la población chilena. *Revista chilena de nutrición*, *43*(3), 271-278. http://dx.doi.org/10.4067/S0717-75182016000300006

Barril García, A. (2021). *Pobreza e indigencia rural: evolución 1990–2000*. ODEPA. https://www.odepa.gob.cl/odepaweb/servicios-informacion/publica/pobreza-indigencia-rural.pdf

BCN (Biblioteca del Congreso Nacional de Chile). (2020). *Proceso constituyente*. BCN. https://www.bcn.cl/leyfacil/recurso/proceso-constituyente

Bellisario, A. (2013). El fin del antiguo régimen agrario chileno (1955–1965). *Revista mexicana de sociología*, *75*(3), 341-370.

Bengoa, J. (1988). *Historia social de la agricultura chilena*. Ediciones Sur.

Campos, J. A., & Polit, E. (2011). Nuevos enfoques para "Chile potencia alimentaria y forestal." ODEPA, Ministerio de Agricultura. https://www.odepa.gob.cl/wp-content/uploads/2011/02/nuevos_enfoques_para_Chile201101.pdf

FAO. (2013). *Panorama de la seguridad alimentaria y nutricional en América Latina y El Caribe* [Panorama of food and nutritional security in Latin America and the Caribbean 2013]. FAO. https://www.fao.org/3/i3520s/i3520s.pdf

FAO, FIDA, OMS, PMA & UNICEF. (2020). *El estado de la seguridad alimentaria y la nutrición en el mundo 2020. Transformación de los sistemas alimentarios para que promuevan dietas asequibles y saludables* [The state of food security and nutrition in the world 2020. Transforming food systems for affordable healthy diets]. FAO. https://doi.org/10.4060/ca9692es

Gobierno de Chile. (1965). *Proyecto de ley de reforma agraria*.

Leporati Néron, M., Binelli Maino, P., & Rojas Olavarría, H. (2010). Seguridad sanitaria de los alimentos y comercio: desafíos para la salud pública y el desarrollo competitivo de la industria. In Barrera, A., & O. Sotomayor (Eds.), *La agricultura chilena en la nueva revolución alimentaria*. Editorial Universitaria.

Llorca-Jaña, M., Ortiz, C. R., Navarrete-Montalvo, J., & Valenzuela, R. A. (2017). La agricultura y la élite agraria chilena a través de los catastros agrícolas, c. 1830-1855. *Historia (Santiago)*, *50*(2), 597-639. http://dx.doi.org/10.4067/s0717-71942017000200597

Meller, P. (1990). Una perspectiva de largo plazo del desarrollo económic chileno. In Blomström, M., & P. Meller (Eds.), *Trayectorias divergentes. Comparación de un siglo de desarrollo económico latinoamericano y escandinavo* (pp. 53–82). CIEPLAN-Hachette.

Ministerio de Agricultura. (1968). *Plan de desarrollo agropecuario, 1965–1980*. Ministerio de Agricultura, Chile.

Ministerio de la Salud. (2010). *Nutrición para el desarrollo. El modelo chileno*. MINSAL, Chile. https://www.minsal.cl/sites/default/files/files/Nutrición%20para%20el%20Desarrollo_%20El%20 modelo%20chileno_%20MINSAL%202010.pdf

Olivares, S., Zacarías, I., González, C. G., & Villalobos, E. (2013). Proceso de formulación y validación de las guías alimentarias para la población chilena. *Revista chilena de nutrición*, *40*(3), 262-268. http://dx.doi.org/10.4067/S0717-75182013000300008

ProChile, 2019. Introducción a Los acuerdos comerciales. https://acceso.prochile.cl/wp-content/ uploads/2019/06/introduccion_acuerdos_comerciales.pdf

Ríos-Núñez, S. (2013). Reestructuración del sector agrario en Chile 1975–2010: entre el proteccionismo del Estado y el modelo económico neoliberal. *Revista de Economia e Sociologia Rural*, *51*, 515-533. https://doi.org/10.1590/S0103-20032013000300006

Riveros, L. A., & Ferraro, R. (1985). La historia económica del siglo XIX a la luz de la evolución de los precios. *Estudios de Economía*, *12*(1), 49-78.

Rojas, A. (2007). Una mirada a 400 años de agricultura en Chile. In Ministério de Agricultura, *Chile: su agricultura, su territorio y su gente*. FUCOA.

Schejtman, A. (1971). *El inquilino de Chile Central*. ICIRA.

Schneider T. (1904). *La agricultura chilena en los últimos cincuenta años*. Barcelona.

Sociedad de Fomento Fabril (SOFOFA). (1983). *100 años de industria (1883–1993)*. Ediciones Patmos.

Villalobos, P., Rojas, A., & Leporati, M. (2006). Chile potencia alimentaria: compromiso con la nutrición y la salud de la población. *Revista chilena de nutrición*, *33*(S1), 232-237. http://dx.doi. org/10.4067/S0717-75182006000300004

Waissbluth, M. (2020). *Origen y evolución del estallido social en Chile*. Centro de Sistemas Públicos, Universidad de Chile. https://www.mariowaissbluth.com/descargas/mario_waissbluth_el_estallido_ social_en_chile_v1_feb1.pdf

World Health Organization. (2015). *Foodborne disease burden epidemiology reference group 2007– 2015*. WHO.

Part III

Recent changes
in national food policies

Chapter 9

Food policies in Argentina: the challenges of implementing an agenda for sustainable systems with a social inclusion component

CECILIA ARANGUREN, ANA MARÍA COSTA,
SUSANA BRIEVA, GRACIELA BORRÁS

▸▸ Introduction

In Argentina and other Latin American countries, issues relating to food production, distribution and consumption have taken on greater significance in public policy, research and development agendas since the beginning of the 21st century. Food issues impact the workings of the economy, government agencies and science and technology institutions, as well as civil society organizations. The complexity of their impact can have direct or indirect consequences on the ability of the population to obtain food and secure a healthy diet. As part of the drive to consolidate its neoliberal development model during the 1990s, Argentina developed targeted policies that adopted a compensatory approach as a strategy to contain poverty, minimizing the two fundamentals of comprehensiveness and universality in the way it addressed food issues. During its neo-developmental period, although Argentina sought to promote a comprehensive approach to food policies, the concepts emerging were many and varied, placing the emphasis on nutritional aspects, marked by significant fragmentation or targeting sectors of high social vulnerability.

From a systemic and constructivist perspective, this chapter pursues a dual objective: first, to analyze the processes underpinning the social construction of food policies in Argentina from the crisis of 2001 through to 2019, and second, to study the capacity of these policies to bring about changes in the food systems and promote sustainability. This analysis places particular emphasis on the processes applied in formulating and implementing national food policies, with the purpose of providing access to food of improved quality, in distribution and consumption. This chapter seeks to answer the following questions: Which visions took precedence in the design and implementation of food policies in Argentina? What repercussions did they have on national production, distribution and consumption systems? What is the role of government, business and civil society representatives in policy-making?

How are the different interests interlinked? What capacities have been developed in technical skills, management and knowledge to address food issues?

To this end, the chapter is organized as follows: first, a non-exhaustive selection and review of studies on food policy measures in Argentina, followed by a description of the analytical and methodological approach applied to this public issue, the social construction of policy, and the technologies for social inclusion. Next, the chapter selects and analyzes the processes applied in the design and formulation of two[1] food policies in Argentina: the National Food Security Plan (PNSA) and the National Plan for a Healthy Argentina (PNAS). The chapter concludes with a number of thoughts and questions.

▸▸ Literature review: a brief summary

Argentina has a long history in the formulation of food[2] programs. Since 1980, it has developed a number of national programs to meet the varied nutritional requirements of the population. Since then, and to a greater extent during the crisis of 2001, the food issue has become part of the public policy agenda. Argentina is a country characterized by its food surpluses but it has faced increasing difficulties in meeting the food requirements of the population, giving rise from 2015 to growing social demands. This has become a public issue of increasing social interest and concern.

Over the years, the design, formulation and implementation of food plans and programs in Argentina have reflected a range of policy conceptions and visions. The question of State intervention, mainly targeting the most vulnerable, has been studied by a number of social scientists and experts employing different theoretical and disciplinary approaches, with varying degrees of scope, highlighting the social, organizational, legal, nutritional and/or economic aspects of social policies. Over the past decade, a series of studies have been carried out on the implementation of food policies in Argentina. Broadly speaking, the key areas of study are: i) food policies over time; ii) the implementation of policies at the provincial and/or municipal level; iii) institutional mechanisms and management in social policies; and iv) conceptualization, food sovereignty, food security and a rights-based approach.

i. Most of the studies looking at food policies over time are analyzed as part of a rights-based approach. In his analysis of food-based interventions over the last 25 years, Maceira (2010) highlights a failure to address the problem of nutritional deficiencies, and a lack of effective mechanisms for transferring capacity to the beneficiaries. He therefore concludes that food should be considered by decision-makers as one of the main health indicators. In a review of food policies implemented over time, Borrás and García (2012) point out that these policies are still adopting a targeted approach, involving only the populations considered to be at nutritional risk, and that they lack a comprehensive approach. Similarly, Salomone (2015) refers to the food policies

---

1. This chapter does not analyze all the rural, production and food policies implemented in Argentina, but focuses on two of the most important and relevant programs of the past two decades.
2. Abeya Gilardon (2016) considers Act No. 12341 of 1936 (known as the "Palacios Act") to be the first attempt to provide a State response to the health, food and nutrition problems of mothers and children living in poverty in Argentina.

implemented in 50 years of intervention as short-term, palliative and targeted. He pleads for a broader, comprehensive and interdisciplinary view of the food problem through joint studies that also encompass the problems of obesity, excess weight and stunted growth (the most prevalent problems). Abeya Gilardon (2016) takes as his basis the conceptualization of food programs and three fundamental components of food security – availability, accessibility and utilization – to analyze the national food programs set up to improve the food security of the families most in need. He also provides an update on the nutritional situation of the target population, putting forward arguments to analyze ongoing programs applying a typology based on the components of food security. He believes that a general review of food programs is necessary, encompassing their conception, content and scope, so that they cease to be mere social aids and start to play a real role in promoting better food and nutrition.

ii. Many authors have turned their attention to the implementation of programs at the provincial and/or municipal level, based on household surveys and studies in different Argentine provinces. Of these, Salvia et al. (2012) make a particularly noteworthy contribution concerning the macro-social conditioning factors faced by the population, based on their area and socio-residential condition, in achieving food security. Other authors address the implementation of plans at subnational levels, such as Aulicino (2012), who shows how political-institutional variables, expressed through practical policies and programs, influence the modes of organization, sharing of responsibilities and intervention practices, as well as the scope and relevance of social interventions. González et al. (2013) follow similar lines in their work, which analyzes the scope and limitations of plans in a number of jurisdictions, as well as the relations between policy makers and recipients of public policies based on negotiation, control and exchange. Based on an analysis of food policies in one of the districts in the province of Buenos Aires, they warn that an approach based on economic change is not sufficient to explain changing consumption habits among the population, just as food education alone cannot bring about a change in diet. Martini (2016) analyzes the effective implementation of public food policies in different regions of the country, highlighting the relations between the State and civil society and looking at the role played by ordinary people in the implementation of public policies. Based on an analysis of the permanent household survey, Colomarde (2018), focuses on nutritional differences at the household level and the methods used for measurement.

iii. Turning to the institutional mechanisms and management necessary to ensure efficiency, governance and equity in the implementation of food policies, Roffler (2010) systematizes and analyzes the main characteristics of the food and nutrition programs implemented by national and provincial governments in recent years, placing particular emphasis on the public spending involved in this type of intervention and incorporating it into an analytical perspective of public policies. He notes that provincial governments (and to a lesser extent municipalities) have been playing a greater role in the oversight of public spending in recent decades, to the detriment of national government, even though they lack the institutional (technical, political) capacity to carry out these functions. In most cases, this has had a negative impact both on management capacity and on the quality of the services and programs. Based on the vision of policy management, Rebon and Roffler (2014) focus their analysis on information systems and on an assessment of these programs. They conclude that the implementation of assessment programs and follow-up initiatives have reached a certain level of development and sophistication in recent decades but are rarely consolidated. At the same time, they point out that most assessment bodies are set up as part of the proposals and/or terms set out by international financing organizations and included in the planning process, making them dependent on the outlay of resources. At the same time, when programs are financed by national resources, it is difficult – with few exceptions – to find studies assessing their results or impact.

iv. Concerning problems relating to the food issue in Argentina, Santarsiero (2012) highlights the growing importance of food policies in welfare policies for poverty alleviation. He notes that the State response to the problem of food and inequality in access to food is characterized by the expansion of targeted food supplementation programs through food modules or food aid in canteens. He points out that State intervention has generally been fragmented, with the main characteristics being a poor institutional framework, a high level of discretion, irregularities in management, and a welfare-based approach. In her analysis of food policies, Basualdo (2018) looks at current tensions in the conceptualization/orientation of the food sovereignty component in social policies, arguing that these tensions are part of an unequal social dynamic and a global systemic crisis of which the food issue is also a part and in which we can also see a social struggle characterized by demand and resistance within the shifting conflict of social policy. At the same time, this struggle prefigures another social horizon that is in conflict with the socially legitimized system of needs.

Taking the standpoint of food sovereignty, Viola and Marichal (2020) analyze the mechanisms applied in coordinating the food emergency with the right to food by the Argentina Against Hunger Program (PACH) in the context of the Argentine legal system. Making emergency food programs the central component of national food policy means that the right to food is limited to its most basic aspect (the right not to die of hunger). While this is clearly a priority, it does not address the food issue in depth, separating it from other systemic aspects such as the production, distribution and consumption of food. The food sovereignty approach of the PACH warrants a more complex approach, coordinating the most fundamental need with the other systemic aspects.

▸▸ An analytical and methodological perspective

An analysis of food policies in Argentina in recent decades shows a range of concepts reflecting different theories and disciplinary bases, providing an explanation for the case under study. Taking as our starting point a systemic, constructivist and interactive approach to processes, we will explore the possibilities for dialog and the integration of concepts from the approach of Policy Analysis and notions relating to approaches in Social Studies in Science, Technology and Society (ESCTyS). According to Serafim and Días (2010), both approaches seek to open "black boxes," in order to understand the processes generated in their respective fields of research. Policy Analysis looks at the phases of public policy (identification of the problem, construction of the agenda, formulation, implementation and assessment). Focusing on the formulation processes, it endeavors to understand how this construction occurs, through the behavior of the social players. In contrast, the approach of the Social Construction of Technology seeks to investigate how technology is socially constructed (Serafin & Dias, 2010, p. 62).

In terms of Policy Analysis, the concept of policy is used in the reconstruction and examination of food regulations and policies (Elmore, 1978; Ham & Hill, 1993; Hogwood & Gunn, 1984). This complex concept involves a range of definitions expressing different positions. This diversity partially reflects the subject. Policy can be considered as a "web of decisions and actions" that allocate value. The processes underpinning the formulation and implementation of public policies always involve different social groups and are a subject of dispute between players and political arguments, all of whom have an interest in issues where the State apparatus is the main space of expression (Thomas, 1999).

Analyzing public policy in terms of processes enables us to understand how players define the problems and agendas, how they are formulated, how decisions are taken and approved, and how actions are implemented (Parsons, 2007). These aspects are considered in the design and formulation of sustainable public food policies.

Analyzing the processes underpinning the construction of sustainable public food policies requires a multidimensional and transdisciplinary approach to the forms applied to the construction, negotiation and implementation of knowledge, strategies and policies. Social constructivist studies in science, technology and society contribute can help to explain these aspects, particularly by identifying problems in the conception and design of the implemented systems, overcoming the determinism that has been implicit until now in most public policy recommendations and forms of governance in the Argentine food system.

From an ESCTyS perspective, technology is an object that is inherently social and therefore political. According to Thomas (2012), technologies – all technologies – play a central role in processes of social change. They stake out the positions and behaviors of players; they determine social distribution structures, production costs, access to goods and services; they generate social and environmental problems; and they facilitate or hinder their resolution. From a constructivist perspective, technology and politics are mutually constituted, like two sides of the same coin (Bijker, 2005). In this way, politics can be understood as a technology of social organization and social intervention (Serafim & Díaz, 2010). Likewise, all technologies are political, albeit with varying levels of visibility, and aim to control some aspect of social practices, processes, behaviors or spaces in the social fabric.

This approach to the food issue is based on a socio-technical analysis, specifically in terms of the Technologies for Social Inclusion. These are understood as a way to design, develop, implement and manage product, process and organizational technologies in order to solve social and environmental problems, generating social and economic dynamics of social inclusion and sustainable development (Thomas, 2010).

With regard to the methodological strategy underpinning the conceptual framework guiding our research, we adopted a qualitative-quantitative explanatory approach. This enables an analytical reconstruction covering the processes of social construction in food policy, as well as changes in the forms of governance and the understanding of the dynamics and processes of policy design and implementation, through a flexible, interactive and continuous process, emphasizing the finding, systematization and collection of information from national plans, programs and policies, along with a critical review of background information and studies on the subject, as well as a comparative and interpretative analysis of public policy initiatives, and the institutions and structures involved over time, particularly those involving participants in the scientific, technological, normative and regulatory system underpinning the agrifood and agro-industrial system in Argentina.

Based on this theoretical-methodological perspective, the study highlights two of the public policies implemented in Argentina over the past 20 years: the PNSA set up in 2003 by Act No. 25724 and the PNAS set up in 2008 by Act No. 27.396. This chapter undertakes a comparative study in stylized form of the processes of design, formulation and implementation applied for both policies.

▸▸ Food programs in Argentina (2003–2019)

Since the crisis of 2001, and following a change in the direction of public policy in 2003, Argentina has promoted new strategies and public policies on food and public health. It has sought to address the problem of socio-economic vulnerability, affecting babies and pregnant women, school-age children and families living in poverty, and to promote the consumption of healthy foods, in order to combat malnutrition and the risks of food-related diseases. The PNSA and the PNAS stand out as two of the main food programs implemented between 2002 and 2019 (table 9.1).

The unprecedented economic and political crisis of 2001 saw unemployment rise to 21%, while poverty increased to 54%, bringing the problem of hunger and malnutrition onto the public policy agenda. In response to strong demands by civil society, journalists, social leaders and non-governmental organizations (NGOs) pooled their efforts to involve and gain the support of politicians and public officials in developing measures to address the problem, through the founding in 2003 of the National Food Security Plan, *El hambre más urgente* (The most urgent hunger) (PNSA), following the approval of Act No. 25.724 referred to as the National Nutrition and Food Plan. With respect to this point, Abeyá Gilardón (2016), talks about the technical proposal put forward by the Sophia Group, along with *Poder Ciudadano* and *Red Solidaria*, under the name *El hambre más urgente.* The main emphasis of this initiative is placed on the nutrition, health care and early learning of all children in poverty from conception through to the age of five, across the country. This initiative, supported by the journalist Luis Majul and the newspaper *La Nación*, gathered more than one million signatures. As a result, it gained the status of popular initiative, a mechanism allowing citizens to submit draft bills for discussion in Parliament, as indicated in Article 39 of the Argentine National Constitution. At the same time, in July 2002, the National Council for the Coordination of Social Policies organized a two-day forum for a National Food and Nutrition Plan in the national library. The event was attended by 300 people: technicians from governmental agencies, legislators, researchers from universities and scientific centers, and members of non-governmental organizations. The Forum's conclusions provided input for parliamentary discussion and the original proposal finally became law. According to Arcidiacono and Carrasco (2012), the list of organizations drawn up for the event included representatives of the food industry such as the Food Industry Coordinator (COPAL) who did not attend.

Table 9.1. Food policy formulation and implementation process

Policy Process	National Food Security Plan (PNSA)	National Plan for a Healthy Argentina (PNAS)
Year of founding	2003	2009
Identification of the problem	Problem of access to food (prices and purchasing power); food monotony; acute and chronic malnutrition, malnutrition (excess weight and obesity); specific deficiencies: iron, calcium, vitamins	Identify nutritional problems, nutrient deficiencies, particularly in the poorest population groups. Nutritional anemia, stunting and increases in excess weight and obesity. Problems in access, distribution (inequalities) and consumption (eating habits); sustained increase in diseases associated with non-communicable diseases
Decision and players involved	Act No. 25.724. Civil society initiative (NGOs, journalists) "Most Urgent Hunger"	Act No. 26.396. National Ministry of Health Initiative (MSN)
Context of implementation	Crisis and social explosion (2001). Increased poverty and social vulnerability. New direction in State action and management (2003–2015)	Results of surveys on health indices in Argentina. WHO Recommendations. Healthy consumption habits. Urbanization, food globalization.
DESIGN AND FORMULATION OF FOOD POLICIES		
Goal of the program	Give socially vulnerable people access to a complementary and sufficient food supply, in accordance with the specific characteristics and customs of each national region	Promote and adopt healthy eating habit and active lifestyles, control obesity and chronic diseases related to malnutrition. Formulate public policy, bring about institutional change, adopt communication strategies and conduct research into diets and physical activity. Implement health promotion and disease prevention strategies
Target population	Families in a situation of socio-economic vulnerability. Children up to 14 years old, pregnant women and adults over 70 (priority: pregnant women and children under the age of five)	General population
Lines of action	Design strategies; establish criteria for access, continuity of the program and control mechanisms. Ensure equity in health care and benefits. Implement a food education program. Establish a permanent system to assess the nutritional status of the population	Promote healthy habits. Coordinate products and services. Promote a healthy environment

Policy Process	National Food Security Plan (PNSA)	National Plan for a Healthy Argentina (PNAS)
IMPLEMENTATION OF FOOD POLICIES		
Enforcement authority	Ministry of Social Development (MDS) with, at the founding stage, the Ministry of Health of the Nation (MSN)	Ministry of Health of the Nation (MSN)
Management system	Decentralized to the provinces and municipalities: the program transfers funds to the provinces and provides technical assistance and training for provincial staff; it aims to coordinate all existing national programs and create a single base of the target population	Centralized in framework agreements with the players at the negotiating table and decentralized in subnational jurisdictions
Program coordination and links	National Commission on Nutrition and Food. Links between the areas of Social Development, Health, Education, Economy, Labor, Employment and Human Resources Training, Production and Non-Governmental Organizations and, for each province, expert advice from scientific entities, universities, welfare and ecclesiastical organizations Missing players: agro-industrial chambers of commerce, which did not specifically take part in the public hearings prior to the approval of the act	National Commission for the Prevention and Control of Non-Communicable Diseases Links with: MSN, Education, Social Development, Economy and Public Finance, National Institute of Social Services for the Retired, Superintendence of Health Services, Universities, Societies and other institutions and players associated with the problem Framework agreements with industry, chambers of commerce and the Ministry of Agriculture Working groups: salt reduction; trans-fat reduction; sugar reduction; front-of-pack labelling; school environments
Type of services	Community and school food and prevention services Food deliveries: bags, tickets, food cards, school and community canteens Education and training	Promote healthy habits Conduct communication campaigns on smoking, healthy eating and an active lifestyle Carry out educational projects at different educational levels, feasibility studies on reducing/replacing sodium, sugar and trans fats in frequently consumed processed foods Agreements with the food industry to reduce sodium and sugar and eliminate trans fats

Policy Process	National Food Security Plan (PNSA)	National Plan for a Healthy Argentina (PNAS)
Complementary programs and links	ProHuerta (INTA/MDS) Local Development and Social Economy Plan *Manos a la Obra*, (Let's get to work) Family plan for social inclusion (MDS); Head of household Plan, work; Young people and first jobs; Social flat-rate tax; Universal Child and Pregnancy Allowance (AUH, AUE); VAT reduction and special prices on basic food items	Campaign: "Trans fat free Argentina." High oleic oil Incorporation of Art. 155 ter* into the Argentine Food Code Campaign: "Less Salt More Life." Salt Act (2013) Agreements with the bakers' trade association Fortification of mass consumption foods (flour, salt, infant formula milk, etc.); "Remediar" program for drug access, Community Doctors Program; Cardiovascular health and diabetes program, etc. AUH, AUE
Monitoring and control tools, training	New Dietary Guidelines for the Argentine Population (oval diagram) Training courses for multipliers carried out mainly by nutritionists (predominantly medical design) Training in market gardens and farms, food preparation in schools, institutions and the community (ProHuerta INTA/MDS program)	New Dietary Guidelines for the Argentine Population (healthy eating plate) National Survey of Risk Factors (ENNFR, 2005,2009,2013, 2018) National Health and Nutrition Survey (NHS, 2007, 2019) Global School-based Student Health Survey (EMSE, 2012, 2019) Training
Assessment mechanisms	Follow-up on compliance with the different indicators of the MDS Internal Audit Units (UAI), the National Audit Office (SIGEN), the Federal Network of Public Control, and the National General Audit Office (AGN)	Monitoring of the compliance of indicators with MSN Survey components
Funding	Special Fund for Nutrition and the National Food Supply	Budgets allocated to the MNS, special allocations per Act and possibly international organizations and other institutions
Demand for Scientific and Technological Innovation	No explicit demand relating to technocognitive problems and human resources training to address the food issue	
Policy orientation	Combines the concept of food security, the idea of hunger and a rights-based approach. Palliative, targeted and welfare-based	It combines the concept of food security and a rights-based approach. Holistic health

* Article 155 ter - (Joint Resolution SPReI No. 137/2010 and SAGyP No. 941/2010) "The content of industrially produced trans fatty acids in food shall not exceed: 2% of the total fats in vegetable oils and margarines intended for direct consumption and 5% of the total fats in all other foods. These limits do not apply to fats from ruminants, including milk fat."

In mid-2000, the National Nutrition and Health Survey (ENNyS) for 2004–2005 revealed worrying rates of obesity, anemia, stunting, nutrient deficiencies, and other nutritional problems, as a result of malnutrition. These figures were an alert for the health system regarding the food and nutrition of the population. In 2008, the MSN set up an inter-agency commission to work on the recommendations of the WHO set out in the "Declaration of Rio de Janeiro – Trans Fats Free Americas." At the request of this commission, agreements were signed between the MDS and the Ministry of Agriculture, Livestock and Fisheries, food control representatives and the main chambers of the Food Industry.

In 2008, under the auspices of the National Ministry of Health (MSN), Argentina passed Act No 26.396 on the Prevention and Control of Eating Disorders. This Act declares the prevention and control of these disorders to be a subject of interest. It encompasses research into causal agents, the diagnosis and treatment of associated diseases, as well as comprehensive care and rehabilitation, including for derived pathologies, and measures to prevent their spread. In 2009, through the Undersecretariat for Risk Prevention and Risk Control, the MSN was the driving force for the Healthy Argentina National Plan (Ministerial Resolution 1083/2009), which aims to encourage healthy living, to oversee products and services and to promote a healthy environment. The plan includes three components: health promotion, the reorientation of health care services for the integrated management of chronic diseases, and greater epidemiological surveillance.

The shift towards neoliberal policies between 2015 and 2019 triggered a new socio-economic crisis with serious social and economic consequences. The national food situation deteriorated further, leading to more public initiatives and strategies in 2020, including the Argentina Against Hunger Plan, which aims to step up the actions carried out as part of the National Program for Food and Nutrition Security. The plan seeks to promote and increase access to the basic food basket.

▸▸ Food policy design and formulation

The goals of the National Food Security Plan are to guarantee supply and access to food for the most vulnerable families, particularly children under five years of age and nursing and/or pregnant mothers. In contrast, the National Healthy Food Plan focuses on the health and dietary habits of the population as a whole. It promotes healthy eating, an active lifestyle, lower levels of obesity and chronic nutrition-related diseases, an affordable supply of healthy food products and reductions in the foods that pose a health risk, such as physical inactivity, a poor diet and smoking.

The target population of the PNSA includes a profile of beneficiaries: children up to the age of 14, pregnant women, disabled and elderly people living in poverty. Within these parameters, the provinces can adjustment the criteria, leading to variations in the characteristics of beneficiaries in each province. In contrast, the PNAS is aimed at the population in general and includes three components giving priority to: health promotion, the reorientation of health care services for the integrated management of chronic diseases and improved epidemiological surveillance. To this end, State agencies are promoting initiatives aimed at developing foods with lower levels of salt and trans fats.

▸▸ Policy implementation

The Ministry of Social Development is the enforcing authority of the PNSA, which is implemented through framework agreements in all national provinces, resulting in a wide variety of implementation procedures. The different levels of government (municipalities, provinces) have dissimilar development contexts and starting points. Reflecting this, we can see processes where the municipal levels of government play a leading role in the implementation of the plan, and others where the provinces have greater levels of participation and decision-making. The Ministry of Health represents the enforcing authority for the PNAS, which is implemented through voluntary agreements with a range of players from business, civil society and various levels of government. Within this framework, the National Commission for the Prevention and Control of Non-Communicable Diseases was set up with the objective of advising the authorities on intersectoral initiatives aimed at the prevention and control of diseases and their risk factors.

Both plans include a range of services to support the intervention strategy: the PNSA organizes food aid, health and school-based prevention programs, compensatory initiatives targeting the age and social groups with the highest levels of nutritional vulnerability (food transfers, modules or bags, tickets or magnetic cards for the direct purchase of food, support for community kitchens). It also promotes the self-production of food. According to Aulicino and Diaz Langou (2012), the benefits of this program are grouped into five categories: direct food and nutritional assistance, assistance for social, children's and community canteens, self-production of food, improvement of food services in school canteens, healthcare for pregnant women and children's health. When it was first set up, the main action of the PNSA involved direct food assistance to families through the distribution of non-perishable food. In some cities, preference was subsequently given to tickets for the purchase of food in shops, and bags containing fresh food (such as with the Family Food Service or SAF). The third phase involved the direct purchase of food with magnetic cards.[3] This service develops a greater sense of independence in selecting food, encouraging family meals and promoting access to fresh food (fruit, vegetables, milk, yogurt, cheese and meat, among others).

According to Carrasco and Pautassi (2015), the reports issued by the AGN make a number of criticisms regarding the effectiveness of benefits, for example: a nutritional value that is insufficient to meet the goals set, the amount provided for the food basket, the monetary contribution necessary to make an effective contribution to reducing food insecurity or at least the poverty gap, the lack of regularity in deliveries and the delays in implementing agreements with the various jurisdictions, as well as a low level of execution in the resources available, and a lack of objective criteria in the selection of benefit holders, meaning that people meeting the conditions for the plan are actually excluded[4]. Although the PNSA has set out new guidelines, the authors point out that the main thrust of the social food policy is targeted and welfare-based, with direct delivery, canteens and market gardens being the main forms of assistance.

---

3. This service preceded the current hunger prevention program implemented with the change of government in December 2019.
4. For more information see Carrasco and Pautassi (2015).

In contrast, the PNAS gives priority to services that seek to: promote healthy living, oversee products and services, e.g., less salt, zero trans fats in food, taxation trade regulations on tobacco products, food and beverages, and the development of a healthy environment, based on draft bills for smoke-free and healthy environments, for example.

In terms of the instruments of intervention, the plan provides for survey-based monitoring,[5] educational projects at different levels of the education system, food guides, communication campaigns and feasibility studies for the reduction/replacement of sodium, sugar and saturated and trans fats in widely consumed processed foods. In this respect, efforts have been made alongside food distributors and services to promote a healthy diet and improve food preparation, for example. At the same time, voluntary agreements have been signed with the players involved (government, business and civil society organizations).

With regard to food education instruments, the PNSA encourages the development of capacities to overcome poverty in the target population, through food and nutrition education or initiatives targeting specific risk groups, together with training and technical support for market gardens and farms, for example. The PNAS organizes communication campaigns to promote smoke-free environments in the home, the workplace and public spaces, alongside initiatives to train, register and certify institutions and companies, educational and health centers.

To combine their material and human efforts, both the PNSA and the PNAS make provision in their implementation strategy to forge inter-institutional links with complementary programs and with the public institutions promoting community action, social development and public health. On the one hand, programs providing tools and training, and on the other, income transfer policies, with plans and programs that have been reformulated and adapted over the years.

In the first case, alongside the economic aid and subsidies, the programs are coordinated with ProHuerta program set up by the INTA (Argentine National Agricultural Technology Institute) in 1990 to promote access to healthy food through the self-production of fresh food for personal, family and community consumption.[6] ProHuerta is a public policy program encouraging agro-ecological growing practices for purposes of self-sufficiency, food education and the promotion of alternative trade points and markets to include producer families.

Although the different programs and institutions are coordinated in the implementation of food policies, the relationship between science, technology and food is poorly explained or completely absent. Juarez and Serafin (2010) argue that ProHuerta is the only program coordinated with the PNSA that effectively develops and implements technologies for organic market gardens and farms.

---

5. As part of the plan, a series of surveys were developed and continued over time, including the First National Survey of Risk Factors conducted every four years (2005, 2009, 2013 and 2018). These surveys have made it possible to diagnose and assess trends in the consumption of critical foods over different periods (daily consumption of fruit and vegetables, use of salt and sugars); the National Nutrition and Health Survey (2005 and 2019) (ENNyS); the Global School-based Student Health Survey (2012, 2018) and in prevalent diseases caused by a poor diet: such as obesity, diabetes, heart disease, and high blood pressure.
6. ProHuerta has an extensive network of educational institutions, canteens, schools, municipalities and NGOs, coordinated with the National Plan for Local Development and the Social Economy, *Manos a la Obra,* the National Families for Social Inclusion Plan, and PROFEDER.

The PNAS seeks to integrate, coordinate and supplement initiatives relating to nutrition and obesity prevention with those developed in other areas as well as with the current MSN programs. To ensure efficiency and effectiveness, the possibility is being considered of tying the program into campaigns promoting healthy eating and lifestyles and access to medical care and drugs. An analysis of the PNAS shows a change of perspective in the design of food policies, from a focus on vulnerable populations to a policy with a universal profile, and from a focus on food welfare to a comprehensive approach to preventing disease and promoting better health promotion, encompassing nutrition, the environment, production, processing and the distribution and consumption of food.

Although these programs entail the participation of all the players involved in the production and distribution of food, the processes underpinning the social construction of policies do not take account of the mechanisms of citizen participation or the guidelines associated with the promotion of sustainable food systems.

With reference to monitoring and evaluation systems, Aulicino and Langou (2012) note that the PNSA makes formal provision for a continuous (monthly) evaluation of processes (management control), results and impact, both from a quantitative and qualitative standpoint, by looking at a set of indicators for: coverage and satisfaction of beneficiaries, frequency of the benefits delivered by the program, nutritional contribution, level of execution of funds, contribution of funds from different levels and sectors, cost of benefits and unit cost of food benefits. They also point out that plans are under way to assess compliance with the agreements signed by the providers on accessibility, quality of benefits and the satisfaction of beneficiaries, as well as to conduct audits with the MDS Internal Audit Units (UAI), the Federal Internal Audit Agency (SIGEN), the Federal Network of Public Control, and the National General Auditor (AGN). The evaluation of the PNAS falls within the remit of the MSN, which monitors indicators relating to the components of the plan, for example, the prevalence of risk factors in its determinants (economic, behavioral, mass communication indicators) and the performance of health services, among other factors.

In terms of financing, the PNSA has seen a considerable increase in resources over the years through the Special Fund for Nutrition and National Food. For the PNAS, the required expenditure is covered by budgetary provisions allocated annually to the MSN and/or by special budget lines voted into law and, in some cases, by international organizations and other institutions.

An analysis of both programs reveals the absence of an explicit technocognitive and human resources training strategy for the solution of food problems. Instead, programs are based on and seek to promote innovations in organization and coordination with the programs set up by public institutions, such as INTA's ProHuerta and complementary programs designed to control obesity, for example. Each of these programs has different objectives, characteristics and implementation processes as part of a policy orientation that includes a focus on rights, food security and hunger. While the first program is characterized by its emphasis on relief and welfare, the second adopts a more comprehensive approach. However, both have a common denominator: the fight against poverty and malnutrition. Considering these programs as part of an approach based social technologies would also enable us to move forward in a comprehensive discussion of the problems of inclusive and environmentally sustainable development.

▸▸ Final thoughts

From an analysis of the agenda and social construction processes of food policy, it is clear that this issue has become a public problem since the crisis of the early 2000s and that it occupies a central position in the Argentine socio-economic policy agenda. A number of institutions are devoting considerable material and human resources to the development of programs aimed at ensuring access to food and a healthy diet.

A review of the process applied to policy design, formulation and implementation shows that, as a general rule and with the exception of the initiatives or demands made by different sectors of society, there has been relatively little participation by citizens, the beneficiaries of these programs and the players responsible for supplying and distributing food as part of these processes, particularly as regards assessments relating to the implementation and fulfilment of the goals of food programs. In strategies set up to promote food and good health, the aim is not only to satisfy the demands of the population in terms of the access, appropriation and use of food, but also to generate sustainable and healthy food systems, incorporating the interests of the players involved in production, supply, transport, storage, processing, distribution and consumption. Given that these systems were designed to promote and encourage initiatives that protect and respect biodiversity, ecosystems and the environment, and to implement development processes based on social equity, they constitute a real challenge for concentrated and globalized economies.

Solving the food problem requires partnerships between the public sector and non-governmental entities, including universities, research and development centers, civil organizations and private entities to jointly identify food-related problems, challenges and vulnerabilities. To this end, it is essential to build stronger ties not only between science, technology and food – currently variable in plans and programs – but also between scientific-technical and academic institutions. This will make it possible to turn food problems into research and capacity-building problems in order to build sustainable food systems and drive forward new agendas for inclusive and sustainable development.

▸▸ References

Abeyá Gilardon, E.O. (2016). Una evaluación crítica de los programas alimentarios en Argentina. *Salud Colectiva, 12*(4), 589–604. http://www.doi.org/10.18294/sc.2016.935

Arcidiácono, P., & Carrasco, M. (2012, September 27–28). *Política alimentaria en la Argentina post crisis 2001/2. Una mirada desde la participación social, los derechos y el rol del Poder Legislativo* [Conference presentation]. III Congreso Internacional en Gobierno, Administración y Políticas Públicas, Madrid, España. http://www.gigapp.org/administrator/components/com_jresearch/files/publications/051%20ARCIDIACONO%20Y%20CARRASCO.pdf

Aulicino, C. (2012). *Una primera aproximación a las políticas de Educación Nutricional en las provincias argentinas* (Documento de Trabajo N°90). CIPPEC. https://www.cippec.org/wp-content/uploads/2017/03/2412.pdf

Basualdo, M. A. (2018). *La soberanía alimentaria en las políticas sociales del periodo 2003- 2012 en Argentina: un análisis en tensión* [Undergraduate thesis, Universidad Nacional de Lujan]. UNLu DSpace. http://ri.unlu.edu.ar/xmlui/handle/rediunlu/435

Bijker, W. E. (2005). ¿Cómo y por qué es importante la tecnología? *Redes, 11*(21), 19–53. http://ridaa.unq.edu.ar/handle/20.500.11807/578

Bonet de Viola, A. M., & Marichal, M. E. (2020). Emergencia alimentaria y derecho humano a la alimentación. Un análisis del Programa Argentina contra el Hambre. *Revista Derechos en Acción, 14*(14), 480–512. https://doi.org/10.24215/25251678e366

Borrás, G., & García, J. (2013). Políticas alimentarias en Argentina, derechos y ciudadanía. *Revista Interdisciplinaria de Estudios Agrarios, 39*, 111–136. https://www.ciea.com.ar/web/wp-content/uploads/2016/11/Páginas-desdeRIEA39-a-imprenta.pdf

Britos, S., O'Donnell, A., Ugalde, V., & Clacheo, R. (2003). *Programas alimentarios en Argentina.* Centros de Estudios Sobre Desnutrición Infantil. https://cesni-biblioteca.org/wp-content/uploads/2018/11/35programas_alimentarios_en_argentina.pdf

Calomarde, B. (2018). *Seguridad alimentaria de los hogares en Argentina: Un análisis a partir de la Encuesta de Gastos* [Undergraduate thesis, Universidad Nacional de Mar del Plata]. Repositorio Digital de la FCEyS-UNMDP. http://nulan.mdp.edu.ar/2933/1/calomarde-2018.pdf

Carrasco, M., & Pautassi, L. (2015). Diez años del "Plan Nacional de Seguridad Alimentaria" en Argentina. Una aproximación desde el enfoque de derechos. *De Prácticas y Discursos, 4*(5). http://dx.doi.org/10.30972/dpd.45805

De Martini, S., Carpintero, K. G., Donzelli, B. E., García Rossi, M. I., Aras, F., & Bohl, F. (2016). *Análisis crítico de la implementación de políticas públicas alimentarias en la Argentina.* Observatorio del Derecho a la Alimentación en América Latina y el Caribe. http://www.fao.org/3/I8961ES/i8961es.pdf

Elmore, R. F. (1978). Organizational models of social program implementation. *Public Policy, 26*(2), 185–228.

González Díaz, J. G., García-Velasco, R., Ramírez-Hernández, J. J., & Castañeda Martínez, T. (2013). La territorialización de la política pública en el proceso de gestión territorial como praxis para el desarrollo. *Cuadernos de Desarrollo Rural, 10*(72), 243–265. http://www.scielo.org.co/scielo.php?script=sci_arttext&pid=S0122-14502013000300013

Ham, C., & Hill, M. (1993). *The policy process in the modern capitalist state.* Harvester Wheatsheaf.

Hogwood, B. W., & Gunn, L. A. (1984). *Policy analysis for the real world.* Oxford University Press.

Juárez, P., & Serafín, M. (2010, July 20–23). *Tecnologías para la inclusión social y políticas públicas en América Latina: el caso de la problemática alimentaria en Argentina y Brasil* [Conference presentation]. VIII Jornadas Latinoamericanas de Estudios Sociales de la Ciencia y la Tecnología, Buenos Aires, Argentina.

Maceira, D., & Stechina, M. (2011). Intervenciones de política alimentaria en 25 años de democracia en Argentina. *Revista Cubana de Salud Pública, 37*(1). http://scielo.sld.cu/scielo.php?script=sci_arttext&pid=S0864-34662011000100006

Parsons, W. (2007). *Política pública: una introducción a la teoría y la práctica del análisis de políticas públicas* (A. Acevedo, Trans.). FLACSO.

Rebon, M. & Roffler, E. (2014). *La gestión de las políticas sociales en Argentina: aprendizajes, avances y desafíos futuros.* VIII Jornadas de Sociología de la UNLP. https://www.aacademica.org/000-099/238

Roffler, E. (2010). *Política alimentaria y gasto público en Argentina Desafíos para la construcción de un sistema de protección social.* V Congreso Latinoamericano de Ciencia Política. http://www.aacademica.org/000-036/357

Salomone, A. (2015). Un repaso de la política alimentaria argentina en los últimos 50 años. *Revista Argonautas, 5*(5), 1–22. http://www.argonautas.unsl.edu.ar/files/02%20Salomone%20Anabella.pdf

Salvia, A., Tuñon, I., & Musante, B. (2012). *Informe sobre la inseguridad alimentaria en la Argentina. Hogares Urbanos. Año 2011.* Observatorio de la Deuda Social/UCA. http://wadmin.uca.edu.ar/public/ckeditor/Informe_Inseguridad_Alimentaria___doc_de_trabajo_.pdf

Santarsiero, L. H. (2012). Las políticas sociales en el caso de la satisfacción de necesidades alimentarias. Algunos elementos conceptuales para su determinación. *Revista Trabajo y Sociedad, 18*(15), 159–176 https://www.unse.edu.ar/trabajoysociedad/18%20SANTARSIERO%20Necesidades%20alimentarias.pdf

Serafim, M., & Dias, R. (2010). Construção social da tecnologia e análise de política: estabelecendo um diálogo entre as duas abordagens. *Revista Redes, 16*(31), 61–73. http://www.redalyc.org/articulo.oa?id=90721346003

Thomas, H. (1999). *Dinâmicas de inovação na Argentina (1970–1995): abertura comercial, crise sistémica e rearticulação* [Doctoral thesis, Universidad Estadual de Campinas]. Repositorio UNICAMP. https://doi.org/10.47749/T/UNICAMP.1999.175418

Thomas, H. (2012). Tecnologías para la inclusión social en América Latina: de las tecnologías apropiadas a los sistemas tecnológicos sociales. Problemas conceptuales y soluciones estratégicas. In Thomas, H., Fressoli, M. & G. Santos (Eds.), *Tecnología, desarrollo y democracia. Nueve estudios sobre dinámicas socio-técnicas de exclusión/inclusión social* (pp. 25–76). Ministerio de Ciencia, Tecnología e Innovación Productiva de la Nación, Argentina. http://www.repositorio.cenpat-conicet.gob.ar/bitstream/handle/123456789/453/tecnologiaDesarrolloYdemocracia.pdf

Thomas, H. (2008). *Tecnologías para la inclusión social y políticas públicas en América Latina.* IESCT/UNQ, CONICET.

Chapter 10

Changes in food programs and consumption patterns in Peru between 2004 and 2018

Carolina Trivelli, Carlos E. Urrutia

▸▸ Introduction

One of the most serious problems facing Latin America and the Caribbean is the lack of food security. The partial or total interruption of access to food affects 187 million people across the region. Among the adult population, a level of gender disparity is visible: 69 million women suffer from food insecurity compared with 55 million men. Undernourishment affects 6.5% of the regional population (about 42 million people), while obesity affects 28% of women and 20% of men over 18 years of age (FAO et al., 2019)

Food security has four main aspects: availability, access, utilization and stability (FAO, 2011). The 1996 World Food Summit defined food security as follows: Food security exists when all people, at all times, have physical, social and economic access to sufficient, safe, nutritious food to meet their daily energy needs and food preferences for an active and healthy life.

The importance of food security lies in its impact on the public health of the population and its links with poverty and inequality. Malnutrition means eating too little, too much or having an unbalanced diet that does not contain all the nutrients necessary for nutritional health (Latham, 2002). One of the clearest examples of this is chronic undernutrition in children. This is diagnosed where the height measured is lower than in previously established patterns (Pajuelo et al., 2000).

Other health problems associated with malnutrition are anemia and obesity. Anemia is a condition in which the hemoglobin count in the blood is lower than normal. Caused primarily by iron deficiency, it predominantly affects pregnant women and children (de Benoist et al., 2008). Obesity is defined as the abnormal accumulation of fat. It is diagnosed using pre-established weight-to-height ratios (WHO, 2016).

Chronic malnutrition, anemia, excess weight and obesity are all public health problems associated with food security. They have a serious impact on the development of young people and constitute an obstacle to social well-being. The characteristics, trends and scale of these conditions have varied over time and in the way they affect different populations. Analyzing their specific characteristics in the Peruvian context contributes to a better understanding of the relevant and urgent need for adequate food security policies.

Chronic malnutrition is considered to be an indicator of national development. A prevention approach contributes to the development of children's physical, intellectual, emotional and social capacities (INEI, 2009). In 2007 chronic child malnutrition affected 28.5% of children under five. By 2019, this figure had been cut to 12.4%, according to ENDES survey data. As well as being considered as a significant achievement in terms of public health, this success also highlights the importance of food security policies that are institutionally supported and results-based (Marini & Rokx, 2017). Despite the progress made, however, the rate remains high and data are even more alarming when the regional factor is taken into account. In 2007 45.7% of rural children under five suffered from chronic malnutrition. Although this figure fell to 24.9% in 2019, it is still more than three times the rate in urban areas.

The inequalities observed in rural populations are not the only significant regional divide in chronic malnutrition. In 2018 according to data from ENDES, the region with the highest rate of chronic malnutrition was Huancavelica (32.0%), followed by Cajamarca (27.4%) and Huánuco (22.4%). In 2007 the four districts with the highest rates of child malnutrition were Bambamarca, Condomarca, Gamarra and Sitabamba, all with figures exceeding 90% (INEI, 2009). Three of these districts are in the region of La Libertad and all are largely rural. Similarly, Hernández-Vásquez and Tapia-López (2017) found that 20% of the 1,834 districts across the country had a high prevalence of chronic malnutrition in 2010 compared with 17.2% in 2016.

Concerning the factors associated with the incidence of chronic malnutrition, Sobrino et al. (2014) identified a number of statistically significant variables: education of the mother, families living in the Sierra or at higher altitudes, households with two or more children, being the third child and having diarrhea in the 15 days prior to the survey. The authors also indicate that the decrease observed between 2005 to 2011 did not affect all children equally. It was faster for girls, children aged between 25 and 59 months, children from urban areas, children with more educated mothers, and children living on the coast.

This decrease can be attributed primarily to a raft of social policies introduced in 2006 with the aim of bringing down the high rates of malnutrition. A high level of political commitment has allowed the measures to be continued and improved through to the present. A new national strategy was introduced with the name of *Crecer* (grow). This evidence-based plan, placed the emphasis on incentives and their results, with the aim of changing parental behavior. All of the above initiatives were reinforced through the institutionalization of measures as well as through comprehensive interventions. (Beltrán & Seinfeld, 2009; Marini & Rokx, 2017).

Many of these policies were social programs. According to Mejía-Acosta and Haddad (2014), one of the most important factors in reducing malnutrition is the coordination and active engagement of civil society in the design and implementation of *Crecer* and *Juntos*. The first program focuses on improving water and sanitation services, while also promoting better food and nutrition practices (Sánchez-Abanto, 2012).

244

The second is a social program based on conditional cash transfers in exchange for taking young children to health checkups. Efforts to monitor and encourage growth acted as a catalyst in reducing malnutrition (Marini & Rokx, 2017).

Another public health problem associated with food security is anemia. According to INEI (2019), the rate of anemia in children aged between 6 and 35 months was 56.8% in 2007, falling to 43.5% in 2018. While this reduction is nationwide, differences remain between urban and rural areas. The rate observed last year in rural areas was 10 percentage points at 50.9%. However, the level of anemia in women of between 15 and 49 years of age remained virtually unchanged for almost ten years, with a rate of 21.0 in 2009 and 21.1% in 2018 (INEI, 2013, 2019).

Drawing on earlier data, Loret de Mola et al. (2014) report a significant reduction in anemia in Peru between 2000 and 2011, particularly in women living in rural areas and children living in urban areas. According to the authors, the reduction in levels of anemia across Latin America was driven by strategies to add iron supplements and fortified foods to daily diets. The fact levels of anemia remain high in Peru can be explained by the low iron intake in daily diets and by the fact that these strategies failed to target the most vulnerable.

Concerning these programs, according to Aparco et al. (2019), the Peruvian government decided in 2010 to give children micronutrient powders. The authors conducted an impact assessment based on a propensity-score approach, and found that this initiative reduced anemia levels by 11 percentage points in Apurimac, in the mountains of Peru. In this sense, changing dietary habits is extremely important, as explained by Banna et al. (2016). The authors found that adolescents on the outskirts of Lima knew that eating beans helped to prevent anemia, but nevertheless based their eating habits on online data, family influences, concern for their physical appearance, and purchasing power.

Finally, a public health problem rarely seen as a consequence of food insecurity is obesity. According to data gathered by the INEI (2019), the obesity rate of women aged between 15 and 49 was 25.1% in 2019. This compares with a rate of 15.7% ten years earlier, in 2009 (INEI, 2014). This increase in obesity has been observed since 1996, although figures have fallen significantly for children (Loret de Mola et al., 2014).

Hernández-Vásquez et al. (2016) found a childhood obesity rate of 1.5%, in 2015. The rate was slightly lower for girls (1.3%) and slightly higher in urban areas (1.7%) and in coastal regions (2.2%). Similarly, Sotomayor-Beltran (2018) found childhood obesity to be significantly higher in the coastal areas of Peru in 2017, particularly in the regions of Lima, Moquegua and Tacna. Further, obesity in young and adult women is associated with socioeconomic status. The likelihood of obesity increases with a higher level of purchasing power (particularly in rural areas) and decreases with a higher level of education (particularly in urban areas) (Poterico et al., 2012).

One of the most important measures in controlling obesity was the approval of mandatory octagonal warning labels in advertising for foods containing high levels of sugar, sodium, saturated fats and trans fats. Introduced in June 2019, Supreme Decree No. 017-2017-SA regulating Act No. 30021 – Act for the Promotion of

Healthy Eating – controls the advertising of certain products. According to studies in Chile, this type of measure is effective in cutting sugar consumption, particularly in children.[1]

All of these three problems arise from inadequate nutrition and they all have long-term effects on the health and development possibilities of those affected, leading to high-impact aggregate effects on the country's development possibilities and on the demands made on the health sector for care and coverage.

▸▸ Food security and public policy

These three major problems of chronic undernutrition, anemia and obesity are directly related to food security. However, they have followed different patterns. While chronic malnutrition has decreased significantly, anemia has stagnated and obesity has increased (figure 10.1). The role of social policies and programs is crucial in bringing about change.

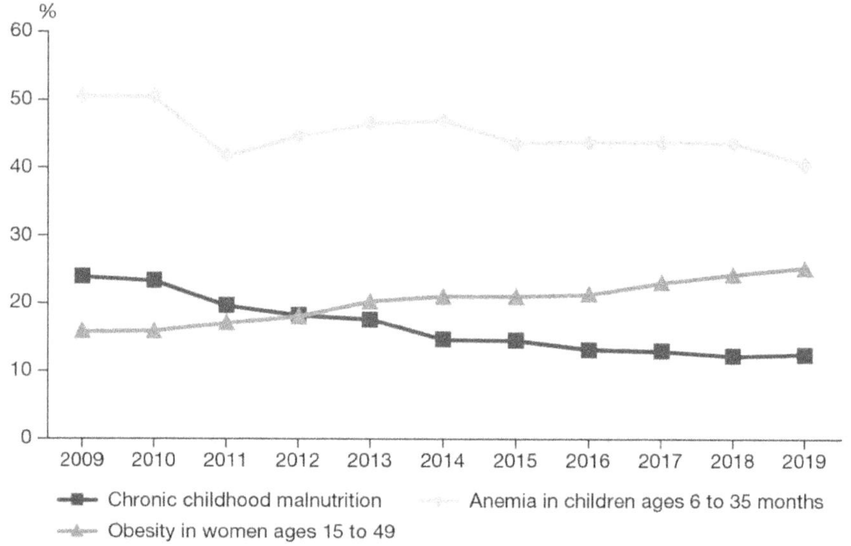

Figure 10.1. Evolution of the main health problems associated with food insecurity in Peru 2009–2019. Source: ENDES 2009–2019.

In the case of Peru, food security policies are primarily the concern of the agricultural sector, while social food programs are the concern of the social sector through the Ministry of Development and Social Inclusion and the Ministry of Health. This prevents a comprehensive approach to food security problems. Instead, as explained below, policies take a specialized and direct approach to each problem, generating a disconnect that does not address food insecurity as a whole.

---

1. https://elcomercio.pe/tecnologia/ciencias/octogonos-advertencia-experiencia-chile-etiquetas-alimentos-procesados-noticia-581259-noticia/

This research focuses on the aspect more closely linked to social policy, with a particular emphasis on food programs. The objective is to characterize changes in food programs between 2004 and 2019 in several stages, and to look at the differences between households in each region. We also aim to observe whether changes in the scale of these programs coincide with changes in consumption patterns for specific product groups, based on their origin.

This chapter is divided into four sections. The first provides a detailed description of the institutional framework developed in Peru in recent years in the area of food security. The second presents a review of the literature on the main food programs of the past 20 years. The third shows changes to the main social food programs in Peru during the period 2004–2019. The fourth details changes in patterns of food consumption by the Peruvian population over the same period. The last part is a summary setting out a number of conclusions, with an agenda for pending research.

▸▸ Public policies on food security

With the return to democracy in 2001, Peru entered a new phase in its republican history and set out to correct mistakes made in the 1990s. In this context, the main political parties decided to establish priorities for public policies that would transcend ideological and political differences. This commitment was referred to as the National Agreement. Signed in 2002, the agreement had four goals: i) strengthen democracy and the rule of law, ii) pursue development based on equity and social justice, iii) increase competitiveness, and iv) establish an efficient, transparent and decentralized state. One of the 35 public policies prioritized was the promotion of food security and nutrition.[2]

As a result of this political commitment, food security took on greater importance in Peru through the creation of the Multisectoral Commission on Food Security and Supreme Decree No. 118-2002-PCM. Chaired by the Ministry of Agriculture, this commission was responsible for drawing up medium- and long-term policies and plans that sought to align all three levels of government as part of efforts to reduce food insecurity. This legal document defined food security as a priority for the State and set up a commission to adapt existing plans and programs to food security objectives.

In 2004 the commission presented the National Food Security Strategy for 2004–2015, approved by Supreme Decree No. 066-2004-PCM. This conceptual document set out the first goals and the main long-term challenges. Its main objective was to "prevent the risks of nutritional deficiencies and reduce levels of malnutrition (...) by promoting a healthy diet and lifestyle, and by ensuring a sustainable and competitive food supply of national origin." The main goals were to reduce chronic malnutrition in children and also micronutrient deficiencies.

At the end of the period covered by the strategy above, a new strategy was developed against a backdrop of growing government concern around food security issues. The creation of the Ministry of Development and Social Inclusion gave new

---

2. http://acuerdonacional.pe/politicas-de-estado-del-acuerdo-nacional/politicas-de-estado%E2%80%8B/politicas-de-estado-castellano/

impetus to social policies leading to a new vision of food security. The National Food and Nutrition Security Strategy for the period 2013–2021 adopted a new, more modern framework concerning the relevance and role of the State in promoting food security. The overall objective was to "ensure that the population is able to meet its nutritional requirements at all times" (MINAGRI, 2013).

The new strategy was backed by the National Food and Nutrition Security Plan 2015–2021 approved by Supreme Decree No. 008-2015-MINAGRI. This plan develops in greater detail the objectives and goals set out by the strategy. At the same time, it specifies which state entities are responsible for their fulfillment and monitoring. The plan sets five specific food security goals (MINAGRI, 2015): i) Availability of food, ii) Access to safe and nutritious food, iii) Adequate intake of safe and nutritious food, iv) Adaptation to food insecurity crises, v) Institutional and program-based framework.

Each goal is under the primary responsibility of one ministry. The Ministry of Agriculture is responsible for the first and fourth goals, while the second and third are mainly under the responsibility of the Ministry of Development and the Ministry of Health, respectively. The fifth is developed by the Technical Secretariat of COMSAN, the food security committee. This means that food policies are designed by the Ministry of Agriculture and social policies are developed by the Ministry of Development and the Ministry of Health. This disconnect is clear when we look at specific details: social policies make no mention of food access and vice versa.

Social issues took on growing importance for the government between 2011 and 2016. Measures developed included initiatives to reduce levels of chronic child malnutrition. At the same time, food assistance programs were renewed and their budgets increased. These measures, along with the creation of the Ministry of Development and Social Inclusion – MIDIS – paved the way for the organization of initiatives and their improved management. The government also passed the Law for the Promotion of Healthy Food for Children and Adolescents and adopted a new official National Strategy for Food and Nutrition Security (Eguren, 2016, Zegarra, 2010).

▸▸ Major social food programs and their impact

One of the essential measures in efforts to improve the nutritional status of the population has been to promote adequate nutrition for Peru's children. Social food programs have played a key role in this respect. These programs were set up to improve the nutritional health of the most vulnerable segments of the population, particularly children. The main programs spearheading this aspect of food security policies over the past 20 years include *Vaso de Leche* (glass of milk) and PRONAA (subsequently replaced by Qali Warma). Both programs first began over two decades ago, but have gone through a number of transformations to reach the point where they are today.

The *Vaso de Leche* program was founded at national level in 1985 through Act No. 24059, after a successful start on the outskirts of Lima in 1984. A political initiative by the then mayor of the capital led to the program gaining funding in all provinces of Peru with the aim of improving food quality.[3] The political context

---

3. https://elcomercio.pe/blog/huellasdigitales/2014/03/las-tres-decadas-del-vaso-de-leche/?ref=ecr

was completely different when Congress passed Act No. 26637 in 1996, taking the program to every corner of the country, and making it one of the few social programs to be truly universal. In 2001 Act No. 27470 set out directives on the content of food rations, the beneficiaries and the administration of resources.

According to Gajate and Inurritegui (2002), *Vaso de Leche* does not have a positive impact on the nutritional status of children benefiting from the program. An impact assessment using the Propensity Score Matching (PSM) method showed the impact to be consistently negative, regardless of the model applied for selecting beneficiaries and the matching method. These findings complement the results of other studies, which indicated that food rations do not meet nutritional requirements, and that health and nutritional quality is not always optimal.

Similarly, Stifel and Alderman (2003) assessed the program's impact on the nutritional status of children from the standpoint of height. The method used public district expenditure on the program as an independent variable, rather than household participation. With various corrections, the results showed that the program has had no significant impact on the standardized height of children, its target population. Furthermore, the results could not be attributed to poor targeting, so the lack of nutrients in the rations could explain the findings.

Other studies found insufficiencies in the way the program works at different stages. Among the most noteworthy findings is the fact that between 40% and 50% of the beneficiaries do not actually belong to the target population, based on either the poverty or age criteria. In addition, studies showed that 71% of the budget allocated by the central government is lost through leakage before reaching the child beneficiary. Finally, the program fails to sufficiently reach the group of beneficiaries owing to problems of filtration (55%) and under-coverage (50%) (Alcázar, 2012; Alcázar et al., 2003; Valdivia, 2005).

Concerning this point, an inspection by the Comptroller's Office[4] in 2011 showed that only 21.6% of municipalities distribute rations for almost the entire year. Moreover, 75.9% of municipalities deliver raw rations and, in many cases, without adequate training for mothers on how to prepare them. Further, the participation of a professional nutritionist was reported in only 12.9% of municipalities, which partly explains why only 27.5% deliver rations with all the nutritional requirements. Finally, in terms of cost, 30.7% reported a real cost that was double or triple the allocated cost.

As a result of the problems above, combined with declining political importance and the expansion of other social food programs, the number of beneficiaries in the *Vaso de Leche* program is falling. Even the criticisms around its operation seem to have come to a standstill, since the last inspection by the Comptroller's Office was in 2011. Currently, the program is facing investigations concerning purchases made from a single large company[5] and its neglect of beneficiaries during the 2020 quarantine.[6]

---

4. https://apps.contraloria.gob.pe/pvl/informes.asp
5. https://ojo-publico.com/1331/familia-lechera-los-millonarios-contratos-del-grupo-niisa-con-el-vaso-de-leche
6. https://rpp.pe/politica/gobierno/coronavirus-en-peru-ivonne-tapia-el-presidente-vizcarra-se-olvido-del-vaso-de-leche-video-noticia-1272060

Another of the main social food programs of the past 20 years is the National Food Assistance Program – PRONAA – set up in 1992 by Supreme Decree No. 020-92-PCM. This program organized actions to combat child malnutrition, with one of its most widely known measures being to provide meals in State schools in both rural and urban areas. The program also delivered food rations to organized women's groups and tuberculosis patients.

In 2002 as part of a reorganization, PRONAA absorbed a number of other food programs with similar objectives. However, there was no real integration since the process took place against a backdrop of decentralization with no transfer of the technical capacities or structural framework. At the same time, the move created overlapping programs targeting the same population, which led to coordination problems and confusion regarding the goals pursued (Alcázar, 2007).

Among the operations managed by PRONAA, the actions targeting younger children had a real and positive impact on nutrition levels and school attendance. However, these effects were diluted by the problems encountered, so operations tended to be short-term only. In children's canteens, the food provided for children under three was found to be unsuitable, while the school breakfasts had no impact on school performance or stature (Alcázar, 2007; Beltrán & Seinfeld, 2010; Cueto & Chinen, 2001; Instituto Cuánto, 2008).

In an attempt to improve the organization of PRONAA's subprograms, it was decided to merge the actions targeting children and schoolchildren with the Comprehensive Nutrition Program – PIN – as part of Directorial Resolution No. 395-2006-MIMDES-PRONAA/DE. The objective of this program is to improve the nutritional status of schoolchildren, with particular emphasis on those suffering from high rates of chronic child malnutrition and poverty, while also increasing educational perfor-mance and school attendance. However, the PIN simply inherited the problems of the old sub-programs, since its design and implementation remained unchanged. For example: deliveries of inadequate rations, a purchasing process favoring local monopolies, a lack of targeting, insufficient rations and unjustified fluctuations in cost (Alcázar, 2016; Beltrán & Seinfeld, 2009).

In view of the persistence of these problems, despite the attempted reforms,[7] PRONAA was brought to an end in 2012 by Supreme Decree No. 007-2012-MIDIS. The newly created MIDIS decided to implement a new program called Qali Warma, to replace the school breakfast program run by the defunct PRONAA. The objec-tive was to continue delivering food rations for schoolchildren, but as part of a new management model. Supreme Decree No. 008-2012-MIDIS created the National School Food Program Qali Warma with the general objective of "guaranteeing food delivery services for children attending State educational institutions at the early education level and (…) at primary level."

In its first months of implementation, Qali Warma achieved high levels of coverage and a higher rate of acceptability since the rations took account of the usual diet of school-children (Romero Lora et al., 2016). In addition, it was set up as a universal program delivering rations for 190 school days, far more than the 90 days budgeted by PRONAA. Moreover, it provides two daily rations for children in districts in the first and second

---

7. https://revistaideele.com/ideele/content/el-cierre-del-pronaa

quintile of poverty. Nevertheless, it also faced difficulties, especially in rural areas, since schools lack the adequate infrastructure for food preparation, and the program makes little provision for training (Defensoría del Pueblo, 2013; MIDIS, 2018).

School food programs have a significant impact at the international level on variables such as school attendance rates, educational achievement, nutritional status, and short-term memory (MIDIS, 2013). In the case of Qali Warma, a preliminary study suggests that it has brought about improvements relating to the short-term memory of girls as well as to school performance (PRISMA, 2019). However, it still seems unlikely that the program will work with local producers, owing to a lack of information and a preference for making purchases in the capital owing to the volume required (Vildoso, 2016).

Despite the problems encountered, Qali Warma has clearly improved the management of food programs. One of the main problems observed in the initiatives reviewed is the lack of monitoring and evaluation mechanisms, or studies on cost-effectiveness (Alcázar, 2016). A program that does not learn from its mistakes will inevitably make them again, and costs will increase if the information is not handled transparently, widening the gap between the program's goals and its results.

▸▸ Changes to social food programs

The National Household Survey – ENAHO – includes a social program module through which it studies members who are beneficiaries of a social food program. This survey is representative at the national and regional levels and by level of urbanization. The content of questions has changed to reflect the termination of some programs and the creation of new ones. Responses have been standardized in order to keep the same categories from one year to the next. Comparable information is available from 2004, with the latest published survey taking place in 2019.

The inference capacity of the survey at the regional level is fundamental to an analysis of the inequalities behind the national averages. Figure 10.2 shows the rates of chronic malnutrition, anemia and obesity by area. Although the figures reflect the patterns of figure 10.1, the levels vary. Rural areas have a higher rate of chronic malnutrition and anemia, while urban area have higher rates of obesity.

The regional inequalities observed in the health problems associated with food insecurity have also had an impact on the organization of food programs. Figure 10.3 shows the households supported by a social food program by level of urbanization. The system of categorization used classifies population centers according to their population: big cities with over 500,000 inhabitants, medium-sized cities with over 50,000 inhabitants, small urban centers with over 2,000 inhabitants and rural areas.

The results show that the number of households supported by social food programs is far higher in rural areas than in urban areas. The bigger the city, the lower the level of participating households. In 2019 the percentage of households with at least one beneficiary was 46.5% at the national level, more than double the figure for big cities and close to half that of rural areas. In 2004 this indicator was slightly higher at 49.1%. From this point, the level of coverage in Peru fell steadily until 2013, when it reached 33.7%, before starting to rise again to its current level.

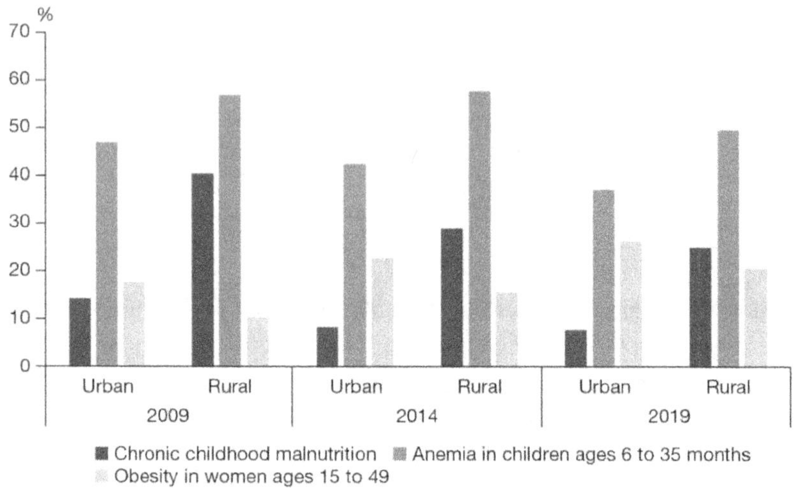

Figure 10.2. Territorial inequality in chronic malnutrition, anemia and obesity in Peru from 2009 to 2019. Source: ENDES 2009–2019.

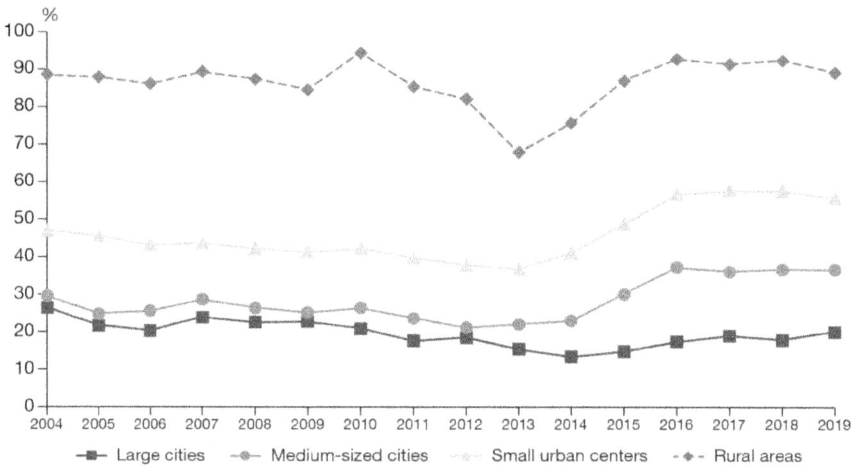

Figure 10.3. Evolution of households benefiting from social food programs from 2004 to 2019. Source: ENAHO 2004–2019.

The pattern of the past 16 years has been similar for the different levels of urbanization, albeit with specific characteristics and phases. We have identified four phases in the coverage of social food programs based on the main trends:

– *First period of stagnation (2004–2010)*: During this phase, the percentage of beneficiary households remained constant for all four levels of urbanization. Overall, little change was observed during this period.

– *Decline (2010–2013)*: Following an increase in the coverage of rural areas in 2010, the trend followed a steady downward path, particularly in rural areas. This downward trend was also visible in big and medium-sized cities, but to a lesser extent than for the other levels.

– *Recovery (2013–2016)*: After reaching a low point in 2013, the number of beneficiaries began to rise once more, peaking in 2016. As in the previous phase, the trend was far more marked in rural areas, but could also be seen in small urban centers and medium-sized cities.

– *Second period of stagnation (2016–2019)*: In recent years, the number of beneficiary households has remained stable overall, with a slight dip in 2019. However, unlike in the first period where no changes were observed, coverage levels are highest in small urban centers and medium-sized cities, showing a more even distribution by level of urbanization.

General trends in the percentage of beneficiaries conceal variations in the figures for each food program. Figure 10.4 shows the proportion of beneficiary households for each major program over the past two decades. The programs with the highest levels of coverage are *Vaso de Leche*, the school breakfasts and lunches provided by PRONAA, soup kitchens and Qali Warma. Programs with only a small number of beneficiaries have been grouped in the "Other" category.

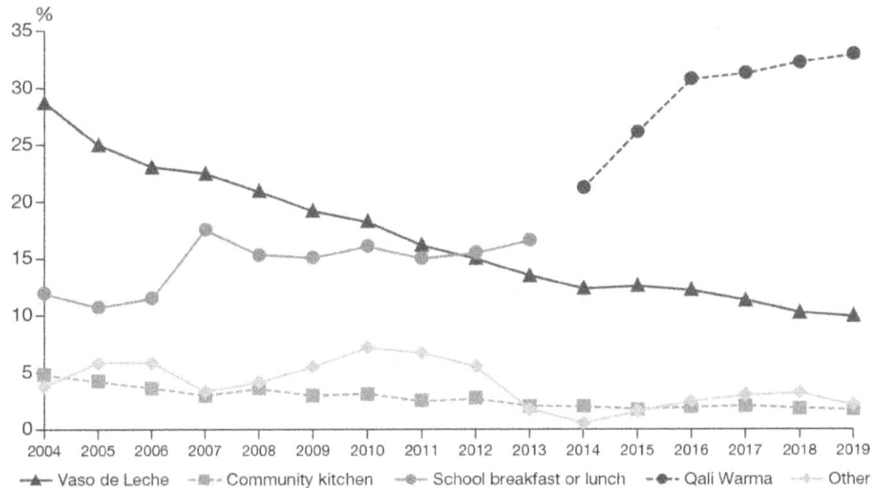

Figure 10.4. Evolution of beneficiaries by food program in Peru 2004–2019. Source: ENAHO 2004–2019.

In contrast with the previous graph, figure 10.4 shows that the trajectory followed by each program is unique and unrelated to the national average. *Vaso de Leche*, for example, has followed a steady downward trend in the number of beneficiaries over time, in the same way, although to a lesser extent, as the soup kitchens. The programs grouped under the label of "Other" rise and fall periodically, possibly as the result of the creation and dissolution of programs with lower coverage. Finally, PRONAA and Qali Warma have reported a steady increase in the number of beneficiaries, albeit with periods of stagnation owing to political and management problems.

Figure 10.4 is based on the same four phases as figure 10.3 but provides an explanation for the general trends:

– *First period of stagnation (2004–2010)*: The continuous fall in the coverage of *Vaso de Leche* was compensated by a steady increase in the meals provided through school canteens. At the same time, a fall in the number of soup kitchens was compensated in some cases by the increased coverage of other programs, leading to overall stagnation at the national level.

– *Decline (2010–2013)*: No program extended its coverage significantly during this period. With the stagnation in the coverage of PRONAA and soup kitchens, the decline of *Vaso de Leche* and other programs became visible at the national level.

– *Recovery (2013–2016)*: The termination of PRONAA and the significant decline in other programs were offset by the rapid growth of Qali Warma, which was the driving force of the recovery observed during this period. Qali Warma offset the stagnation of *Vaso de Leche* and the decline in other programs.

– *Second period of stagnation (2016–2019)*: The change of government marked a slowdown in the expansion of Qali Warma, exactly offsetting the decline of *Vaso de Leche*. This new balance was maintained until 2019, a year before the Covid-19 pandemic led to the closure of schools. The *Vaso de Leche* program is still working with relevant questioning and investigations, but has not been closed.

The coverage of social programs is an indicator that has allowed us to observe trends in social food programs and to understand how these trends correspond to changes in public food security policies. However, it is important to emphasize that the characteristics of the programs cannot be summed up purely on the basis of the number of households involved. The quality of the food provided, its cost, the characteristics of the beneficiary population and the overall impact also stand as key indicators for which the trends have not been considered in this section.

Breaking down the period into phases enables us to summarize the main points in the changing coverage of food programs and to explain trends based on the expansion or reduction of each program. The four phases identified coincide with the policies marking these changes. However, they also show that these policies are not linked with the other programs. On the contrary, although all the programs share the goal of food security, each one follows a different path, reflecting its political importance and results.

It is important to remember that the target population of each program is different, so differences in the rates observed may reflect a smaller target population. In the case of Qali Warma, children are the target population, so the percentage of households without children are not covered by this program. The opposite is true for *Vaso de Leche*, a program with virtually no restrictions.

Complementing this information, data relating to the budgets of social programs show trends in the public funds invested in each program. This gives us a better idea of their importance at State level and in the eyes of the political players concerned. Figure 10.5 shows the total budget for social food programs. In order to fully understand the graph, it is important to remember that the national budget has increased steadily since 2004.

The changes in the total amount budgeted each year for social food programs partially coincide with the phases identified in program coverage. In the first phase, the PRONAA budget increased steadily, but a lack of comparable data prevents the identification of a clear trend for the other programs. The following phases show a stagnation in the *Vaso de Leche* program and a slight but steady decrease for the soup kitchens. This contrasts with a steady increase for PRONAA, carried on by Qali Warma, further accentuating the positive trend, except for a period of one year.

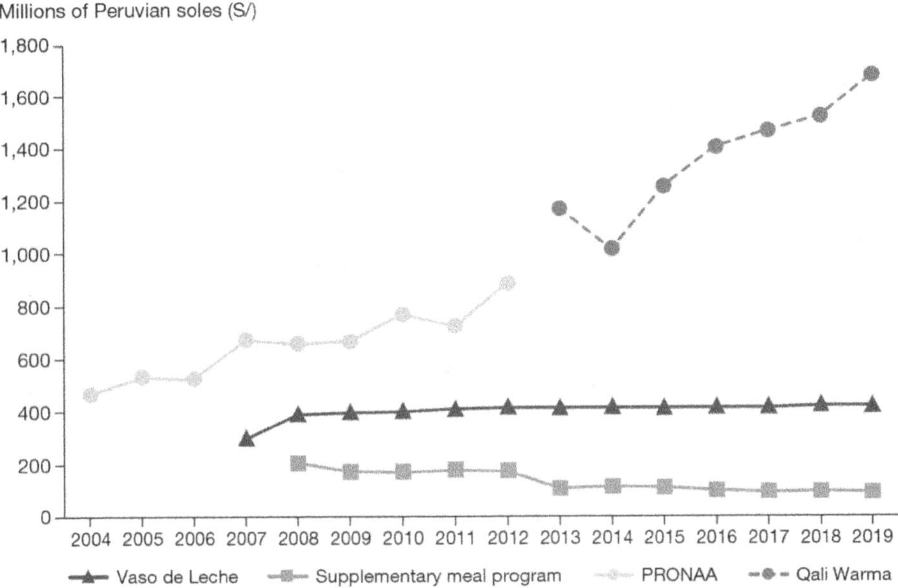

Figure 10.5. Budgetary evolution of social food program in Peru 2004–2019. Source: Economic Transparency – MEF.

These budget data show that the State has given priority to school-based programs in its food security strategy, while programs with a more universal approach have been downgraded over time. The level of coverage and the budgets both show that *Vaso de Leche* and soup kitchens were not a priority in food policy, this being illustrated by a gradual reduction in their capacity.

▸▸ Changes in consumption patterns

The world's food systems are increasingly interdependent and globalized. Information on trends and characteristics play a fundamental tool in public policy decisions on food security. At the international level, we can see a convergence in the consumption of foods of animal origin and sugar and, at the same time, a divergence in the consumption of vegetables, seafood and oil crops. A better understanding of these trends would help to improve food systems and contribute to the goal of ensuring food security at the national level (Benthamet al., 2020).

In relation to this point, the impact of food security on public health depends on the consumption of micronutrients such as iodine, iron and vitamin A. A suitable diet is essential to providing people with the nutrients necessary for their development. In this respect, trends in food expenditure highlight the nutritional status of the population. At the same time, changes in the importance of some products indicate whether policies have an impact on the population's diet (Gómez & La Serna, 2005).

Figure 10.6 shows trends in the cost of the food basket by household between 2004 and 2019. The cost of the basket for each household is equal to the items purchased plus the donated food and beverages consumed inside and outside the household by all its members. This means that food delivered through the social programs described earlier is included in the following analysis. The levels of urbanization are the same as those applied in the previous section and the average cost of the basket at the national level in 2019 was 696 soles, or 208 US dollars.

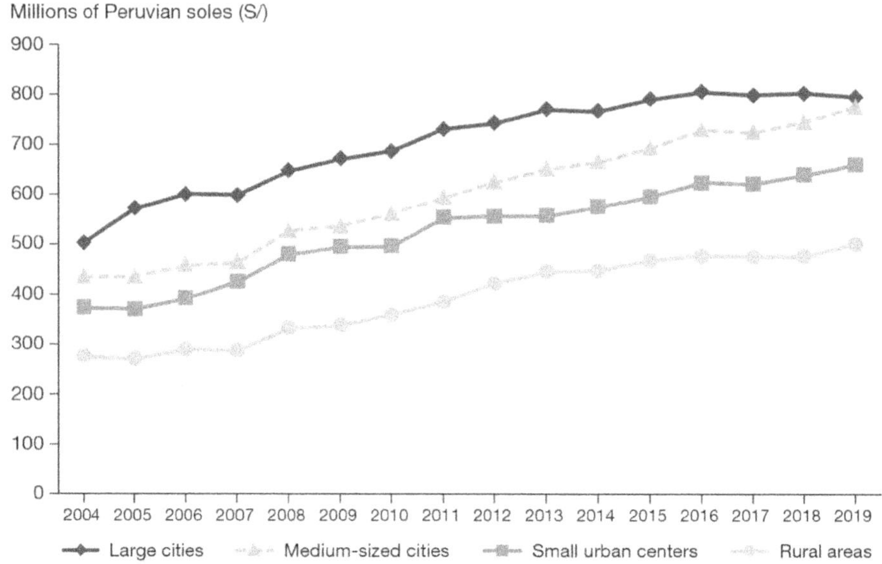

Figure 10.6. Evolution of the value of the food basket in Peru from 2004 to 2019 (value in current soles deflated in each year). Source: ENAHO 2004–2019.

The value of the food basket has increased steadily since 2004, as shown in figure 10.6. This increase has been constant for all levels of urbanization, with the cost increasing according to the size of the city. In 2016, however, the cost remained constant in big cities and the rate of increase for the rest of the country was far lower. As a result, in 2019, medium-sized cities consumed almost the same amount of food as big cities, although the gap remained between small urban centers and rural areas.

The cost shown, however, does not have the same relative importance in total household expenditure. In 2004 the value of the food basket accounted for 25% of total gross household expenditure in big cities, compared with 51% in rural areas. By 2019, the figure was 23% for large cities and 43% for rural areas. This figure of 43% was 10 percentage points higher than in small urban centers and 15 percent higher than in medium-sized cities.

This food basket includes the products consumed by all members of the household, grouped into 50 categories by the ENAHO. For purposes of simplification, these data have been split into seven product groups based on the categories listed in Appendix 10.1. The product groups chosen have been grouped according to origin as shown in table 10.1. Figures show the value of each group as a proportion of the total amount spent on the household food basket for 2004 and 2019.

The first two product groups account for over half of the value of the shopping: cereals, flour and bread, and meat, dairy products and eggs, based on the information in table 10.1. These two categories accounted for 51.9% of the total basket in 2019, compared with 55.2% in 2004. Further, the lowest value in 2004 corresponded to tubers at 7.2%, compared with a slightly lower 7.1% in 2019, equaling the percentage of beverages, sugar and sweets.

Table 10.1. Relative importance of expenditure on the household food basket by product group in Peru in 2004 and 2019

Product group	2004	2019	Difference
Cereals, flour and bread	26.0%	19.3%	–6.7%
Meat, dairy products and eggs	29.2%	32.6%	3.4%
Tubers	7.2%	7.1%	–0.1%
Beans, pulses and vegetables	8.1%	9.0%	0.9%
Fruit	8.2%	11.7%	3.5%
Beverages, sugar and sweets	9.4%	7.1%	–2.3%
Other	12.0%	13.2%	1.2%

Source: ENAHO 2004, 2019. Percentage of expenditure on each product group in the household food basket.

The differences in the percentages shown for each product group between 2004 and 2019 reflect the main changes in consumption patterns in Peru. The sharpest drop is seen in the group made up of cereals, flour and bread, with a fall of 6.7 percentage points. This reduction was offset by an increase of 3.5 percentage points for fruit, and 3.4 points for meat, dairy products and eggs. The group made up of beverages, sugar and sweets, also fell significantly by 2.3 percentage points.

Although these general trends indicate changing consumption patterns at the national level, we can also see some differences by level of urbanization. Figure 10.7 shows the percentages by product group for 2004 and 2019, but by level of urbanization. In both 2004 and 2019, the consumption of meat, dairy products and eggs follows a downward trend with the level of rurality. This trend is reversed for cereals, flour and bread, and to a lesser extent for tubers, while the figures remain similar for other product groups.

In terms of trends between 2004 and 2019, the biggest changes shown in table 10.1 can be seen in both large cities and rural areas. The difference in figures for cereals, flour and bread is even greater in small urban centers, falling by 7.3 points, compared with 5.8 points in rural areas. The increase in the percentage value of fruit can also be seen at all levels, reaching 3.9 points in rural areas. Similarly, the product groups accounting for the lowest percentages show only minor changes for all levels of urbanization.

However, not all trends follow the same path for all the levels of urbanization. The increase observed for meat, dairy products and eggs was far greater in rural areas, although the distribution of animal protein by Qali Warma may have had an impact in this respect. While the percentage value rose by 0.3 points in big cities, it jumped by 4.8 and 5.1 points in small urban centers and rural areas, respectively. Similarly, the decrease observed for beverages, sugar and sweets was also greater in rural areas, with a percentage fall of 3.1 points compared with 1.7 points in big cities.

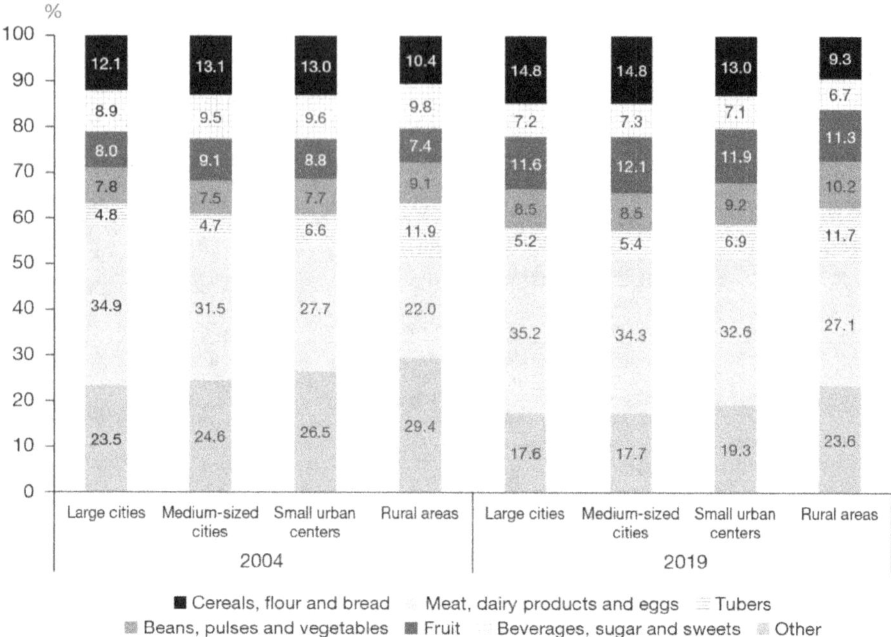

Figure 10.7. Relative importance of food basket expenditure by product group and level of urbanization in Peru for 2004 and 2019. Source: ENAHO 2004, 2019.

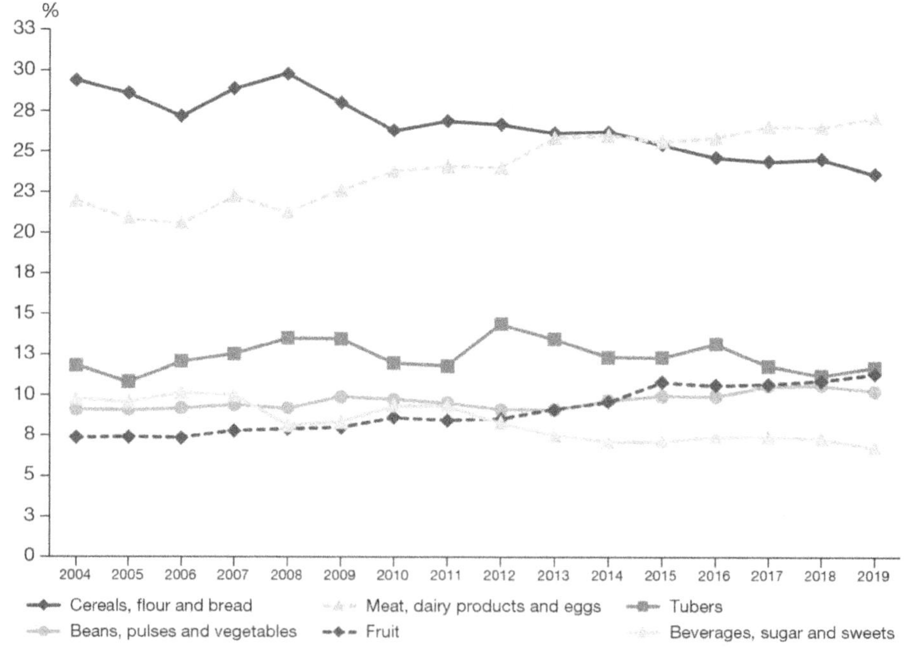

Figure 10.8. Evolution of the value of the food basket by groups of products in rural Peru from 2004 to 2019. Source: ENAHO 2004–2019.

Figure 10.8 shows trends between 2004 and 2019 in the value of each product group for rural areas, where the coverage of social food programs is highest. The graph also shows the four phases identified in the previous section. The only product group excluded is "Other" since it contained no relevant information.

In figure 10.8, we can see that the main changes in consumption patterns in rural areas concern the product groups of cereals, flour and bread, meat, dairy products and eggs, and tubers. In the case of the first group, we can see a constant decrease with periods of fluctuation in the first years. The opposite is true for the second group, where a steady increase can be seen including, as in the first group, periods of fluctuations in the first years. In the case of the tubers group, the change between 2004 to 2019 is not significant, but the rises and falls over the period as a whole merit detailed analysis.

As in the previous section, the definition of phases enables a far better characterization of the changes taking place. They also allow us to view the main changes in food programs against the changes in consumption patterns in rural areas, as shown in figure 10.8. In this way, we can contrast the analyses conducted in the previous sections, taking the four phases as our basis, in order to explore a possible link.
— *First period of stagnation (2004–2010)*: Values remained relatively stable for the various product groups, with the exception of cereals and meat, which fluctuated from year to year. The tubers group showed a slight increase, while a fall was observed in 2009 for beverages, sugar and sweets.
— *Decline (2010–2013)*: The trends observed in the year preceding this period became more pronounced for some product groups during these three years. This is particularly true of the falling and rising trends observed for cereals and meat respectively. In the case of tubers and beverages, sugar and sweets, we can see a fall followed by an increase that returns the value to its starting point at the end of the period.
— *Recovery (2013–2016)*: During this period, the falling and rising trends observed respectively for cereals and meat continue to gather pace. At the same time, tubers begin a steady slide, as do beverages, sugar and sweets, while fruit accentuates the slight increase observed during the previous period.
— *Second period of stagnation (2016–2019)*: In the same way as in the first period, little change can be seen in the product groups. All the values seen in figure 10.8 remain stable, except for tubers, which show a tendency to fluctuate, although figures remain largely stable compared to other periods.

The trends observed in the changing consumption patterns of rural areas highlight two main types of movement: stable and fluctuating. The first category refers to the gradual changes taking place over the long term with little variation from one period to the next. The decrease in cereals, flour and bread falls into this category, as does the increase in meat, dairy products and eggs. These changes may be more closely linked to changes in the purchasing power of households, with incomes rising steadily in rural areas. For lower-income households, especially households living in poverty, cereals are the main source of nutrients rather than meat as in higher-income households (Calderónet et al., 2005; Martínez & Villezca, 2005; Morón & Schejtman, 1997).

The second category refers to fluctuating changes. These are the trends observed for tubers, beverages, sugar and sweets and, to a lesser extent, fruit. The changes seen in these product groups can be characterized on the basis of the phases defined above.

In the case of tubers, the changes in trends are more visible during the decline and recovery periods of social food programs. To a lesser extent, the consumption of beverages, sugar and sweets increased during the decline of PRONAA and decreased with the development of Qali Warma. The consumption of fruit followed a similar trend over the same periods, although figures were more stable.

▶▶ Conclusions

One of the main aims of this paper is to characterize trends in social food programs and consumption patterns in Peru between 2004 and 2019. At the same time, it sets out to review the main public health problems associated with food security and the public policies developed to address them.

The first conclusion to be drawn from the literature review and the compilation of the main public policies on food security is that these policies are highly disjointed in their respective goals. On the one hand, food policies with the emphasis on access to food, are designed by the Ministry of Agriculture – MINAGRI. On the other, social policies with the emphasis on health issues, are designed by the Ministry of Development and the Ministry of Health.

Reflecting the above disconnect, food security policies with a social focus, including food programs or strategies to address chronic malnutrition, make no mention of food-related issues, and vice versa. This disconnect is also visible in the trends observed for chronic malnutrition, anemia and obesity. Given that they are associated with a lack of food security, they have not received the same level of attention.

Changes in the coverage of social food programs have ntalso followed an uneven course. For the second part of our conclusion, we can make a distinction between four phases: a first period of stagnation (2004–2010) characterized by the downgrading of the *Vaso de Leche* program and the expansion of PRONAA; a decline (2010–2013) marked by the stagnation of PRONAA through to its closure; a recovery (2013–2016), with the rapid expansion of the newly created Qali Warma; and a second period of stagnation (2016–2019), marked by the slow growth of Qali Warma and the continuing decline of *Vaso de Leche* (table 10.2).

Table 10.2. Summary table of the main changes by phase between 2004 and 2019 at the national level

Phases	Chronic child malnutrition	Food program coverage	Spending on cereals, flour and bread
First period of stagnation (2004–2010)	Fall from 30%* to 23%	Fall from 49% to 45%	Fall from 26% to 22%
Decline (2010–2013)	Fall from 23% to 18%	Fall from 45% to 34%	Fall from 22% to 21%
Recovery (2013–2016)	Fall from 18% to 13%	Increase from 34% to 47%	Fall from 21% to 20%
Second period of stagnation (2016–2019)	Fall from 13% to 12%	Stable at 47%	Fall from 20% to 19%

*Data from the study by Sánchez-Abanto (2012).

Moreover, the trends in changes observed are far more pronounced in rural areas, where rates are highest for the programs described. For the third part of our conclusion, we can say that consumption patterns in this area have been marked by a steady increase in the value of the food basket. Further, between 2004 and 2019, we can see an increase in the values observed for meat, dairy products and eggs, as well as for fruit, with a decrease for cereals, flour and bread, and beverages, sugar and sweets. Figures can be observed to rise by 4.5% and 3.5% for meat and fruit respectively, and to fall by 4.8% and 2.5% for cereals and beverages.

The research agenda identified will involve further studies on the quality of the food products delivered by the programs. Looking beyond coverage, the services they provide and their impact are determined by their nutritional relevance and the frequency with which food rations are delivered. At the same time, it would also be useful to establish which targeting strategies are the most effective for distributing resources, and to decide whether these programs should work with local producers as suppliers of the food delivered. This last point is particularly relevant in rural areas.

The changes described for consumption patterns focus on monetary value rather than nutritional value. Another relevant point for the agenda would be to study the nutritional value of the food consumed by households, along with any changes and inequalities. Similarly, although a number of coinciding trends were identified in the changes observed for some product groups and food programs, it is difficult to establish a causal link between the two. Impact evaluations that take account of spillover effects in rural areas would help to clarify the extent to which programs contribute to food security beyond their target population.

▸▸ References

Alcázar, L. (2007). ¿Por qué no funcionan los programas alimentarios y nutricionales en el Perú?: riesgos y oportunidades para su reforma. In GRADE, *Investigación, políticas y desarrollo en el Perú* (pp.185–234). http://www.grade.org.pe/upload/publicaciones/archivo/download/pubs/InvPolitDesarr-5.pdf

Alcázar, L. (2012). *Algunas reflexiones en torno a los riesgos y oportunidades de la reforma de los programas alimentarios en el Perú*. GRADE.

Alcázar, L. (2016). Algunas reflexiones sobre los programas alimentarios y nutricionales: cambios y retos durante la última década. In GRADE, *Investigación para el desarrollo en el Perú. once balances* (pp. 251–296). https://nbn-resolving.org/urn:nbn:de:0168-ssoar-51787-3

Alcázar, L., López-Calix, J. R., & Wachtenheim, E. (2003). *Las pérdidas en el camino. Fugas en el gasto público: transferencias municipales, Vaso de Leche y sector educación*. Instituto Apoyo. http://www.grade.org.pe/upload/publicaciones/archivo/download/pubs/LA-perdidas%20en%20el%20camino.pdf

Aparco, J. P., Bullón, L., & Cusirramos, S. (2019). Impacto de micronutrientes en polvo sobre la anemia en niños de 10 a 35 meses de edad en Apurímac, Perú. *Revista Peruana de Medicina Experimental y Salud Pública, 36*(1), 17–25. https://doi.org/10.17843/rpmesp.2019.361.4042

Banna, J. C., Buchthal, O. V., Delormier, T., Creed-Kanashiro, H. M., & Penny, M. E. (2016). Influences on eating: a qualitative study of adolescents in a periurban area in Lima, Peru. *BMC Public Health, 16*, Article 40. https://doi.org/10.1186/s12889-016-2724-7

Beltrán, A., & Seinfeld, J. (2009). *Desnutrición crónica infantil en el Perú: un problema persistente* (Documento de Discusión DD/09/14). CIUP. http://repositorio.minedu.gob.pe/handle/123456789/1514

Beltrán, A., & Seinfeld, J. (2010). Desnutrición crónica infantil en el Perú: un problema persistente. In Portocarrero, F., Vásquez, E., & G. Yamada (Eds.), *Políticas sociales en el Perú: nuevos desafíos* (pp. 141–199). Fondo Editorial, Universidad del Pacífico.

Bentham, J., Singh, G. M., Danaei, G., Green, R., Lin, J. K., Stevens, G. A., Farzadfar, F., Bennett, J. E., Di Cesare, M., Dangour, A. D., & Ezzati, M. (2020). Multidimensional characterization of global food supply from 1961 to 2013. *Nature Food, 1*, 70–75. https://doi.org/10.1038/s43016-019-0012-2

Calderón A, M., Moreno P, C., Rojas D, C., & Barboza del C, J. (2005). Consumo de alimentos según condición de pobreza en mujeres en edad fértil y niños de 12 a 35 meses de edad. *Revista Peruana de Medicina Experimental y Salud Publica, 22*(1), 19–25.

Cueto, S., & Chinen, M. (2001). *Impacto educativo de un programa de desayunos escolares en escuelas rurales del Perú* (Documento de Trabajo 34). GRADE. https://www.grade.org.pe/wp-content/uploads/ddt34.pdf

de Benoist, B., McLean, E., Egli, I., & Cogswell, M. (Eds.). (2008). *Worldwide prevalence of anaemia 1993–2005.* WHO. https://apps.who.int/iris/bitstream/handle/10665/43894/9789241596657_eng.pdf

Defensoría del Pueblo. (2013). *Primer reporte de supervisión al funcionamiento del Programa Nacional de Alimentación Escolar Qali Warma en instituciones educativas ubicadas en zonas rurales.* Defensoría del Pueblo. https://www.defensoria.gob.pe/wp-content/uploads/2018/07/Reporte-Supervision-Qali-Warma.pdf

Eguren, F. (Ed.). (2016). *Seguridad alimentaria en el Perú. Compendio de artículos publicados en La Revista Agraria 2010–2015.* CEPES. https://cepes.org.pe/wp-content/uploads/2019/01/seguridad-alimentaria-en-el-perú.pdf

FAO. (2011). *Una introducción a los conceptos básicos de la seguridad alimentaria.* FAO. http://www.fao.org/docrep/014/al936s/al936s00.pdf

FAO, OPS, WFP, & UNICEF. (2019). *Panorama de la seguridad alimentaria y nutricional en América Latina y el Caribe 2019.* FAO. https://www.fao.org/3/ca6979es/ca6979es.pdf

Gajate Garrido, G., & Inurritegui Maúrtua, M. (2002). *El impacto de los programas alimentarios sobre el nivel de nutrición infantil: una aproximación a partir de la metodología del "Propensity Score Matching."* CIES.

Gómez Gamarra, R., & La Serna Studzinski, K. (2005). Gestión pública y seguridad alimentaria en el Perú. In S. Salcedo Baca (Ed.), *Políticas de seguridad alimentaria en los países de la Comunidad Andina.* FAO. https://www.bivica.org/files/seguridad-alimentaria_comunidad_andina.pdf#page=120

Hernández-Vásquez, A., Bendezú-Quispe, G., Santero, M., & Azañedo, D. (2016). Prevalence of childhood obesity by sex and regions in Peru, 2015. *Revista Española de Salud Pública, 90*, 1–10. https://www.mscbs.gob.es/biblioPublic/publicaciones/recursos_propios/resp/revista_cdrom/VOL90/ORIGINALES/RS90C_AHVingles.pdf

Hernández-Vásquez, A., & Tapia-López, E. (2017). Desnutrición crónica en niños menores de cinco años en Perú: análisis espacial de información nutricional, 2010–2016. *Revista Española de Salud Pública, 91*, 1–10. http://www.redalyc.org/articulo.oa?id=17049838032

INEI (Instituto Nacional de Estadística e Informatica). (2009). *Mapa de desnutrición crónica en niños menores de cinco años a nivel provincial y distrital, 2007.* INEI.

INEI. (2013). *Perú, Encuesta demográfica y de salud familiar 2012.* INEI. http://proyectos.inei.gob.pe/web/biblioineipub/bancopub/Est/Lib1075/index.html

INEI. (2014). *Perú: Enfermedades no transmisibles y transmisibles, 2013.* INEI. https://www.inei.gob.pe/media/MenuRecursivo/publicaciones_digitales/Est/Lib1152/libro.pdf

INEI. (2019). *Perú: Encuesta demográfica y de salud familiar 2018 – nacional y departamental.* INEI. https://www.inei.gob.pe/media/MenuRecursivo/publicaciones_digitales/Est/Lib1656/pdf/Libro.pdf

Instituto Cuánto. (2008). *Evaluación de impacto del subprograma escolar – Nivel primaria desayunos – Fase II.* Instituto Nacional de Salud. https://www.ins.gob.pe/insvirtual/BiblioDig/MISC/PDE07/IF030308.pdf

Latham, M. C. (2002). *Nutrición humana en el mundo en desarrollo* (Colección FAO: Alimentación y Nutrición n. 29). FAO. https://www.fao.org/3/W0073S/w0073s00.htm

Loret de Mola, C., Quispe, R., Valle, G. A., & Poterico, J. A. (2014). Nutritional transition in children under five years and women of reproductive age: A 15-years trend analysis in Peru. PLoS ONE, 9(3), Article e92550. https://doi.org/10.1371/journal.pone.0092550

Marini, A., & Rokx, C. (2017). *Dando la talla: El éxito del Perú en la lucha contra la desnutrición crónica.* Banco Internacional de Reconstrucción y Fomento/Banco Mundial. https://documents1.worldbank.org/curated/en/891441505495680959/pdf/FINAL-Peru-Nutrition-Book-in-Spanish-Oct-11.pdf

Martínez Jasso, I., & Villezca Becerra, P. A. (2005). La alimentación en México. Un estudio a partir de la encuesta nacional de ingresos y gastos de los hogares y de las hojas de balance alimenticio de la FAO. *Ciencia UANL, 20*(1), p. 196–208. http://eprints.uanl.mx/id/eprint/1659

Mejía-Acosta, A., & Haddad, L. (2014). The Politics of Success in the Fight Against Malnutrition in Peru. *Food Policy, 44*, 26–35. https://doi.org/10.1016/j.foodpol.2013.10.009

MIDIS (Ministerio de Desarrollo e Inclusión Social, Peru). (2013). *Nota metodológica para la evaluación de impacto del Programa Nacional de Alimentación Escolar Qali Warma.* MIDIS. https://repositorio.minedu.gob.pe/bitstream/handle/20.500.12799/3969/Nota%20metodológica%20para%20la%20evaluación%20de%20impacto%20del%20Programa%20Nacional%20de%20Alimentación%20Escolar%20Qali%20Warma.pdf

MIDIS. (2018). *Síntesis de los estudios realizados por la Dirección General de Seguimiento y Evaluación del MIDIS en relación al Programa Nacional de Alimentación Escolar Qali Warma (PNAEQW), 2013–2017.* https://glocalevalweek.org/sites/default/files/2021-06/Documento-de-Trabajo-Final4_QW_Sintesis_Estudios.pdf

MINAGRI (Ministerio de Agricultura y Riego, Peru). (2013). *Estrategia nacional de seguridad alimentaria y nutricional 2013–2021.* MINAGRI. https://www.midagri.gob.pe/portal/download/pdf/seguridad-alimentaria/estrategia-nacional-2013-2021.pdf

MINAGRI. (2015). *Plan nacional de seguridad alimentaria y nutricional 2015–2021.* MINAGRI. https://www.midagri.gob.pe/portal/download/pdf/seguridad-alimentaria/plan-acional-seguridad-2015-2021.pdf

Morón, C., & Schejtman, A. (1997). Evolución del consumo de alimentos en América Latina. In Morón, C., Zacarías, I., & S. de Pablo (Eds.), *Producción y manejo de datos de composición química de alimentos en nutrición* (pp. 57–76). FAO. https://www.fao.org/3/ah833s/AH833S00.htm

Pajuelo, J., Villanueva, M., & Chávez, J. (2000). La desnutrición crónica, el sobrepeso y la obesidad en niños de áreas rurales en el Perú. *Anuales de La Facultad de Medicina, 61*(3), 201–206. http://hdl.handle.net/20.500.12799/3637

Poterico, J. A., Stanojevic, S., Ruiz-Grosso, P., Bernabe-Ortiz, A., & Miranda, J. J. (2012). The association between socioeconomic status and obesity in Peruvian women. *Obesity, 20*(11), 2283–2289. https://doi.org/10.1038/oby.2011.288

PRISMA. (2019). *Evaluación de impacto del Programa Nacional de Alimentación Escolar Qali Warma.* PRISMA. http://disde.minedu.gob.pe/bitstream/handle/20.500.12799/6585/Evaluación%20de%20Impacto%20Programa%20Nacional%20de%20Alimentación%20Escolar%20Qali%20Warma.pdf

Romero Lora, G., Riva Castañeda, M., & Benites Orjeda, S. (2016). *Crónica de una reforma desconocida; experiencia de implementación del Programa Nacional de Alimentación Escolar Qali Warma* (Documento de Trabajo n. 229). Instituto de Estudios Peruanos. http://repositorio.minedu.gob.pe/handle/MINEDU/5319

Sánchez-Abanto, J. (2012). Evolución de la desnutrición crónica en menores de cinco años en el Perú. *Revista Peruana de Medicina Experimental y Salud Publica, 29*(3), 402–405.

Sobrino, M., Gutiérrez, C., Cunha, A. J., Dávila, M., & Alarcón, J. (2014). Desnutrición infantil en menores de cinco años en Perú: tendencias y factores determinantes. *Revista Panamericana de Salud Pública, 35*(2), 104–112.

Sotomayor-Beltran, C. (2018). Obesity among Peruvian children under five years of age in 2017: a Geographic Information System analysis. *Acta Informatica Medica, 26*(3), 207–2010. http://doi.org/10.5455/aim.2018.26.207-210

Valdivia, M. (2005). Peru: Is Identifying the poor the main problem in reaching them with nutritional programs? In Gwatkin, D. R., Wagstaff, A., & A. S. Yazbeck (Eds.), *Reaching the poor with health, nutrition, and population services: what works, what doesn't, and why* (pp. 307–334). The World Bank.

Vildoso, C. (2016). *Contribución de las compras del Programa Nacional de Alimentación Escolar Qali Warma en la dinamización de la economía local.* MIDIS.

World Health Organization. (2016). *Informe de la Comisión para acabar con la obesidad infantile [Report of the commission on ending childhood obesity]*. World Health Organization. https://apps. who.int/iris/handle/10665/206450

Zegarra Méndez, E. (2010). Seguridad alimentaria: una propuesta de política para el próximo gobierno. In Rodríguez, J. S., & M. Tello (Eds.), *Opciones de política económica en el Perú: 2011 – 2015* (pp. 71–106). PUCP. http://repositorio.pucp.edu.pe/index/handle/123456789/173163

▸▸ Appendix 10.1. Correspondence table between the products and their groups

Product groups	Products
Beverages, sugar and sweets	Sugar
	Coffee, tea and similar
	Candy, chocolate and honey
	Alcoholic beverages
	Sodas
	Mineral water and juices
Meat, eggs and dairy products	Milk
	Eggs
	Red meat
	White meat
	Giblets
	Meat by-products
	Beef liver
	Beef tripe
	Other offal
	Fresh fish
	Tuna and similar
	Seafood
	Fresh cheese
	Margarine
	Butter
	Other dairy products
Grains, flour and bread	Bread
	Pastries
	Rice
	Corn
	Wheat and wheat products
	Quinoa and quinoa products
	Pea flour and similar
	Noodles
	Hominy

Product groups	Products
Fruits	Tomatoes
	Lemons
	Mandarin oranges, oranges and papayas
	Bananas
	Other fruits
Pulses, beans and vegetables	Pulses
	Onions
	Carrots or squash
	Vegetables and beans
Tubers	Potatoes
	Sweet potatoes, cassava and ulluco
Other	Oil
	Iodized salt
	Chili peppers
	Seasonings
	Prepared meals
	Other
	Pet food
	Other meals outside the home

Source: National Household Survey (ENAHO).

Chapter 11

Food sovereignty and security policy in Paraguay: a study from the standpoint of policy, politics and polity

SILVIA APARECIDA ZIMMERMANN, DIANA JAZMIN BRITEZ COHENE,
NOELIA RIQUELME

▸▸ Introduction

Paraguay is highly dependent on exports but far less so on imports of primary commodities (FAO, 2019, p. 194). It was South America's fastest growing economy between 2012 and 2017, with gross domestic product (GDP) increasing by around 28%, driven primarily by the country's natural resources, particularly electricity (through the binational hydroelectric power plants of Itaipu and Yacyretá), agriculture and livestock production. These key economic activities accounted for over 60% of Paraguayan exports in 2017 (Villalba, 2019). However, Paraguay is also one of the Latin American countries known for its social inequalities, with high rates of poverty and extreme poverty.

Paraguay has a population of just over seven million. Of this total, 33% of those living in poverty or extreme poverty – the poorest among the population – live in rural areas. Some 73% of the population considered as being in extreme poverty are in rural areas (DGEEC, 2020a). While government programs in the 1990s already included policy proposals for peasant agriculture, the actual institutionalization of public policies for Peasant Family Farming (AFC) in Paraguay began in the 2000s, influenced by the regional discussions of the Mercosur Special Meeting on Family Farming (REAF) and the input from institutions including the Inter-American Institute for Cooperation on Agriculture (IICA) and the Food and Agriculture Organization of the United Nations (FAO) (Wesz Jr., Zimmermann & Ríos, 2018). This marked the start of efforts to provide a formal definition for this public. At the same time, real progress was made in the programs aimed at this social group, both from a legal standpoint (acts, decrees, resolutions), and with reference to the development of public policies, programs and projects, for the benefit of Paraguayan family farmers.

In response to the social, political, and food-related situation, both civil society and the international agencies present in Paraguay have put pressure on the executive and legislative branches to build public policies for food and nutritional sovereignty and security. The first steps were the creation of the National Plan for Food and Nutritional Sovereignty and Security (PLANAL), and the resumption of the process to pass the Framework Act on Sovereignty, Food and Nutritional Security and the Right to Food (FNSS Framework Act). In this way, the choices made by the government through national public policies have driven and influenced sustainable food systems across the country. It is not easy to summarize all of these processes in a few pages as this paper will attempt to do. It is nevertheless an exercise that groups a range of public policies, in some cases overlapping and in others with little or no dialogue between them.

The purpose of this chapter is to describe the main public policies implemented by the government in Paraguay in the areas of food sovereignty and security from the standpoint of the three dimensions of public policy: polity, politics and policy (Secchi, 2014; Jaime et al., 2013). Polity refers to the institutional aspect of politics. It is understood to refer to the institutions and rules setting out the way in which power is organized and shared and how it can be exercised as part of the political order. Politics refers to the inherently conflictual dynamics of politics, encompassing aspects such as power struggles, the behavior of players (political, state, economic and social), and the processes of negotiation and cooperation between political players. Finally, policy refers to the actions or decisions made by the government (public policies), along with their objectives, results and impact (Jaime et al., 2013).

To this end, a literature review was conducted, based on official documents, analyses of statistical data, and semi-structured interviews with the political players involved in implementing and studying food sovereignty and security policies in Paraguay. It took place between November 2019 and August 2020, using two online platforms, WhatsApp and Skype. The study is therefore qualitative in nature, in that interviewees were asked about the policies they consider to be most relevant to food sovereignty and security in their country, including aspects relating to conflict and progress. Seven experts were interviewed, including three technical officers from the FAO; three representatives of organized civil society (Paraguay Peasant Movement, the National Coordination of Rural and Indigenous Women – CONAMURI, and the National Peasant Organization – ONAC), together with one government representative (Senator Hugo Richer, member of the Parliamentary Front against Hunger). The Covid-19 pandemic made access to the interviewees more difficult.

This chapter comprises three parts, in addition to this introduction, the final thoughts and the bibliography. The first part describes the food context in Paraguay, from production to consumption, including local data on the agrarian structure, production conditions and other items. The second part describes the public policies on food sovereignty and security implemented in Paraguay since the end of the 2000s, in the different fields of public policy, including the government initiatives put in place to support the population during the Covid-19 pandemic. The third part describes the main political players involved in food sovereignty and food security policies in Paraguay. It also includes the final thoughts and bibliography.[1]

---

1. We would like to thank the interviewees for their time and to acknowledge the support of the CNPq as part of Universal Call MCTI/CNPq No. 01/2016, and of the Federal University of Latin American Integration through Call PRPPG/UNILA No. 110/2018, Call PRPPG/UNILA No. 137/2018 and Call PRPPG/ No. 80/2019.

▸▸ The food context in Paraguay: from production to consumption

In Paraguay, agriculture has been closely linked to economy and society since the arrival of the Spaniards in the sixteenth century. Agricultural products became a driving force in exports and have been a source of currency income at different periods of history (UGP, 2015). In the early 20th century, for example, exports of tobacco and cotton – the two main crops – made Paraguay part of the regional and global economy.

Paraguay is divided into the Western Region or Chaco (departments of Alto Paraguay, Boquerón and Presidente Hayes) and the Eastern Region (departments of Central-Asunción – the capital district – Amambay, Canindeyú, Concepción, Cordillera, Guairá, Itapúa, Misiones, San Pedro, Alto Paraná, Caazapá, Ñeembucú, Caaguazú and Paraguarí). In 2020, the population numbered around 7.2 million: 3.6 million women and 3.6 million men. Sixty-three percent of the population live in urban areas and 38% in rural areas (DGEEC, 2020b).

The characteristics of each region have a direct impact on the type of agriculture. The Chaco has the most land, but it is sparsely populated and characterized by a scarcity of water. The Western Region has a bigger population, easier access to water, and vast plantations in each department. As a result, the Western Region has a higher concentration of wealth than the Eastern Region, where land ownership is primarily concentrated in the hands of national and foreign rural businesses. This has an impact on the AFC and indigenous populations (Guereña & Villagra, 2016).

Paraguay has two main two agricultural subsectors: industrialized agriculture on large tracts of land using state-of-the-art technology to produce soybean, corn, beef cattle and rice grown under irrigation, and family farms on smaller holdings with poor soil fertility and lower levels of crop production and productivity, primarily for purposes of self-consumption (Adib & Almada, 2017). This subsector rarely uses technology, receives little technical assistance and credit, and has limited marketing capacity. Over the past 15 years, agribusiness crops have expanded over most of the country (Schmalko & Sarta, 2018). According to figures provided by the Ministry of Agriculture and Livestock (MAG), the land used by agribusiness has expanded by almost 3 million ha (118%) nationwide during this period, primarily for soybean, corn, wheat, rice, sugarcane and sunflower, in addition to livestock production.

In 2017 soybean crops were grown in 13 of the 17 departments, covering an area of just over 3 million ha (Schmalko & Sarta, 2018). Soybean is the main export crop – in grain, flour, pellet or oil form – and its derivatives are used for an increasingly wide range of purposes. Nevertheless, this form of monoculture is primarily responsible for the absence of food diversity in traditional agriculture, in the areas where it has expanded over the past two decades (Wesz Jr., 2020). Concerning maize crops, the number of hectares cultivated in 2017 grew most strongly in the departments where agribusiness has been established for longest. Paraguay provided 2% of the world's total maize exports between 2013 and 2016, when it was one of the leading maize exporters (Vázquez, 2017; Schmalko & Sarta, 2018, p. 21). Imas (2018) points out that the maize crop is partly produced and sold by peasants. However, the MAG does not specify the production of agribusiness peasants (Schmalko & Sarta, 2018).

Wheat is cultivated in nine departments around the country and this has not varied over the past 15 years. Nevertheless, production is now rising, based on an increase in the number of cultivated ha (Schmalko & Sarta, 2018). Irrigated rice is grown throughout the Eastern region, primarily for Brazil, which acquired 79% of exports between 2013 and 2017, but also for other countries including Chile, Argentina and Belgium (Schmalko & Sarta, 2018). Rice is a "dangerous" crop owing to its impact on stream beds and biodiversity, to the extent that a number of companies have been sanctioned (Ortega, 2018, p. 18). Sugarcane production is concentrated in the Eastern Region. In 2017 the department of Paraguarí grew almost six times more sugarcane than in 2002. The expansion and industrial production of sugarcane can be attributed in part to the demand for organic sugar on the international market and for the alcohol used in Brazil as fuel. Another factor is the support provided through government policies, which have contributed to the development of the business infrastructure (Schmalko & Sarta). These authors also state that sunflower production grew by 65% between 2002–2017 at the national level, but with a lower impact on exports than other crops. At the same time, livestock production increased in the Western Region, where the department of Alto Paraguay reported an eightfold increase in the number of heads of cattle compared with 2002, a rise of 735%. It was followed by the departments of Boquerón, with an increase of 89% and Presidente Hayes, with an increase of 49%. This last department has been the biggest producer for two decades (Schmalko & Sarta, 2018).

The expansion in agribusiness activities and crops for export in Paraguay is also of concern for smallholdings. Between 2013 and 2017, peasant production continued to decline, with 9,000 ha lost in five years, a fall of 3% (Ortega, 2018, p. 19). The same author points to a fall in the production of fruit and vegetables by smallholdings, which do not have the same capacity for expansion as agribusiness. Two factors have been identified: on the one hand stagnating production in terms of surface areas and volume, and on the other, high imports of commodities (Imas, 2018).

The main crops cultivated by peasants are: garlic, peas, sweet potatoes, onions, strawberries, beans, peppers, manioc, cassava, peanuts, potatoes, beans, sesame, tomatoes, carrots, bananas, lemons, tangerines, oranges and pineapples. In 2012–2013, production of these crops totaled 3.4 thousand tons for around 360 thousand ha. For the 2016–2017 harvest, the figures were 3.8 thousand tons and 375 thousand ha, up by 10% and 4% respectively (Imas, 2018, p. 78).

Despite the difficulties and recognized characteristics of the AFC, agrifood production for markets remains diverse, covering almost the entire country, primarily the Eastern Region where the main area of cultivation is Caaguazú (Imas, 2018). However, based on a crop production sequence of over 21 varieties, the order varies in Central, Alto Paraná, Itapúa and Caaguazú. The demand for food staples is satisfied through imports. Imports of food traditionally produced by the AFC (citrus fruit, other fresh fruit, vegetables and legumes) totaled around 89 thousand tons in 2013, rising to 153 thousand tons in 2017, an increase of 42% (Imas, 2018). These foods made up 47% of the import of products and by-products of plant origin. At the same time, the number of Phytosanitary Import Authorizations required for products entering the country grew by 51% between 2013–2017. With respect to imports of food staples: "In 2016, 98% of potatoes came from Argentina, 96% of garlic from China and Argentina, 78% of locoto from Brazil,

60% of onions from Argentina and Brazil, 56% of tomatoes from Argentina, 50% of watermelons from Brazil, 30% of oranges from Brazil and 10% of pineapples from Brazil" (Imas, 2018, p. 82). Only cassava and banana were produced exclusively by Paraguay.

The author states that the food dependence of Paraguay can be attributed to the lack of a food security and sovereignty policy, as well as to the failure by the State to back the AFC through the implementation of comprehensive public policies for support and protection. At the same time, agribusiness activities use high levels of agrochemicals, leading to the contamination of natural resources and damaging public health. They also take up thousands of hectares for the production of monocultures to meet external demand, eliminating diversity in food production and affecting sustainable food systems. Changes in the material conditions of production have also altered food consumption. Outlining the origins of the Paraguayan smallholding system, Dougham (2011) describes how the indigenous heritage allowed the development of a food system in tune with the ecosystem, particularly in the northern part of the Rio de la Plata basin, based on the production and harvesting of food in quantities that easily exceeded the needs of the population. As a result, the indigenous diet was largely vegetarian, reflecting their farming activities, their ample knowledge of wild fruit, and an abundance of fish in the rivers and streams. They did not depend to any great extent on hunting. The main crops grown by peasants were always based on corn, cassava, peanuts and beans, pumpkin, sweet potato and rice, all with little technical input and based largely on manual cultivation. Today, around 11% of the total population – some 800 thousand people – suffer from malnutrition. We can also observe a high rate of obesity, impacting around 19% of the adult population. At the same time, 12% of the child population is overweight (FAO et al., 2019). Agro-exporting Paraguay has a population with significant rates of excess weight, infant mortality and childhood stunting, owing to food insecurity and other food-related issues.

In May 2019, 70% of the products in the basic food basket had risen in price, particularly vegetables (cabbages, carrots, tomatoes, etc.). In some cases, this increase was close to 40%. Meat was also more expensive (ABC Color, 2019). During the Covid-19 pandemic, the price of the basic food basket fell by 0.2% in April 2020, compared with the previous month, as a result of lower consumption figures (AIP, 2020a). The most significant decrease was in the consumption of beef, owing partly to an oversupply of cattle, and also to a fall in both internal and external demand during this period, following the social isolation health measures adopted by Paraguay and beef importing countries.

▸▸ Public policies for food sovereignty and security in Paraguay[2]

Around 20 public policies relating to food sovereignty and security were identified in Paraguay (Imas, 2019), with research interviewees highlighting the ones they

---

2. The Republic of Paraguay is a unitary state. Its National Constitution stipulates the main characteristic of unitarianism to be state centralism, in which government decisions are made from a central nucleus. This characteristic is extremely important for an analysis of national public policies, which were overly influenced by governmental fluctuations during this period.

considered to be most important. This paper will therefore set out to show how these policies were shaped and modified, terminated or maintained to this day as part of the political process and across the three dimensions of public policy.

The debate around the need for public policies to combat poverty and ensure food sovereignty and security in Paraguay began in the 2000s. At this time, Paraguay had signed up to the Millennium Development Goals, with the purpose of eradicating hunger. The most relevant public policy implemented during this period was the Program of Nutritional Supplements for Schools, *Vaso de Leche* (glass of milk), replaced in 2014 by the Paraguay School Meals Program. At that time, the country was governed by Nicanor Duarte Frutos (2003–2008), of the Colorado Party, who was considered to be behind the first social protection initiatives. He "promoted a non-contributory social protection system, systematically developed from 2003 with the formulation of the first National Strategy to Combat Poverty, Inequality and Social Exclusion, Decree No. 8152/2006 (…)." With this strategy, social assistance was structured around an emblematic program: *Tekoporã* Conditional Cash Transfer, Tekoporã meaning "good life" in the Guarani language. This program was set up in 2005 (Duarte-Recalde, 2018, p. 48).

Also during this period, Paraguay passed Act No. 10559/2000, creating the National Food and Nutrition Commission. In 2006 a need was expressed for a National Food Security Plan with the support of the FAO technical cooperation program. Consequently, in December 2007, work began on the first draft of what would become the PLANAL, with the participation of the FAO, coordinated by the Technical Planning Secretariat for Planning of the Social Cabinet of the Republic, with the participation of social organizations, public and private institutions, NGOs and supporting agencies. Three programs created at that time are still in place today and continue to play an important role: *Tekoporã*, the Comprehensive Nutritional Food Program (PANI) and *Abrazo* (hug), the National Program to Prevent Child Labor.

Tekoporã is based on the experience gained through previous studies between 2003 and 2004. It was first implemented in 2005 as a pilot program in five districts. The Social Action Secretariat was set up through Decree No. 9235/1995 as an agency with the role of identifying, coordinating, managing and supervising the plans, programs, projects and activities put in place to advance social policies with the emphasis on fighting poverty (Fukuoka, 2011). Through Act No. 6137 of 2018, the Secretariat set up the Ministry of Social Development (MDS), which currently oversees the program. The main aim of *Tekoporã* is to improve the quality of life of beneficiaries by helping them to enforce their rights to food, health and education (MDS, 2020)[3]. Minister Mario Varela said that the Paraguayan State is investing 63 billion Guaraní (USD 9,500,000 thousand)[4] every two months to protect the 167 thousand families in the program, of whom 28 thousand indigenous families, 25 thousand disabled people and about nine thousand people with severe disabilities. He also pointed out

---

3. Cash transfers are made up of a fixed amount (food allowance) and a variable amount (family allowance), linked to the number of eligible persons in the household, i.e. children aged between 0 and 18, pregnant women, older adults and people with disabilities. In the case of families belonging to indigenous communities, a single amount is paid (MDS, 2020).
4. Values in dollars are converted on the basis of the rates listed on February 16, 2021 on the website of the Central Bank of Paraguay. Available at: https://www.bcp.gov.py/webapps/web/cotizacion/monedas

that *Tekoporā* has a female focus since, for 82% of the families in the program, the head of the household is female (Foco, 2019).

The National Program for Nutritional Food Assistance (PANI) was set up in 2005 and changed its name in 2011 (UTGS, 2018). Through Act No. 4698/2012 on the Nutritional Guarantee for Early Childhood, the program became the main instrument of public policies relating to long-term nutritional guarantees, highlighting national efforts to promote mechanisms aimed at reducing malnutrition. The National Food and Nutrition Institute of the Ministry of Public Health and Social Welfare is the institution with responsibility for PANI, which delivers nutritional supplements to families (two kilos of whole milk enriched with iron, calcium, zinc, copper and vitamin C). Investment has increased significantly over the years, rising from around 17 billion Guaraní (USD 2,565,000) in 2011 to 81 billion Guaraní (USD 12,222,000) in 2015 (IMAS, 2019). In 2019 the number of users totaled around 56 thousand children, 17 thousand pregnant women and 664 Exceptional Cases, making a total of almost 74 thousand people from the 18 healthcare regions. At the same time, the Ministry awarded National Public Tender 035/2019 to buy 1.3 million kilos of enriched milk powder (MSPyBS, 2020).

The *Abrazo* program began in 2005 as an initiative implemented by non-governmental organizations. In 2007 it was transferred to the Secretariat of Social Action (currently MDS) and since the end of 2008, it has been based in the Ministry of Children and Adolescents. The purpose of the program is to eradicate dangerous child labor and to prevent its development in vulnerable communities (MINNA, 2020). *Abrazo* provides conditional cash transfers, health and nutrition services, food aid and care facilities for children and adolescents up to age 17 (IMAS, 2019). In 2019 the program was present in 10 departments and 26 districts across the country, serving 10 thousand children and adolescents and three thousand families, of whom almost two thousand receive cash transfers and just over one thousand receive basic food baskets (IMAS, 2019). No overall data has been found at national level concerning the budget.

The arrival of the Patriotic Alliance government headed by Fernando Lugo (2008–2012) marked the start of a new period in food sovereignty and security policies in Paraguay. With the support of the FAO, the government launched a plan for the Nutritional Sovereignty and Security of Paraguay (PLANAL), through Decree 2789/2009. For the Senator interviewed as part of this research: "(…) The Lugo Government made a real start on the process of consolidating social protection policies…there was a real strategy for fighting poverty." In his inaugural speech, the President made a personal commitment to eradicate hunger (Riquelme, 2014). In principle, PLANAL brought together a significant number of organizations and institutions, led by the Social Cabinet of the Presidency of the Republic between 2008 and 2013 (Riquelme, 2014), and coordinated by the Technical Secretariat of Planning, managed by the Social Cabinet. All associated programs were led by the Ministry of Agriculture.

The projects generated through PLANAL gave new impetus to various programs in different ministries: i) the Family Agriculture Food Production Program, run by the Ministry of Agriculture, ii) the Indigenous Agriculture and Economy Program, run by the Department of Technical Assistance for Indigenous Communities, under

Department of Agricultural Extension/Ministry of Agriculture, iii) the Comprehensive Nutritional Food Program, run by the Ministry of Public Health and Social Welfare; iv) the National School Meals Plan, run by the MEC, with the creation of the School Meals Directorate. All of these programs set out projected figures through to 2015. PLANAL ran into problems owing to a lack of political support. This prevented any practical cooperation between Ministries. As a result, we could say that: "Although PLANAL adopted an intersectoral and inter-institutional approach, it failed to meet its intended goals, making it difficult to pursue initiatives outside the institutional framework" (Riquelme, 2014, p. 81).

The Program to Promote Food Production through Family Farming (PPA) was set up in 2008, implemented in 2011 and continued until 2016. The PPA sought to promote the supply of institutional services (technical assistance in organization, production, marketing), rural education and the transfer of incentives. Aimed at families in the AFC sector, it gave priority to achieving food security and sovereignty, through a number of initiatives such as: increasing the production of safe, high-quality food, opening new opportunities for the marketing of products from the AFC; increasing the availability (access) of food in terms of both quality and quantity, placing the emphasis on sustainable and diversified production (MAG, 2020).

The Pension Program for Older Adults living in poverty was set up in 2009 through Act No. 3728/2009. It is based on a cash payment equivalent to 25% of the applicable minimum wage, around 550 thousand Guaraní (USD 83) in 2020 (MH, 2020a). The Ministry of Finance is responsible for its implementation with, as of 2016, the Technical Secretariat for Economic and Social Development Planning. This support is available to any Paraguayan aged 65 or older, living in poverty, who does not receive a salary or any other financial aid and who is not currently facing any criminal charges. At the end of February 2020, the program had around 208,000 active beneficiaries (AIP, 2020b).

Following the ousting of President Lugo in a parliamentary coup in 2012, Vice-President Federico Franco of the Liberal Party took office until mid-2013. The subsequent elections were won by Horacio Cartes (2013–2018) of the Colorado Party, who took a number of new measures including, for example, the National Program for the Reduction of Extreme Poverty *Sembrando Oportunidades* (sowing opportunities) and the National Development Plan Paraguay 2030. At the same time, some initiatives were expanded through existing policies and advances were made with legal frameworks.

Sembrando Oportunidades was a government strategy implemented between 2013 and 2018. It enabled families in situations of vulnerability, inside or outside the agricultural sector, to increase their income and access the full range of social services (Plataforma CELAC-SAN, 2020). The Technical Secretariat for Economic and Social Development Planning had responsibility for this strategy. Over five million people benefited from this initiative, for which the government invested more than USD 400 million through to November 2018 (STP, 2018). However, the program did not meet its goal of reducing poverty (IMAS, 2019).

Launched in 2014, Paraguay 2030 pursues three main lines of action: i) poverty and social development; ii) inclusive economic growth; iii) Paraguay's global integration (PND, 2014). The Technical Secretariat for Economic and Social Development Planning is responsible for this plan. In 2019 the plan set out the main processes as part of public policies, with the next stage being to send out a call to the private sector and organized civil society[5]. The Cartes government also set up a Joint Program on Food and Nutritional Security (PC-SAN), implemented with the financial support of the Spanish Agency for International Development Cooperation, which benefited around 3,000 indigenous families and 600 peasant families (STP, 2018). The role of PC-SAN was to protect the vulnerable, to improve food and nutritional security and to reduce maternal and child malnutrition in the departments of Presidente Hayes, Caazapá and Caaguazú.

In late 2013 Paraguay began debating the FNSS Draft Framework Act, No. 6175. The draft was passed by the various commissions in the Chamber of Deputies and Senators. It was approved by the National Parliament in June 2018 but rejected in November by the new president, Abdo Benítez, who argued that it required additional resources. He also argued that Paraguay already had other social programs as part of a comprehensive protection system (COPROFAM, 2019). During the Ordinary Session of May 22, 2019, the Chamber of Deputies accepted the total veto of the Executive with 30 votes in favor, 1 against, 13 blank and 25 abstentions. Consequently, the Project was shelved. This annihilated the efforts made to put in place a formal, coordinated national policy on food sovereignty and security. It meant that all initiatives in this area had to start once again, practically from scratch.

Abdo Benítez of the Colorado Party, took office in 2018 and will remain in office until 2022. The interviewees from his government drew attention to the creation of the Ministry of Social Development (MDS), which has since taken over social welfare and protection initiatives, including *Tekoporã*. They also mentioned the creation of the Integrated Social Information System through Decree No. 4509/2015, as a tool to ensure the proper implementation of social protection programs and projects. The system includes data on the public policies described in this section. An interviewee representing FAO supposes that the Framework Act on Food Sovereignty and Security was not approved precisely because the current government was more interested in consolidating the Social Protection System and rarely talks about food sovereignty and security.

In 2020 to address the impact of the Covid-19 pandemic, Paraguay introduced the food security program *Ñangareko* ("take care of" in Guaraní) and the program *Pytyvõ* ("help" in Guaraní). *Ñangareko* was set up to deliver food packages. However, with the implementation of social distancing measures, a decision was made to implement one-off cash transfers of 500,000 Guaraní (USD 75.44), while the food packages were intended mainly for the Chaco region (MH, 2020b). By May 2020, 260,000 families had benefited from food vouchers and 15,000 families from food packages (MH, 2020b). *Pytyvõ* is a temporary welfare initiative aimed at informal workers or self-employed people who were economically affected by Covid-19.

---

5. Alongside, Paraguay 2030, the government set up an Inter-institutional Commission on Food and Nutritional Security (FNS) with the aim of promoting the Paraguayan Institute of Indigenous Affairs by linking and coordinating inter-institutional initiatives with the indigenous communities as part of the Food Security Program, acting in an advisory capacity and monitoring the commitments made by public and private institutions and indigenous organizations in this area (PARAGUAY, 2016, p. 47). We recommend another article by the authors in this same book, which deals with the policies of food sovereignty and security for indigenous peoples in Paraguay.

It involves a cash transfer of 550,000 Guaraní (USD 83), with up to two payments per person for up to two members per family. Until May 2020, *Pytyvõ* involved about 1.1 million beneficiaries (MH, 2020b). *Ñangareko* is run by the National Emergency Secretariat (SEN) in coordination with other institutions. Its budget comes from the SEN of the Presidency of the Republic. *Pytyvõ* is the responsibility of the Ministry of Finance. Its budget comes from part of Paraguay's debt with the World Bank (USD 1.6 billion), of which USD 426 million has been allocated to social protection measures, including the two programs (MH, 2020b).

Table 11.1 below shows the main National Programs already detailed, as well as the main Acts, Decrees and organizations with responsibilities or involvement in the institutional framework of food sovereignty and security in Paraguay. The programs are described in the order in which they were set up, alongside the associated Acts and Decrees. The Institutions responsible are listed in alphabetical order, some of them being responsible for more than one program, as described earlier. Figure 11.1 shows the timeline of public policies. This highlights, with even greater clarity, the correlation between governments, their initiatives and changes in national public policies on food sovereignty and food security over time. The diagrams show the three dimensions of public policy in Paraguay.

Table 11.1. Components of national policies on food sovereignty and security in Paraguay, 2020

Policy components		
Policy	*Polity*	*Politics (governmental bodies only)*
Paraguay School Meals Program of (PAEP): 1995/ Act of Universality in 2014 to present	Act No. 10559/2000, establishing the National Commission on Food and Nutrition	Parliamentary Front Against Hunger (Senate and Congress)
Tekoporã Program: 2005 to present	Decree No. 8152/2006, National Strategy to Combat Poverty	Social Cabinet
Comprehensive Nutritional Food Program (PANI): 2005 to present	Decree No. 2789/2009, creating the PLANAL	Driving Group (various government and civil society institutions)
National Program to Prevent Child Labor, *Abrazo*: 2005 to present	Act No. 3728/2009, funded by the Directorate of Non-Contributory Pensions	National Institute of Food and Nutrition of the Ministry of Public Health and Social Welfare
PLANAL: 2009–2015	Act No. 5210/2014, upholding student rights to food and health	Ministry of Agriculture and Livestock
Program to Promote Food Production through Family Farming: 2009–2016	Act No. 6175/2018, Framework for Food Sovereignty, Food and Nutritional Security and the Right to Food (passed by the Senate in September 2018 and vetoed in 2019)	Ministry of Education and Science
Pension Program for Older Adults in Poverty: 2009 to present		Ministry of Finance
Paraguay 2030 National Development Plan: 2014 to present	Act No. 6286/2019, Defense, Restoration and Promotion of Family Farming	Ministry of Children and Adolescents
National Program for the Reduction of Extreme Poverty, *Sembrando Oportunidades*: 2013–2018	Act No. 6137/2018, founding of the MDS	Secretariat of Social Action, now the Ministry of Social Development.
Ñangareko program: 2020		Technical Secretariat of Planning
Pytyvõ program: 2020		Technical Secretariat for Economic and Social Development Planning

We cannot list the names of all the civil society representatives involved here, so they are mentioned further down.

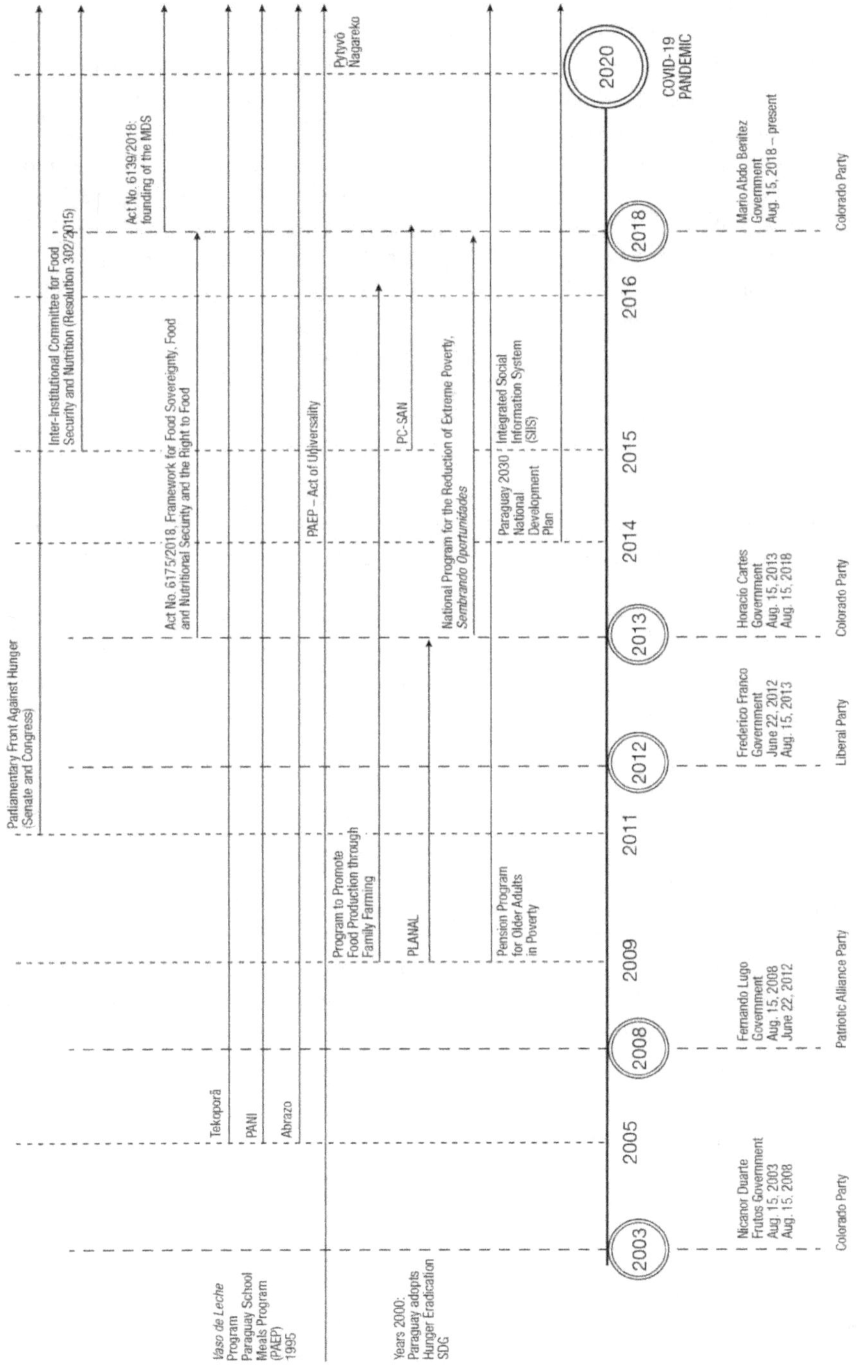

Figure 11.1. Timeline of food sovereignty and security policies in Paraguay, 2000–2020.

▸▸ The political players

The main players include the institutions of the executive branch, with an important role played by: the Secretariat of Social Action of the Presidency of the Republic until 2018 and now the MDS; the Technical Secretariat for Economic and Social Development Planning; the Ministry of Agriculture and Livestock, through the Directorate of Agrarian Extension, the National Institute for Rural and Land Development, the Department of Technical Assistance for Indigenous Communities, the Paraguayan Institute for Indigenous Affairs, the Crédito Agrícola de Habilitación (providing loans to farmers), the Ministry of Finance, the Ministry of Labor and Social Security; the Ministry of Women's Affairs, the Ministry of Public Health and Social Welfare through the National Secretariat for Children and Adolescents, the Ministry of Education and Science; the Secretariat for the Environment and, more recently, owing to the pandemic, by the National Emergency Secretariat, in addition to the support of the government offices and municipalities in the area of influence of the programs. It should also be noted that the Technical Secretariat for Economic and Social Development Planning is responsible for the international agreements that Paraguay signs in other countries and is in close contact with bilateral and multilateral organizations.

The role of the legislative institutions was fundamental in the presentation of the FNSS Draft Framework Act No. 6175 and in the related procedures, particularly as regards the action of the Parliamentary Front against Hunger, formally set up in 2011 in the Chambers of Senators and Deputies. The process nevertheless began in 2009, on the initiative of the Latin American and Caribbean Parliament (Parlatino) and FAO (FAO, 2017). In 2015, on the basis of Resolutions 2842 and 21056, the mandates of the eight deputies and eight senators of the Front were renewed. The Parliamentary Front played a key role in the approval of: i) Act No. 5210/2014 "On School Meals and Health Control," extended by Act No. 6277/2019; ii) the creation and approval of the FNSS Framework Act; iii) Act No. 6286/2019 on the Defense, Restoration and Promotion of Family Farming, approved and published on May 25, 2019. In the same way as the executive branch, the legislative branch also saw a number of changes following the national elections, which brought about a broad (but not complete) renewal of the representatives of the Parliamentary Front and National Parliament[6]. In the opinion of an FAO representative interviewed, the Act for the Promotion of Family Farming is controversial because it entails the creation of a vice-ministry. Given that the FNSS Framework Act and the Act for the Promotion of Family Farming were being discussed at practically the same time, she believes that the reason the government chose only one is that new budgets would be needed or the budgets of the existing ministries would need to be redistributed for the two new vice ministries.

Bilateral and multilateral organizations are just two of the players involved in public policy discussions on food sovereignty and security. One of the main players is the FAO, an active participant in surveys, in the FNS Inter-institutional Commission,

---

6. The changes in Parliamentary forces and interests are considered to have played a role in the acceptance of the president's rejection of the SSAN Framework Act, since the Parliament that passed the Act was not the same as the one that agreed to it being thrown out. At the same time, the Abdo Benitez government seems determined to institutionalize the social protection system, recognizing the public policies of food sovereignty and food security as a part of this.

supporting the Parliamentary Front, etc. Other players fund projects and programs through specific initiatives, such as the Spanish Agency for International Development Cooperation, the Pan American Health Organization/World Health Organization, the World Food Programme, the United Nations Children's Fund, UN Women and the Office of the United Nations High Commissioner for Human Rights, IICA and IFAD.

The political players involved in social organizations and movements have maybe been the most fragile component in recent years. One of the FAO representatives interviewed pointed out: "social organizations in Paraguay are weak or fragmented" in the field of food sovereignty and security. In her opinion, this is partly why there was no real social pressure around the veto of the FNSS Framework Act. In the early stages of the development of the Draft Act, a number of civil society organizations were closely involved: "With the support of the FAO, coordinated by the Vice-Ministry of Agriculture and the Technical Secretariat of Planning, a preliminary draft of the framework act was developed as part of a participatory construction process between February and April 2012" (Fukuoka, 2014, p. 75). The interviewees said that these were different times, different circumstances.

Most of the social movements were against the parliamentary coup that ousted Lugo and, after 2012, they maintained their opposition albeit with less force. The victory of Horacio Cartes "led to a new dynamic in the social struggle with a re-emergence of coordinated action against neoliberal policies. However, the players failed to maintain their alliances and forums of concertation over time, and were unable to influence the policies they challenged" (Palau et al., 2018, p. 68). This gave greater prominence to the role of the Parliamentary Front. With reference to this point: a representative of the FAO said that: "The participation of organized civil society itself is not really representative at present although we do not really know why." She continued: "At the moment there are no high-profile figures promoting actions in this area. Rather, there are projects linked to food and nutrition security initiated by NGOs or international cooperation agencies, but they do not seek to establish public policy (…) they are in day-to-day contact with certain communities but not at the national level." Interviewed about the FNSS Framework Act, a representative of the Association of Women Farmers of Caaguazú (AMUCAP) said: "(…) And about the Law (…) I already talked about it, the organizations had a lot of discussions, but then it was dropped halfway and I think the parliamentarians took it, changed everything and then they pushed us out of the Act. We no longer had information on its scope and we didn't know if our proposal had been incorporated into this draft Act. These are issues for new discussions that we can update and present to the government" (Balbuena, 2020). The SSAN Framework Act changed over the years. In the words of an interviewee representing the FAO: "It was mutilated, because it had no tool to read the workings of the State, such as a council. There was no council, no discussion table, no coordination." She points out: "The laws on food security and family agriculture were strongly supported by cooperation agencies in coordination with public institutions dialoguing with Congress, but the weight of the social organizations is not specified."

In 2013, the platform *Ñamosẽke Monsanto* (Monsanto Out in Guaraní) was set up. Bringing together around 30 urban and rural organizations and institutions, it rejects the extractivism model and the cultivation of transgenic soy, seeking rather to promote food sovereignty, seeds and agroecology. It also involves NGOs for research, the dissemination of information on the situation in Paraguay, and the training of social organizations. The platform has requested support to promote the AFC (Palau et al., 2018).

▸▸ Final thoughts

In the 2000s, the Paraguayan government began to implement national public policies to fight hunger and malnutrition, setting up three of the best-known national programs to date, *Tekoporã*, PANI and *Abrazo*. These programs continue to play a fundamental role in improving the lives of many Paraguayan families. They also marked the start of efforts to develop a non-contributory social protection system. Another highlight is the school meals policy initiated in the 1990s, which became universal in 2014. Over time, a range of public policies on food sovereignty and security have been implemented through the three dimensions of public policy (polity-politics-policy). The FAO, the political entrepreneur of food and nutritional security at multiple levels in Latin America, has played a particularly important role (Beduschi, Faret & Lobo, 2014).

The Lugo government initiated new actions such as PLANAL, a policy coordinating different actions as part of a Plan, "a major attempt to achieve institutional coordination between 2009 and 2012" (Imas, 2019). PLANAL ended in 2013, but other actions continued until 2016, such as the PPA. Lugo's government also began debating the FNSS Framework Act, with strong social participation from civil society organizations, and the creation of the Parliamentary Front against Hunger in the Chamber of Deputies and Senators.

Over the years, social participation has followed a downward path. According to our interviewees, the reasons lie with changes in the government and the loss of opportunities for participation, as well as with changes that were incorporated into the Framework Act and that altered its objectives. This led to a new stage in the action of social organizations, with new priorities. Demands around FNSS took other forms, such as a demand for policies supporting the AFC. Looking at the situation of the Framework Act, we can observe a continuity of the legal process with longer-term action by the Parliamentary Front and the FAO, as well as the more remote position of the non-governmental organizations of civil society.

The Cartes government implemented two broad initiatives, *Sembrando Oportunidades* and the SAN joint program. However, both of these public policies lasted no longer than the government that implemented them. There were some long-term actions, such as Paraguay 2030 and a special program for the indigenous communities, which began in 2015 and is currently being finalized. When Abdo Benitez came to power, he vetoed the SSAN Framework Act, turned the Secretariat of Social Action into a ministry and created an Integrated Social Information System. For one of the interviewees, this reflects a clear choice by the government to give greater visibility to social protection and social assistance actions, while at the same time failing to address these issues from the perspective of food sovereignty and security. No doubt because this would mean addressing issues relating to national food production and consumption, land occupation and the treatment of the indigenous communities. For the current government, talking about social assistance and protection seems to be less complex and less likely to lead to political conflict over the country's development model than talking about food sovereignty and security. Poverty, especially rural poverty, becomes a social welfare problem rather than a development problem. The government has decided to deal with the consequences of social issues, rather than confronting the underlying causes, which are in the policies guiding the country's current development model.

Although both *Pytyvõ* and *Ñangareko* were designed to address the problems of people economically affected by the Covid-19 pandemic, many people in real need received nothing owing to a lack of access to the information necessary for application or for other unknown reasons. Interviewees from social organizations said that few, if any, people belonging to their organizations received benefits. They even believe that only urban dwellers were considered. Nevertheless, they said that the *Olla Popular* (soup kitchen), in which neighbors cook and share food, continues to be a fundamental community-based initiative that ensures a daily meal for many.

The diversity of actions described in this paper has not been sufficient to eradicate poverty in Paraguay. Further, it can be seen that the best coordinated initiatives were implemented only for a short time or were unable to make progress, such as PLANAL. At the same time, some public policies remain on the government agenda to this day and have been altered through decrees or new laws. Finally, a question that we were unable to study in greater detail: is it possible to guarantee food sovereignty and security while maintaining the current development model, which is still growing strongly through the two key components of land concentration and the intensive production of commodities for export? Reality shows that there are limits to an amicable coexistence, and that breakthrough changes will be necessary if governments really want to eradicate hunger and guarantee the sustainable development of Paraguay.

▸▸ References

ABC Color. (2019, May 14). *Setenta por ciento de los productos suben de precio*. Editorial AZETA. https://www.abc.com.py/tv/abc-noticias/setenta-por-ciento-de-los-productos-suben-de-precio-1814783.html

Adib, A., & Almada, F. (Eds.). (2017). *Políticas públicas y marcos institucionales para la agricultura familiar en América Latina*. IICA. https://repositorio.iica.int/bitstream/11324/4114/1/BVE17089182e.pdf

Agencia de Información Paraguaya (AIP). (2020a, May 5). *Productos de la canasta básica familiar se abarataron en 0,2% en abril, según BCP.* Ministerio de Tecnologías de la Información y Comunicación. https://www.ip.gov.py/ip/productos-de-la-canasta-basica-familiar-se-abarataron-en-02-en-abril-de-2020-segun-bcp/

Agencia de Información Paraguaya (AIP). (2020b, March 5). *Un total de 207.712 adultos mayores reciben pensión alimentaria al cierre de febrero*. Ministerio de Tecnologías de la Información y Comunicación. https://www.ip.gov.py/ip/un-total-de-207-712-adultos-mayores-reciben-pension-alimentaria-al-cierre-de-febrero/

Agencia de Información Paraguaya (AIP). (2020c, April 4). *Gobierno benefició hasta la fecha a 41.662 familias por medio de Ñangareko*. Ministerio de Tecnologías de la Información y Comunicación. https://www.ip.gov.py/ip/gobierno-beneficio-hasta-la-fecha-a-41-662-familias-por-medio-de-nangareko/

Balbuena, M. (2020, July 9). *Ciclo de charlas "Pandemia del Covid-19," Reflexiones de mujeres campesinas, Es hora de crear un nuevo mundo* [Facebook Live dialogue]. https://www.facebook.com/CentroDeDocumentacionYEstudios/videos/810283739503497

Beduschi, L. C., Faret, P., & Lobo, L. (2014). Un marco conceptual para el análisis de experiencias de promoción de políticas públicas de seguridad alimentaria y nutricional en América Latina y el Caribe. In FAO, *Cooperación internacional y políticas públicas de seguridad alimentaria y nutricional* (pp. 37–56). FAO.

Confederación de Organizaciones de Productores Familiares del Mercosur Ampliado (COPROFAM). (2019, May 29). *Diputados aceptan el rechazo del Ejecutivo a la Ley de Derecho a la Alimentación.* https://coprofam.org/2019/05/29/diputados-aceptan-el-rechazo-del-ejecutivo-a-la-ley-de-derecho-a-la-alimentacion/

Dirección General de Estadística, Encuestas y Censos (DGEEC). (2020a). *Principales resultados de pobreza monetaria y distribución de ingreso – 2019.* DGEEC.

Dirección General de Estadística, Encuestas y Censos (DGEEC). (2020b, January 2). *La población de Paraguay en el año 2020.* DGEEC.

Doughman, R. (2011). *La chipa y la soja: la pugna gastro-política en la frontera agroexportadora del este paraguayo.* BASE Investigaciones Sociales.

Duarte-Recalde, L. R. (2018). La institucionalidad de la política de asistencia social en Paraguay. In Andrenacci, L., Campana, M., & M. Servio (Eds.), *La asistencia social en Argentina y América Latina. Avances, problemas y desafíos* (pp. 46–65). PEGUES Rosario.

FAO. (2017). *Frentes parlamentarios contra el hambre e iniciativas legislativas para el derecho a una alimentación y nutrición adecuadas [Parliamentary fronts against hunger, and legislative initiatives for the right to adequate food and nutrition].* FAO. https://www.fao.org/3/i7872s/i7872s.pdf

FAO, FIDA, OMS, PMA & UNICEF. (2019). *El estado de la seguridad alimentaria y la nutrición en el mundo 2019. Protegerse frente a la desaceleración y el debilitamiento de la economía [The state of food security and nutrition in the world 2019. Safeguarding against economic slowdowns and downturns].* FAO. https://www.fao.org/3/ca5162es/ca5162es.pdf

Foco. (2019, December 13). *Tekoporã alcanzó a más de 167.000 nuevas familias.* Grupo Nacion. http://foco.lanacion.com.py/2019/12/13/tekopora-alcanzo-a-mas-de-167000-nuevas-familias/

Fukuoka, M. P. (2011). *El Estado y la garantía del derecho a la alimentación adecuada en Paraguay.* BASE Investigaciones Sociales.

Fukuoka, M. P. (2014). *Derecho a la alimentación en la administración Cartes.* BASE Investigaciones Sociales.

Guereña, A., & Rojas Villagra, L. (2016). *Yvy jára: Los dueños de la tierra en Paraguay.* OXFAM.

Hoy. (2020, May 8). *Ya suman más de 1.150.000 beneficiarios del programa Pytyvô: acreditaciones van en aumento.* Grupo Multimedia. https://www.hoy.com.py/nacionales/ya-suman-mas-de-1.150.000-beneficiarios-del-programa-pytyvo-acreditaciones-van-en-aumento

Imas, V. J. (2018). Producción e importación de alimentos en el Paraguay: aumentando la dependencia alimentaria. In M. Palau (Ed.), *Con la soja al cuello 2018. Informe sobre agronegocios en Paraguay* (pp. 78–83). BASE Investigaciones Sociales. http://www.baseis.org.py/wp-content/uploads/2018/11/2018Nov_ConLaSojaAlCuello-2018.pdf

Imas, V. J. (Ed.). (2019). *Seguridad y soberanía alimentaria en Paraguay. Sistema de indicadores y línea de base.* Centro de Análisis y Difusión de la Economía Paraguaya (CADEP). https://www.conacyt.gov.py/sites/default/files/upload_editores/u294/seguridad_soberania_alimentaria_cadep.pdf

Jaime, F. M., Dufour, G., Alessandro, M., & Amaya, P. (2013). *Introducción al análisis de políticas públicas* (1st ed.) Universidad Nacional Arturo Jauretche. https://www.unaj.edu.ar/wp-content/uploads/2017/02/Pol%C3%ADticas-públicas2013.pdf

Ministerio de Agricultura y Ganadería, Paraguay. *Programa de fomento de la producción de alimentos por la agricultura familiar.* MAG. http://www.mag.gov.py/index.php/programas-y-proyectos/programas-y-proyectos-cerrados/ppa

Ministerio de Hacienda, Paraguay. (2020a). *Programa pensión alimentaria para adultos mayores en situación de pobreza.* Ministerio de Hacienda. https://www.hacienda.gov.py/web-sseaf/index.php?c=181

Ministerio de Hacienda, Paraguay. (2020b). *Paraguay ante la pandemia. ¿Qué se está haciendo?* Presupuesto, Ministerio de Hacienda. https://ppr.hacienda.gov.py/wp-content/uploads/2020/06/Informe-PARAGUAY-COVID.pdf

Ministerio de la Niñez y la Adolescencia (MINNA), Paraguay. *Abrazo*. MINNA. http://www.minna.gov.py/pagina/229-abrazo.html

Ministerio de Salud Pública y Bienestar Social, Paraguay. (2020, February 12). *Más de 56 mil niños asistidos por desnutrición gracias al PANI*. Ministerio de Salud Pública y Bienestar Social. https://www.mspbs.gov.py/portal/20405/mas-de-56-mil-nintildeos-asistidos-por-desnutricion-gracias-a-pani.html

Ortega, G. (2018). La agricultura campesina durante el gobierno de Horacio Cartes. In M. Palau, (Ed.), *Con la soja al cuello 2018. Informe sobre agronegocios en Paraguay* (pp. 18-23). Base Investigaciones Sociales. http://www.baseis.org.py/wp-content/uploads/2018/11/2018Nov_ConLaSojaAlCuello-2018.pdf

Palau, M., Coronel, C., Irala, A., & Yuste, J. C. (2018). *Canalización de demandas de los Movimientos Sociales al Estado paraguayo*. Base Investigaciones Sociales, 2018.

Riquelme, N. (2014). *Políticas públicas de soberanía y seguridad alimentaria en Paraguay: PLANAL (2008–2013)* [Undergraduate thesis, Curso de Desarrollo Rural y Seguridad Alimentaria, Universidad Federal de la Integración Latino-Americana]. UNILA. https://dspace.unila.edu.br/bitstream/handle/123456789/648/TCC.Noelia.Silvia.FINAL.pdf

Schmalko, C. A., & Sarta, A. M. (2018). *Mapeando el agronegocio en el Paraguay*. Base Investigaciones Sociales.

Secchi, L. (2014). *Políticas públicas. Conceptos, esquemas de análise, casos práticos* (2nd ed.). Cengage Learning.

Secretaría Técnica de Planificación del Desarrollo Económico y Social (STP). (2014). *Plan nacional de desarrollo Paraguay* 2030. Presidencia de la República del Paraguay. https://www.stp.gov.py/pnd/wp-content/uploads/2014/12/pnd2030.pdf

Secretaría Técnica de Planificación del Desarrollo Económico y Social (STP). (2018a). *Apuntes de experiencias y lecciones aprendidas: Esta es la historia que construimos en el periodo 2013–2018*. Dirección de Comunicación, Presidencia de la República del Paraguay. https://www.stp.gov.py/v1/wp-content/uploads/2020/03/STP_-Apuntes-de-experiencias-y-lecciones-aprendidas-2013-2018.pdf

Secretaría Técnica de Planificación del Desarrollo Económico y Social (STP). (2018b, January 2). *Más de USD 400 millones fueron invertidos en el programa Sembrando Oportunidades*. Presidencia de la República del Paraguay. https://www.stp.gov.py/v1/mas-de-usd-400-millones-fueron-invertidos-en-el-programa-sembrando-oportunidades/

Secretaría Técnica de Planificación del Desarrollo Económico y Social (STP). (2019, September 26). *Avanza construcción del Plan Nacional de Pueblos Indígenas en Paraguay*. Presidencia de la República del Paraguay. https://www.stp.gov.py/v1/avanza-construccion-del-plan-nacional-de-pueblos-indigenas-en-paraguay/

Unidad Técnica del Gabinete Social (UTGS). (2018). *Programa Alimentario Nutricional Integral (PANI)*. Presidencia de la República del Paraguay. https://www.gabinetesocial.gov.py/archivos/documentos/PANI-docu-eva_uhfqt18i.pdf

Unión de Gremios de la Producción (UGP). (2015). *Agricultura y desarrollo en Paraguay*. UGP.

Villalba, A. M. (2019). *Políticas públicas diferenciadas para la agricultura familiar campesina. Paraguay (2004–2017)*. CLAEH.

Wesz, Jr., V. J. (2020). A rentabilidade dos produtores de soja no Paraguai: concentração e exclusão. *Estudos Sociedade e Agricultura, 28*(1), 156–179. https://doi.org/10.36920/esa-v28n1-7

Wesz, Jr., V. J., Zimmermann, S. A., & Carreras Rios, F. D. (2018). La institucionalización de políticas públicas para la agricultura familiar en Paraguay. *Raízes: Revista de Ciências Sociais e Econômicas, 38*(1), 80–97. https://doi.org/10.37370/raizes.2018.v38.40.

Part IV

Building food policies and actions in regions and cities

Chapter 12

The process of politicizing food sustainability in the city of Brasilia: towards a transition of the local food system?

STÉPHANE GUÉNEAU, MAURO G. M. CAPELARI,
JANAÍNA DEANE DE ABREU SÁ DINIZ, JESSICA PEREIRA GARCIA,
TAINÁ BACELLAR ZANETI

▶▶ Introduction

There is widespread recognition in the literature that food systems are responsible for greenhouse gas emissions, the depletion of groundwater resources, the irreversible loss of biodiversity, and other socio-environmental damage (Clark et al., 2019; Béné et al., 2020). Unsustainable food systems lead to the destabilization and loss of the regulatory functions on which human life depends (Campbell et al., 2017; Springmann et al., 2018). Many studies have thus highlighted the need for a transition to sustainable food systems (Stassart et al., 2012; Hinrichs, 2014; Jurgilevich et al., 2016; Caron et al., 2018; Springmann et al., 2018; Vermeulen et al., 2020). Given this need, cities are regarded by several authors as places of social mobilization and as important drivers of the innovation that can make this transition to sustainable food systems a reality (Brand et al., 2019).

Historically, urban food policies have aimed, above all, to guarantee the supply of food to large cities, thus guaranteeing urban consumers a sufficient and diverse food supply (Daviron et al., 2019). In Brazil, this issue has gained in importance since the 1970s due to the acceleration of the rural exodus, which has led to the concentration of its population in cities. According to data from the last demographic census, in 2010, the urban population accounted for 85% of Brazil's total population (IBGE, 2011).

Largely the responsibility of the federal government, the supply logistics of Brazil's major cities were given a structure with the creation of the National Supply System (SINAC) and the Supply Centers (Ceasa) network, which had been planned for the nation's cities as early as the 1970s (Cunha, 2006). The municipalities were responsible for only part of the food policies. Compared to other issues such as transport, housing and the environment, the food system was considered far more of a rural concern than an urban one (Pothukuchi & Kaufman, 1999).

The situation changed in the 1990s with the advent of high prices for agricultural commodities. As a result of these price rises, Brazil's poorest urban populations had serious problems accessing food (Hoffmann, 1995), leading municipalities to develop their own public food policies (Hawkes & Halliday, 2017). The city of Belo Horizonte was a pioneer in this area, implementing actions to combat food insecurity, which, in part, served as a model for the Zero Hunger Program, launched in 2003 by the Brazilian government with the aim of eradicating hunger and extreme poverty in Brazil (Rocha & Lessa, 2009).

In recent years, urban food issues seem to have become more diverse and complex in nature. Though the need to supply urban populations with sufficient quantities of healthy food continues to be an issue of concern for policymakers, public action seems to be more expansive in nature, as it relates to the sustainability of the entire food system (Dury et al., 2019). Specifically, these challenges result in a desire to transform a system that, on the one hand, involves the reduction of the waste, by-products and excessive pollution generated by all processes linked to the production and consumption of food (Alexander et al., 2017) and, on the other, to the reduction of the food system's contribution to climate change (Vermeulen et al., 2012), and lead ultimately to the improved management of natural resources used in food production (Westhoek et al., 2016), which equates to influencing agricultural production models to make them more sustainable (Eakin et al., 2017). Moreover, in many cities dependent on distant food supplies, the stated desire of a growing number of their inhabitants, especially the wealthier and educated classes, to reduce the environmental and social impact of their food consumption involves a desire to reconnect the city with the countryside (Bricas, 2019).

Although Brazilian cities account for a large part of the country's population and have acquired considerable economic and political weight, the question of how urban actors face these new challenges is still little researched in Brazil, albeit with some exceptions (Preiss et al., 2017; Niederle & Schubert, 2020). Nevertheless, Brazil's agrifood system, which is largely dominated by agribusiness and supported by federal public policies, is strongly criticized for its disastrous socio-environmental effects (Sauer, 2018).

The issue that we propose to address in this chapter is the transformation of the initiatives promoting the transition to more sustainable food systems that are emerging in Brazilian cities into policy, all within a context of the dominant position agribusiness enjoys in the country. We chose to address this issue through a case study of the city of Brasília, including the urban area of the Federal District. This area is home to 3 million people and has witnessed sustained demographic growth (IBGE, 2019). As the capital of Brazil, Brasília has considerable symbolic and political weight. It is also one of the cities with the highest income per capita in Brazil (IBGE, 2020).

On the basis of this case study of Brasília, we aim to answer the following two questions in this chapter: Which actors are working to transform food systems in Brasília? What actions are they engaging in to turn the issue of food system sustainability into the subject of local public action? Through these research questions, we aim to provide a wider analysis as to whether the politicization of the food issue in one of Brazil's major urban agglomerations is helping to bring about a specific and structural transformation of the food system and make it more sustainable as a result.

▶▶ Methodology

In order to explain the processes of development and change in public action as it relates to food systems in Brasília, a methodology was applied that involved, first of all, analyzing the main documentary sources: academic literature, opinion articles, and the websites of different groups of actors involved in debates on food.

Next, a round of interviews based on the sociology of public action was conducted. This approach analyzes the interactions between actors, the ideas they hold, and the places where these ideas, which are essential elements of the process of politicizing a social fact, are debated and disseminated (Hassenteufel, 2011; Lascoumes & Le Galès, 2012). Public action is, therefore, approached as a coalition of actors who seek to create alliances around causes, groups moved by a system of shared beliefs or by a shared perception of a public problem and the solutions that are best suited to address it (Jenkins-Smith et al., 2014). This analysis grid was applied in 36 semi-structured interviews conducted between 2017 and 2019 with actors involved in the food sustainability debate in Brasília: producers, chefs, members of cooperatives and Community-Supported Agriculture (CSAs) initiatives, social movements linked to consumption, technical services, and local administrations in charge of food-related issues. The sample of interviewees was selected, firstly, on the basis of their participation in events, seminars and district councils discussing food-related issues and, secondly, using the snowball method, through which a key respondent provides access to other contacts for interview (Noy, 2008).

This research work was complemented by a workshop held in Brasília in May 2019 and attended by several of these actors. The workshop provided an opportunity to analyze the innovations and actions implemented with the aim of making the food system more sustainable, using a methodology developed as part of the URBAL project (Valette et al., 2020).

▶▶ Stakeholders' involvement in the food policy-making process in the city of Brasília

In Brasília, the process of putting the sustainability of the food system on the policy agenda has been a gradual one. It began with work that was initially carried out by actors who supported the implementation of alternative approaches to agriculture, such as agroecology. It is only in more recent times that downstream actors of the value crains – i.e., consumers and intermediary consumers (restaurants and schools) – have come together to create a network of actors bound by certain ideas and positions relating to food sustainability.

The boost provided by the advocacy coalition on agroecology

The actors who first put the issue of transforming Brasília's food system on the political agenda were members of another alliance built around the agroecological transition. This alliance was the initiative of a small group of alternative-agriculture-supporting agronomists who created the Association of Ecological Agriculture (AGE) in Brasília in 1988. Some of these AGE activists entered the public sphere in the 1990s, occupying positions of responsibility in the Ministry of Agriculture, Livestock and Supply, in the technical assistance and rural extension services of the Federal District (Emater-DF), and as district deputies (Sabourin et al., 2019). Some of these actors maintained their links with agriculture, owning agroecological or organic plots.

The actors in this group shared a particular vision of agricultural science, based on the principles of non-use of pesticides and respect for natural ecosystems (known as ecological agriculture). Acting across a range of decision-making spheres, these actors stimulated a slow process of institutional change in the 1990s by spreading agroecological ideas. Under their influence, a strong advocacy coalition was formed around agroecology, including many representatives of public bodies, producers, and associations. The activism of this coalition and the circulation of agroecological ideas within political authorities led to the development and implementation of public action in support of agroecology in the Federal District (Sabourin et al., 2019).

Given Brasília's status as the consumer market for most of this organic and agroeco-logical production, a number of urban actors, such as the Ceasa, the Federal District's State Secretariat of Agriculture, Supply and Rural Development (Seagri-DF), the Federal District's Secretary of State for Education (SEE-DF), and other local services joined the advocacy coalition on agroecology to create fresh markets for organic and agroecological products and to supply canteens and restaurants in public schools and other public establishments (hospitals, nursing homes, and shelters for the homeless) in the Federal District. In doing so, they contributed significantly to making the food issue a subject of local public debate and to triggering the first public measures towards a transition to more sustainable food systems.

In 2014, these actors formed a working group on food procurement for schools of the Federal District, which allowed for the circulation of ideas on alternative food systems among new groups of actors, in particular nutritionists and SEE-DF officials responsible for the supply of school meals. These ideas led to a gradual change in the way food is supplied to the city.

In 2012, the government of the Federal District launched the Agricultural Produc-tion Procurement Program (Papa/DF) under District Law No. 4.752, of February 7, 2012, in support of other federal food procurement measures, such as the National School Meals Program (PNAE) and the Food Acquisition Program (PAA). The Papa/DF enables the government of the Federal District to purchase food directly from small rural producers and social organizations operating in the agricultural sector. It includes a mode for the procurement of agro-ecological and organic products for distribution at public establishments, such as school and nursery school canteens (Sabourin et al., 2019).

Farmers' markets provide another cornerstone of public policy in support of sustain-able food and have grown significantly in size in Brasília, under the leadership of Emater-DF, Ceasa and Seagri-DF. In 2019, there were about 50 farmers' markets in the Federal District, the best known of them being the Ceasa's family farming and agroecological market. According to Emater-DF, the city's high levels of schooling and income account for the high demand among its consumers for healthy food, which thus explains the success of these markets. Health concerns are greater in higher income brackets. Enjoying greater wealth, they are better able to buy more expensive organic and agroecological products. Given that the farmers of the Federal District produce on a small scale, farmers' markets are their main distribution channel, as they do not produce in sufficient quantity to access institutional markets.

Emater-DF, a central actor in the transition of the local food system

As the body responsible for implementing most of the policies in support of agroecology and organic production, Emater-DF has played its part in bringing together the rural world and the urban food actors in Brasília.

A significant proportion of the technical support that Emater-DF provides for agroecology is focused on the smallholdings of family farmers. This public organization has a large number of technicians covering the area devoted to family farming in the Federal District, which has resulted in close ties between technical support staff and family farmers. On interview, an Emater-DF official said: "What perhaps makes the difference here is that there is a larger number of technicians. As well as providing focus, you get to give this personalized service, to go to the smallholding. At other Emater-DF offices, they do a lot of work with the collective."

This in-depth knowledge of the actors involved in family farming allows Emater-DF to provide a key link between those responsible for overseeing urban food policies and farmers. As a result, cooperation between Emater-DF and the agencies responsible for family farming support programs and federal and local procurement, among them the National Supply Company (Conab), Seagri-DF and SEE-DF, has been strengthened. According to an Emater-DF program manager, these organizations have strong, interdependent ties:

> We ourselves depend on their resources, on the signing of the agreement between the Secretariat and the Ministry, which gives us the ability to take this policy out to the farmer. (...) And the Secretariat depends on us to reach out to farmers and engage with them.

Cooperation between Emater-DF and various actors, including Seagri-DF and the social family farming and retailer movements, led the government of Cristovam Buarque to launch the Program for the Vertical Integration of Small Family Production (PROVE) in 1995. The objective of PROVE was to sell local products in large supermarkets in Brasília. However, the program was quickly shelved following a change of government in the Federal District (De Carvalho, 2005).

Emater-DF operates directly in the urban space through various programmes. First, Emater-DF implements an urban agriculture program. According to its manager, this program aims "to promote, first and foremost, food production in urban areas." In reality, however, the volume of food produced in urban areas is very small and the program focuses more on educational activities, offering support for community gardening, mainly in schools:

> We essentially work with schools, health centers, hospitals, youth detention centers, Papuda Prison, government bodies, be it education, health, etc. (...) Our biggest client is schools. We go to around 100 to 120 state schools in the Federal District every year. The work we do in schools is mainly teaching-related. That's the focus.

According to Emater-DF, this action is underpinned by the belief that by educating students at urban schools about agroecology, they can influence their consumption patterns and encourage them to consume healthier and more natural products: "We saw that as educators we can train new consumers. So for rural producers to have people who eat more vegetables and fruits (...) It's interesting for us to have good consumers."

Second, Emater-DF's Center for Technological Training and Rural Development (Centrer) offers a series of food technology courses given by specialists. This center was initially located outside Brasília and mainly dealt with requests from the rural population. Its main objective is to increase the ability of farmers to add value by enhancing their capacity to process basic agricultural products. Centrer later moved to the central area of Brasília, close to the main organic food market, in the Ceasa complex. The training courses offered to family farming communities were gradually made available to the urban population, to directly attend to the requests of city dwellers looking to produce their own processed food and be less dependent on the food industry, or to sell healthy, homegrown processed products. In the words of Centrer's head:

> We have the housewives, the people who have very large gardens and want to use that space in some way and fill it with fruit trees, vegetables or medicinal plants. And then there are the social welfare institutions. We receive a lot of requests from churches, support centers, spiritual centers, and from organizations working with vulnerable people. These are where requests are popping up and where people are getting involved. (...) We try to bring farmers and city dwellers together at our training sessions, where they can interact with each other, and where they can promote what they have to offer and strike up relationships. Farmers' markets provide a place where the producers can sell their products and build up their networks of consumers.

Through the Centrer's actions, Emater-DF is also involved in safeguarding food heritage, an issue that urban populations are becoming increasingly aware of through the consumption of non-conventional food plants (PANCs) (Kinupp & Lorenzi, 2014). According to one of the Emater-DF officials interviewed, many city dwellers want to reconnect with traditional culinary practices: "There are many community vegetable gardens scattered around and they want to work in this way rather than follow the conventional practices showcased at the markets. They want to bring back the culture of our grandparents and parents in working with these types of products." In response, Centrer held a training course on cultivating food products based on PANCs (mainly preserves and seasonings).

Third, although more sporadic and indirect in nature, the central role played by Emater-DF has also helped bring together rural and urban actors involved in the transformation of food systems. For example, a number of chefs who work at restaurants in Brasília and are looking to create cuisine based on local, agroecological and regional products regularly go through Emater-DF to contact agroecological producers and access information on the Community-Supported Agriculture (CSAs) initiatives in the Federal District. One of the heads of the Emater-DF program said: "The most common thing is for people from restaurants to get in touch with Emater-DF to find out who the farmers are (...) Then they call the farmers and buy from them at the gate. It is a niche area that is growing."

Several restaurants that have adopted a more alternative approach by offering, for example, a range of dishes based only on agroecological and/or local products, confirm the key role Emater-DF has played in making a wide variety of food available in Brasília. One of them said: "We had help from Emater-DF. They took us to Ceasa, to rural seetlements and agroecological producers."

▸▸ Citizen initiatives that strengthen the advocacy coalition

Some Federal District inhabitants with an awareness of the issue of sustainable food have initiated actions that go beyond individual consumption choices of agroecological and organic food. For example, Brasília's first collective vegetable gardens were created 20 years ago by citizens who wanted to carry out agroecological experiments in urban spaces. The manager of Emater-DF's urban agriculture program said: "Our oldest community vegetable garden, in São Sebastião, must have been created around 15 to 20 years ago. It was the first community vegetable garden in the Federal District. (...) So, they took action, but they didn't have the program."

Emater-DF believes that citizen investment in collective vegetable gardens in the Federal District is gradually increasing, especially among Brasília's wealthier classes, as they look to create new social ties:

> In terms of urban agriculture, it is the middle and upper-middle classes that have taken action. (...) These are people who are not looking for a financial return, for cheap food. They are concerned about the city and the environment, and they have money to spend and time on their hands, which is the most difficult thing. Something else that motivates a lot of people is the chance to interact.

Some citizens involved in these urban agriculture initiatives have created urban agriculture groups as part of the Nossa Brasília Movement.[1] Although it is not particularly well organized, the group is led by some prominent figures with strong links to the advocacy coalition on agroecology, which successfully campaigned for the introduction of a law in 2012 (Act No. 4.772/2012) that provides specifically for policies supporting urban agriculture in the Federal District. This law was complemented by Act No. 6.671/2020, permitting the use of underutilized public and private urban spaces in the Federal District for the development of urban agriculture activities. The law prohibits the use of pesticides and the planting of genetically modified plant species in these spaces.

In terms of consumption, food safety issues are seen as a major topic of public interest, not least because of health scandals, such as "Operação carne fraca,"[2] which are frequently brought to light by health control agencies. Though still somewhat limited today, the political activism of Brasília's consumers can be traced back to the late 2000s and their participation in organizations such as the CSAs and the international Slow Food movement.

Created in November 2009 in Brasília, the Convivium SlowFood Cerrado is made up of actors closely involved in food issues (chefs, consumers, academics, members of

---

1. http://www.movimentonossabrasilia.org.br (accessed on 19 May 2020).
2. In 2017, "Operação carne fraca" (Operation "weak meat") revealed a meat adulteration scheme involving at least 30 slaughterhouses.

cooperatives, urban farmers, etc.). The group devotes its energies to highlighting the biodiversity of the Cerrado (a vast tropical biome covering more than 20% of Brazil, where natural vegetation is gradually being replaced by cash crops and pastures) and activities implemented by traditional communities to conserve it through its sustainable use as food (Duarte et al., 2020). The Convivium organizes food festivals that promote the Cerrado's biodiversity products, undertakes projects supporting the sustainable use of the biodiversity of the biome, and carries out actions that aim to combat food waste. These actions take the form of "Disco Xepa" festivals, during which volunteers are invited to gather and cook food that would otherwise be disposed of because it does not meet trading standards. The meals are then served free of charge. In Brasília, these events are carried out with the support of Emater-DF and Ceasa, which lays on the facilities and equipment of Centrer, next to the main organic and agro-ecologic farmer's market.

Although the public authorities do not develop projects directly with the Convivium SlowFood Cerrado, they do support its political actions. In the words of one manager of Emater-DF:

> SlowFood is seen as a way of shaking people up and saying: It can be done. We can change mindsets and habits. They have the expertise for this. Centrer provides SlowFood with the physical support it needs in terms of machinery and equipment. In that respect, we do meet demand.

In recent years, the inhabitants of Brasília have engaged in many other initiatives in an effort to transform the way in which food is distributed and consumed, such as CSAs, the number of which have grown exponentially. The conferences organized by the European members of CSAs at the University of Brasília since 2012 have prompted some NGOs, among them Mutirão Florestal, to create their own CSAs in Brasília, the first three of them founded in 2015. The political activity of these pioneering initiatives led to the formation of a support network for the creation and consolidation of CSAs in Brasília, through communication channels on social networks, emails, and round tables. Within a short time, working groups were set up and the CSA Brasília website was created,[3] while newspapers and TV and radio programs reported on the initiative. The objectives of the CSA Brasília Network are "to promote a supportive and healthy culture of food production and consumption." By 2020, the city had 36 CSAs, the largest such community in Brazil.

However, several of Brasília's CSAs have a very unstable and small membership base (members are known as *co-agricultores* or "farmer partners") and have problems in finding the money to continue operating. On interview, CSA members said the movement's political activity is limited, with each CSA focusing mainly on its own production and marketing activities.

In the final analysis, many of the initiatives set up to create direct supply chains between producers and consumers lack sufficient influence to obtain political support. In the words of one of Seagri-DF's program managers:

> A lot of things happen without the support of the state. A lot of farmer's markets operate without any government regulation. There are CSAs that operate outside

---

3. https://CSAbrasilia.wordpress.com/

government control. In terms of consumption, though, we can see that the biggest support the market can give is to strengthen the institutional market, which I believe is the main instrument that we now have in public policy for promoting sales and distribution.

These comments highlight the weakness of networks and the relative lack of political power that urban consumers have in transforming food systems. Consumer demands for change remain relatively unstructured and depend heavily on the activism of other groups. In this whole issue, the alliance around agroecology and organic farming is undoubtedly the most active, though there are two other groups of actors that also influence the political process as it relates to sustainable food: nutritionists and chefs.

▸▸ Changing attitudes among nutritionists

Nutritionists are regarded by producers as key actors that can truly bring their influence to bear in the transition towards more sustainable urban food systems, as explained by a representative of a farmer cooperative of the Cerrado:

> The challenge is to put what is ours on the market. Everyone's heard of soda, right, but no one knows about *coquinho*[4] juice. So, the interesting thing here is to change this idea that "what is ours is exotic, and what in the industry is commonplace." If chefs and nutritionists don't want to put it on their menus, they don't. We need to start doing is to change that mindset.

In the past, many nutritionists responsible for implementing food procurement programs were reluctant to include a certain proportion of family farming products on menus, particularly in school canteens. The food issue is widely addressed in Brazilian cities and links are made between food and health, with the focus on the nutritional characteristics of food being understood in a broad sense. In other words, food safety characteristics of these family farming products are taken into consideration. Generally speaking, family farming and agroecology products are often seen by nutritionists as posing more of a health risk than food industry products, despite the fact that agro-industrial production involves the use of a large number of pesticides and chemical additives. A Seagri-DF program manager said:

> Nutritionists do not generally put typical family farming products on menus. They tend to go for cornstarch biscuits, margarine and bread, for example. But they have a problem putting homegrown things like a jam from the Cerrado biome on school menus and in school lunches.

As part of the working group on food procurement, the members of the advocacy coalition on agroecology started to put forward their ideas for closing the gaps between family farmers and nutritionists. According to the heads of Emater-DF's food procurement programs:

> When we were in the working group, we raised the nutrition team's awareness of this. (…) When we were putting the menu of vegetables and fruits together, we worked as one to decide which ones to order. They also made a commitment to put seasonal products on the menu too.

---

4. *Coquinho azedo* is a fruit of the Cerrado biome.

Many nutritionists have agreed to come on board as a result of this effort to change mindsets, with many industrially processed food products now being omitted from menus at public establishments. A school nutrition head at the Federal District's Ministry of Education (SEE-DF) said:

> Some 80% of our items are now based on natural products or are minimally processed. And only three of 62 items are ultra-processed. (…) We now serve 30 family farming items out of these 62, including fruits and vegetables. We have seen exponential growth in supply from family farming since 2015. We've been serving products since 2013, but since 2015 we've seen exponential growth, which has also come about because of the strength of the Federal District decree, and because the ministry's representative – the director – has formed a monitoring group with the Ministry of Agriculture's head of institutional procurement and with Emater-DF's head of public procurement.

Passed in 2015 and known as the School Canteens Act, Federal District Decree 36.900 prohibits, within the Federal District school network (commercial canteens located in schools, areas next to educational establishments, parties, events, etc.), the sale of certain foods that are particularly harmful to the food and nutritional safety of children (soft drinks, chocolate, sweet biscuits, fried food, etc.).

As a result of their interaction with the other actors that make up the working group on food procurement, the scope of action of the SEE-DF's nutritionists was expanded to include issues that extend beyond nutrition. Though their main focus initially was the health and nutritional quality of food intended for supply to school canteens in the Federal District, they gradually began implementing sustainable food education projects. A SEE-DF nutritionist said:

> When we talk about school food, we always talk about school lunches and whether food is nutritional enough or not. But what we forget is that it's all a question of education and that we have to work on educating children. That's why we include food and nutrition education activities on the school timetable for a few weeks. Last year, for example, we focused on food waste. There's a week devoted to cultural heritage, which is when we include food heritage.

▸▸ Chefs as intermediaries

Restaurant chefs are seen as key players in the transformation of urban food systems because they play a direct role as intermediaries between producers and end consumers. Restaurants in Brasília that have said they are part of the alternative movement have shown an interest in buying agroecological produce from local producers. Among other things, this creates ties with CSAs.

Some restaurants in Brasília have set up their own CSAs, such as Girassol restaurant, which now supplies itself. Due to a wide range of organizational constraints, however, the relationship between CSAs and restaurants in Brasília is not straightforward, as confirmed by the representative of a CSA:

> I think the most difficult thing is the organization of the restaurant. A lot of restaurants also need to take a more flexible approach to keep up with production because production is seasonal and there are different periods. (…) A restaurant might have the same set dishes, but we can't always have the same food, because of seasonality.

Some restaurants and CSAs have tried to bring in and establish organizational innovations, not least through the creation of Communities Supporting Restaurants (CSR). The CSR principle is based on trust, with restaurant owners agreeing to buy a basket of produce, on a regular basis, from a community of farmers whose products and production methods they are familiar with. This innovation has not been successful, however, as the representative of a CSA pointed out:

> We were trying to make the CSR the same kind of partnership as we have with the CSA. So, you pay at the end of the month. (…) We knew that to turn the CSA into the CSR, into a restaurant supporting farming, or something like that, that we would need them to give us a list of essentials that they use every week, on average. The problem was none of the restaurants ever sent us a list. We had a basic list that we talked about in a meeting, and which was written down, but we didn't know exactly what they wanted from one week to the next.

The way in which the restaurants of Brasília influence the transformation of food systems is not so much related, therefore, to the direct impact of their culinary activities than to the things they say in the public sphere. Some Brazilian chefs with big reputations play, in fact, major roles as political entrepreneurs, as one of the representatives of the Central do Cerrado cooperative pointed out:

> None of the restaurants that take this approach have the capacity to use these products in any great volume. It's different with the most famous chefs (…) In terms of actual consumption, it doesn't have a major direct impact. (…) If a chef is making dishes with Kalunga monkey pepper, they'll only buy a kilo every six months, but when they talk about monkey pepper in a feature in the local newspaper *Correio Braziliense*, it generates a lot of indirect sales.

This has been confirmed by a number of well-known gourmet restaurants in the city, which use their fame to get certain messages across, as the owner of one of them said:

> We understand that we are not just a restaurant, and I've spoken about us wanting to be a place where we are free to create – both in the kitchen and in terms of concepts – a place that needs to have an appeal, that is really focused on the producer, on sustainability, on the fairest way of doing business with partners, and which is fair with customers too. (…) Sometimes you might not use the product, but you can try to have an impact by talking about the Cerrado and promoting local products, for example.

Some of these restaurants are taking part in an initiative called "Panela Candanga." A network of restaurants created in 2016 by a group of leading chefs, it aims to "discover and rediscover the role of Brasília in national cuisine."[5] The network organizes events showcasing local cuisine, including regular Panela Candanga fairs in Brasília. According to its website, these actions help establish a narrative, a story, that gives local cuisine solid roots:

> Besides extolling the virtues of dishes made with local ingredients, the chefs go a step further and draw attention to what is behind this creative process, where these products are from and how they reach Brasília's tables. The aim is to provide a different culinary experience, to present, through cuisine, a Brasília unknown to its people, and to bring to each dish the history of its ingredients. It is to offer new ways of understanding food and to identify each mouthful with our own culture.

---

5. http://panelacandanga.com.br/sobre/ (accessed on 5 September 2020).

Brasília's chefs also take part in other events such as food festivals organized by the Convivium SlowFood Cerrado, cooking demonstrations in shopping malls and schools, and more exclusive initiatives. These include "Cerrado no Prato," which is described as a collective of chefs and other culinary professionals "that understands cuisine and tourism as levers for promoting the sustainable use of the Cerrado's sociobiodiversity and, at the same time, contributing to the biological and cultural recognition of Brazil's second largest biome."

The chefs and restaurant owners involved in activities and discussions on the transformation of the food system in Brasília operate in very different ways, however. While many of these actors limit their role to speaking about the transformation of food systems, a few put their words into practice, mainly by preparing food that respects certain principles that correspond to a certain vision of food sustainability, using recyclable and non-polluting packaging (some restaurants operate a zero-plastic policy, for example), setting up waste recycling programs, and taking part in food education activities, mainly in schools.

For example, the Ecozinha Institute brings together about 15 restaurants in Brasília and has as its objective: "to implement actions that result in the economic, social and cultural development of society and its institutions, with an emphasis on environmental education and conservation, promoting a new low-carbon, circular economy through food."[6] The institute has rolled out a kitchen-waste management project, which involves setting up recycling containers near restaurants for the collection and treatment of waste, including glass, that is normally consigned to local landfill sites.

Another example of these actions involves a restaurant that does not have a set menu and prepares different dishes every day, depending on the products offered to them on the day by producers with whom they have long-standing relationships. The restaurant's chef and owner said: "It's a menu that evolves from bottom to top. It's not my head that makes the decisions; it's the food supply, which makes the restaurant more environmentally friendly, right? We don't have a set menu, which means we're not asking producers to provide the same food all year long." The chef is also an activist, both with Convivium SlowFood and her own restaurant, where she tells customers about the restaurant's guiding principles. The chef sees this stance as a political act: "I think that educating people through cooking is absolutely wonderful (…) I see it as a revolutionary act. There is something political about what you eat."

The owner of another establishment confirms this political stance, which prompts the restaurants of Brasília to promote ideas as part of their aim of transforming food systems:

> The restaurant was born with the idea of giving talks and running courses. We're not just about selling things; this needs to be a place of transformation too. This is a learning process for us and for our customers. Our mission is to show that it is possible to eat better and pollute less.

Some chefs are backing up the actions taken by the advocacy coalition on agroecology by taking part in events that aim to bring urban and rural communities closer

---

6. https://www.institutoecozinha.org.br/

together. For example, the head of the Centrer spoke about a chef who regularly participates in Emater-DF training workshops held at the farmer's market:

> We organize workshops from time to time and he comes here on a Saturday, which is the day of the farmer's market, and he brings all his expertise, picks up what he sees at the market, makes all these dishes, and shows them to the urban community. (...) The chef already knows who the producers are. He already knows with who he's buying the herbs from here. (...) The proximity between chefs, the urban community and the producers at the market is what creates this relationship of trust. They come into contact with each other at the market, and producers ask them to come and visit their properties.

A SEE-DF nutritionist spoke of the participation of a chef, who is also a professor of gastronomy at the Brazilian Institute of Higher Education (IESB), in the projects they are implementing to encourage school canteens to use more natural produce from family farming and fewer ultra-processed products. The project, which goes by the name of "Chef e Nutri na Escola" (Chef and Nutrition at School) aims, for example, to have a chef going into schools and working with cooks to create recipes using food procured through institutional purchasing. One of the nutritionists responsible for this project explained the role of the chef:

> The aim is to use the food that we have in our school meals to develop cooking techniques with the canteen cooks and work with them on putting dishes together. (...) The chef works with the menu of the day and tries to come up with more innovative ways of preparing it. (...) He focuses on the issue of waste, because he saw for himself that the cooks were throwing a lot of food away.

According to one of the managers of SEE-DF's Food Procurement Program, the project is yielding tangible results in terms of the transformation of food systems, as it is connecting the culinary world with school meals:

> It totally deconstructs the glamorous image that chefs have. (...) Chef e Nutri na Escola is the star of our program. It attracts the most attention, it's very Instagrammable, and we can see that it generates a lot of positive coverage in the media. We have a lot of other things going on. We have the Ministry of Health's public policies, and we have partnerships, etc. But it's the success of Chef e Nutri na Escola that drives the others and helps them all achieve success.

▸▸ Discussion

Analysis of the politicization of the food system transition in Brasília reveals that some urban populations (mainly those with higher levels of education and income) want to see change with regard to issues such as access to healthy food, the environmental and social conditions of food production (linked to the environmental degradation of the Cerrado or the protection of traditional communities, for example), waste management (especially packaging), and food waste. As it was described in Brazil, the social demand for the transformation of food systems is similar to that studied by authors in other continents, who have shown that urban actors – once mere consumers of food products supplied by the rural population – are now responsible for a large number of innovations. These urban actors take action and organize themselves in transforming the entire food system, from production to waste management (Blay-Palmer et al., 2016; Debru et al., 2017; Hawkes & Halliday, 2017; Kropp, 2018).

The desire for change among actors relates to a demand to regain local control of public problems that are administered at national and international levels and where there is great deal of interference by powerful private actors in the agrifood industry. In this respect, the politicization of food issues in the city of Brasília is similar to that observed by other authors (Fouilleux & Michel, 2020). In fact, the actions taken by actors in the urban area of Brasília is clearly presented as a counterweight to food systems based on agro-industrial modes of production and distribution, described by Jan Douwe van der Ploeg (2008) as "food empires."

One of the peculiarities of Brazil is that, in recent times, this movement has been linked to the large-scale dismantling of public policies directly connected to the food system (Sabourin et al., 2020). This process began in President Dilma Rousseff's second term in office, when Brazilian Congress representatives defending the interests of agribusiness instigated a series of measures aimed at limiting public support for family farming, which depends heavily on programs combatting food insecurity. The dismantling of these policies gathered pace under the President Temer administration and culminated in the closure of the National Council of Food and Nutrition Security (CONSEA), one of the first actions of the President Bolsonaro administration on January 2, 2019, the first working day of the year.

In the face of these policy changes, some actors in Brasília have sought to continue with a process that began in the 1990s and organize resistance and opposition to a specific vision, widely promoted by federal government, of food systems based on the production of standardized and industrially processed food and promoting the consumption of ultra-processed products. The increased presence of family farming and agroecological products in the Federal District's public food procurement programs, the emergence of programs supporting urban agriculture, the sudden growth of agroecological markets, and the advent of programs aimed at the development of chains of processed family farming products, and food education initiatives implemented with the help of chefs and nutritionists are all clear signs of a dynamic that aims to change public action in favor of sustainable food systems in the city of Brasília.

The shift towards more sustainable food systems was initially brought about by actors from the alternative farming world making their way into the institutions of the Federal District and circulating their ideas there. Our results show that the process of turning the issue of food system transformation into policy depends, in large part, on the actions of an advocacy coalition devoted to agroecological transition, in which Emater-DF plays an important role in linking the urban and the rural worlds. These results echo the findings of studies that have shown that alternative food networks are emerging in Brazil and are bringing family farmers excluded from agricultural modernization into contact with urban markets (Cassol & Schneider, 2015; Darolt et al., 2016).

However, while complementing these studies, our results also highlight the connection that has grown between networks traditionally formed by urban consumers and family farmers and other categories of actors, such as chefs and nutritionists.

In the case of nutritionists in Brazil, the actions taken by these actors were previously limited to issues of health quality control and improving the nutritional quality of food. We have shown that they now work within the policy framework that they are part of (school food programs) in supporting the movement of resistance to the dominant food system, contributing to the development of various measures aimed at increasing the use of family farming and agroecological produce in schools.

The strengthening of ties between the world of nutrition and alternative food systems in the urban space has invigorated the broader political struggle waged by Brazilian nutritionists against the excessive consumption of processed and ultra-processed products, which leads to increased problems of overweight and obesity often associated with diabetes, cardiovascular diseases and certain types of cancer (Martins et al., 2013). Nutritionists have thus manifested their opposition to the pressure imposed by major agribusiness groups, which have worked their way into the public domain in an effort to minimize the nutritional problems caused by junk food (Azevedo, 2019). At the end of 2020, for example, the Ministry of Agriculture, Livestock and Supply issued a technical note requesting that the Ministry of Health remove the term "ultra-processed foods" from the Food Guide for the Brazilian Population, a document that provides the people of Brazil with guidance on eating a healthy diet and warns against the consumption of ultra-processed foods. Nutritionists reacted by issuing a resolution in support of the Food Guide.[7]

For their part, chefs and restaurant owners in Brasília are taking action through various channels and initiatives to ensure that food system sustainability becomes a public issue on the political agenda. In terms of the principles regulating meal preparation, chefs are expanding the boundaries by showing a preference for organic produce, prohibiting the use of monoculture products, and using only products purchased directly from producers. Some have drastically changed their culinary practices, for example, and are creating daily menus based on products supplied directly by farmers, making investments in remote rural communities, and taking part in social movements promoting sustainable food. Others are supporting initiatives on recycling and combating waste and the use of polluting products such as plastics. Our findings thus back up those of Niederle and Shubert (2020) in the city of Porto Alegre, which show that some restaurants involved in the veganism movement are engaging in practices, such as establishing close ties with local producers, that go far beyond the decision not to eat animal products.

Our results also indicate that some chefs promote local produce from the Cerrado biome to create a unique culinary identity for the city, in line with the conclusions of other studies (Zaneti & Brumano, 2019). They seek to spread the message about food system transformation more through their fame and presence at media events than through their purchasing and culinary practices.

This diverse positioning generates a certain tension among Brasília's chefs. While there are some who look to radically transform the food system and engage in activism, there are others who believe that their identity as chefs gives them the freedom to become political entrepreneurs. These tensions are reminiscent of those reported on by Duarte et al. (2020) as a result of the elitism that stems from the "gastronomization" of local species. Similarly, the activist chef Tainá Marajoara, denounces the expropriation of Amazonian identities resulting from this "gastronomic spectacularisation" created by some famous chefs (Granchamp, 2019).

---

7. https://www.fca.unicamp.br/portal/pt-br/comunic-2/comunicacao-noticias/comunicacao-not-socie-dade/1527-mocao-de-apoio-ao-guia-alimentar-para-a-populacao-brasileira.html

▸▸ Conclusion

While the creation of an extensive coalition of actors of the city of Brasília on food sustainability suggests that the issue of food system transition has gradually become a public problem, the actions undertaken by these urban actors have not translated into a coherent and structured public policy framework. The initiatives that have been implemented remain too fragile and poorly connected to bring about a shift in public action leading to sweeping reform of the food system.

The main reason for these limited changes is that the coalition of actors is still very weakened by diverging viewpoints. On the one hand, among the chefs, the representations of food and the modes of action to transform the system are not homogeneous. On the other hand, the relationship between the urban and rural world is still far from harmonious, as shown for example by the difficult relationship between restaurants, who have their own economic and commercial approaches, and CSAs, which seemingly cover a wide range of initiatives and are insufficiently robust and coordinated with each other. There is no question that there is a greater dynamic of change towards a sustainable food system in the city of Brasília, but it is still way from being a movement that could prevail against food industry and distribution actors, who are more powerful and organized.

▸▸ References

Alexander, P., Brown, C., Arneth, A., Finnigan, J., Moran, D., & Rounsevell, M. D. A. (2017). Losses, inefficiencies and waste in the global food system. *Agricultural Systems*, *153*, 190–200. https://doi.org/10.1016/j.agsy.2017.01.014

Azevedo, E. (2019). Lobbies alimentares. *Revista Ingesta*, *1*(1), 53–67. https://doi.org/10.11606/issn.2596-3147.v1i1p53-67

Béné, C., Fanzo, J., Prager, S. D., Achicanoy, H. A., Mapes, B. R., Alvarez Toro, P., & Bonilla Cedrez, C. (2020). Global drivers of food system (un) sustainability: a multi-country correlation analysis. *PloS ONE*, *15*(4), Article e0231071. https://doi.org/10.1371/journal.pone.0231071

Blay-Palmer, A., Sonnino, R., & Custot, J. (2016). A food politics of the possible? Growing sustainable food systems through networks of knowledge. *Agriculture and Human Values*, *33*(1), 27–43. https://doi.org/10.1007/s10460-015-9592-0

Brand, C., Bricas, N., Conaré, D., Daviron, B., Debru, J., Michel, L., & Soulard, C.-T. (Eds.). (2019). *Designing urban food policies: concepts and approaches*. Springer. https://doi.org/10.1007/978-3-030-13958-2

Bricas, N. Urbanization issues affecting food system sustainability. (2019). In Brand, C., Bricas, N., Conaré, D., Daviron, B., Debru, J., Michel, L., & C.-T. Soulard (Eds.), *Designing urban food policies: concepts and approaches* (pp. 1–25). Springer. https://doi.org/10.1007/978-3-030-13958-2

Campbell, B. M., Beare, D. J., Bennett, E. M., Hall-Spencer, J. M., Ingram, J. S. I., Jaramillo, F., Ortiz, R., Ramankutty, N., Sayer, J. A., & Shindell, D. (2017). Agriculture production as a major driver of the Earth system exceeding planetary boundaries. *Ecology and Society*, *22*(4), Article 8. https://doi.org/10.5751/ES-09595-220408

Caron, P., Ferrero y de Loma-Osorio, G., Nabarro, D., Hainzelin, E., Guillou, M., Andersen, I., Arnold, T., Astralaga, M., Beukeboom, M., Bickersteth, S., Bwalya, M., Caballero, P., Campbell, B. M., Divine, M., Fan, S., Frick, M., Friis, A., Gallagher, M., Halkin, J.-P., … Verburg, G. (2018). Food systems for sustainable development: proposals for a profound four-part transformation. *Agronomy for sustainable development*, *38*, Article 41. https://doi.org/10.1007/s13593-018-0519-1

Cassol, A., & Schneider, S. (2015). Produção e consumo de alimentos: novas redes e atores. *Lua Nova*, *95*, 143–177. https://doi.org/10.1590/0102-6445143-177/95

Clark, M. A., Springmann, M., Hill, J., & Tilman, D. (2019). Multiple health and environmental impacts of foods. *Proceedings of the National Academy of Sciences, 116*(46), 23357–23362. https://doi.org/10.1073/pnas.1906908116

Cunha, A. R. A. D. A. (2006). Dimensões estratégicas e dilemas das Centrais de Abastecimento no Brasil. *Revista de Política Agrícola, 15*(4), 37–46. https://seer.sede.embrapa.br/index.php/RPA/article/view/516

Darolt, M. R., Lamine, C., Brandenburg, A., Alencar, M. D. C. F., & Abreu, L. S. (2016). Redes alimentares alternativas e novas relações produção-consumo na França e no Brasil. *Ambiente & Sociedade, 19*(2), 1–22. https://doi.org/10.1590/1809-4422ASOC121132V1922016

Daviron, B., Perin, C., & Soulard, C. T. (2019). History of urban food policy in Europe, from the ancient city to the industrial city. In Brand, C., Bricas, N., Conaré, D., Daviron, B., Debru, J., Michel, L., & C.-T. Soulard (Eds.), *Designing urban food policies* (pp. 27–51). Springer. https://doi.org/10.1007/978-3-030-13958-2

De Carvalho, J. L. H. (2005). PROVE-DF. A descontinuidade apesar do amparo legal. *Revista de Agricultura Urbana, 16,* 109–113. https://ruaf.org/assets/2006/11/rau16_total.pdf#page=109

Debru, J., Albert, S., Bricas, N., & Conaré, D. (2017). *Politiques alimentaires urbaines. Actes de la rencontre internationale sur les expériences en Afrique, Amérique latine et Asie.* Chaire Unesco Alimentations du Monde. https://hal.archives-ouvertes.fr/hal-01995460

Duarte, L. M. G., Guéneau, S., Diniz, J. D. A. S., & Passos, C. J. S. (2020). Sistemas agroalimentares alternativos, construção social de mercados e gastronomização de produtos agroextrativistas do cerrado brasileiro. In Guéneau, S. Diniz, J,. D. A. S., & C. J. S. Passos (Eds.), *Alternativas para o bioma Cerrado. Agroextrativismo e uso sustentável da sociobiodiversidade* (pp. 405–446). Editora Mil Folhas. http://cds.unb.br/images/CDS/8.publicacoes/2020/Gueneau_et_al_2020.pdf

Dury, S., Bendjebbar, P., Hainzelin, E., Giordano, T., & Bricas, N. (Eds.). (2019). *Food systems at risk: new trends and challenges.* FAO, CIRAD, and European Commission. http://www.doi.org/10.19182/agritrop/00080

Eakin, H., Connors, J. P., Wharton, C., Bertmann, F., Xiong, A., & Stoltzfus, J. (2017). Identifying attributes of food system sustainability: emerging themes and consensus. *Agriculture and Human Values, 34*(3), 757–773. https://doi.org/10.1007/s10460-016-9754-8

Fouilleux, E., & Michel, L. (Eds.). (2020). *Quand l'alimentation se fait politique(s).* Presses Universitaires de Rennes.

Granchamp, L. (2019). Penser l'alimentation d'un point de vue décolonial. Entretien avec Tainá Marajoara (Bélém, Brésil). *Revue des Sciences Sociales, 61,* 132–141. https://doi.org/10.4000/revss.3611

Hassenteufel, P. (2011). *Sociologie politique: l'action publique.* Armand Colin. https://doi.org/10.3917/arco.hasse.2011.01

Hawkes, C., & Halliday, J. (2017). *What makes urban food policy happen? Insights from five case studies.* IPES-Food. https://www.ipes-food.org/_img/upload/files/Cities_full.pdf

Hinrichs, C. C. (2014). Transitions to sustainability: a change in thinking about food systems change? *Agriculture and Human Values, 31*(1), 143–155. https://doi.org/10.1007/s10460-014-9479-5

Hoffmann, R. (1995). Pobreza, insegurança alimentar e desnutrição no Brasil. *Estudos Avançados, 9*(24), 159–172. https://doi.org/10.1590/S0103-40141995000200007

IBGE. (2011). *2010 Demographic Census.* https://www.ibge.gov.br/en/statistics/social/education/18391-2010-population-census.html

IBGE. (2019). *IBGE discloses the population estimates of the municipalities for 2019.* https://censoagro2017.ibge.gov.br/en/2185-news-agency/releases-en/25283-ibge-divulga-as-estimativas-da-populacao-dos-municipios-para-2020.html

IBGE. (2020). *IBGE divulga o rendimento domiciliar per capita 2019.* https://censos.ibge.gov.br/2013-agencia-de-noticias/releases/26956-ibge-divulga-o-rendimento-domiciliar-per-capita-2019.html

Jenkins-Smith, H., Nohrstedt, D., Weible, C. M., & Sabatier, P. A. (2014). The advocacy coalition framework: foundations, evolution, and ongoing research. In Sabatier, P., & C. Weible (Eds.), *Theories of the policy process* (pp. 183–224). Routledge.

Jurgilevich, A., Birge, T., Kentala-Lehtonen, J., Korhonen-Kurki, K., Pietikäinen, J., Saikku, L., & Schösler, H. (2016). Transition towards circular economy in the food system. *Sustainability, 8*(1), Article 69. https://doi.org/10.3390/su8010069

Kinupp, V. F., & Lorenzi, H. (2014). *Plantas alimentícias não convencionais (PANC) no Brasil: guia de identificação, aspectos nutricionais e receitas ilustradas.* Instituto Plantarum de Estudos da Flora.

Kropp, C. (2018). Urban food movements and their transformative capacities. *The International Journal of Sociology of Agriculture and Food, 24*(3). https://doi.org/10.48416/ijsaf.v24i3.6

Lascoumes, P., & Le Galès, P. (2012). *Sociologie de l'action publique* (2nd ed.). Armand Colin.

Martins, A. P. B., Levy, R. B., Claro, R. M., Moubarac, J. C., & Monteiro, C. A. (2013). Participação crescente de produtos ultraprocessados na dieta brasileira (1987–2009) [Increased contribution of ultra-processed food products in the Brazilian diet (1987–2009)]. *Revista de Saúde Pública, 47*(4), 656–665. https://doi.org/10.1590/S0034-8910.2013047004968

Niederle, P., & Schubert, M. N. (2020). How does veganism contribute to shape sustainable food systems? Practices, meanings and identities of vegan restaurants in Porto Alegre, Brazil. *Journal of Rural Studies, 78,* 304–313. https://doi.org/10.1016/j.jrurstud.2020.06.021

Noy, C. (2008). Sampling knowledge: The hermeneutics of snowball sampling in qualitative research. *International Journal of Social Research Methodology, 11*(4), 327–344. https://doi.org/10.1080/13645570701401305

Ploeg, J. D. V. D. (2008). *Camponeses e impérios alimentares: lutas por autonomia e sustentabilidade na era da globalicação* (R. Pereira, Trans.). UFRGS. https://edepot.wur.nl/424203

Pothukuchi, K., & Kaufman, J. L. (1999). Placing the food system on the urban agenda: the role of municipal institutions in food systems planning. *Agriculture and Human Values, 16,* 213–224. https://doi.org/10.1023/A:1007558805953

Preiss, P., Charão-Marques, F., & Wiskerke, J. S. C. (2017). Fostering sustainable urban-rural linkages through local food supply: a transnational analysis of collaborative food alliances. *Sustainability, 9*(7), Article 1155. https://doi.org/10.3390/su9071155

Rocha, C., & Lessa, I. (2009). Urban governance for food security: the alternative food system in Belo Horizonte, Brazil. *International Planning Studies, 14*(4), 389–400. https://doi.org/10.1080/13563471003642787

Sabourin, E., Grisa, C., Niederle, P., Leite, S., Milhorance, C., Ferreira, A., Sauer S., & Andriguetto-Filho, J. (2020). Le démantèlement des politiques publiques rurales et environnementales au Brésil. *Cahiers Agricultures, 29,* Article 31. https://doi.org/10.1051/cagri/2020029

Sabourin, E., Tadeu Da Silva, L. R., & De Avila, M. L. (2019). A rede de ação pública em torno da agroecologia e produção orgânica no Distrito Federal. In Sabourin, E. Guéneau, S., Colonna, J., & L. R. Tadeu da Silva (Eds.), *Construção de políticas estaduais de agroecologia e produção orgânica no Brasil: avanços, obstáculos e efeitos das dinâmicas subnacionais* (pp. 219–241). Editora CRV. https://hal.archives-ouvertes.fr/hal-02545007

Sauer, S. (2018). Soy expansion into the agricultural frontiers of the Brazilian Amazon: the agribusiness economy and its social and environmental conflicts. *Land Use Policy, 79,* 326–338. https://doi.org/10.1016/j.landusepol.2018.08.030

Springmann, M., Clark, M., Mason-D'Croz, D., Wiebe, K., Bodirsky, B. L., Lassaletta, L., de Vries, W., Vermeulen, S. J., Herrero, M., Carlson, K. M., Jonell, M., Troell, M., DeClerck, F., Gordon, L. J., Zurayk, R., Scarborough, P., Rayner, M., Loken, B., Fanzo, B., … Willett, W. (2018). Options for keeping the food system within environmental limits. *Nature, 562*(7728), 519–525. https://doi.org/10.1038/s41586-018-0594-0

Stassart, P. M., Baret, P., Grégoire, J.-C., Hance, T., Mormont, M., Reheul, D., Stilmant, D., Vanloqueren, G., & Visser, M. (2012). L'agroécologie: trajectoire et potentiel. Pour une transition vers des systèmes alimentaires durables. In Van Dam, D., Streith, M., Nizet, J., & P. M. Stassart (Eds.), *Agroecology. Entre pratiques et sciences sociaux* (pp. 25–51). Éducagri.

Valette, E., Schreiber, K., Conaré, D., Bonomelli, V., Blay-Palmer, A., & Bricas, N. (2020). An emerging user-led participatory methodology. Mapping impact pathways of urban food system sustainability innovations. In Blay-Palmer, A., Conaré, D., Meter, K., Di Battista, A., & C. Johnston (Eds.), *Sustainable food system assessment: lessons from global practice* (pp. 19–41). Routledge. https://doi.org/10.4324/9780429439896

Vermeulen, S. J., Campbell, B. M., & Ingram, J. S. I. (2012). Climate change and food systems. *Annual Review of Environment and Resources, 37*, 195–222. https://doi.org/10.1146/annurev-environ-020411-130608

Vermeulen, S. J., Park, T., Khoury, C. K., & Béné, C. (2020). Changing diets and the transformation of the global food system. *Annals of the New York Academy of Sciences, 1478*(1), 3–17. https://doi.org/10.1111/nyas.14446

Westhoek, H., Ingram, J., Van Berkum, S., Özay, L., & Hajer, M. (2016). Food systems and natural resources. United Nations Environmental Programme. https://www.resourcepanel.org/sites/default/files/documents/document/media/food_systems_summary_report_english.pdf

Zaneti, T. B., & Brumano, C. N. (2019). Um olhar sobre o cenário gastronômico de Brasília. *Cenário: Revista Interdisciplinar em Turismo e Território, 7*(13), 42–53. https://doi.org/10.26512/revistacenario.v7i13.25513

Chapter 13

Building an urban food policy: the case of Cali, Colombia

RUBY ISAZA CASTELLANO, GUY HENRY, SARA RANKIN

▸▸ Introduction

Faced with the major sustainability challenges raised by modern society and growing urban populations, an increasingly clear need is emerging for science-based public policies, and for tools developed by academia to support decision-making. Food and nutrition security, for example, depend largely on local and national food policies.

Food and nutrition security is a topic of global importance, as are the challenges affecting the sustainability of agricultural and food systems. The need to focus on sustainable food systems has been recognized as the way to ensure food security and nutrition for everybody, without compromising the economic, social and environmental needs of future generations. This is also the only way to end malnutrition in all its forms (undernutrition, micronutrient deficiencies, excess weight and obesity).

According to a major review by Popkin and Reardon (2018), in Latin America and the Caribbean (LAC), widespread obesity in children and adults, along with poor diets, and inadequate physical activity, are causing high levels of diabetes, high blood pressure, and other noncommunicable diseases. At the same time, a high proportion of children in many countries around the region are suffering from undernourishment and stunting owing to poor nutrition in the first 1,000 days of life.

Many organizations have pointed to the need for measures to make food systems sustainable, fair and inclusive, through regulations that give people access to nutritious, safe, varied, affordable and environmentally responsible products. A major report by the EAT/Lancet Commission (2019) notes: "The absence of globally agreed scientific targets for healthy diets and sustainable food production has hindered large-scale and coordinated efforts to transform the global food system." This commission brings together 19 commissioners and 18 co-authors from 16 counties in the fields of human health, agriculture, policy, science, and environmental sustainability to develop global scientific targets based on the best available evidence relating to healthy diets and sustainable food production. These global targets are incorporated into a framework that acts as a safe operating space for food systems, allowing us to assess which diets and food production practices will help us to achieve the UN and Paris Sustainable Development Goals (SDGs). In addition to this, a number of international frameworks are emerging, such as

Agenda 2030, the Food Plan for the Community of Latin America and Caribbean States (CELAC), the Milan Pact and the EAT-Lancet Commission, among others. These pacts, movements and frameworks are beginning to transform conventional agrifood dynamics and concepts of urban development. They have also attracted the attention of decision-makers and raised new challenges for the many different players involved in big cities. This has led to the development of new programs and policies in response to these new objectives.

According to Arciniegas and Peña (2017), the overall nutritional situation in Colombia is characterized by a significant increase in malnutrition by excess and a significant fall in malnutrition by deficiency; a situation that is reflected in anthropometric indicators. In Cali, the rate of chronic malnutrition in children under five was 5.3%, meaning that five out of every 100 children were underweight for their age. Concerning low weight-for-age, 0.9% of children are affected. Overall, in terms of malnutrition indicators, Cali (as mentioned above) is one of the cities with the lowest indicators compared with other cities in Colombia.

In Colombia, the introduction of the National Food and Nutritional Security Policy (CONPES 113, 2008) gave the issue national relevance, setting out clear approaches for food insecurity in the country and in turn creating an enabling context for talking about these issues across the regions, particularly in capital cities such as Cali, which are the destinations of large migratory movements. However, Cali did not have its own Food and Nutrition Security (FNS) policy.

Returning to the national level, Colombia's food policies have undergone a number of changes from certain standpoints. In 1996 it implemented a National Food and Nutrition Plan (PNAN), focusing on access to food and a greater supply of micronutrients for the population. Following this, food issues were integrated to a greater extent into national development plans, as well as into CONPES 91 for the achievement of the Millennium Development Goals. In 2007, as part of CONPES 113, the National Food Security Policy was established and implemented through the National Food and Nutritional Security Plan (PNSAN), during its formal period of implementation between 2012 and 2019.

The FNS policy in Colombia was assessed by the National Planning Department (DNP), with particular emphasis on its institutional capacities. The assessment noted that some goals had been met but that problems had increased in other cases. In relation to this, the National Survey of the Nutritional Situation (ENSIN) in 2015 showed that the indicator for national food insecurity fell from 57.7% in 2010 to 54.2% in 2015. Over the same period, the level of stunting in children under the age of four fell by 2.4%, while acute malnutrition increased from 0.9% to 1.6%. Another rising trend in Colombia is excess weight, which rose from 51.2% in 2010 to 56.5% in 2015 among adults of 18 or over, and from 15% to 18% among schoolchildren over the same period. These are just a few of the indicators tracing the food situation and providing an overview of the need to strengthen public policy on food security in Colombia.

This chapter discusses the process of building an FNS policy in Cali. The conclusion, asks a number of key questions as part of an analysis of the main driving forces behind the successful construction and formulation.

▸▸ Public policy: major efforts

The world is talking about the challenges of sustainability, how to address them and, more specifically, how to reduce food insecurity through the construction of sustainable food systems. Academia must generate answers.

In 2014, the CIAT launched "FoodLens," a strategy aimed at delivering new knowledge and interventions, through research into changing patterns in food production, distribution, and consumption. The purpose is to enable countries to provide urban and rural consumers with easy access to healthy food, through three strategic aims: sustainable farming adapted to the climate, higher income in rural areas, and affordable, high-quality food. As part of this approach, the project *Cali Come Mejor* (Cali eats better) was developed between 2015 and 2017, with the support of the Ford Foundation, CIRAD and the CGIAR Water, Land and Ecosystem (WLE) program. The aim was to analyze Cali's food system, and its social, political, cultural and physical context, in order to deliver strategies that could optimize operation.

Identifying the main players in the food system, the project reviewed and gathered available data on three main topics: the food supply system of vulnerable people, the main food problems, and the public policies and strategies involved in this system. Some of the findings were related to the prevalence of malnutrition caused by nutrient deficiencies, particularly in children under the age of five; the prevalence of obesity in the adult population; insufficient access to healthy and fresh food; deficiencies in market infrastructure and management, pushing up food costs and reducing the competitiveness of small farmers in rural areas; and the inadequate implementation and monitoring of agricultural policies to promote food availability and access to food, among other factors.

The new questions raised by all the local urban players and the new urban dynamics relating to changing food production models or even positive changes in consumption, are just a few of the enabling factors that helped to facilitate a dialogue and to build this policy, as well as transforming the conception of what the regional food system is and the importance of understanding it in order to act.

The analysis of this information and other evidence gathered from players in a range of sectors provided the basis of efforts to develop a Municipal Public Policy for Food and Nutritional Security and Sovereignty, designed, recognized and developed as part of a participatory approach, in order to guarantee access to food as a right. A comparison with cities such as Quito and Medellin showed that, although city programs or initiatives relating to food sovereignty and food security have diverse starting points, they share a common need: to scale up actions for the consolidation of public policies establishing a framework for comprehensive and intersectoral action, and providing resources for the planning and implementation of long-term activities.

Cali as part of the city-region: Strategic Territorial Planning

The city-region is an academic concept of decentralization that has taken on greater importance owing to its link with globalization processes, as well as with regional and local development.[1] It includes the complex network of players, processes and

---

1. Friedmann and C. Weaver (1981): 154, Territorio y Función, Madrid, Instituto de Estudios de Administración Local. Note the debatable Spanish translation of city-region by *región ciudadana*.

relationships involved in the production, processing, marketing and consumption of food within a given geographical area (e.g., an urban center, its peri-urban environment and the surrounding rural areas). The concept is very important when it comes to understanding "city-region links."

The Resource Centre for Urban Agriculture & Forestry (RUAF) foundation and other organizations have documented several case studies of city-region food systems projects, programs and policies developed around the world. In their words: "city-region food systems offer practical policy and program opportunities to address multiple development objectives and to build direct links between rural and urban areas and communities in a given city-region"[2]. Activities focusing on city-region food systems also provide an opportunity for local governments to streamline their efforts and develop common agendas for integrated regional planning, considering cities as hubs within a single system.

Instruments and tools

An agrifood system, with all its components (see figure 13.1), designed to be sustainable and resilient in the face of global challenges and climate change, will undoubtedly involve doubling or tripling efforts to coordinate players who can work from different perspectives as part of a complementary approach, in order to effectively facilitate a dialogue between the different components of the system.

To this end, a two-pronged process was adopted for formulating the suggested FNS Policy (figure 13.2):
– The Food and Nutritional Security and Sovereignty Roundtable of Cali, set up through a municipal decree in 2009 by the Municipal Mayor's Office as an inter-institutional, advisory and coordinating body to formulate, socialize, manage and monitor the Municipal Plan for this issue. The Roundtable is made up of representatives of government agencies, NGOs, private companies in the fields of food, social organizations, trade unions and civil society. Its functions include: formulating policies aimed at eradicating hunger and malnutrition and achieving sustainable food security for the municipality, promoting research into food and nutritional security as a basis for decisions on community interventions, and encouraging the production, use and consumption of cultural food crops; promoting clean agriculture through the use of natural resources with sustainable technologies. The Roundtable is divided into working groups based on the lines of action under research; food assistance with the emphasis on education, rural and urban agricultural production, agroindustries and consumer organizations, and the food supply and distribution system. Over a number of years, the Roundtable has provided a space to coordinate players and a space for discussions around the food issue. Prior to this, players conducted their own individual analysis, based on their skills area, without a holistic view of the food context in Cali and its region. The Roundtable has played a greater role in some periods than in others, but it nevertheless produced the first technical document for a food security policy proposal,

---

2. Dubbeling, M.; Bucatariu, C.; Santini, G.; Vogt, C.; Eisenbeiß, K. 2016. City region food systems and food waste management: linking urban and rural areas for sustainable and resilient development. Available at: http://edepot.wur.nl/413114

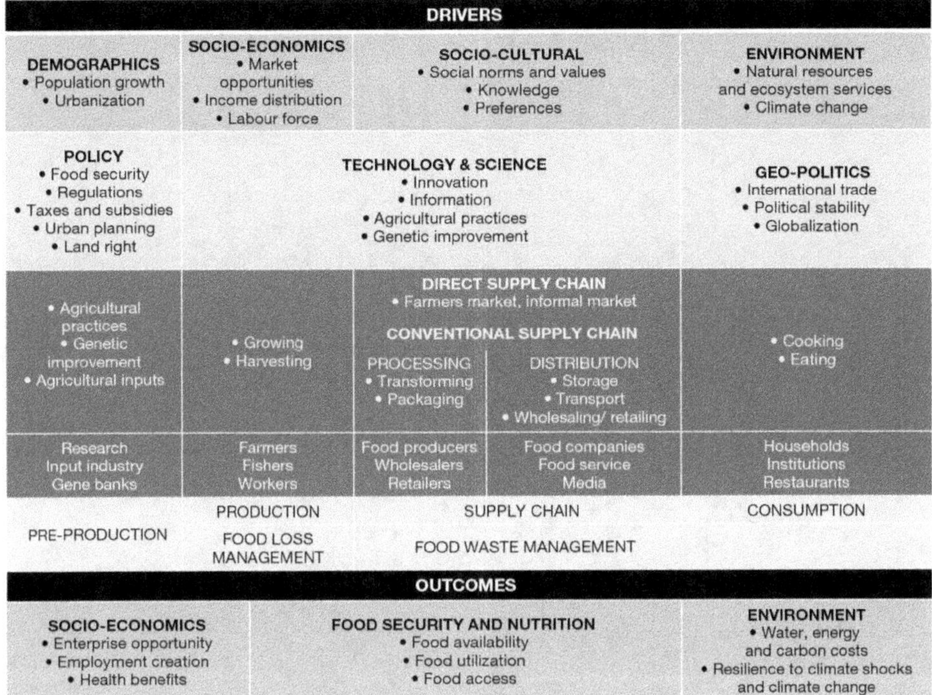

Figure 13.1. CIAT's conceptualization of what a sustainable food system is. Source: Bioversity-CIAT Alliance, 2018.

with the participation of the following players: Life, Justice and Peace Commission, DAGMA, CEDECUR, Economic Development and Competitiveness, ICBF, market places, representatives of the communities and surrounding municipalities, CIAT, Departmental Health Secretariat, Municipal Planning, CAVASA (central market), the Association of community kitchens, and companies from the food industry, among others. Since its inception, this process has been led by the Municipal Health Secretariat. Efforts to strengthen the process are continuing in order to generate sufficient debate and a public agenda to consolidate an updated proposal to social and emerging situations in the field of food and nutrition.

– The Platform for academic dialogue on food and nutritional security and sovereignty was set up in 2016 by CIAT, with the University of Valle, the World Food Programme, the Municipal Secretariat for Public Health, the BIOTEC Corporation, the National University of Colombia, the Autonomous University of the West, Harvest Plus,[3] the Network of Agroecological Markets, Valle en Paz[4], the Center of Education and Research for Urban and Rural Community Development (CEDECUR), and the World Wild Fund for Nature (WWF), to encourage integration between

---

3. HarvestPlus is a program co-managed by IFPRI and CIAT, dedicated to ending hunger by providing nutrient-rich foods through biofortification.
4. Valle en Paz is a private not-for-profit corporation set up to create development opportunities in rural communities affected by the armed conflict.

2009	2010	2015	2016	2017	2018	2019
Building of a municipal public policy through Municipal Working Group on Food Security (Mesas SAN)	Policy proposal not-approved by the Council	CIAT starts its work with the Mayor's Office to boost the development of a new Policy and participates simultaneously in the Food Security Working Group. The "Cali better food" Project began and generated knowledge about the city's food system. This project was funded by FORD Fundation.	Call to Municipal Secretariats to build the Policy, in order to create multisectoral interest. Creation of the Academic Platform to support the process aiming to build the public policy. CIAT research on food consumption informed by gaps identified in the Working groups established by the municipal authorities.	Validation of policy proposals and guidelines. Food and Nutritional Security Municipal Forum, that brought together 400 people and 42 public and private initiatives on production and waste management.	Validation of policy proposals and guidelines. Food and Nutritional Security Municipal Forum, that brought together 400 people and 42 public and private initiatives on production and waste management.	Food Policy approval. Implementation of pilot project on the establishment of a Food and Nutrition security observatory.

Figure 13.2. Timeline formulation of municipal public policy for food and nutritional security and sovereignty of Cali. Source: Bioversity-CIAT Alliance, 2018.

representatives of the academic, public, and NGO sectors, to provide scientific and technical input regarding the food system of the city-region, and to provide scientific support for the construction of the FSN policy. A multi-player group with a common agenda to analyze, discuss and deliver research proposals and interventions related to healthy eating, environmental sustainability, climate change, food sovereignty, waste management and consumer awareness, among other areas, able to contribute to more coordinated actions.

The collaborative approach makes it possible to identify problems and information gaps, to analyze and submit solutions and recommendations to improve Cali's food system, to organize actions to mobilize public opinion and encourage political advocacy. The platform took on the role of providing technical support for the construction of the policy proposal, which was approved by the Roundtable as it moved forward, and adjusted to reflect new recommendations.

The interaction between players and academics generated extensive discussions and provided an interesting learning experience on new issues and aspects relating to the concepts of sustainable food systems and the city-region. The main challenge lay in the increasing importance of aspects related to good nutrition, high quality food and healthy diets. The interactions between urban consumers and, for example, agroecological producers have proved valuable for both parties. Successful implementation of the new FNS policy would involve maintaining these spaces and continuing the dialogue. In order to strengthen interaction and integration between urban and rural players, stakeholders have suggested two types of actions: i) raising awareness of the issues through documentation/ pamphlets, fairs, workshops, etc. and ii) implementing projects to improve/expand/support more local farmers' markets (primarily as part of an agroecological approach), along with urban and peri-urban farming projects.

Milan Pact: roadmap and strategic framework for action

The creation of Cali's FSN policy was led by the Municipal Health Secretariat. The policy establishes a legal framework for action and coordination around the initiatives to be developed over the next 10 years in six strategic areas: availability, access, consumption, safety and quality, biological utilization and knowledge management and communication. The first five areas were based on the framework for action set out by the Milan Urban Food Policy Pact.

The Milan Pact (MUFPP) is a "voluntary treaty signed by cities that have made a commitment to working towards the development of sustainable, inclusive, resilient, safe and diversified food systems, to ensure a healthy and accessible food supply for all people, as part of a rights-based framework for action, in order to reduce food waste and protect biodiversity while mitigating and adapting to the effects of climate change" (MUFPP, 2015).

This pact gives visibility to successful initiatives relating to sustainable, fair, inclusive and environmentally friendly urban food systems, suggesting a framework and a guide for planning. As part of the process to build the proposal, the CIAT has also encouraged the identification and subsequent tracking of the food system monitoring indicators suggested by the RUAF, the FAO and the Milan Pact[5], structured along six main lines: governance, sustainable diets and nutrition, social and economic equity, food production, food supply and food waste. The use of gender indicators has been suggested, in so far as possible, with the aim of highlighting some of the differences associated with the dynamics and characteristics of the role of women in these contexts. If this monitoring is conducted systematically and on a formal basis, it will allow us to establish the real state of the food system and to monitor future trends, while measuring the impact of the actions initiated as part of the FNSS Policy.

▸▸ What do we already know and where are the information gaps?

The causes of food problems are complex and multidimensional, encompassing not only economic aspects, but also the social, political, cultural and physical environment. For this reason, it is essential to work together in developing conceptual frameworks and decision-making tools for the food system in general. To do this, we need to continuously gather information that helps us to understand the specific characteristics of the system in Cali, within the city-region. In Cali, for example, the options open to consumers are often limited to cheap food of low nutritional value. This is particularly true in low-income neighborhoods, where healthy food is either expensive or non-existent.

According to a study conducted as part of the *Cali, come mejor* project (Arciniegas & Peña, 2017), an explanation for the problems of malnutrition by deficiency and excess can be found partly at the individual level, since the nutritional status of the person is the result of a balance between their food intake, nutrient requirements and their relationship to the individual's physiological status. However, the quality and quantity of food consumed does not only depend on individual factors related to biological determinants and lifestyle, but also on other economic, social, cultural, and environmental factors.

Decisions relating to food consumption are strongly influenced by food standards. However, the food security of the most vulnerable socio-economic groups is impacted by economic determinants that prevent them from meeting these standards. In contrast, for the more stable socio-economic groups, the situation depends on the weight of these concerns and also on the relationship between health and food.

As a result of the socio-economic changes associated with higher incomes, households are more open to including new food products in their diets, although these may be perceived in a positive or negative way. Although domestic cooking is hugely important for all socio-economic groups, people with a higher socio-economic status are less likely to prepare food within the home, primarily for reasons relating to household organization, such as changes in the role of women in the home or an improved financial status enabling or facilitating access to new foods.

Improving the economic situation or purchasing power of households does not solve the food problems of the poorest households, since a number of other non-individual

---

5. Milan Urban Food Policy Pact Monitoring Framework Indicators. Available at: https://www.milanurbanfoodpolicypact.org/wp-content/uploads/2020/12/Pubblications_Menu-of-actions-to-shape-urban-food-environments-for-improved-nutrition_compressed.pdf

factors also play a role. The easy availability and affordability of cheap unhealthy food can translate into risk factors for nutritional excess. Consumers shift between these two conditions in accordance with their perceptions and the dietary norms.

Food environments play a major role in explaining differences in food consumption. The high availability of ultra-processed foods and the environments promoting their consumption have significant effects on the national diet, primarily affecting children and young people, who play a key role in choosing what is eaten at home.

Using qualitative research methodology, this study gathered extensive knowledge on eating behavior and the reasons behind it. Given that this research and most other studies based on the same type of methodology have sought to explain trends, we have been able to gain a broad understanding of household food patterns in Cali, looking more deeply not at 'what' and 'how', but rather at 'why' concerning changes and transitions in food.

As part of this same project, a descriptive/qualitative study was carried out on the operation of the food supply and production subsystem (Temple & Dury, 2015), including the supply process for the most vulnerable. This made it possible to identify the major difficulties encountered by this segment of the population in gaining sufficient access to healthy food with fresh produce.

One major problem lies in the very high prices of these products in an urban context where the monetary poverty rate stands at 15.7%.[6] These high prices can be explained by the oligopolistic structure of the food distribution system, which is dominated by supermarket chains. Another problem is the lack of access to markets in areas lower down the social hierarchy, resulting in transport and logistic costs that are higher for vulnerable populations than for other populations. Shops are the main supply source for vulnerable populations, but these shops obtain some of their supplies from markets with health quality problems and the rest from supermarket chains selling at high prices. Some circuits have been identified where local producers supply food to the city, but they remain few and far between. The only existing agroecological market enabling small producers to market their produce directly is experiencing difficulties caused by a lack of access to the point of sale. These findings have highlighted the need to model the food systems of vulnerable communities in order to focus on the main challenges for public policies and on the objectives of any necessary complementary studies.

These studies identified information gaps regarding the nutritional status and nutritional monitoring of the population of Cali, particularly children under the age of five, who are most at risk of malnutrition. Although a nutritional status survey exists at the national level, the results are only published two years later and concern just a small segment of the population. Problems relating to economic access are clearly the main barrier faced by households in seeking a healthier diet. Unemployment and informal labor are determining factors in the choice of food. In Colombia, as in several other Latin American countries, people are eating more highly processed food, and outside sources are playing a growing role in the household diet. People know what sort of food they should be eating, based on the promotion of good eating habits by health

---

6. Cali in figures 2018-2019. Available at: https://www.cali.gov.co/planeacion/loader.php?lServicio=Tools2&lTipo=descargas&lFuncion=descargar&idFile=41162

institutions and the media, but there are a number of barriers preventing them from achieving this. At the same time, they are also influenced by the globalization process.

These studies are an approach to the understanding of food patterns. However, we must continue to study the driving forces behind change as well as the permanent factors relating to the supply, processing and consumption of food by households. By understanding the motivations and determinants of consumers, we will be able to generate strategies aimed at transforming eating habits and reducing the barriers that limit adequate nutrition.

Neighborhood shops and shopkeepers both play a major role in the supply of food and in the relationships woven around food. They are a major source of support for food patterns, particularly with respect to the most vulnerable socio-economic groups. Although it is important to analyze individual food patterns, it is increasingly necessary to relate them to the household and to the city, since each level has a determining impact on the other and they are strongly related when it comes to food. This clearly highlights the need to continue in-depth studies encompassing each of the spheres that influence food. Although quantitative studies are necessary, the in-depth results of this study and many other current studies show the need to associate both research paradigms. Gaining an in-depth understanding of food-related trends is a necessary task that must be addressed as part of a comprehensive approach in order to continuously feed into the programs and policies related to these issues.

Some of the conclusions/recommendations of these studies that served as input for developing the municipal policy are set out below:
– Given that the supply of natural food is becoming more complex for the most vulnerable socio-economic groups, we need to find strategies to support the food distribution chain, i.e., to strengthen mobile markets and marketplaces in order to facilitate access to natural foods.
– Preference must be given to short food distribution circuits, for the benefit of both producers and (ultimately) consumers, since lower prices would make it easier for them to access natural foods.
– Health authorities must organize inspections of fast food stalls, in order to ensure the safety and quality of the food available.
– Given the decisive influence of the urban environment on the diets of people in all socio-economic groups, we must encourage measures to regulate the high availability of fast food in public spaces.
– We must also promote alternative forms of nutritional education, not only to provide knowledge and information, but also to raise awareness on the importance of proper nutrition in preventing health problems.

▸▸ Municipal Public Policy Proposal for Food and Nutritional Security and Sovereignty (PSSAN) 2019

Today, Santiago de Cali has a public policy on food and nutritional sovereignty and security, following a long process that began in 2009 with the signing of municipal decree No. 411.0.20.0072 setting up the Municipal Roundtable on Food and

Nutritional Security. The Roundtable began consolidating the process of formulating a local food policy. In 2011 it produced a technical document with policy guidelines, although this was not approved by the City Council at the time.

During the period 2012–2015, the formulation of public policy in Cali in the field of FNSS was part of the remit of the Secretariat of Social Welfare in the Municipal Development Plan of the municipality. The Secretariat worked on the financial component and the responsibility of each agency, but was unable to formally set out a policy within this period, owing to the prioritization of other issues. During the period 2016–2019, the process was resumed, this time through coordination between the Secretariats of Public Health and Social Welfare with the support of the FNS Roundtable, backed up by Cali's new Academic Platform for FNS.

A work schedule was put in place with the methodology necessary to gather contributions through meetings based on a range of perspectives, with input from various players as part of plenary sessions of the FNS Roundtable and the FNS Academic Platform. At the same time, academic events were organized with international and local speakers, and activities developed with local communities in which the initiatives of civil society were recognized. The next step was to define the main problems impacting each area of focus in the FNSS plan and the alternative approaches for addressing them. An agreement was then made concerning a series of actions to address food and nutritional security and sovereignty, alongside the guarantee of the human right to food, within the context of Cali.

This process was described in a document, which was then widely socialized to provide input for the analysis, approach and guidelines. The technical document set out the basis for the explanatory memorandum, a technical requirement for presentation to the Municipal Council. The draft agreement was reviewed and further developed by the different agencies of the Cali administration that would be responsible for policy implementation and evaluation. Once this process was completed, the document was ready to be presented to the Municipal Council for approval. Five councilors took up the issue for presentation and began the process of analysis and plenary debate, with the participation of a range of players from civil society. Finally, on December 19, 2019, the Municipal Public Policy on Food and Nutritional Sovereignty and Security was approved by municipal agreement 0470.

The FNSS policy was structured by 18 articles in three chapters: i) General Provisions, Scope, Principles, Approaches, definitions and objectives, ii) Strategic areas of focus, iii) Action plan, monitoring, follow-up and assessment. The action plan comprises six areas of focus, 14 lines of action and 26 actions. The six areas of focus include the following (see table 13.1):

Based on a regional approach, the policy seeks to recognize the relations and scale of the dynamics with neighboring departments as part of the food system. It emphasizes the need for coordination with sectoral and population-based public policies, providing a comprehensive approach to the goals set.

Table 13.1. Areas of focus and actions included in the FNSS plan

Areas of focus	Actions
Food availability	Implement agroecological training processes, support the production of traditional foods, strengthen the organization of local and rural producers, transfer technology for the management and control of processes in the production and processing of food, generate processes to reduce intermediation with marketplace traders, shopkeepers, mini markets and agroecological markets with food producers. Strengthen the sustainable operation of food distribution centers, implement and assess strategies and activities at the institutional level to increase breastfeeding
Food consumption	Implement measures to control advertising for food of low nutritional content in educational institutions in Cali. Inspect food vending establishments with respect to consumption risks, develop actions to protect exclusive and complementary breastfeeding, provide care and monitoring regarding micronutrient deficiencies in infants, pregnant women and schoolchildren, develop strategies to promote a culture of responsible consumption and the rights and duties of consumers
Access to food	Strengthen the service coverage of soup kitchens and school meals programs. Generate strategies to boost the commercial development of farmers' markets in order to promote and encourage a sense of ownership regarding the farming, preparation and use of local native and ancestral products, strengthening the recognition of cultural identities
Quality and safety	Adopt human capital technologies (knowledge, equipment, inputs) to better control food risks, enforce compliance with the standards relating to good manufacturing practices and food handling throughout the food system cycle, formulate and develop intersectoral and community strategies for healthy environments, ensure monitoring of the quality and safety of food supplementation programs for schoolchildren, make a proposal for the design, formulation and implementation of a "Circular Economy" model encompassing the chain of food production, distribution and consumption, as well as the disposal and processing of organic waste from coordinated programs and projects, establish sustainability strategies and actions relating to the issue of public services (energy, water and sewerage) as part of the programs and projects associated with the FNSS plan
Biological use of food	Ensure coordinated implementation of a comprehensive care program to address malnutrition, reinforce the system in place for the nutritional and food surveillance of children under five, schoolchildren, pregnant women and older adults. Provide technical assistance for the regulations and guidelines of Comprehensive Healthcare Programs (RIAS) and the protocols in force relating to malnutrition prevention and care and the monitoring of low birth weight, strengthen the strategy around Healthy School Shops
Knowledge promotion and management	Increase the visibility of successful experiences in self-consumption and short-circuit supply chains, food security and the reduction of malnutrition, in order to promote processes encouraging healthy food environments in areas such as work, school and the community, implement actions to prevent food loss and waste

▸▸ Discussion, lessons and conclusions

Several aspects need to be taken into account in formulating an FNS policy for Colombia and its regions. Among them are the new health trends such as the prevalence of chronic non-communicable diseases and the double nutritional burden of low weight/micronutrient deficiencies and excess weight. Other factors are: the peace agreement resulting from several years of dialog, and which includes agrarian reform, migration from Venezuela, which has increased in recent years with Cali being one of the cities with the highest number of migrants, and the current Covid-19 pandemic. These factors highlight historic weaknesses in the national food system, such as an economic inability to access a balanced diet on a permanent basis. This crisis nevertheless provides an opportunity to reorient the National Food Safety Plan and its conception, by considering food to be a human right, with the role of policy being to implement adequate measures to guarantee this right. Its formulation, implementation and assessment must be considered as part of a participatory, inclusive and differential approach. We must also take account of other emerging alternative conceptual frameworks for food security, such as Food Sovereignty, Food Justice and Food Citizenship, which provide a wider range of alternatives for addressing situations and promoting systemic developments in all public policy processes.

Nevertheless, for food policies to be implemented and to achieve the desired impact in the various regions, institutional changes must be made to the coordination, linkage and participation of management and implementation entities.

Public policies are being developed to provide solutions to the social, economic, environmental and cultural problems requiring State action in the short, medium and long term, as described by Aguilar Astorga and Lima Facio (2009). Policies are the result of deliberate collective action; the course taken by the action as a result of decisions and interactions are the facts produced by the action. Policies are crystallized into decisions with the participation of society, with a higher level of participation indicating a higher level of democracy. On this basis, a number of aspects such as participation and coordination are required as essential steps in developments. They are taken into consideration in theory but are confronted by deficient processes for their implementation in reality.

In Colombia, coordination between the center and the regions is extremely weak. This leads to separate actions and separate efforts by the various institutional structures. At the same time, technical support for the regions is lacking. Further, as CONPES 113 has taken over the coordination of FNS through the creation of a central committee with no regional participation, developments cannot be aligned at the different levels of government.

▸▸ What's next? Prospects for cross-sectoral work

Today, significant progress has been made with a public policy on food sovereignty and security for Cali, built on the participation and support of organized groups and entities. It is essential for this policy to be implemented and for it to have an impact on the food and nutritional wellbeing of local communities. To this end, it is important to continue with the processes of social participation, increasing the participation of different schools of thought based on the same conceptual framework of rights, and for the State to be present in this exercise, protecting the various perspectives. Where citizens recognize food and participation as a right, the development of governance will be broader and more efficient for supporting both the implementation and the assessment of public policy.

At the same time, work must continue on regional autonomy and on the technical and financial support required from central government in order to fully develop the FNSS action plan for the policy proposal with its ten-year timeframe.

Additionally, a number of specific conditions need to be taken into consideration when developing actions to promote environmental conversation and sustainability in the regional food system in such a way as to reflect the specific conditions of Cali. Much of this region is classified as an environmentally protected area since it is part of the Los Farallones national park. The seven rivers running through Cali contribute to water supplies but require special protection and last, the monoculture of sugarcane is expanding in the valley. All these factors constitute a major challenge for the development of food production areas.

The discussions on policy construction have highlighted the need for food producers in Cali to transform their traditional agricultural practices or unsustainable agricultural techniques to agroecological practices, based on respect for the environment and for consumers. This is because the intensive use of pesticides and herbicides affects not only natural resources, but also the health and welfare of those consuming these products.

These land-related constraints combined with environmental and health requirements, are increasingly compelling citizens and governments to raise their gaze and to adopt a regional vision. Most of the food consumed in Cali is produced regionally and, without this vision, we cannot envisage food security in the city. The policy formulated includes a regional focus, but its development involves working with neighboring municipalities and departments, in order to design comprehensive policies that strengthen the components of food production and transportation in particular. The next step is to move from cooperation to the pooling of efforts as part of a fully integrated approach to food policy with a regional emphasis on south-west Colombia, aligning the needs of producer areas with the requirements of consumer areas as part of sustainable food systems.

Finally, in terms of "What comes next in Cali to better guarantee the right to food?", it is important to manage knowledge more effectively, as this is essential to making accurate and effective decisions, "There are gaps at the local level with respect to the generation of information on FNS. There is less research than in other regions and this makes it difficult to take technical decisions." Another point is also mentioned: "Unfortunately in Colombia, government decisions are based not on facts and data, but on the convictions of the ruling party" (Castellanos, 2018, p. 69). This was one of the most common views expressed when a number of FNS players in Cali were asked for their views on the existence of technical information to support decision-making. Although some progress has been made in this area, it is important to continue strengthening research into the most complex aspects of food security, starting from the standpoint of food production, particularly as one of the aims of the policy approved in Cali is to transform its agriculture.

Furthermore, access to food is one of the main problems for local communities. In 2018, 15.7% of the population in Cali was affected by monetary poverty, without the income necessary to access a basic basket of food and non-food items (housing, clothing, education, health, transport, etc.) or to meet their primary needs. The extreme poverty, rate in Cali was 3.5%, meaning that around 85,000 inhabitants lacked the resources to access even a minimal supply of food, according to the report Cali in Figures 2018–2019 (Municipal Planning 2019). With the current pandemic, the situation is likely to get worse. Concerning the consumption of food, in Cali and Colombia as a whole, a transformation is under way with interventions spanning all life stages, in order to prevent excess weight and chronic non-communicable diseases. To gather the data necessary for technical, social, economic and environmental intervention, knowledge management must be a constant in the implementation of public policies on food sovereignty and security in Santiago de Cali.

▸▸ Lessons learned about the construction process

If we were to think of the process of constructing a municipal food security policy for Cali as a recipe for a functional menu, recognizing the importance of trust would be the nutritional load provided by a vitamin-rich dressing.

While it is true that this construction process would not have been possible without the enabling context, the commitment of local government and the technical support of academia, we have seen the significant difference made by efforts to build trust among the stakeholders in this process. Trust helps to unlock synergies among players, nurturing the collective effort in working towards a common goal.

The presence of research organizations and NGOs as mediators in the relationship between the public and private sectors, their scientific contributions, the technical support for constructing the proposal, the international frameworks for guidance, the globality of the topic addressed, the opportunity to pursue different objectives and goals, and the chance to take part in a well-coordinated study with input from everybody, are just a few of the factors that made it possible to address the issue from a different perspective that reduced resistance to change.

Cali started a study aimed at improving its understanding of the food system and identifying needs from the standpoint of different players who are interested in contributing, rather than criticizing or making demands. The idea is to recognize the work done by each of the sectors working directly on the food issue, respecting their independence as a way to promote innovation, learning from each other and sharing the information available, with each person or entity contributing to the process.

Establishing a list of these ingredients as an experience rather than as a recipe, could contribute in some way to the needs of other medium-sized and small cities that do not have a policy of this type, but need to build one on a scientific basis. We are open to analysis and discussion.

Cali needs to continue consolidating its public food policy. Gaining approval was just a step along the way. Now it needs to strengthen the role of the Territorial Council of Food and Nutritional Sovereignty and Security (COTSSAN), the entity that created the policy, in coordinating civil society and the State and overseeing the suggested initiatives. It will also be necessary to implement a system to monitor the commitments made through the FNSS policy, reporting to the community on the development of the processes and enabling them to hold the council accountable for what it does or does not do in the field of food and nutritional security.

At the same time, the process showed us the need to develop processes for positioning FNSS issues in the public agendas of the governors and municipal council as a strategic topic for the development of the district of Santiago de Cali. For the moment, it rare to find governors with an understanding of the importance and complexity of this issue or a conceptual grasp of food sovereignty and security.

The introduction and implementation of the Milan Pact as an international reference and strategic framework for constructing the FNS policy was a key factor. Members of the government of Cali, players in the food chain and other partners recognized the relevance of the Milan Pact with its guidelines and the valuable experience gained in many cities around the world. For the local government, it is a source of pride knowing that its FNS policy uses the same framework as cities such as London, Sao Paulo, Vancouver, Paris and Copenhagen.

When thinking about a sustainable agrifood system, it will be necessary for Cali to extend its viewpoint and to embrace a regional vision based on stronger ties with the departments and municipalities that produce the food consumed in Cali. This relationship must go hand-in-hand with technical support to achieve environmental sustainability, technology transfer, and improved logistics for the transportation and distribution of food.

To develop a response to the content of the FNSS plan of Cali, with actions that translate into greater wellbeing for local communities, a relevant step would be to strengthen institutional problem-solving capacities.

▸▸ Final conclusions and outlook

As part of the discussions that took place for the construction of this policy, the following challenges were identified for Cali. Hopefully, the city will be able to overcome them through the new dynamics created and the participation of the various players. These challenges can also be understood as recommendations for any municipality seeking to initiate this type of process:
— Understand the food chain as a system within the city-region approach.
— Better integrate the issue of FNSS and agrifood systems in urban planning (e.g., land use planning).
— Obtain the participation of a wide range of players. Planning a sustainable agrifood system is a collective initiative requiring the collective action of society, and the participation of all players.
— Institutionalize comprehensive food and nutritional security policies and programs.
— Align efforts with international frameworks. Implementing a *référentiel* and aligning initiatives with international agreements such as the Milan Pact is an opportunity to harmonize objectives and adjust them to local realities.
— Adopt and implement indicators for monitoring the progress, results and impact of public food policies.

– Overcome information gaps and the lack of data or social mapping.
– Recognize the time required to meet economic and social development goals in cities.

▸▸ References

Aguilar Astorga, C., R., & Lima Facio, M. A. (2009). ¿Qué son y para qué sirven las políticas públicas? *Contribuciones a las Ciencias Sociales*, *9*. www.eumed.net/rev/cccss/05/aalf.htm.

Arciniegas, L., & Pena, J. (2017). *La transición alimentaria y nutricional en el modelo alimentario de los hogares caleños (Cali – Colombia)*. CIAT. https://hdl.handle.net/10568/97839

Castellanos Peñaloza, R. E. (2018). *Lineamientos de política pública de seguridad alimentaria y nutricional desde la perspectiva de diferentes actores para el municipio de Santiago de Cali* [Master's thesis, Universidad del Valle, Santiago de Cali]. Biblioteca digitale Universidad del Valle. http://hdl.handle.net/10893/14967

Concejo de Santiago de Cali. (2019). *Acuerdo No. 0470 de 2019. Por el cual de adopta la política pública de soberanía y seguridad alimentaria y nutricional del distrito especial de Santiago de Cali y se dictan otras disposiciones.* http://www.concejodecali.gov.co/Documentos/Acuerdos/acuerdos_2019

Departamento Administrativo de Planeación Municipal de Santiago de Cali. (2019). *Cali en cifras 2018–2019*. http://www.cali.gov.co/planeacion/publicaciones/137803/documentos-de-cali-en-cifras/

Departamento Nacional de Planeación, República de Colombia. (1996). *Plan nacional de alimentación y nutrición 1996–2005* (CONPES DNP-2847 UDS-DISAL). https://www.icbf.gov.co/cargues/avance/docs/conpes_dnp_2847_1996.htm.

Departamento Nacional de Planeación, República de Colombia. (2015). *Evaluación institucional y de resultados de la Política Nacional de Seguridad Alimentaria y Nutricional (PSAN)*.

Ministerio de Salud y Protección Social & Instituto Colombiano de Bienestar Familiar. (2019). *Documento general de análisis Encuesta Nacional de la Situación Nutricional en Colombia – ENSIN 2015*.

Popkin, B. M., & Reardon, T. (2018). Obesity and the food system transformation in Latin America. *Obesity Reviews 19*(8), 1028–1064. https://doi.org/10.1111/obr.12694

Temple, L., & Rippke, U. (2015). *Identificación de los problemas alimentarios de la población vulnerable de Cali: análisis de las cadenas de abastecimiento y de producción* (Cali Come Mejor, Reporte n°4). CIRAD, CIAT.

Willett, W., Rockström, J., Loken, B., Springmann, M., Lang, T., Vermeulen, S., Garnett, T., Tilman, D., DeClerck, F., Wood, A., Jonell, M., Clark, M., Gordon, L. J., Fanzo, J., Hawkes, C., Zurayk, R., Rivera, J., De Vries, W., Sibanda, L. M., … Murray, C. J. (2019). Food in the Anthropocene: the EAT–Lancet Commission on healthy diets from sustainable food systems. *The Lancet*, *393*(10170), 447–492. https://doi.org/10.1016/S0140-6736(18)31788-4

Chapter 14

The sustainable transformation of food systems and their impact on food and nutritional security: the case of the municipality of San Ramón, Nicaragua

JAIRO ROJAS MEZA, PEDRO PABLO BENAVÍDEZ, FRANCISCO CHAVARRÍA ARÁUZ

▸▸ Introduction

Few studies have been carried out in Nicaragua on the local transformations brought about by trade and social organizations, or their interaction with local government initiatives and national public policies for the sustainable transformation of food systems and their impact on the FNS of the population (families and communities).

San Ramón, where the case study was conducted, is one of Nicaragua's 154 municipalities. It is located in the central region, 12 km from the departmental capital of Matagalpa and 142 km from the capital Managua. It covers an area of 424 km², equal to 7% of the department and 0.33% of the national territory. No up-to-date population data are available, although the forecast provided by the association of municipalities for northern Nicaragua AMUPNOR (2010) gives a figure of around 45,150 inhabitants for 2020. Of this total, 85% are rural, a higher proportion than the national average of 40%.

This municipality was founded in 1904, when José Santos Zelaya was President of the Republic. Its origins are linked to the coffee plantations that were colonized by parts of the bourgeoisie through the appropriation of stretches of national lands and/or communal lands previously occupied by small mestizo or indigenous producers (Maldidier and Marchetti, 1996). As is common in the coffee-growing areas of Nicaragua, communities of small producers sprang up around the plantations, working on their own farms for part of the year and selling their labor to the coffee plantations during periods of low demand. These communities are also home to the families of permanent and seasonal agricultural workers, working for these large production units.

According to data from the CENAGRO national agricultural survey (2011), the municipality covers an area of 46,370 *manzanas* (32,460 ha), with cultivated and natural pastures occupying most of the agricultural land in the municipality for a total of 18,214 mz (12,750 ha). This is followed by the areas of land occupied by forest (9,228 mz or 6,460 ha), coffee (7,835 mz or 5,485 ha), and basic grains (5,169 mz or 3,618 ha). The municipality has three separate agroecological zones: in the north, a zone of higher altitude and rainfall, which is where most of the coffee is grown and, in the south-east, a dryer intermediate zone, which is where the activities relating to livestock and basic grain crops are based. The third zone is located in the southern part of the municipality; it is considered a dry corridor, specialized in the production of basic grains

The municipality is home to 9,131 families, of whom 2,550 are urban and 6,581 rural. Of these rural families 2,154 own land and 4,704 do not. Of the families in this last category, some rent land to grow crops, while others sell their labor for agricultural or non-agricultural activities. In other words, agricultural production in the municipality is maintained by 2,154 families, mainly relying on conventional production systems, with little use of farming technology. According to a study by Bonilla, Gutiérrez, Ramírez and Rojas (2010), 76% of the food consumed by farming families is purchased from outside sources. In other words, they are deficient production units. The study also revealed shortfalls in protein consumption (37.74%) and in kcal consumption (22.88%).

At the same time, landless rural families are even more exposed to food insecurity since they depend on farm work, particularly coffee growing, which is an unstable activity owing to the continuous fluctuations in prices. As a result, coffee producers often dispense with their services. In addition, farming activities in general depend on services such as the loans granted by the national financial system, which are more difficult to access in periods of socio-political instability, such as in 2018 or during the current pandemic.

Environmental damage is another aspect of the problems faced by the municipality. Forty percent of the soil is overused, suffering continuous damage that affects its productivity (AMUPNOR, 2010). Production units tend to be located in areas with gradients of over 30%, where the main productive vocation is agroforestry or conservation. Deforestation is a problem in the higher part of the municipality, where the main water sources originate, primarily owing to the actions of the big coffee producers. Their drive to increase profits is leading to a gradual decrease in water flows. Contamination by chemical pesticide residues and coffee by-products such as honey water and pulp, pose a risk to the water supply used by urban populations and communities.

In recent decades, this complex, multifactorial reality has affected the food and nutritional security of families, as well as the natural assets of the municipality and of the productive units in particular.

The first response to this crisis came from the National Union of Farmers and Ranchers (UNAG) through the farmer-to-farmer program *Campesino a Campesino* (PCaC) in the early 1990s. Other local social organizations subsequently became involved, together with the municipal government, through public policies and specific projects. Two emblematic programs have been developed on the initiative of central government: The Productive Food Program or *Hambre Cero* (Zero hunger), which has benefited 1,190 small farmers in the municipality and a school meals program that feeds 7,458 children in kindergarten and primary education.

Against this problematic backdrop, the study will answer the following questions:
– How can we best describe the development and current state of the Food System in the municipality of San Ramón?

– What has been the impact of national public policies (regulatory framework, programs, projects and actions) on Food and Nutritional Security (FNS) and sustainability in the municipality of San Ramón?

– What has been the role of local players in building the institutional framework for the transformation of the Food System, overcoming food insecurity and environmental, social and economic unsustainability?

– Are there any links between national and local policies, particularly those related to production, education, health and the environment?

▸▸ The conceptual and methodological framework for analyzing the problem

The conceptual framework is based on the definitions of food systems, life strategies in farming homes and agroecology as a basis for sustainability, recovery, the preservation of natural resources, and governance (coalitions and institutional structure) in the Food System transformation process. The study used the methodological approach of complex systems, socio-historical analysis and constructivism as the epistemology to build scientific knowledge. The tools used were: a documentary review (research into Food and Nutritional Security in the municipality, case studies on the impact of agroecological farming models, municipal diagnoses, plans, strategies, the local regulatory framework developed by social, trade union, community and public sector players, particularly the municipal government), and interviews with key players.

The conceptual framework is briefly detailed below:

Food systems: A food system is the sum of the various components, activities and players which, through their interconnections, enable the production, processing, distribution and consumption of food. Owing to their multidimensional nature, food systems include economic, environmental, political and socio-cultural aspects, with a wide range of players and dynamic environments (FAO, 2017).

Nutritional Food Security: Defined as the state in which all people have physical, economic and social access to the food they need, in a timely and permanent way, in sufficient quantity and quality for their consumption needs and biological utilization, ensuring a state of general well-being that contributes to their successful development (INCAP, 2012). The *Via Campesina* (International Peasants' Movement) describes the concept of Food Sovereignty as: "The right of peoples, communities and countries to define their own food and agriculture policies, as part of an approach that is ecologically, socially, economically and culturally appropriate for their unique situation." One of the ways to address food security is through its four components: availability, access, biological utilization and stability.

Sustainability of food systems: Sustainability is a multidimensional concept referring to the need to maintain the natural capital, production base, social development and the economy over the long term, with particular emphasis on Food and Nutritional Security. Agrifood sustainability encompasses production and consumption systems with continuous, regenerative cycles, reducing dependence on external inputs and energy, and minimizing the production of waste and pollutant emissions (HLPE, 2014). However, the concentration of land and resource ownership, the damage to soil, water and biodiversity, the rapidly growing population and the economic factors linked to the food trade have a determining impact on the food security of the most vulnerable, as well as leading to high environmental costs, inequality and a rise in diseases linked to malnutrition. The proposed idea is to move towards sustainable food systems, improving process efficiency and, more particularly, including all players, in order to generate more equitable results and benefits (FAO, 2017).

Reproduction strategies in domestic peasant units: Strategies are conceived as production and social practices carried out consciously or unconsciously to maintain or change a situation. The viability of a farming production strategy depends on the smallholding's capacity for self-sufficiency as a unit of production, consumption and salaried labor (Ramírez & Méndez, 2007).

Martinez and Rendón (1978) describe two types of reproduction according to production capacity and based on the labor force and production resources available to satisfy the needs of the production unit. These are:
– Farming units that are characterized by their inability to reach a sufficient level of production for their survival, using their own resources. The reproduction of these production units depends on labor market conditions.
– Farming units that manage to satisfy basic reproduction needs through agriculture. However, their vulnerability is immediately laid bare by any unfavorable changes, no matter how small, in the terms of trade, climate conditions or even their own family composition.

Peasant farming in Nicaragua today is a complex mix of market and subsistence strategies, international capital flows, global, national and local institutions, the relationship with the environment and the controversial use of resources (Holt, 2008). In the region considered, small-scale farms coexist with medium and large farms, giving rise to a highly heterogeneous agrarian structure that reproduces a pattern of inequality in the distribution of assets, perpetuating and accentuating productivity gaps (FAO, 2017).

Agroecology as a strategy for sustainability and rural development: This paper defines agroecology as an approach that seeks to integrate ecological science with other sciences, such as agronomy, sociology, economics and history, along with local and indigenous knowledge systems, in order to guide research and action towards the sustainable transformation of our agrifood systems (Gliessman, 2015). This concept supports the notion that agroecology is a framework expressed through science, practice and social movements, and that it is most effective when these three dimensions converge. From this perspective, agroecology is the most appropriate way to promote processes leading to sustainability and rural development, this last aim being understood as a process of productive and institutional transformation in a given area, with the purpose of improving the living conditions of local families.

Governance, coalitions and the institutional framework in the dynamics of Food System transformation: the analysis of the transformation of local food systems, from the standpoint of governance and local players, can draw upon the Advocacy Coalition Framework (ACF). This theoretical model aims to explain how changes in public policy take place. The underlying message is that policy changes are not only the result of power relations or competition between interests, but that the process of learning within the coalitions and alliances formed by players is also important. Sabatier and Weible

(2007) state that the purposes behind the development of this conceptual framework of the policy process was to simplify and explain the complexity of public policy.

Method of complex systems, socio-historical and constructivist analysis: to understand how a complex system works, it is necessary to analyze the history of the processes that led to the type of organization (structure) in its existing form at a given point in time. From this perspective, we can affirm, as Piaget once said: "There is no structure without history, nor history without structure."

To approach food systems and their sustainability, we need to represent a set of situations and phenomena that can be addressed as an organizational whole with a characteristic form of operation. The structuring of reality is based on two principles that are characteristic of complex systems: an arrangement of components by level of organization, each with their own dynamics but interacting with each other; and a form of progression relying on successive reorganizations rather than continuous development (García, 2006). In the subject that concerns us, the levels are: i) the municipal level with its components and relations; ii) the set of national policies on FNS and sustainability, and; iii) the international level, coffee and other markets, and global policies on development and the environment. Consequently, the transformations undergone by the system depend to a great extent on the interactions within and between these levels.

This approach makes it possible to generate systemic properties such as resilience and sustainability. Both concepts refer to the system's capacity to adapt to disturbances of a certain magnitude, i.e., that do not exceed the characteristic threshold of the system at any given time. When this threshold is exceeded, the system is destabilized, making it vulnerable to these disturbances.

The constructivist perspective adopted in this paper (figure 14.1) is based on the work of García (2006):
— *First, a process of differentiation and integration in the construction of knowledge*: in the first instance, whole entities are defined with a level of imprecision, with

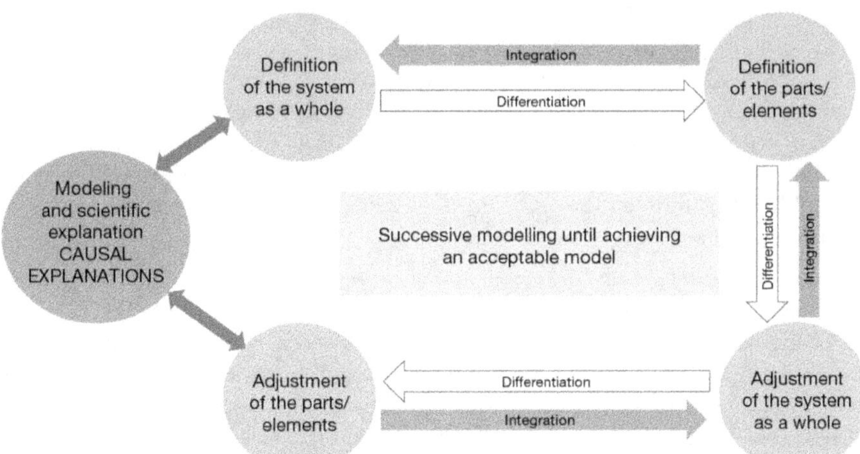

Figure 14.1. Constructivist methodological model. Source: Rojas (2020), adapted from García (2006).

subsequent and gradual differentiation of their constituent parts (also with a level of imprecision) through a process of analysis. Studying these components enables better characterization for a more clearly determined whole. The process begins and continues in successive stages, with adjustments being made at each one. These may involve incorporating factors that have been left out or eliminating those that are seen to be unnecessary or secondary. The dialectic of differentiation and integration forms the basis of the construction of knowledge.

– *Second, modelling and a scientific explanation*: Research involves successive modelling until an acceptable model is reached, i.e., one that provides a causal explanation of the phenomena under study. In other words, the set of relationships involved constitutes the explanation of how the system works.

▸▸ The municipality of San Ramón combines old coffee plantations and peasant farming

The governors of the late nineteenth and early twentieth centuries were keen to enter the modern era by gaining access to international markets. For this reason, they introduced a range of public policies on agriculture, including coffee production. Presidents Fruto Chamorro (1853), Tomás Martínez (1858) and Evaristo Carazo (1889) issued decrees rewarding coffee growers and offering up to 500 manzanas (350 ha) of "vacant" land free to any foreigner or national who planted more than 25,000 coffee trees (Choza, 2008). In the period between 1893 and 1910, over 28,725 manzanas (20,233 ha) of land were taken from the indigenous people.

Following the victory of the Sandinista Popular Revolution in 1979, most of these large properties were confiscated, becoming part of a group of public companies called the *Área Propiedad del Pueblo* or Area of People's Property (APP). The lands were not handed over to the peasants, however. The reason given was the need to maintain productive levels and obtain the foreign exchange income necessary to invest in the new government's economic and social development programs. With the FSLN's electoral defeat in 1990, these companies followed different paths: some were returned to their former owners, others were turned over to workers and cooperatives. In this last case, the state compensated the former owners for the value of the confiscated land. A few years later, however, with the rollout of the neoliberal policies of the 1990s and the limited capacity of worker-owned companies for technical, administrative and financial self-management, some of the cooperatives sold their lands to the old and new owners, reconfiguring the large coffee plantations once again.

The large plantations concern producers whose main aim is to maximize the profits from their investment. Following the same approach, they also invest in other sectors such as livestock or other agricultural activities, when coffee is in crisis. A small portion of producers in this sector invest in non-agricultural trade and agribusiness activities such as coffee processing, slaughterhouses or services. Unlike peasants, they have more advanced means of production such as barns, farmhouses, mills, haciendas, vehicles, etc. (Rojas & Blandino, 1996; Maldidier & Marchetti, 1996) This sector is also characterized by its semi-settled nature (meaning that producers live in the city rather than on the farm and delegate management to an administrator and/or other supervisors). Labor contracting, tenant farming, the possession of multiple farms and the dependence on credit for the production process, are the most common production relations.

The coffee-growing sector has a secure labor force in landless peasants, agricultural workers and even a number of peasants whose family members, in addition to working on their own farms, sell their labor for a few months every year, particularly during the coffee harvest. This provides them with additional income to acquire the food and basic necessities that are not generated by the farm. It is also important to point out that the municipality has around 200 small coffee growers, implementing peasant reproduction strategies, who not only produce coffee but also set aside small areas for basic grains (corn and beans) for self-supply, and who sell their labor, particularly to the large coffee plantations. These peasant groups are vulnerable to the recurrent crisis in coffee prices in international markets. In consequence, their food security depends on the international market in two ways: the income generated through the sale of their own coffee and the employment possibilities for family labor in this area.

▶▶ The peasant production crisis and the emergence of alternative technological and political advocacy options

The *Campesino a Campesino* program (PCaC) was set up by the National Union of Farmers and Ranchers (UNAG) in 1992 in the four communities of San Ramón, under the leadership of four developers. Its first initiatives focused on Soil and Water Conservation (SWC) in corn and bean plots. To this end, the program set up experimental plots, using green manure and implementing reforestation practices to protect water sources (Cuadra & Vásquez, 2016). The main purpose of this work was to improve soil fertility, and to reduce production costs and external dependence on inputs.

In the mid-1990s, the PCaC decided after a process of reflection to expand its initial scope in order to address issues related to integrated production systems, soil fertility, the ecological management of pests and diseases, large and small livestock, and animal traction, among other issues. At the end of the decade, in the wake of Hurricane Mitch that swept through several Central American countries, a regional study (El Salvador, Honduras and Nicaragua) found that the plots implementing agroecological practices and PCaC methodology were better placed to recover from the losses and destruction caused by the hurricane. This demonstrates their capacity for resilience.

Since the beginning of the 2000s, the PCaC has expanded its coverage of local communities to enter public spaces, becoming part of the coordination of the Municipal Government (GM) and particularly the Environmental Commission. As part of this role in 2007, working with civil organizations, the public institutions of central government and the GM itself, it promoted the rollout of an environmental ordinance to regulate and counter the environmental damage caused by agricultural activities, particularly the deforestation driven by the expansion of the coffee plantations in the highlands of the municipality. The PCaC members involved talked about the difficulties of the advocacy work involved. Prior to approval: "There were serious problems, because of the deforested areas created by the production and extraction practices of the large coffee landowners, who were opposed to the municipal ordinance as it would prevent them from continuing to damage the environment."

The ordinance was approved and now protects watersheds, aquifers, genetic resources and the genetic diversity of wild flora and fauna. It provides the basis for promoting environmental education, scientific research and the study of ecosystems, as well as furthering sustainable local development, and encouraging the implementation of clean processes and technologies for the improvement and rational/sustainable use of natural ecosystems.

According Cuadra and Vásquez (2016), in late 2007, the Humboldt Center, a national non-governmental organization, supported a nutritional study of children in San Ramón schools. The children who were receiving food distributed as humanitarian aid, were experiencing vomiting, diarrhea and skin allergies. A number of food samples were collected and sent to Genetic ID, Inc., a laboratory in the United States. The result showed that the food consumed by the children contained seven types of GMOs at high levels of concentration.

This situation prompted social players, unions and the Municipal Government to discuss an ordinance to declare the municipality a GMO-free zone. In August 2008 the "Ordinance for the Protection of Biodiversity in the Municipality of San Ramón" was approved. Its stated goals are to: i) promote food production within the community; ii) ensure the establishment of GMO-free crops in order to secure nutritional food of sufficient quality and quantity; iii) rescue and conserve native biological diversity with particular emphasis on heirloom and hybrid seeds of crops and species of nutritional interest, as a strategic component in strengthening the Food Sovereignty and sustainable development of the municipality of San Ramón; iv) promote the establishment of centers for the conservation of heirloom seeds (Seed Banks) at the community and municipal level; v) harmonize initiatives in the area of municipal Food and Nutritional Sovereignty and Security with the current national legal framework.

In 2012 social players and the municipal government further expanded the local framework for promoting Food Security in the municipality by approving the "Ordinance for the Promotion of Food and Nutritional Sovereignty and Security (FNSS) in the Municipality of San Ramón." The ordinance states that food security in the municipality is impacted by factors associated with the productivity of agricultural production systems and family income, coupled with an external dependence on food, expressed in a high rate of food purchases (76%) and a dietary pattern based on eight foods. It also identifies agroecology as an appropriate model to increase and diversify the production and consumption of nutritious and safe food and the protection of natural resources.

▶▶ Flagship public programs and their impact on FSN and the sustainability of production systems

With the return of the Sandinista National Liberation Front (FSLN) to government, the issue of nutritional food security and agriculture, particularly family agriculture, took on greater importance once again. In this context, the government outlined a new Food Production Program (McBain-Hass & Wolpold-Bosien, 2008) based on international experience, particularly in Brazil, and on the lessons learned by national civil society organizations, particularly the Center for Promotion, Research and Rural and Social Development (CIPRES).

The ideas underpinning the CIPRES Food Production Program were formed in the wake of Hurricane Mitch. The objective was for every Nicaraguan family to have the same products on their table as families in developed countries, such as milk, meat, eggs, fruit, vegetables and cereals. The general aim of the program was to provide capital and support through agroecological technologies, targeting peasants as the main players in food production, as well as supporting the organization, integration and participation of the rural population in its management.

The purpose of the *Bono Productivo Alimentario* (Food Production Support or BPA) or *Hambre Cero* was to build sustainability in food production. This means recovering the natural resources (biodiversity, soil, water) used in the production process and reconstituting the original properties of these systems through the use of eco-technologies, the manufacture and use of bio-inputs, the generation of domestic energy through biodigesters, nutrient recycling and water harvesting, to give just a few examples.

A number of requirements and mechanisms were established for the allocation of the BPA, as follows: i) the woman of the household would be the direct beneficiary; ii) the family was required to have access to a plot of land of between one and five manzanas (3.5 ha), this land did not necessarily have to belong to the woman, but could belong to another member of the family; iii) the family did not already own the assets provided by the BPA; iv) the woman accepted the assets and made a commitment on their appropriate management; v) she signed a document in which she agreed to take part in training workshops, not to sell the assets provided, to take part in organizing a members' group and to return 20% of the value of the BPA, for payment into a savings fund that is part of the group's assets.

In the case of the municipality of San Ramón, 1,190 BPAs have been awarded since 2008. According to one of the technicians who took part in the running of the program and who is currently a member of the Municipal System of Production, Consumption and Trade, an internal assessment showed that only 628 (52%) of the families are able to maintain or increase the assets provided. This means that the program is relatively inefficient. The main causes are thought to be: the selection of beneficiaries who do not meet the requirements set, and assets that do not reflect the socio-economic circumstances of local families. For example, the poultry and pigs provided belonged to improved breeds requiring concentrated feed that the families could not afford. The beneficiaries therefore sold the assets or used them for food before they were able to reproduce. In the case of BPAs that included cattle, if the family did not have enough land for grazing, the beneficiaries took the option of selling the assets.

The thinking behind the program was to organize groups of 50 women, who would make the leap in the medium-term to higher forms of organization, such as cooperatives. In the case of the municipality, among the groups of women who received pigs, a decision was made to organize a cooperative to coordinate the chain and add value to the meat. The municipal government provided support in the form of land for the infrastructure, while the Ministry of Family, Community, Cooperative and Associative Economy built the infrastructure. The cooperative operated for a relatively short time before disintegrating.

Nevertheless, among the women and groups who are still part of the program, there is some evidence as to the program's success. Some families have managed to build up and maintain a herd of cattle from a cow they received. As a result, they have milk and cheese for their own food security, and they can sell what they don't need in local markets. Similarly, concerning the poultry and pigs supplied, some evidence points to the role of these assets in improving food security and family income. The municipality has conducted no assessments on the impact and sustainability of this program, but tangible achievements can be seen concerning the families that manage to keep the assets. The Municipal Government of San Ramón built a municipal market for the agroecological production of rural families, with produce from this program. This has undoubtedly improved not only the food security of farming families, but also the supply of products available to rural and urban families who are unable to produce their own food. The challenge now is to assess the benefits of the *Hambre Cero* program in terms of production, income, sustainability and well-being for families in the municipality of San Ramón.

In the public sector, through the Ministry of Education (MINED), the government has set up a Comprehensive School Nutrition Program (PINE) comprising four components: i) Education in Food and Nutritional Security; ii) School kitchen gardens; iii) School meals and iv) School shops. PINE-MINED has responsibility for monitoring and supervising school meals through regional technicians and the school management office. In each school, a School Meals Committee (CAE) has responsibility for the program. Parents are able to take part in the committee, which is coordinated by the school principal. Since they were first set up, the CAEs have played an important role in keeping the PINE program going, since they oversee the process of preparing school meals. The rations are received, stored, prepared and delivered to each child.

In the case of San Ramón, the municipality has 90 nursery and primary schools, with 7,458 children receiving school meals. An assessment of the program conducted by the WFP for the period 2013–2019 concluded that the PINE adopted a cross-cutting approach to the issues of nutrition, resilience and climate change with specific workshops, but that there is no evidence that the program has had any impact on changing capacities or behavior.

It is clear that some mistakes in the implementation of public-sector agricultural development programs and projects can be attributed to the lack of a comprehensive vision. The municipality of San Ramón could be considered as a reference in the production of *Passiflora edulis* (passion fruit). Its origins date back to a program set up the Rural Development Institute (IDR) in 1994. Many producers saw greater economic benefits in passion fruit, so they turned their plots over to this crop, reducing or stopping their production of basic grains (maize and beans). This affected the basic food security of families, leaving them vulnerable to market prices. The same occurred in other areas, where the main emphasis was on profitability and using the income to acquire the basic foodstuffs that they used to produce.

▸▸ Players, leadership training and the institutional framework in place to address food insecurity, environmental damage and rural development in San Ramón

The current situation in San Ramón in terms of the Food System, Food Security and sustainability is the result of a complex history, with interactions between global, national and local processes. Among the most important global processes with local impact are those relating to international markets, and particularly coffee. The decisions made through the institutional framework concerning this crop, have had a major impact and continue to influence production, income, employment and therefore the nutritional food security of the municipality.

At the same time, national events such as the triumph of the Sandinista Revolution in the 1980s, led to the emergence of new social and trade union players such as the National Union of Farmers and Ranchers (UNAG), founders of the program *Campesino a Campesino*, which arrived in the municipality in the early 1990s. The Union of Agricultural Cooperatives (UCA – Augusto Cesar Sandino) of this municipality was formed through this process, and most of the land owned by cooperative members was gained through the Agrarian Reform led by the revolution. Although they have failed to bring about major transformations in the production systems of their members, as a productive diagnosis of this organization shows, they have nevertheless made pertinent efforts in environmental, social and economic sustainability, and food security in particular.

Other social organizations, with national, departmental or local coverage, have played a role in the socio-productive and environmental dynamics of the municipality over the last thirty years. We could say that the most tangible result of the work of the social and trade union players lies in the promotion of sustainable production systems based on the postulates of agroecology. At present, five players with relevant environmental experience are present in the communities of the municipality, as follows:

– The program *Campesino a Campesino* (PCaC) has been present in San Ramón for almost 30 years without interruption. It currently has 2,000 families involved in its labor networks. The program is present in 65 communities out of the 96 in the municipality, with eight Community Seed Banks of heirloom and hybrid seeds, produced by recovering local genetic material. It works with a network of 120 leaders and developers, 65 of whom hold a range of community and public management responsibilities at the municipal level. For example, the current mayor and vice-mayor of the municipality belong to this movement.

– The *Unión de Cooperativas Agropecuarias (UCA – Augusto Cesar Sandino)* is a local trade union organization that has been present in the municipality for 30 years. Its main leaders have held positions of responsibility in municipal public administration and in the departmental and national agricultural public sector. In the municipality of San Ramón, the UCA has eight member cooperatives. In terms of Food Security, the UCA focuses primarily on production diversification, the improvement of agroecological practices to increase the availability of food for member families, the sustainability of local food systems and the strengthening of the cooperative movement, with the participation of young people and women.

– The *Denis Ernesto Gonzáles López Foundation (FUDEGL)* is a non-profit organization that has been present in the municipality for 15 years. Its first actions were aimed at promoting conservationist practices in agriculture. It then expanded its focus to include permaculture, the saving of heirloom seeds and agroecology in general. It has helped to set up five Community Seed Banks in the municipality, as well as contributing to the characterization of 20 heirloom varieties. These varieties are part of the genetic material of peasant families and offer a number of advantages in fighting climate change. This organization has demonstrated leadership in the areas of coordination and advocacy, encompassing a range of issues in order to promote agroecology and sustainability in the agenda of the municipal government.

– The *Movement of Agroecological and Organic Producers (MAONIC)* emerged from the Organic Roundtable set up in 2007. It has successfully developed synergies with a range of social, production and public-sector organizations. Its main action has been to contribute to the development of national proposals to promote agroecological production. It has developed technical competencies in the sustainable recovery of soils. Although it is a national organization, it is also a useful support for local organizations and producers.

– The *Organization for Economic and Social Development in Urban and Rural Areas (ODESAR)* is a non-governmental organization (NGO) present in several municipalities in the department of Matagalpa, including San Ramón. Its objectives concern the need to develop programs, projects and actions focusing on the most impoverished and historically marginalized population groups. It prioritizes projects to reduce extreme poverty and to promote the food and nutritional security of peasants and landless rural workers, the priority being to establish direct contact with the women in the community.

The emergence of community, inter-community and municipal leadership is one of the main results of several decades of work in San Ramón. This leadership has contributed to a number of agreements in the form of ordinances targeting Food and Nutritional Security, environmental protection and sustainable production. The municipal government has enabled these organized movements to become part of working commissions that have taken different names at different times, but that nevertheless serve the same purpose.

The influence of national institutions and public agricultural, environmental, health and educational policies has been more diffuse in terms of Food Security and Sustainability. The municipality has applied neither the Food and Nutritional Sovereignty and Security Act (Act 693) nor the Act for the Promotion of Agroecological and Organic Production (Act 765), particularly from the standpoint of organizational mechanisms. Act 693 establishes the organization and operation of the Municipal Commissions of Food and Nutritional Sovereignty and Security, but the municipality already has a commission called the "Cabinet of Production, Environment and Water." This commission brings together the leaders of the organizations above, together with councilors from rural communities with experience and commitment to the development of sustainable agriculture. Through this commission, the social and trade union organizations develop their coordination and advocacy mechanisms.

The dispersion of public-sector organizations has been one of the characteristics observed at the municipal level. In 2020 municipalities across the country, and San Ramón in particular, organized the National System of Production, Consumption and Trade (SNPCC). This entity brings together the delegates or technicians serving the municipality, among others: the Ministry of Agriculture (MAG), the Ministry of Family, Community, Cooperative and Associative Economy, the Nicaraguan Institute of Agricultural Technology (INTA), the Ministry of the Environment and Natural Resources, the National Forestry Institute (INAFOR), and the Institute of Agricultural Protection and Health (IPSA). This coordinating entity holds monthly meetings to oversee and assess the work carried out in the field of agriculture. For the purposes of these meetings, it regularly invites three members of the municipal government, particularly those occupying the most senior positions in the Mayor's Office and the Presidency of the Cabinet of Production, Environment and Water.

As Ayala (2004) points out, furthering institutional innovation and setting up efficient institutions in particular, is a difficult task, since it involves complex negotiations and – frequently – conflicting demands. Nevertheless, the municipality of San Ramón could build on the experience acquired by its players in the formation of social capital to make more effective use of these mechanisms in order to address the challenges of transforming the food system and its sustainability.

▸▸ Some evidence relating to agroecology and the sustainable transformation of the food system in the municipality of San Ramón

Despite the complexity of the food issues in this municipality, the technical, social and organizational foundations have been laid for the large-scale implementation of agroecological practices. The municipality has a network of community seed banks (heirloom and hybrid), as well as flagship or reference farms showing the results of the model in real conditions, and organizational mechanisms at the institutional level to further progress in this direction.

The last municipal study on Food and Nutritional Security was conducted in 2010. It found that food patterns were made up of eight foods in rural areas and twelve in urban areas. Case studies carried out on agroecological farms in the municipality illustrated the potential of agroecology for achieving food sufficiency at the household level as well as producing surpluses (40%) for the local market (table 14.1). These farms produce an average of 20 items, which allows them to diversify their diet and to supplement their income by selling surpluses. The families involved in the case studies buy a wider range of foods than those that they produce on their farms.

In terms of environmental sustainability, the cases studied show the transformation of the conventional farming model, through the use of 14 agroecological technologies and practices, including: plant diversity (around 20 on average), the incorporation of organic fertilizers to improve soil fertility, pest and disease management through the manufacture of natural bio-inputs, the use of heirloom and hybrid seeds according to local climate projections, soil protection using living and dead matter, windbreaks, technologies for water harvesting and the application of biofertilizers. This change in the technological matrix contributes to the restoration of farms and the surrounding environment, while also helping to reduce diseases caused by poisoning, improve food quality and cut production costs.

Table 14.1. Food patterns in urban and rural areas and agroecological farms

Sector	Number of meals in the pattern	Type of food consumed	
		Produced	Purchased
Urban	12		Corn tortilla, beans, rice, coffee, chili, onion, curd cheese, basic bread, sugar, oil, salt and egg
Rural in general	8	Corn, beans, rice, coffee, egg (with 60% produced on the farm on average)	Sugar, oil, salt
Agroecological farms	20	Corn, beans, bananas, cassava, bananas, eggs, citrus fruits, avocado, vegetables, fruit, chicken (with the exception of the chicken, which is purchased in part, the average surplus of the other items is 40%)	Rice, cheese, curd cheese, cream, oil, salt, sugar, meat, oatmeal bread

Source: Based on the diagnosis of the Food and Nutritional Security study conducted by the municipality in 2010 and the study by Cuadra & Vásquez (2016).

Although the results are encouraging concerning efforts to improve the food system and its sustainability in the municipality, a number of questions will require further study: What progress has been made in the agroecological transformation of farms supported by the local organizations described in this paper? What are the bottlenecks limiting the adoption of the agroecological model? What national and local public policies could accelerate this transformation? What are the possible options for rural families who have no land and who depend on the sale of their labor to acquire their food supply?

▸▸ Conclusions

Based on the questions raised in this paper, we have reached the following conclusions:

– The existing food system in the municipality of San Ramón is the result of a historical process of agricultural public policies and relations with the national and international market. This food system is deficient in terms of food production, both for peasant families and for the rural and urban population. The land tenure structure, in which large properties coexist with smallholdings, is the result of historical dynamics with a determining impact on food production.

– The access of landless rural families to food depends on their income, and this is uncertain, particularly in sectors such as coffee growing, owing to instability in international markets, socio-political crises and exceptional events such as the Covid pandemic.

– Public policies expressed through programs of capital investment in peasant units such as Food Production Support (*Hambre Cero*) have played a regular role, even though almost 50% of the families concerned have been unable to maintain and develop the assets provided. However, those who were able to maintain and develop the assets were able to provide a wider range of food for their families and to supply products for the local markets. The PINE – school meals program has provided stable support for children in nursery and primary education, complementing the existing food supply of families, which as we said earlier, is insufficient. The participation of parents in the management of this program has been fundamental to its success.

– The agroecological transformation of peasant production units has been mainly driven by trade union and social organizations, which have shown steady commitment over a period spanning almost 30 years. These entities have created an organizational base of sustainable agroecological technologies that can be expanded to include a greater number of farms in the municipality. They also have the leadership and skills to influence local public policies to improve the food system. There is strong evidence showing that the agroecological model of diversified farms is superior to that of conventional farms, in terms of production sufficiency.

– Findings have shown that the coherence between national and local policies, coordination mechanisms and intersectoral cooperation is still insufficient. Neither Act 693 on Food Security nor Act 765 are implemented at the municipal level, and local coordination mechanisms such as the Cabinet of Production, Environment and Water do not involve delegations from central government entities, such as MINED and MINSA, both of which are key to a strategy for Food and Nutritional Security. The coordination of the Public Agricultural Sector, through the System of Production, Consumption and Trade with the participation of the Municipal Government, is an important step, but it must bring in other key players, such as those mentioned above.

– This research highlighted weaknesses in the management of information and knowledge of the food system, and therefore of Food and Nutritional Security. The team carrying out this work needed to bring together a number of widely dispersed components in order to fully understand the subject. It will be necessary to generate a system of indicators around the food system, with monitoring and follow-up strategies, in order to generate knowledge for decision-making in this area.

▸▸ References

AMUPNOR (Asociación de Municipios Productivos del Norte). (2010). *Diagnóstico Municipal de San Ramón*.

Ayala, J. (2004). *Mercado, elección pública e instituciones: una revisión de las teorías modernas del Estado* (2nd ed.). Universidad Nacional Autónoma de México.

Bonilla Escoto, H. J., Gutiérrez, M., Ramírez, J., & Rojas, J. (2012). Estado de la seguridad alimentaria y nutricional en el municipio de San Ramón, departamento de Matagalpa, durante el segundo semestre del 2009. In Rojas Meza, J., & J. Ramírez Juárez, *Aportes para el desarrollo rural en la región centro norte de Nicaragua* (pp. 289–305). UNAN. https://repositorio.unan.edu.ni/1829/1/2004.pdf

Cuadra, M., & Vásquez, J. I. (Eds.). (2016). *De la práctica agroecológica en la finca a la incidencia política. Seis experiencias en San Ramón, Nicaragua*. UNAG-PCac, SIMAS. https://cluster-nicaragua.net/media/publicaciones/archivos/EstudiosCasosAgroecoliaAxA_fYsBHqi.pdf

García, R. (2006). *Sistemas complejos: Conceptos, método y fundamentación epistemológica de la investigación interdisciplinaria*. Editorial Gedisa. http://secat.unicen.edu.ar/wp-content/uploads/2020/03/GARCIA-Sistemas-complejos1.pdf

González Dufau, G. I., Santamaría Guerra, J., & Rojas Meza, J. (2019). *Soberanía y seguridad alimentaria nutricional en la Comarca Ngäbe-Buglè, Panamá: Escalamiento de la agricultura agroecológica para aumentar la producción de alimentos*. IDIAP. http://www.idiap.gob.pa/?wpdmdl=3944

FAO. (2017). *Reflexiones sobre el sistema alimentario y perspectivas para alcanzar su sostenibilidad en América Latina y el Caribe.* https://www.fao.org/3/i7053s/i7053s.pdf

Holt-Giménez, E. (2008). *Campesino a Campesino: Voces de Latinoamérica, Movimiento Campesino a Campesino para la agricultura sustentable* (L. Hurtado, E. Botella, D. Mac, A. Núñez, & C. Nilsen, Trans.). SIMAS. (Original work published 2006).

McBain-Hass, B., & Wolpold-Bosien, M. (2008). *El derecho a la alimentación y la lucha para combatir el hambre en Nicaragua. Un año del Programa Hambre Cero.* FIAN Internacional.

Maldidier, C., & Marchetti, P. (1996). *El campesino-finquero y el potencial económico del campesinado nicaragüense.* Universidad Centroamericana. http://repositorio.uca.edu.ni/id/eprint/572

Martinez, M. P. L., & Rendón, T. (1978). Fuerza de trabajo y reproducción campesina. *Revista Comercio Exterior, 28*(6), 663–674.

Ramírez Juárez, J., & Méndez Espinoza, J. A. (2007). *Transformaciones agrarias y estrategias de reproducción campesina en el Soconusco, Chiapas.* Colegio de Postgraduados en Ciencias Agrícolas, Campus Puebla.

Rivas Choza, Edgard. (2008). *San Ramón: Indígena y fértil.* Editorial Apante.

Rojas Meza, J., & Blandino Granados, L. (1996). *Dinámica de la renovación y fomento cafetalero en el municipio de San Ramón, Matagalpa, período 1990–1995* [Graduate thesis, Universidad Nacional Agraria]. Repositorio UNA.

Sabatier, P. & Weible, C. (2007). The advocacy coalition framework: innovation and clarifications. In P.A. Sabatier (ed), *Theories of the policy process* (pp.189–220). Westview Press. https://doi.org/10.4324/9780367274689

Development of and changes in public food and nutrition security policy in the department of Antioquia, Colombia

José Anibal Quintero Hernández

▶▶ Introduction

The research presented here on the development of and changes in public food and nutrition security (FNS) policies in the department of Antioquia (Colombia) has been undertaken within the framework of the Planning, Development and Education (Plan D+E) research group of the University Institution Colegio Mayor de Antioquia, which looks at issues relating to agrifood areas in the department of Antioquia.

Food security is an issue of great concern in Colombia. Chapter II of the Colombian Constitution of 1991 states that a balanced diet is a fundamental right of children and sets out the duties of the state, from production to research support, raw materials, and production technologies. However, a report compiled by the Attorney General of the Republic in Colombia and recently published on the website of the Food Bank of Colombia reported that five million people suffer from hunger in the country and 54.2% of households suffer from chronic malnutrition, hence the high level of concern surrounding the issue.

The department of Antioquia is made up of 125 municipalities and covers an area of 63,612 km² (5.5% of Colombia's total area). It has a population of 6,407,102, of which 4,972,941 (77.61%) live in urban areas and 1,434,161 (22.38%) in rural communities, according to preliminary data from the 2018 census. The country sits mainly on the Central Andes and has a diverse ecosystem that allows agricultural production for export and consumption.

On the issue of food security, the department has played a leading role in Colombia. It introduced the Food and Nutrition Improvement Program of Antioquia (MANA) in 2001. Forming part of the Departmental Development Plan, it was recognized in 2003 as a departmental public policy (Government of Antioquia, 2011). Significant strides have been made thanks to the policy, and reference to the literature, this is shown throught some important research conducted in assessing public policies (Mazo & Imasato, 2018; Mazo, 2015; Oviedo, 2015;

Quintero, 2011; Arboleda et al., 2008); from nutrition (Alvarez et al., 2006) and participation (Zuliani, 2010), but without, as yet, identifying aspects specific to the region, which is the aim of this research.

In terms of food and nutrition security, sub-regions have been established for the constant search of maintaining food production in that sense sustain the possibility to solve the need at hand, in the cast round of the integration of the FNS plan on both local and national levels. This text is thus divided into five sections, in addition to this introduction. The first section deals with the theoretical approach and also considers the relational approach. The second describes the methodology used. The third describes the course taken by public policy since its creation, highlighting five four-year periods of government in the department of Antioquia and identifying the strategies employed in implementing the MANA policy. The fourth section looks at emerging experiences in the region in relation to public school feeding procurement and the development of rural social organizations in the municipalities of Cáceres, Granada, and Carmen de Viboral. The fifth section offers some final considerations.

▸▸ Public policy from a relational perspective

The theoretical approach of this research paper is based on the recent contribution made by Tirelli (2020), which takes into account two fields of study: regional studies, and analysis of the implementation of public policies as a theoretical approach to them, based on a relational approach of the region. According to Reis (2005, 2015) and Favareto et al. (2015), the region has an important contribution when considering public policies and, in particular, how they are implemented in different spaces. As such, the region is not seen as a mere recipient of public policies. It is possible to find different results that hinge on the diverse relations between the region's actors, the positions they occupy in the relational sphere, and the resources they have in order to change or strengthen the policy (Tirelli, 2020). The aim is to contribute to a dialogue that provides an understanding of how the region acts on the agency of policy implementers at a local level, with the result that the outcomes of public policies are contingent upon and sometimes deviate from the proposed objectives.

In studies on the implementation of public policies, the region has not been looked at as an analytical category, with the exception of studies in which the focus is on regional planning and development policies (Favareto et al., 2015). According to Tirelli (2020), the concept of region is being used to analyze public policies and their design as a specific space in which the policy seeks to exert its influence in responding to a specific problem. In the case of regional policies, the literature argues that some policies reduce the concept of region to a recipient of government investment for its development (Lotta & Favareto, 2016). When regional public policies are drawn up, the region is seen in many cases as a homogeneous space for their implementation. As seen, and according to the relational approach, it is based on the emergence of social realities set in motion by both individual and collective subjects that are in some way linked to them (Donati, 1993).

This study defines the concept of region as the result of a combination of discourses, technologies, alliances and original modes that are not pre-existing, a blueprint of heterogeneous elements that are linked together to create interwoven routes, uses, conflicts, and effects (Deleuze & Guattari, 1994; De Landa, 2006).

Similarly, Reis (2015) argues that regions can be seen as being uncertain, contingent and unexpected. It is a question of moving from an organicist vision to a one that recognizes their polymorphism (Reis, 2005).

With regard to the notion of food and nutrition security in terms of human and sustainable development, agroecology, food sovereignty and the human right to adequate food, a commitment to address FNS from multidimensional perspectives has been reaffirmed. According to some authors, food sovereignty, the human right to adequate food, and the promotion of development are *référentiels* inseparable from FNS (Burlandy, Magalhães & Maluf, 2006).

In taking into account these approaches to public policies and the region in relation to food and nutrition security, the following questions can be addressed: How has the process of emergence come about, and which mechanisms have triggered social, political, environmental and other changes in public policies? How do the relationships between institutions and through which actors build their own visions of the world, giving new meaning to the relationship with public food and nutrition security policies in the department of Antioquia, come about on a regional level? Can the disparities between communities and centers of power in relation to food and nutrition security be resolved?

In asking these questions, this research paper aims to address the development of and change in public food security and nutrition policy in the department of Antioquia, from its creation to the present day, and to identify relationships between local actors in order to gain an understanding of the construction and emergence of their visions of public policy.

▸▸ Methodology

The subject of research is the MANA public policy in the department of Antioquia, which has been central to discussion of the policy at a regional level. The research is of a qualitative nature, with primary and secondary data having been collated during the public policy period running from 2000 to 2020. Primary data was gathered in open interviews with participants involved in implementing public policy at both regional and municipal level. Follow-up questions were asked in these interviews, which expanded on baseline information. The actors in question include public organizations at different levels of municipal and departmental government. A total of 16 interviews were conducted in 16 municipalities in the department's nine sub-regions, the aim being to account for regional, social, economic, environmental, historical and cultural variations in both urban and rural areas. Interviews were also conducted with participants in various projects supporting the work of social organizations and NGOs in the department's sub-regions.

▸▸ The development of public policy for the improvement of food and nutrition in Antioquia

In charting the development of public policy since its creation, five four-year periods of government in the department of Antioquia have been analyzed to enable identification of MANA policy strategies. The period 2000 to 2003 saw their emergence;

2004 to 2007 their continuation and consolidation; 2008 to 2011 the sustaining of some strategies and the breakdown of others; 2012 to 2014 the reintegration and creation of new strategies; 2015 to 2019 – the most recent period – again saw the breakdown of some strategies, the sustaining of others, and a new planning process (table 15.1). Analysis also takes in the period from 2020, which coincides with the ongoing pandemic and begins with a new 2020–2023 development plan. These aspects have been linked to national policy and other developments that may be important in relation to the policy under discussion.

There is extensive literature on public food and nutrition security policies in Colombia. The National Food and Nutrition Security Policy (PSAN) remains in force at this moment in time. Part of the National Council for Economic and Social Policy document (CONPES, No. 113 of 2008), it links with the policy experiences of the department of Antioquia as a *référentiel* for formulation and subsequent implementation (Government of Colombia's National Development Plan, 2006–2010).

As regards the department of Antioquia, the Food and Nutrition Improvement Program (MANA) addresses food problems in the region and focuses in particular on children under the age of 14, who occupy an important place in the local, regional and national government agenda (Nussio & Pernet, 2013; Quintero, 2011).

MANA emerges in first period – from 2000 to 2003 – through its development plan "Por una Antioquia Nueva" (For a New Antioquia). It created an epidemiological profile to contextualize the problem of Antioquia, with a rate that stood at 32.9 in the foundation years (Garrett, 2011). In that same year, some 7.6% of Antioquia's population suffered malnutrition and 3.6% chronic malnutrition, compared to the national average of 6.7% and 2.8% respectively (Demographic Health Survey, 2000). The government considered the improvement of the food policy implemented by developing a FNS program estructured by several sector and led by the Antioquia Sectional Health Directorate.

MANA has the following six core facets: community alternatives for supplemental food assistance, introduction to health services, properly handled food, the Antioquia Food and Nutrition Monitoring System (SISVAN), the development of production projects, and educational projects (Government of Antioquia, 2007). These core facets are designed to be comprehensive and complementary in terms of food and nutrition security by taking organizational, economic, social, health, promotional, prevention, and training aspects into consideration. Thanks to these core facets, the Antioquia Food and Nutrition Security Department (GSAN) was created. It embarked on a funding management process that brought in other public, private, academic and non-profit actors.

Introduced by the political party in power in the previous period, the policy "Antioquia Nueva, un hogar para la vida" (New Antioquia, a home for life) did not result in any changes in the MANA structure from 2004 to 2007. Greater attention was paid to introduction to health services; early detection; and specific protection and child, sexual and reproductive health care, mainly for children under the age of six and their families and for pregnant and breastfeeding mothers and their families. Another key aspect was the consolidation of SISVANA in the absence of adequate official information for the monitoring of malnutrition in the department with official sources,

Table 15.1. The development of public food and nutrition security policy (MANA) in each of its core facets and strategies proposed across various periods of government (years)

2000–2003	2004–2007	2008–2011	2012–2015	2016–2019	12-year Plan (2019)	2020–2023
Introduction to health services	Introduction to health services	Partly abandoned	Support and prevention to reduce risk of food insecurity among vulnerable families	Partly abandoned	Social protection of the guarantee of the right to healthy food	5: Social protection of the guarantee of the right to healthy food
Properly handled food	Properly handled food	Partly abandoned	Management of public food and nutrition security policies	Partly abandoned	Sustainable food distribution	3: Governance as a regional food and nutrition security strategy
Antioquia Food and Nutrition Monitoring System (SISVAN)	Antioquia Food and Nutrition Monitoring System (SISVAN)	Partly abandoned	Antioquia Food and Nutrition Monitoring System (SISVAN)	Partly abandoned	Healthy and sustainable food environments	6: Science, technology and innovation for the food and nutrition system. Food and Nutrition Security Observatory (ODSAN)
Development of production projects	Development of production projects	Partly abandoned	Support and prevention to reduce risk of food insecurity among vulnerable families	Development of production projects	Sustainable food production	2: Sustainable food production and distribution
Educational projects	Educational projects		Management of public food and nutrition security policies	FNS professorship	Science, technology and innovation for the food and nutrition system	1: Regional management of food and nutrition security
Supplemental food assistance	Supplemental food assistance	Supplemental food assistance	Support and prevention to reduce risk of food insecurity among vulnerable families	Partly abandoned	Ethnic chapter: Food autonomy and regional sovereignty for Antioquia's ethnic groups	4: Healthy and sustainable food environments

developing an evaluation that could lead to a diagnosis expected by the department (Zuliani, 2010). With the assessment of the pilot plan introduced in the 30 municipalities in the previous period of government, it was found that although the policy prevented children from dying of hunger, the problem of chronic malnutrition had not gone away. In response, a new program entitled "Care for children suffering from severe malnutrition" was created. Agreements with hospitals were signed and protocols for nutritional recovery were introduced (ICBF, 2009). Rolled out in the department's 125 municipalities, the strategy was implemented by municipal officials in some cases and by state school teachers in others, with information being generated by unsuitable team (Mazo & Imasato, 2018). Although an agreement was signed with the 125 municipalities for participation in the policy, only 95 of them ended up taking part (Government of Antioquia, 2007). On a national level, the 2006–2010 Development Plan entitled "estado comunitario: desarrollo para todos" (community state: development for all) identified the need for a food security policy that reflected the national reality (CONPES, 113 of 2008).

A change in government led to the introduction of the development plan "Antioquia para todos, manos a la obra" (Antioquia for all. Get down to work) in the period 2008 to 2011, which spelled change and a break with previous policy. More attention is given in this period to Indigenous peoples and those displaced by the armed conflict in the early 2000s (Government of Antioquia, 2011). Most of the lines of action proposed in the public MANA policy were renamed, while the production projects and the SISVANA were partly abandoned. At a national level, meanwhile, the National Food and Nutrition Security Council was created, in line with similar developments elsewhere in the world.

Introduced in the 2012–2015 period, the development plan "Antioquia la más educada" (Antioquia the most educated) saw the government reach an agreement with the mayor of Medellin in this same period (Governor 2004–2007). The plan allowed for the resumption of MANA policy initiatives embarked on in 2001–2007 and the creation of new strategies harnessing the transformative power of education for the very young. Regarding the six thematic lines of public policy since 2003, these were grouped into three core facets (Government of Antioquia, 2013): care and prevention as means of reducing the risk of food insecurity in vulnerable families (dovetailing with the core facets of supplemental food assistance, introduction to health services, agricultural production projects and nutritional education for healthy eating habits); the management of public food and nutrition security policies (encompassing properly handled food and educational projects); and the food and nutrition monitoring system (corresponding to SISVAN's fourth line of action) (Departmental Development Plan of Antioquia, 2012–2015). On a national level, a food survey (ENSIN) was conducted in 2015 and revealed that 54.2% of Colombian households suffered food insecurity, not least Indigenous families and families headed by women (Ministry of Health, 2017). The survey also revealed shortcomings in access to food and its disposal and use, highlighting the fact that public policy created in 2008 was unable to tackle the problem on a national level. National policy was centered on availability and quality, at the expense of access, consumption and use.

In the fifth and most recent period, from 2016 to 2019, the development plan "Antioquia Piensa en Grande" (Antioquia, Think Big) focused mainly on updating departmental policy through Ordinance 046 of 2016 and set out a 12-year plan for food and nutrition security in Antioquia. It also enhanced the food and nutrition security IT system (SISAN) through Ordinance 038 of 2018 and created the anti-aging ordinance or Ordinance 17 of 2019, a policy designed to improve the quality of life and life expectancy of Antioquia's inhabitants. In the FNS education core facet, FNS professorships were created in the department's nine sub-regions, with food and nutrition security being integrated with basic and core subjects on the syllabus.

As this is the most recent policy period, it is worth focusing on the 12-year plan, the main objective of which was to identify and prioritize aspects impacting on the food system and to identify the various actors involved in the process. The aim was that this would lead to the creation of a prospective strategic design that would conclusively point the way to a solution to these problems, through the identification of six strategic lines of action. (Gobernación de Antioquia, 2019a).

The first strategy involves governance as a regional strategy for food and nutrition security in Antioquia: government and society in tandem. It aims at promoting innovation and institutional development through democratic, pooled governance as a regional initiative that encourages social, public and private actors to work together. The second strategy deals with sustainable food production and seeks to create the conditions for achieving a sustainable, diverse and quality food supply. It aims to enhance the conditions and capabilities of local, family and community farming (ACFC), with an emphasis on agroecological production, marketing and agro-industrial processes.

The third strategy – sustainable food distribution – aims to bring about a move towards supply and distribution systems that enable the procurement of healthy food with institutional, domestic and wholesale consumers. It also seeks to strengthen ACFC organizations. The strategy supports both public procurement and local farmers' markets. The fourth strategy is aimed at healthy and sustainable food environments and seeks to make them safe, culturally acceptable and fair, and enable good food habits through diets that tackle malnutrition in a conscious and responsible manner. The strategy encompasses informative food environments, breastfeeding, and school environments. The fifth strategy, regarding social protection guaranteeing the human right to healthy food, is aimed at people with limited access to sufficient, quality food. It links up with the School Feeding Program (PAE) and care for the elderly, among other areas. The sixth strategy, which deals with science, technology and innovation as they relate to the food and nutrition system, allows for the creation of projects that drive the food and nutrition system forward through knowledge management and which adapt to the needs of regions. It takes into account the development of the food and nutrition profile; the Food and Nutrition Security Observatory (ODSAN); research initiatives; agroecology, production and infrastructure technologies; and innovation and community ownership for welfare and local markets.

Finally, an ethnic chapter was created. Entitled "Autonomía alimentaria y soberanía territorial para los grupos étnicos de Antioquia" (Food autonomy and regional sovereignty for the ethnic groups of Antioquia), it seeks to respect the identities, practices and culture of ethnic groups. It aims to revive, strengthen and preserve traditional production practices, and help these group develop the skills they need to plan their development and healthy living. As can be seen, these strategies seek to consolidate the core facets proposed in the previous periods.

Although the 12-year plan provided key elements in the restructuring of the policy, the 2020–2023 development plan "Unidos por la vida" (United for life) has a difficulty being outdated and outof form based on the 12-year plan proposed by the previous government closely, with only some aspects being taken from it. And based on the fact that in 2019, 27.9% of households in Antioquia were suffering moderate to severe levels of food insecurity, with that figure increasing to over 50% in the regions of Urabá and Bajo Cauca (DANE, 2019). Even more revealingly, more than half of the households (60.9%) suffer some form of food insecurity (Antioquia, Quality of Life Survey, 2019). In response, the food security development plan provides for the following six programs: Program 1: Regional management of food and nutrition security that takes into account the availability of food. The program is mainly supported by the department's Secretary of Agriculture and includes educational production projects. Program 2: Sustainable food production and distribution, consolidating the supply and distribution system. Program 3: Governance as a regional strategy for food and nutrition security in Antioquia, with the search for joint solutions based on participation, innovation and institutional development. Program 4: Healthy and sustainable food environments, enabling the recovery of food culture identity in the transformation of healthy environments. Program 5: Social protection for guaranteeing the human right to healthy food, through the PAE. Program 6: Science, technology and innovation for the food and nutrition system, which the Food and Nutrition Security Observatory aims to implement through research and monitoring. It should be noted that each of these programs takes the pandemic and post-pandemic scenarios into account.

▸▸ Integration and participation of institutions in food and nutrition security

An effective methodology was used in the policy planning stage. On creation of the policy, the secretaries of agriculture, government, finance, planning, and social policy, held joint meetings with the Colombian Institute of Family Welfare (ICBF). These meetings helped create a culture of food and nutrition security and health promotion to address the problems of hunger. Zuliani (2010) points out that the formulation of the MANA policy, which involved the governor, secretaries, the academic sector and the community affected by the problems of hunger and child mortality, did not so much involve the creation of a solid group as raise awareness for the government, even though the situation of the country was rash.

Among the actors emerging at a regional level are community organizations, regional ethnic organizations, children's foundations that work for children, community action boards, community mothers' groups, social movements, and ethnic groups (Kujar Lozano et al., 2008).

Formed by a large group of universities in the department of Antioquia, academic actors provide a link between society and the government and contribute to the development and strengthening of the quality of life of the population by creating instruments for monitoring food and nutrition (SISVAN). In institutional and political terms, the governor is supported by the departmental assembly. It was essential, therefore, to create a public policy that was linked to the government program, was indefinite in nature, was results-based, and also had the support of the national government.

At a national level, further investigation has been achieved thanks to the commitement of some institutions, such as the Colombian Institute of Family Welfare, the Colombian Agricultural Institute, the Colombian Institute of Rural Development, and the Department of Statistics (DANE), not to mention local television and the press. These actors and institutions all provide support at national, departmental and local levels and act as a platform for the public policy. In terms of international actors, the United Nations (UN), the World Food Programme (WFP), the World Health Organization (WHO), and other international cooperation agencies such as France's Family Households have played central roles.

Private actors include compensation funds, marketplaces, food cooperatives, producers' and transporters' cooperatives, religious organizations, food production companies, and clinics. Significant progress was made in the development of the diagnosis in the early years by bringing these actors together (Oviedo, 2015), with local subregional and municipal FNS committees being formed. This integration continues to continues to act as a linchpin of the policy, bringing MANA to life and bringing society into the process of creating spaces for participation. The conjuction of institutes and officials became so strong that it was invited to take part in the 40th session of the Committee on World Food Security (CFS) and share its experience at an international level.

Community participation has gradually declined as the policy has been implemented, with institutions leading the process. At this moment in time, the department's social organizations continue to make demands with regard to other structural problems, such as land ownership, the growth in mining and energy megaprojects, and attacks on social leaders. Similarly, when the 12-year plan was being drawn up in 2018 and 2019, there were no clear signs of the involvement of the department's social organizations. Institutions and the local actors that had been invited were more involved. Thanks to the constant support of many sectors, significant progress has been made in terms of political participation, which is dependent on a process that strengthens communities with political experience of the problems afflicting their regions.

▸▸ The emergency in public procurement for school feeding

In the department of Antioquia there have been three experiences of public procurement for school feeding: in the sub-region of Bajo Cauca, the municipality of Cáceres, and the municipalities of Granada and Carmen de Viboral, in eastern Antioquia. The most commonly referenced of these experiences in the literature is Granada, with authors such as Valderrama (2019); Valderrama et al. (2018) and Giraldo (2018). These programs have all been led by social organizations and have been consolidated thanks to the support of institutions. The Granada program restarted in 2020.

Colombia's long history of public procurement is documented in the literature and spans 83 years, from the PAE to the present (Government of Antioquia, 2019). It dates to 1934, when it was managed by certain government agencies. Colombia's Ministry of National Education in Colombia is currently in charge, providing funding for approved regional bodies and the municipalities, who in turn allocate the funding needed to put the program into practice (Arboleda, et al. 2016; FAO, 2013; Vargas, 2013).

Public procurement operations in eastern Antioquia involve the following social organizations: the Association of Small Producers of Granada (ADEPAG) and the Association of Producers of Tafetanes in the municipality of Granada; and the Association of Fruit Growers of Carmen de Viboral (ASOFRUCAR) and the Agricultural Association of Carmen de Viboral (AGROPOCAR) in Carmen de Viboral. In the sub-region of Bajo Cauca, it involves the Vereda el Deseo Association of Farmers and Growers (AAVED), the Association for Rural Progress (APROCAMPO), and the Association of Rubber Growers and Farmers of the Municipalities of Cáceres and Caucasia (ASOCAUCE).

Public procurement in the municipality of Granada began in 2004 and 2005 at the initiative of the mayor's office and the community action boards represented by the Association of Community Action Boards (ASOCOMUNAL). Problems arose when the University of Antioquia and MANA management, which were responsible for monitoring the process, requested food safety records. This was incompatible with purchases made directly in rural areas, with produce for school feeding being bought directly from the parents of families in the school environment. Other products not produced in the municipality were purchased from community stores and in urban areas. Unfortunately, ASOCOMUNAL was also unable to continue operating, as its statutes did not allow it to engage in such activity, which led to it being run by private operators.

In 2011, Granada's institutions adopted public procurement on their political agendas with a view to supplying school canteens with locally produced food. Sadly, no progress was made and they had to continue with a private operator, buying food from outside the municipality. It was only in 2015 that the municipal administration, through ADEPAG, began to buy directly from farmers again. Public procurement also involved other organizations such as a bakery group and an organization of female victims of the armed conflict, who supplied processed bakery products. Support was also provided by the municipality's cooperatives, which are solidly founded and have a long tradition behind them (Valderrama et al., 2017).

Institutional support can have an important part to play in the initiatives of social organizations, in partnership with public administration. However, in the 2016–2019 period, which saw a new administration, the program ceased to operate with ADEPAG and the contract was awarded to an organization that once again procured food from outside the municipality. The goal of feeding of children with local produce and improving the economic lot of local farmers and growers was consigned to history.

In response, some of the people who have led the public procurement process in Granada have created other mechanisms in an effort to consolidate the processes they began. Some organization leaders and a number of other actors have joined forces with the Tejipaz corporation and a group of professionals to market their products at fair prices and with international support. At the beginning of 2020, the people of municipality re-elected the mayor who years earlier had introduced public procurement. As a result, public procurement resumed, again through Tejipaz. In early 2021, however, it was handed over once more to an external operator, bringing an end to procurement from the municipality's own organizations.

Leaders gain strength through the mechanisms used in search of the constant strive, to respond to new challenges, negotiate to solve the obstacles they face along the way, and stick to the task, at hand, as is the case with the secretariat of the government of the municipality of Granada, which is currently working with a large number of actors. Public procurement aims to improve the income of local farmers and growers and to solve the situation related to the diets of local children. The region becomes a place of action and interaction involving actors operating in networks over time and in different areas, the result being that their outcomes are not predetermined.

In Carmen de Viboral, ASOFRUCAR and AGROPOCAR embarked on projects with the government of Antioquia in 2008, projects that were strengthened by the second "Peace Laboratory" in 2011–2012. The organizations faced a complex situation with the operator of the public procurement program and did not sign agreements with it or the institutions at the time. At this moment in time, the organizations devote their energies more to individual activities, though they do come together in some cases to market their products.

In Cáceres, AAVED, APROCAMPO and ASOCAUCE joined forces between 2011 and 2013 in the Forest Ranger Families project, promoted by the United Nations Office on Drugs and Crime as part of a rice-growing project. Sadly, the municipality's public procurement program was undermined by the armed conflict and has been discontinued.

In 2015 and 2016, the two sub-regions, the three municipalities and the seven organizations involved in a joint FAO-WFP and Brazilian Embassy initiative came together in a national pilot project entitled "Abriendo mercados para la agricultura familiar en Colombia" (Opening markets for family farming in Colombia) and which also involved organizations in the department of Nariño. These regions were chosen for the roll-out and development of a public procurement model for school feeding and for the supply of food for ICBF programs in hospitals and prisons (Giraldo, 2018). The project identified and acted on the problems potentially faced by family farming organizations (OAF) in accessing institutional markets (FAO & WFP, 2016). It functions throught an organized rollout which shows that family farming organizations can access the institutional market and supply food in the municipalities of Cáceres, Carmen de Viboral, and Granada in the department of Antioquia (Giraldo, 2018).

As Reis (2015) and Favareto et al. (2015) pointed out, in this case the region has an important contribution to make to public policies, in the way they are implemented in different areas. Public policies acquire their own characteristics and effects when they are implemented in different regions. As such, the region is seen as more than just a recipient of public policies, as a passive entity.

▶▶ Conclusions

Although the Food and Nutrition Improvement Program of Antioquia (MANA) aims to support the most disadvantaged, especially children under the age of 14 – as was initially planned – it has not been able to solve the problem of hunger once and for all. This is possibly due to the fact that the successive governments that have sought to implement the policy since 2003 have made major changes to it without giving it the proper attention it requires.

The public MANA policy succeeds in bringing major academic, political and economic sectors together around the problem of food security, allowing the department's young people to solve, in part, the problem of food and nutrition security. Nevertheless, social organizations still need to be given a stronger role in this process.

Analysis of public policies from a regional perspective is important in gaining an understanding of how they are implemented in different areas. It is especially true, however, that public policies can have their own unique effects when implemented in different regions, as is the case with public procurement for school feeding in Granada, Carmen de Viboral, and Cáceres, with each municipality having its own social organizations. As such, a region is more than just a recipient, a passive entity for the implementation of the public MANA policy. Such an approach is problematic, particularly for the organizations involved in the food procurement process.

▶▶ References

Álvarez, M. C., Estrada, A., Montoya, E. C., & Melgar-Quiñónez, H. (2006). Validación de escala de la seguridad alimentaria doméstica en Antioquia, Colombia. *Salud pública de México, 48*(6), 474–481. https://saludpublica.mx/index.php/spm/article/view/6722/8395

Arboleda, J., García, L. M., Agudelo, Y., & Orrego, C. (2008). *Política de seguridad alimentaria y nutricional.* Facultad Ciencias Sociales y Humanas, Departamento de Trabajo Social, Universidad de Antioquia.

Burlandy, L., Magalhães, R., & Maluf, R. S. (Eds.). (2006). *Construção e promoção de sistemas locais de segurança alimentar e nutricional no Brasil: aspectos produtivos, de consumo e de políticas públicas.* CERESAN.

CONPES. (2008). *Política nacional de Seguridad Alimentaria y Nutricional (PSAN)* (Documento 113). Ministerio de la Protección Social.

De Landa, M. (2006). A new philosophy of society: assemblage theory and social complexity. Continuum.

Deleuze, G., & Guattari, F. (2004). *Mil mesetas. Capitalismo y ezquizofrenia* (J. Vásquez Pérez, Trans.). PRE-TEXTOS. (Original work published 1980).

Departamento Nacional de Planeación. (2007). *Plan nacional de desarrollo 2006–2010. Estado comunitario: desarrollo para todos.* Gobierno de Colombia. https://www.dnp.gov.co/Plan-Nacional-de-Desarrollo/PND%202006-2010/Paginas/PND-2006-2010.aspx

Donati, P. (1993). Pensamiento sociológico y cambio social: hacia una teoría relacional. *REIS: Revista Española de Investigaciones Sociológicas, 63*, 29–52. https://doi.org/10.2307/40183648

FAO. (2013). *Alimentación escolar y las posibilidades de compra directa de la agricultura familiar, estudio nacional de Colombia* (Proyecto GCP/RLA/180/BRA). https://www.fao.org/docrep/field/009/as513s/as513s.pdf

FAO & WFP. (2016). *Sistematización de la Experiencia: Articulación de la agricultura familiar con las cadenas de abastecimiento de alimentos a los programas institucionales en Colombia.*

Favareto, A., Kleeb, S., Galvanese, C., Magalhães, C., Moralez, R., Seifer, P., Buzato, H., & Cardoso, R. (2015). Territórios importam–Bases conceituais para uma abordagem relacional do desenvolvimento das regiões rurais ou interioranas no Brasil. *Revista em Gestão, Inovação e Sustentabilidade, 1*(1), 14–46.

Garrett, J. (2011). MANA: Improving food and nutrition security in Antioquia, Colombia. In Garret, J., & M. Natalicchio, (Eds.), *Working multisectorally in nutrition: principles, practices, and case studies* (pp. 100–149). IFPRI. http://dx.doi.org/10.2499/9780896291812

Giraldo Calderón, P. E. (2018). *Alimentación escolar y compras públicas en perspectiva comparada: un análisis Colombia-Brasil* [Master's thesis, Universidad Federal Rio Grande do Sul]. UFRFS Digital Repository. http://hdl.handle.net/10183/178579

Gobernación de Antioquia. (2007). *Ordenanza No 17 de 24 de noviembre de 2003. Por medio de la cual se adopta la política pública de seguridad alimentaria y nutricional para los menores de 14 años y sus familias, en el departamento de Antioquia.*

Gobernación de Antioquia. (2011). *¿Dónde nace Mejoramiento Alimentario Nutricional de Antioquia-MANA?*

Gobernación de Antioquia. (2012). *Plan de desarrollo departamental de Antioquia, 2012–2015. "Antioquia la más educada."*

Gobernación de Antioquia. (2019). *Manual para la supervisión integral del PAE en el territorio.*

Gobernación de Antioquia, Gerencia de Seguridad Alimentaria y Nutricional. (2007). *Informe de gestión pública MANA 2004–2007.*

Gobernación de Antioquia, Gerencia de Seguridad Alimentaria y Nutricional. (2011). *Informe de gestión pública MANA 2007–2011.*

Gobernación de Antioquia, Gerencia de Seguridad Alimentaria y Nutricional. (2013). *Programa Antioquia de seguridad alimentaria y nutricional-MANA: Lineamientos técnicos MANA.*

Gobierno de Colombia. (2012). *Plan nacional de seguridad alimentaria y nutricional (PNSAN) 2012–2019.* https://www.icbf.gov.co/sites/default/files/pnsan.pdf

Instituto Colombiano de Bienestar Familiar (ICBF). (2009). *Lineamientos técnicos administrativos de las unidades de atención integral y recuperación nutricional para la primera infancia.* Ministerio de Protección Social, Dirección Técnica, República de Colombia.

Kujar Lozano, N. Y., López Meneses, Á., A., & Meneses Riascos, A. J. (2016). *Percepciones del programa de alimentación escolar por parte de los actores en el corregimiento de San Cristóbal y de a comunas 13, 15 y 16 del Municipio de Medellín* [Degree thesis, Corporación Universitaria Lasallista]. Repository Lasallista. http://hdl.handle.net/10567/1766

Lotta, G., & Favareto, A. (2016). Desafios da integração nos novos arranjos institucionais de políticas públicas no Brasil. *Revista de Sociologia e Política, 24*(57), 49–65. http://doi.org/10.1590/1678-987316245704

Mazo Henao, L. M. (2015). *Processo de formação de estratégias na política publica de segurança alimentar e nutricional do departamento de Antioquia, Colombia* [Master's thesis, Universidade Federal do Rio Grande do Sul]. UFRGS Digital Repository. http://hdl.handle.net/10183/127230

Mazo Henao, L. M., & Imasato T. (2018). Uma análise organizacional da política pública de segurança alimentar e nutricional do departamento de Antioquia, Colombia (2001–2014). *Revista Brasilera de Estudos Organizacionais, 5*(2), 277–301. http://hdl.handle.net/10183/194259

Ministerio de Salud (2017, November 21). *Gobierno presenta Encuesta Nacional de Situación Nutricional de Colombia (ENSIN) 2015.* https://www.minsalud.gov.co/Paginas/Gobierno-presenta-Encuesta-Nacional-de-Situación-Nutricional-de-Colombia-ENSIN-2015.aspx

Nussio, E., & Pernet, C. A. (2013). The securitisation of food security in Colombia, 1970–2010. *Journal of Latin American Studies, 45*(4), 641–668. http://doi.org/10.1017/S0022216X1300117X

Quintero, J., & Zuluaga C. (2018, September 17–21). *Territorios agroalimentarios en el oriente de Antioquia* [Presentation]. Third International Conference on Agriculture and Food in an Urbanizing Society, Porto Alegre, Brazil.

Quintero Velásquez, Á. M. (2011). Modelos de políticas públicas de Colombia, en beneficio de las familias. *Revista Katálysis, 14*(1), 116–125. https://doi.org/10.1590/S1414-49802011000100013

Reis, J. (2005). Uma epistemologia do território. *Estudos sociedade e agricultura, 13*(1), 51–74. https://revistaesa.com/ojs/index.php/esa/article/view/258

Reis, J. (2015). Território e políticas do território. A interpretação e a ação. *Finisterra, 50*(100), 107–122. http://hdl.handle.net/10316/41869

Tirelli, C. (2014). As contribuições da sociologia relacional para as análises das organizações sociais do campo da assistência: o caso da rede parceria social/RS. *Redes, 19*, 25–43.

Tirelli, C. (2020). Conectando políticas públicas e território: a contribuição da perspectiva relacional. In Leite Lima L., & L. M. Schabbach (Eds.), *Políticas públicas: questões teórico-metodológicas emergentes* (pp. 242–265). UFRGS Editora.

Zuliani, L. (2010). *El proceso de participación de los actores en la construcción del caso MANA Antioquia-Colombia 2001–2003.* [Master's thesis, Universidad de Antioquia]. Repositorio Institucional Universidad de Antioquia. http://hdl.handle.net/10495/1381.

Part V

The challenges of specific food policies

Chapter 16

The politicization of public food procurement in Brazilian state governments: actors and ideas in the creation of sustainable food systems

CATIA GRISA, MÁRIO AVILA, RAFAEL CABRAL

▸▸ Introduction

The food supply, healthy eating, and the creation of sustainable food systems have become central issues in the 21st century (Fouilleux & Michel, 2020; IPES Food, 2020; Brand et al., 2019; HLPE, 2014). Recurrent scandals, threats, and food and economic crises had already triggered debate of the world's major food challenges and have been brought into even sharper focus by the Covid-19 pandemic. The growth of urbanization, long-distance supply chains, the threats posed by global warming, the trend towards concentration in the food industry, the growing problem of obesity and related diseases, and the permanence/intensification of hunger and food and nutrition insecurity have become central themes in recent years and, in particular, in 2020. In effect, the current production and consumption model is largely founded on the intensive use of nature (deforestation, the expansion of farmland, and the exhaustive use of soil resources); the intensive use of chemical inputs and pesticides; the reduction of biodiversity; the transportation of products and food over long distances; the ultra-processing and artificial preservation of food; the concentration of the means of food production and distribution; and the intensification of social inequalities (Fouilleux & Michel, 2020; FAO, 2019; Schneider & Gazolla, 2017). These aspects place food and its forms of production and consumption at the center of public and political debates and demand changes in the organization and dynamics of the prevailing food system.

Although the politicization of food is not a recent phenomenon, it has intensified over the last two decades precisely because of criticism of the prevailing food system (Fouilleux & Michel, 2020; Portilho, 2020). By the politicization of food, we understand the following:

> (...) the construction of food as a public problem on the basis of the actions of different actors striving to remove it from its habitual spaces – the private sphere of consumption, the technical sphere of production and agricultural transformation,

the economic sphere of trade relations between different actors in the production chains – and place it in public and political spaces, or, conversely, to keep it in depoliticized spaces, considering it a private, technical, social or economic problem (Fouilleux & Michel, 2020, p. 12).

In other words, the politicization of food implies making it a political issue, one that is rendered problematic and disputed by different interpretations and practices with regard to its forms of production, distribution, marketing, preparation, consumption and waste management (Fouilleux & Michel, 2020; Portilho, 2020; Portilho, Castañeda & Castro, 2011). More than a biological or technical need, the act of eating and the dynamics that permeate food have become an expression of political and social positions (Barbosa, 2009).

The politicization of food has manifested itself in different ways and in different spheres. We could mention individual actions that evidence themselves in political consumption and the adoption of different diets and eating habits, such as vegetarianism, veganism and climatarianism (Niederle & Schubert, 2020; Portilho & Micheletti, 2018). We could also point to the actions of social and civil society movements in guiding new ways of producing, exchanging/trading and consuming, such as the Slow Food movement, agroecological produce fairs, consumer groups, Communities Supporting Agriculture (CSAs), locavorism (Feldmann & Hamm, 2015), and other forms of short supply chains (Ribeiro, 2019; Preiss, 2017). We could also mention the initiatives of the private sector and the solidarity economy, through the actions of chefs, restaurants and other private commercial establishments that seek to establish a dialogue and source food from family farming, sociobiodiversity, and regions (Dorigon, 2019; Zanetti, 2017). Some of these manifestations often seek to make problems visible and put them on the government agenda.

Moving towards the politicization of food oriented to the government agenda, this paper addresses the politicization of public food procurement. Though these activities are already present and highly recurrent in government action, they are often looked upon as technical and administrative procedures that hamper access to food among certain population groups. In building on a wide range of studies on the topic, the article seeks to explore the action of actors in bringing public food procurement out of its usual environment – where it is dealt with as a routine, operational and impartial instrument of public administration, guided essentially by price – and placing it front and center in the political sphere, exploring its repercussions and oversights in terms of the dynamics of food systems (Grisa, Schneider & Vasconcellos, 2020; Sonnino, 2019; Soldi, 2018; Goggins, 2018; Goggins & Rau, 2016). As Goggins (2018) states, public food procurement gives government agencies the opportunity to directly influence food systems by using its purchases to promote social, economic, and environmental objectives.

In Brazil, the politicization of public food procurement has already resulted in some significant changes and innovations. With the creation of the Food Acquisition Program (PAA) in 2003, various social and political actors drew attention to the role of public procurement in boosting family farming, promoting food and nutrition security, and regional development. As a result, the National School Feeding Program (PNAE) was also amended, resulting in the introduction of a requirement that at least 30% of school meal budgets be devoted to the purchase of family farming products and of guidelines for the promotion of agroecology and sociobiodiversity, obesity control, and food education (Teo & Triches, 2016). Furthermore, Decree No. 8.473/2015 – introduced in 2015 – stated that of the total budget allocated by Federal Public Administration agencies and bodies to food procurement, at least 30% must be set aside for the purchase of produce from family farming and their organizations.[1]

In addition to these actions, we have also observed the mobilization of various social and political actors in the creation of similar actions in subnational governments (states and municipalities). In fact, since 2007 the public procurement of food from family farming has been institutionalized in several state governments, with the instruments used in the process, which are not neutral in relation to food-system dynamics, embracing a concern for socioeconomic, environmental and nutritional inclusion.

This chapter seeks to explore the emergence of these experiences of public food procurement in the state governments. More specifically, the article analyzes how the politicization of food was incorporated into state public food procurement policies, the actors and alliances involved in the creation of these initiatives, the issues they brought to light, and the public policy *référentiels* established (Fouilleux, 2011, 2003).

There are two main reasons for this emphasis on the actions of state governments. Firstly, there is a difference in the way these experiences are handled in Brazilian rural and food studies.[2] The recognition of various family farming policies by federal government, their emergence, and the increasing centralization of public policies at this level in the 2000s are factors that may have contributed to the overshadow food policies created at other levels of governance (state and municipal). Secondly, several studies point to the return (Daviron et al., 2017) and the growing prominence of cities and local and regional governments in the creation and implementation of food policies (Fouilleux & Michel, 2020; Brand et al., 2019; Raja et al., 2018; IPES Food, 2017). The enhanced role played by cities and local and regional governments in food policy stems from the intensification of urbanization and the overlapping of food, environmental and social issues that challenge and demand responses from policy actors in local spaces (Fages & Bricas, 2017; IPES Food, 2017); the continuation of the local as a space for innovation; collaborative efforts and a primary sphere of influence and operation for actions supporting social justice (Reece, 2018); and the opportunities that the local and regional offer for bringing actors and sectors together and the creation of systemic actions that replace, complement or compete with other scales (Guéneau et al., chapter 12) linked and aligned with regional characteristics.

Framed in such a way and aiming to address the objectives presented, the chapter has been split into four further sections. The next section contextualizes the politicization of public procurement at a national level (the view being that it has driven the creation of similar actions at state government level), maps the initiatives

---

1. Even before food procurement from family farming became mandatory, it is important to point out that under the National Food and Nutrition Security System (SISAN) and its associated legal framework, Article 17 of Law 12512/2011 already authorized federal bodies to purchase family farming produce by waiving bidding requirements.

2. When analyzing state agroecology and organic production policies in Brazil, Sabourin et al. (2019, p. 18) reach the same conclusion: "Although several studies have already analyzed the path taken in the institutionalization of agroecology at federal level (...), few have examined the processes by which state agroecology and organic production policies are drawn up."

implemented by state governments, and presents the methodological approach. The third section addresses the politicization of public food procurement at state government level, analyzing the temporary nature of actions, and the actors and the dynamics involved in creating the government agenda. The fourth section discusses the issues that guided the public policy *référentiels* outlined here, while the fifth section systematizes some of the reflections provided here.

▸▸ The politicization of public food procurement at a national level, the actions of state government, and the methodologies open to research

Public food procurement gained fresh momentum with the creation of the Food Acquisition Program (PAA) by the federal government in 2003. In rendering the limits of regulations linked to the economy and isonomy – which had until then largely been considered neutral and impersonal –, the program revealed the potential of public procurement in transforming food systems and making them more inclusive, nutritional and sustainable. Innovative in nature, the PAA enables the procurement of food (and seeds) from family farming, allocating them to public food and nutrition facilities and meeting the demands for food by government agencies and for price regulation and added value by social organizations. Built on the demands of political and party actors, food and nutrition security (FNS) organizations, social and family farming union movements, and agroecological organizations, the PAA is structured by means of the six modes presented in the table 16.1 (the configuration of which inspired action at subnational government level).

Table 16.1. Summary of the PAA's modes of implementation in May 2020

Mode	Features
Purchase with donation	Purchases food and donates it to organizations in the social welfare network and to public food and nutrition facilities
Stock formation	Provides financial support for the creation of food stocks by supplier organizations, for subsequent marketing and the return of resources to the public authorities
Direct purchase	Intended for the purchase of products with a view to supporting and regulating prices
PAA milk	Enables the purchase of milk which, after processing, is donated to consumer beneficiaries
Institutional purchase	Purchases from family farms to meet the consumption demands of the federal government
Seed procurement	Intended for the purchase of seeds, seedlings and materials for growing food for human or animal consumption from supplier beneficiaries for its donation to consumer or supplier beneficiaries

Source: Based on consolidated legislation.

The dynamics involved in the operating of the modes, combined with other PAA regulations, have produced and led to important changes in the organization of the contemporary food system. Several studies have detailed the PAA's contributions to diversifying production; promoting agroecology and sociobiodiversity; increasing the visibility of social identities; enhancing the socioeconomic and cultural diversity of family farming and the recognition of rural women; strengthening cooperativism and associativity; promoting healthy eating, making fruit and vegetables an integral part of people's diets, generating good practices and food education actions; and promoting short supply chains and local development (Procedi, 2019; Siliprandi & Cintrão, 2014; Avila, Caldas & Avila, 2013). Although marginal in nature given the configurations of the dominant food system and the program's financial and political weaknesses (Porto, 2014; Mielitz, 2014), the PAA's results have added to a set of initiatives that drive change in the continuing quest for the creation of sustainable food systems (Bricas, 2019).

Based on the experience and lessons learned through the PAA, government actors and/or those linked to FNS and family farming began to question and advocate for change in school food procurement. As a result of their actions, the PNAE began to raise food quality through guidelines for respecting local eating habits, the consumption of fresh food, and the control of the supply of ultra-processed food containing sodium, sugars and fat. It provided for different treatment for school food from traditional communities; ensured the participation of family farming in food supply, with priority given to local farmers, farming reform landholders, indigenous and quilombola communities, agroecological producers, and the members of cooperatives and associations; encouraged the procurement of food produced in line with sociobiodiversity practices; strengthened short supply chains; and brought producers and consumers closer together (Teo & Triches, 2016). Criticism of the dominant food system was incorporated into the operating dynamics of the PNAE, which saw it become a catalyst for the creation of sustainable food systems, thanks to its impact in shaping healthy and sustainable eating habits, addressing social inequalities, and promoting rural and local development (Triches, 2015).

Representing a step further in that direction, Federal Decree No. 8,473, which became law in 2015, boosted the PAA's institutional procurement mode, and set out that at least 30% of the total resources allocated to the procurement of foodstuffs by the bodies and entities of federal government, municipalities and foundations must be allocated to the procurement of family farming produce. Deconstructing the supposed neutrality, the PAA once again sought to put the "public plate" (Morgan & Sonnino, 2008, 2010) at the service of building more inclusive food systems that value local farmers and products.

In assessing the contributions and results of these actions, some state governments have begun to set up similar programs. Table 16.2 presents an overview of the 26 states of the federation and the federal district, indicating the existence of family farming food procurement programs and their operational status as of May 2020 (non-existent, suspended or running).[3] As can be seen, only in five states were

---

3. This mapping was carried out up to the end of May 2020. Only family farming food procurement programs or actions with state government resources, regulations and guidelines were included here. It is important to note that we did not include the following in the mapping: i) complementary or specific state government school feeding programs that favour family farming, or the redirecting of PNAE resources to family farming during the pandemic; ii) the implementation of the PAA by state governments through the Binding Agreement with the Federal Government; iii) milk procurement and distribution programs that favour family farming.

Table 16.2. Overview of state family farming food procurement programs

Regions	Federation Unit	Year law published[1] – name of program or draft bill (PL)	Regulation[2]	Other documents[3]	Status in May 2020
North	Acre (AC)	2008 – State Program for the Incentivizing of Family Forestry and Agroforestry Production	2009		Suspended
	Amazonas (AM)	–	–		Non-existent
	Amapá (AP)	–	–		Non-existent
	Pará (PA)	Draft bill presented	–		Non-existent
	Rondônia (RO)	2017 – Rondônia State Food Procurement Program – PAA Rondônia	2018		Running
	Roraima (RR)	2013 – Roraima State Food Procurement Program – PAA Roraima	2020		Suspended
	Tocantins (TO)	–	–		Non-existent
Northeast	Alagoas (AL)	2017 – State of Alagoas Family Farming Food Procurement Program – PAA/AL	2017		Suspended
	Bahia (BA)	Draft bill presented	–		Non-existent
	Ceará (CE)	2015 – State of Ceará Family Farming Food Procurement Program	2017		Running
	Maranhão (MA)	2015 – Family Farming Procurement Program – Procaf	2016		Running
	Paraíba (PB)	2020 – Emergency Procurement and draft bill presented	–	2020	Running
	Pernambuco (PE)	2020 – Local Procurement Program and State Family Farming Food Procurement Program – PEAAF	2020	2020	Running
	Piauí (PI)	Draft bill presented	–		Non-existent
	Rio Grande do Norte (RN)	2019 – State Program for Government Procurement from Family Farming and the Solidarity Economy (PECAFES)	2019		Running
	Sergipe (SE)	–	–		Non-existent

Regions	Federation Unit	Year law published[1] – name of program or draft bill (PL)	Regulation[2]	Other documents[3]	Status in May 2020
Southeast	Espirito Santo (ES)	2007 – Direct Food Procurement	2007		Running
	Minas Gerais (MG)	2013 – State Policy for Family Farming Food Procurement – PAA-Family	2015		Running
	Rio de Janeiro (RJ)	2018 – State Policy for Family Farming Food Procurement – PAA-Family	–		Suspended
	São Paulo (SP)	2011 – São Paulo Social Farming Program – PPAIS	2012		Running
South	Paraná (PR)	2020 – Paraná Direct Procurement	–	2020	Running
	Rio Grande do Sul (RS)	2012 – State Policy for Government Procurement from Family Farming and Rural Family Enterprises and from the Grassroots and Solidarity Economy – Collective Purchasing/RS	2013		Suspended
	Santa Catarina(SC)	2020 – Food Procurement Program (PAA) and draft bill presented	–	2020	Running
Midwest	Distrito Federal (DF)	2012 – Program for the Procurement of Agricultural Production – PAPA/DF	2012		Running
	Goiás (GO)	2017 – State Family Farming Food Procurement Policy	–		Suspended
	Mato Grosso do Sul (MS)	–	–		Non-existent
	Mato Grosso (MT)	2017 – State Policy for Government Procurement from Family Farming and Rural Family Enterprises – Collective Purchasing/MT	2017		Suspended

1. Refers to the year of publication of the Law creating the state program for the procurement of food from family farming food.
2. Refers to the year of publication of the program regulation.
3. Refers to Council Resolutions, Manuals, Public Calls for Tender or Terms of Reference.

proposals or actions for the procurement of food from family farming with their own resources or regulations/guidelines not identified; five states had draft bills (PL) passing through their legislative assemblies; six states had established their own laws, though these had yet to be or were not being enforced; and 12 states had structured programs for the procurement of food from family farming or had created emergency initiatives in response to the pandemic.

Based on this mapping, the researchers sought to broaden their analysis and achieve the proposed objectives by focusing on the 12 states with programs in place. To this end, documentary analysis of the legislation enacted was conducted along with 19 semi-structured interviews with actors (governmental and/or non-governmental) involved in creating and implementing governmental actions. The interviews sought to address the reasons why the actions were created, the objectives and institutional arrangements giving structure to the programs, and the main actors involved.

▸▸ The politicization of public food procurement: actors and dynamics in the public agenda

As table 16.2 shows, the creation of state programs for the procurement of food from family farming has followed different timelines. We have identified three key periods. The 2007–2013 period sees the emergence of some state government initiatives (4 cases), following on from the success of federal programs. Several interviewees spoke of the influence of the design, results and opportunities marked out by the PAA and the change in the PNAE as factors leading to the creation of state initiatives. In the 2014–2019 period (4 cases), in addition to the influence of the national experience and actions that were already being undertaken in some states, a tailing-off in the implementation of the national PAA from 2013 and the dismantling of national family farming policies from 2016 also promoted the creation of state programs (Sabourin et al., 2020; Grisa, 2018). Political changes at national level and "windows of opportunity" (Kingdon, 2006) at state-government level acted as catalysts in terms of the politicization of public procurement at these levels. In 2020, the coronavirus pandemic led to more resources being made available in some cases to strengthen ongoing initiatives (including international sources, as in the case of Rio Grande do Norte state with the World Bank), and also resulted in changes in public calls for tender and pressure to create new modes. In other cases, the pandemic spurred the creation of additional or emergency actions (4 cases).[4]

Three different processes also stood out during analysis of the way in which the issue found its way on to the public government agenda, albeit with common elements between them (table 16.3). One stems from the actions and food activism of social movements in politicizing the issue and building institutional spaces promoting the issue and/or dialogue with managers for the creation of public policies. In line with Portilho (2020) and Preiss (2017), we understand food activism to mean the actions and discourses of social actors and groups that aim to change the food system and

---

4. Our attention was also drawn to the fact that in some states nearly two years elapsed between the passing of the Law and its regulation by Decree, or that some legislation was not implemented or had been suspended. Problems with setting out budgets, implementing regulations, and political changes (new elections) contributed to these circumstances.

make it more democratic, sustainable, healthy, ethical (particularly in relation to non-human animals), focused on the regions and on other related issues, with such activism entering the sphere of the state and public policies. The cases of the PAA-Family in Minas Gerais, Procaf in Maranhão, PEAAF and the Local Procurement Program in Pernambuco, and the emergency action in Santa Catarina are all illustrative in this respect.

Table 16.3. Processes for building the public agenda with regard to family farming food procurement at state-government level

Type of process	Cases
Food activism of social family farming movements in politicizing the issue, building institutional spaces and/or dialoguing with managers for the creation of public policies	MG, MA, PR, SC, CE and PE
Transit of representatives of social family farming movements into institutional policy spaces	RN, RO
Institutional activism of mid-level and street-level bureaucracy	SP, DF, ES and PB

In the case of Minas Gerais, although the election of Antônio Anastasia (Brazilian Social Democratic Party) ensured political continuity with the previous administration, it did create an opportunity for institutional change, manifested by the creation of the Under-secretariat of Family Farming in 2011. As observed in the interviews and also cited by Schmitt and Barbosa (2019, p. 84), the new under-secretariat "was influenced by actors linked to family farming organizations, FNS, and agroecology networks." Several of these actors were active in the State Council for Sustainable Rural Development (CEDRAF) and the State Council for Food and Nutrition Security (CONSEA) and lobbied for the creation of specific family farming institutions, focusing their efforts in this respect on the president of CONSEA in particular. Once the under-secretariat had been created, the institutional transit (Silva & Oliveira, 2011) of actors with links to social movements to government became more significant. Indeed, "the official who took charge of the secretariat had a long history of action, both in the agroecological movement (since the 1980s) and in the Brazilian Forum for Food Sovereignty and Security (FBSSAN), having also been a councilor of the National CONSEA" (Schmitt & Barbosa, 2019, p. 84).

Through the under-secretariat, these actors with links to family farming, agroecology and FNS oversaw the creation of legal frameworks (such as the State Policy on Agroecology and Organic Production and the State Policy for Sustainable Development of Traditional Peoples and Communities) and public policies, such as the Family PAA, created in 2013. The program became law in 2015, by which time a new government had taken office, thus broadening the institutional transit: "a number of other professionals and members of the political community with long-standing ties to civil society organizations also began to occupy positions in the state government" (Schmitt & Barbosa, 2019, p. 84).

In the case of Maranhão, political change in 2014 paved the way for institutional changes and the creation of new public policies. At the start of his term in office :

> (...) the governor (Flávio Dino – Brazilian Social Democratic Party) created the State Secretariat of Family Agriculture (SAF), meeting a demand of the social movements.

> This measure ensured a place at the government for the agendas and claims of rural workers, extractivist populations, quilombola communities, and indigenous peoples, which had been referred until then to the State Secretariat of Agriculture (Sagrima) and were largely watered down or deemed secondary in importance to the agendas and claims of large-scale farming (Guéneau, Neto & Braga, 2019, p. 203).

Among these agendas was the commercialization of family farming, understood as a lever for production. Procaf was established on the basis of the SAF dialogue with representatives of the rural workers' unions, quilombola communities, landless rural workers, the babassu nut breaker movement, agroecological organizations and social inclusion councils.

The governments of Santa Catarina and Pernambuco came up with emergency actions in response to the pandemic. The latter also set up the State Family Farming Food Procurement Program (PEAAF) in June 2020, meeting the demands of family farming organizations and social movements. In Santa Catarina, milk producers demanded a government response to the drought and the problems the pandemic had caused them in selling their produce. When a claim was received, the government (Governor Carlos Moisés – Social Liberal Party) put together a proposal with the Agricultural Research and Rural Extension Company of Santa Catarina (Epagri), which, given its experience of other institutional markets (PAA and PNAE), proposed that municipalities procure food from farming, with 50% of the budget being allocated to milk and derivatives, and 50% to products listed in the Food Guide for the Brazilian Population.

In the case of Pernambuco, organizations and social movements linked to the Extended Countryside Committee demanded that the state government address the problems caused by the pandemic (Governor Paulo Câmara – Brazilian Socialist Party). In response to these demands, the state's Secretariat of Economic Development created the Local Procurement Program, which linked the demands of family farming with the distribution of food to the socially vulnerable. Meanwhile, Pernambuco's Institute of Agronomy, following dialogue with social movements and members of state congress, put together a proposal for the PEAAF. A little more than a month later, that idea had become law.

The politicization of public food procurement also found its way on to the government agenda as a result of representatives of social movements transitioning to positions in the government, notably to the legislative branch. According to Silva and Oliveira (2011, p. 98) :

> (...) institutional transit is characterized by the continuous movement of social-party activists through different spaces of action (social organizations, parties, institutional forums and government positions), which is largely made possible by interpenetration between parties and movements. As a result, electoral victories and defeats tend to produce a significant transit of activists from civil society to government and vice-versa, creating rapid and intense change in terms of opportunities for accessing institutions (both with regard to levels and form).

Unlike the dichotomy between state and society or characterizations such as the cooptation of social movements or state apparatus, this concept emphasizes the permeability between social movements and states. In our analysis, particularly in the cases of Rondônia and Rio Grande do Norte, this institutional transit and permeability was essential to the politicization of public procurement and led to the creation of family farming food procurement programs.

The creation of the Rondônia State Food Procurement Program (PAA Rondônia) in 2017 was the result of extensive work carried out by the Federation of Rural Farm Workers and Family Farmers of the State of Rondônia (Fetagro) and the extent to which this work was applied across the state. As one of the interviewees pointed out, the issue of short supply chains and direct sales to consumers has long been an important issue for the union and is central to its demands for public policies, having been driven by the creation of the PAA in 2003 and the state government's implementation of the program through the Binding Agreement from 2011 onwards. The change in government in 2010 (Confúcio Moura – Brazilian Democratic Movement) saw Fetagro representatives take up positions in the State Secretariat of Agriculture and other elected positions, such as members of state congress. The implementation of the PAA and this institutional transit provided an opportunity for a member of congress with links to Fetagro (the then president of the Agriculture Commission in the Legislative Assembly) to propose a bill that formally became the PAA-Rondônia in 2017.

In the case of Rio Grande do Norte, the creation of the State Program for Government Procurement from Family Farming and the Solidarity Economy (Pecafes) in 2019 was also the result of the demands made by the social movements of the solidarity economy and family farming combining with the transit of actors to congress. The 2018 elections brought about changes in the state government (the election of Fátima Bezerra – Workers' Party – as governor) and saw the election of deputies with links to social movements. As in other states, various social movements and organizations in Rio Grande do Norte (several of them operating in the northeast region and even nationwide) had already been advocating the creation of institutional markets for family farming and the solidarity economy for some years. The election of a member of congress with ties to the landless workers' movement and who had taken part in the Women's World March and was active in the Xique Solidarity Economy Network provided the opportunity to translate these demands into law. The new government took office in January 2019 and one month later the idea was put forward in the form of a bill. The law was passed that June.

Finally, we ascertained that the politicization of public procurement found its way on to the agenda through the institutional activism of mid-level bureaucrats[5] (Lotta, Pires & Oliveira, 2015) and street-level bureaucracy[6] (Lipsky, 2019). In accordance

---

5. According to Lotta, Pires and Oliveira (2015, p. 25), mid-ranking bureaucracy corresponds to the "intermediate bureaucracy, which manages street-level bureaucrats and acts as a link between these policy implementers and formulators (...). This mid-ranking bureaucracy includes managers, directors, supervisors and agents in charge of putting into operation the strategies formulated by the upper echelons of the bureaucracy."

6. According to Lipsky (2019, pp. 36-37), street-level bureaucrats are all "those public service workers who interact directly with citizens in the course of their work and who have substantial power in the execution of their work (...). Typical street-level bureaucrats are teachers, police officers and other law enforcement officials, social workers, judges, public defenders and other court officials, health care workers, and many other public employees who grant access to government programs and enable the delivery of services within them. (...) They are constantly torn between the demands of service recipients, who want greater effectiveness and responsiveness, and the demands of citizens, who want public services to be more efficient and effective."

with Abers (2015) and Cayres (2015), we understand institutional activism as the militant actions of members of the governing bureaucracy. Whether or not previously linked to social movements, these bureaucrats commit themselves to certain causes and start "promoting political and social projects perceived by the actor as public or collective in nature" (Abers, 2015, p. 148). In the case under analysis, these are projects designed to consolidate family farming, its socioeconomic inclusion, and the promotion of FNS, delivered here by public procurement. The cases of the Federal District and São Paulo are illustrative of these situations.

In the case of the Federal District, the creation of the Program for the Procurement of Agricultural Production (PAPA – 2012) was managed at institutional level by mid-level bureaucrats who had been involved in managing the PAA at national level and, following the election of Agnelo Queiroz (PT) as governor in 2011, began to occupy management positions at this level (the FNS board at the Supply Center (Ceasa) and the board of Emater-DF). Possessing experience of institutional markets at a national level and seeking to promote FNS and family farming initiatives, these actors, in dialogue with the rest of the bureaucracy and Consea, helped create the PAPA.

The São Paulo Social Farming Program (PPAIS), which came into being in 2011, was founded on the actions of officials at the José Gomes da Silva State of São Paulo Land Foundation Institute (ITESP), who were both mid-level and street-level bureaucrats (rural extension workers with links to the body). Based on the results of institutional markets (PAA and PNAE) in land reform settlers and with a view to expanding market opportunities for family farmers, the ITESP bureaucracy took innovative steps in drawing up regulations that provided for social participation in government food procurement by government. This proposal was accepted by the governor at the time (Geraldo Alckmim – Brazilian Social Democracy Party).

In analyzing these processes, three factors attract our attention. The first is the importance of elections as a catalyst for political change and opportunities for the institutionalization of new ideas. With the exception of actions undertaken in response to the pandemic, most cases were driven by the advent of new political alliances and/or the elections of members of congress. Indeed, several authors view elections as opportunities for change and "windows of opportunity" for public policy (Kingdon, 2006; Sabatier & Jenk-Smith, 1999). A second key factor is that policies, programs and actions were created across governments of different political and party persuasions. The politicization of food and its manifestation in public food procurement is not restricted to certain political and ideological positions. However, even though elections are an opportunity for change and different parties can influence changes in public procurement, the creation of initiatives is not an automatic process. The food activism of social movements and the institutional activism of the bureaucracy is fundamental in putting the issue on the agenda, ensuring it remains there, and building institutional frameworks and instruments made possible by current economic and political conditions. The third factor is the centrality of social actors and movements with links to family farming in the politicization of public procurement, whether through food activism or the institutional transit of their members to positions in government. Even in cases where the creation of programs was based on institutional activism, mid-level or street-level bureaucrats generally had strong ties with social movements and family farming unions. We note the absence of actors and organizations that engage in activities focused more on the urban space or on other issues related to food and consumption. This centrality of the social group impacts, in turn, on the public policy *référentiels* outlined herein, as will be discussed below.

▸▸ Issues and *référentiels* in state-government public food policy

This section discusses the various issues, institutional frameworks and objectives that have guided state government in the creation of family farming food procurement programs and actions. This analysis is based on the concept of the public policy *référentiel*, which, according to Fouilleux and Michel (2020) and Fouilleux (2003), refers to ideas and interpretations of public food problems that are institutionalized in public policies. Such ideas and interpretations are constructed, contested and negotiated in specific actor communities and arenas, the aim being to construct or change public policies. The *référentiel* is a snapshot of public policy at a certain moment in time, and conveys its organization, objectives and the image of the disputes and negotiations at stake, subject at all times to the tensions that arise in public arenas (Fouilleux, 2003).

As shown in table 16.4, the PAA's institutional frameworks influenced the creation of state food procurement programs designed to meet public demand. In general terms, there are two main frameworks, although some state governments announced, through their laws, the creation of the set of modes found in the national program (table 16.1). In our analysis, we noted that seven state governments set up actions similar to the purchase with simultaneous donation mode, by virtue of which the state government (based on the allocation of specific funding for this purpose) procures food or produce from family farming and distributes it to social welfare organizations, schools or socially vulnerable populations. In addition, five state governments established regulations that tie in with institutional procurement and Decree No. 8.473/2015, ensuring a place for family farming in existing public food procurement for which funding is already guaranteed. Given the financial difficulties experienced by some state governments, in some cases the social movements of family farming and/or the bureaucracy in question deemed that, in strategic terms, bringing about changes in actions that were already under way (and which had a guaranteed budget) would be more likely to succeed and less likely to create conflict than demanding the creation of new public policies and fighting for funding where it was scarce.

A number of objectives and issues have shaped the politicization of public food demand and the creation of institutional frameworks for family farming. An analysis of the objectives set out in legislation shows that, generally speaking, the various actions implemented have bound together the consolidation of family farming; the socioeconomic inclusion of certain social groups, such as land reform settlers, quilombola communities, indigenous populations, artisanal fishing communities, and women and young people; guaranteed prices and income generation for family farming; the supply of food for public food and nutrition facilities; and the promotion of food and nutrition security for socially vulnerable populations. In addition to these factors, our analysis of the legal documents also identified objectives relating to the promotion of cooperativism, agroecology, sociobiodiversity and local development dynamics. The predominance of family farming-related issues in the creation of state-government public procurement actions in detriment to other issues concerning food systems is noteworthy.

In addition to this documentary analysis, we asked the interviewees to point out the main objectives that shaped the creation of programs. These were prioritized as follows: promoting agroecology, promoting sociobiodiversity, meeting public food demand, combating hunger, improving the quality of food, reducing obesity, promoting short supply chains, promoting local development, promoting the marketing of family farming, and promoting cooperativism and associativity. The five objectives most frequently mentioned by the interviewees were as follows and in this order: promoting the marketing of family farming, improving the quality of food, promoting local development, promoting short supply chains, and meeting public food demand. Promoting cooperativism and associativity, combating hunger, promoting agroecology, and promoting sociobiodiversity were cited less frequently among the first five objectives.

It is noteworthy that in none of the interviews did reducing obesity appear among the top five objectives. It is also noteworthy that although the laws and decrees under analysis refer to the promotion of healthy eating, none of them make specific reference to a concern with overweight and obesity which, as mentioned in the introduction, is one of the main problems facing society today and has already been recognized as a pandemic (Swinburn et al., 2019). The concern with obesity, the dialogue with the Food Guide for the Brazilian Population, and regulations regarding food quality for consumers were discussed and observed with greater intensity in three states (Espírito Santo, Paraná, and Santa Catarina), where the actions and involvement of nutritionists in creating and implementing programs was noted.

Based on analysis of the objectives and issues shaping the politicization and creation of programs, there are two considerations that are worth mentioning. The first concerns the importance given to the promotion of family farming through guaranteed markets, prices and income; the socioeconomic inclusion of the social group's diversity; and the strengthening of markets and local development. These guidelines owe their existence to the prominent role of the social group in shaping changes in public food procurement. As stated in the previous section, in the vast majority of cases, the prominence, food activism and institutional transit of actors with links to family farming organizations were crucial to the establishment of state programs. In several cases, the creation of the programs depended on dialogue with and support from state food and nutrition security councils and agroecological and solidarity economy organizations. There was, however, a noticeable absence of action, interaction and dialogue with other actors and organizations with important roles to play in the issues of food and public food procurement (urban consumer organizations, and organizations and professionals from other areas).

The second consideration, which is closely linked to the first, relates to the public policy *référentiel* promoted by these state programs. Analysis of the ideas underpinning the creation of the actions shows that this is a *référentiel* that seeks to increase the prominence of family farming in food production, guarantee food supply based on the organization of regions and rural communities, and promote healthy eating based on food that is local and linked to the social group. In line with the national experience and results, the aim is to come up with a separate approach for family farming; encourage productive, economic, social and food transformations between and as part of the (re)connection of production and consumption; and create tension in the food system and food policies. As stated, there is a prevalence of family farming and rural issues in the public food procurement *référentiel*.

Despite the importance of these actions and *référentiel*, we emphasize the fact that public procurement of food could aid the creation of integrated food policies. As stated by Sonnino et al. (2020), we increasingly need :

> (...) integrated food policies that build interdependencies between environmental, socioeconomic and health issues, focusing on the whole population, from consumers in general to vulnerable populations (...). These policies require multi-stakeholder engagement, and vertical and horizontal coordination in their implementation.

In view of food challenges and the dynamics of the hegemonic food system detailed in the introduction, it is important for existing experiences to engage in wider dialogue on issues such as overpricing and obesity, decreasing biodiversity, the destruction and pollution of nature, and global warming. Public procurement of food :

> (...) has become increasingly recognized as an important tool to address some of the main challenges of unsustainability in the food system (...). There is growing recognition of the power of public procurement to promote a healthy food system that integrates issues of public health, economic development, democracy, and the environment (Sonnino, 2019, p. 12).

To this end, it is also important to bring other actors, organizations, professional groups and areas of public management into the dialogue between family farming and bureaucracy, such as consumer organizations, social and environmental organizations, and professionals/bureaucrats with links to health, nutrition, social welfare, the environment and urban planning. According to Fouilleux and Michel (2020, p. 14), the politicization of food results in changes in agricultural and food policy arenas that "must open up to new actors from outside the agricultural and industrial world, who bring new ideas, new demands, and who give a new political meaning to food. The food activism of family farming social movements and institutional activism of the bureaucracy could enhance their capacity to influence and transform the food system to the extent that they complement similar and converging demands made by other actors, groups and initiatives, some of which are mentioned above and in the introduction of this paper.

Table 16.4. Objectives, institutional framework, and other information on the public procurement of family farming food by state governments

State	Program	Institutional framework	Other information
Rondônia	PAA Rondônia	Procurement with food donations, implemented in a decentralized way in municipalities. The state government issues a public call for tender, and cooperatives and family farmers present proposals for implementing it in their municipalities, supported by the municipality and Emater-DF	Higher pricing of organic or agroecological production; priority for marketing proposals from women's cooperatives, land credit and land reform settlers, agroecological production, and municipalities with more socially vulnerable people
Maranhão	Procaf	The program acquires produce from family farming and uses it to supply the social welfare network, food and nutrition security programs, and schools offering alternative teaching	Public call for tender for family farmers in general and a specific one for indigenous communities to meet their specific needs; higher pricing of organic or agroecological production; priority for marketing proposals from women's cooperatives and from quilombola and traditional communities, with greater food diversity and marketing experience in other institutional or solidarity economy markets
Ceará	State of Ceará Family Farming Food Procurement Program	Purchase modes with simultaneous donation, direct purchase, incentive for milk production and consumption, stockpiling and institutional purchase, only the last of which had been regulated by May 2020, setting out the procurement of foodstuffs directly from family farming (at least 30% of the budget) or contracting food supply services	Higher pricing of organic or agroecological production; priority for marketing proposals from women, youth, traditional communities, indigenous populations and quilombola communities
Rio Grande do Norte	PECAFES	Direct purchase, aimed at simultaneous donations to educational and social welfare institutions or for the composition of menus to be prepared by education centers in the state public network (family farming produce should account for at least 50% of the budget in 2022); and indirect purchase, in cases where the state government budget is used to purchase prepared food, guaranteeing that family farming produce accounts for at least 30% of menus through a parallel public call for tender	The Program gives priority to projects run by cooperatives or associations, organizations with a higher percentage of women, young people, agroecological production, and quilombola and indigenous communities

State	Program	Institutional framework	Other information
Paraíba	Emergency Procurement	Emergency purchase of food from farming (ten types of fruit, tubers and roots exclusively from the social category, as well as fish and chicken) for social welfare referral centers (CRAS) and associations supporting the socially vulnerable	
Pernambuco	Local Procurement* and PEAAF (June 2020)	Public call for tender for the procurement of 20,000 packs with 13 different products to be distributed to the socially vulnerable, guaranteeing that at least 30% of the budget is invested in family farming	
Espirito Santo	CDA	Rolled out in a decentralized manner in the municipalities, the project procures food from family farming and delivers it to public food and nutrition facilities, public structures providing meals, justice and security networks, the public health network, and private health establishments that have the Charitable Social Welfare Organization Certificate (CEBAS)	Family farmers in the Single Registry (CadÚnico) receive assistance, and the projects proposed by the municipalities must include 40% of the people supported by the Family Grant Program, settlers, quilombola communities, indigenous populations, and other traditional communities; 40% women; and 5% organic/agroecological producers. The Project provides for the practically exclusive purchase of fresh food and up to 15% processed food and provides guidance on the higher pricing of organic or agroecological production
Minas Gerais	PAA Family	The Program ensures that at least 30% of the budget allocated to the institutional purchase of foodstuffs and seeds are channeled to the procurement of family farming produce. The program also ensures that a minimum of 30% of the budget is allocated to the procurement of food and prepared meals from family farming	The program gives priority to marketing projects run by family farmers in the municipality where the food will be consumed, traditional and quilombola communities and indigenous populations, land reform settlements, women's groups and agroecological or organic production. It provides guidance on the higher pricing of organic or agroecological production
Federal District	PAPA	The program ensures the procurement of agricultural and extractivist produce and handicrafts from family farming (at least 30% of budget) by state agencies	Provides guidance on the higher pricing of organic or agroecological production

State	Program	Institutional framework	Other information
São Paulo	HIPC	The program ensures the procurement of food from family farming (at least 30% of budget) by state agencies	
Paraná	Paraná Direct Procurement*	The Program purchases food from family farming to serve the social welfare referral centers, the specialist social welfare referral center, the specialist referral center for the homeless, shelters, nursing homes, halfway houses, public and philanthropic hospitals, public restaurants, community kitchens, and food banks	The Program gives priority to marketing projects (separated by aggregated administrative regions) that include the following: agroecological production, traditional peoples and communities, settlers, young people and women. The sales projects must list the groups of foods to be marketed, with priority being given to foods in line with the Food Guide for the Brazilian Population.
Santa Catarina	Food Procurement Program/ Decentralization of FDR resources from SC*	Municipalities with a Human Development Index (HDI) below 0.7, councils (COMSEA and CMDR), and an economy based on the "agro sector" can submit proposals that link family farming food packs with donations for families suffering from food insecurity, preferably those listed in the Single Registry	50% of the budget for the packs is allocated to milk and derivatives, and 50% to other foods listed in the Food Guide for the Brazilian Population. The initiative gives priority to local farmers, indigenous and quilombola communities, organizations, and agroecological production. It provides guidance on the higher pricing of organic and agroecological food

Source: Based on legal documents and interviews conducted.

First three objectives cited in the legislation in force. Generally, more than three objectives influence the implementation of these public policies.

* Temporary programs created in response to the coronavirus pandemic.

▸▸ Final considerations

In returning to the subject of the 12 state family farming food procurement programs, we note that the temporalities, contexts and processes that put the topic on the governmental public agenda were distinct from each other. The innovations and results of the PAA and PNAE at national level, the PAA's financial and political crisis (2013 onwards), the dismantling of public policies as they relate to family farming and FNS, political changes at national and state level, and the coronavirus pandemic were all factors that contributed to the setting of the public agenda and the politicization of public food procurement. It is essential to recognize that hunger is on the public agenda once again and that federal government has done little to address it. In addition to these factors, a number of social actors mobilized, took center stage and took action. The social and union movements of family farming engaged in food activism, their representatives engaged in institutional transit to the government, and mid-level and street-level bureaucracy engaged in institutional activism. In line with Fouilleux and Michel (2020), we can thus see that the politicization of food and its manifestation in public procurement is the result of various processes and actors that together seek to problematize and change the food system.

Family farming organizations have a dominant position among the actors involved in the creation of state programs. They pressure, interact and engage with the bureaucracy and government actors. In the course of our analysis, we did not observe the involvement of consumers, consumer organizations, other professional groups or areas of public management in changing the food order in force in Brazilian states. The centrality of family farming organizations is reflected in public policy *référentiels* that mainly address and respond to issues that impact on the social group (income generation, guaranteed markets, the socioeconomic inclusion of specific social groups, local development, the promotion of cooperatives, etc.) and the supply of better-quality food to public facilities and in response to demand. Although these factors are fundamental to promoting change in the dominant food system, so urgent is the need to build sustainable food systems that other ideas, issues and actors must be incorporated into the *référentiels* to create integrated food policies. In this respect, it is important to recognize and strengthen policy networks that promote agreements and weaken centralizing elements. Family farming organizations could strengthen their activism and agendas by interlinking them with other intersectoral issues (environmental, social, nutritional, economic, political, and educational).

▶▶ References

Abers, R. N. (2015). Ativismo na burocracia? O médio escalão do Programa Bolsa Verde. In Cavalcante, P. L. C., & G. S. Lotta (Eds.), *Burocracia de médio escalão: perfil, trajetória e atuação* (pp. 143–175). ENAP. https://repositorio.enap.gov.br/handle/1/2063

Avila, M. L., Caldas, E. L., & Avila, S. R. (2013). Coordenação e efeitos sinérgicos em políticas públicas no Brasil: o caso do Programa de Aquisição de Alimentos e do Programa Nacional de Alimentação Escolar. In Del Grossi, M. E., & D. R. Kroeff (Eds.), *PAA: 10 anos de aquisição de alimentos* (pp. 96–113). Ministério do Desenvolvimento Social e Combate à Fome, Brasil. https://aplicacoes.mds.gov.br/sagirmps/ferramentas/docs/PAA.pdf

Barbosa, L. (2009). Tendências da alimentação contemporânea. In Pinto, M. de L., & J. K. Pacheco (Eds.), *Juventude, consumo & educação* (pp. 15–65). ESPM.

Brand, C., Bricas, N., Conaré, D., Daviron, B., Debru, J., Michel, L., & Soulard, C.-T. (Eds.). (2019). *Designing urban food policies: concepts and approaches*. Springer. https://doi.org/10.1007/978-3-030-13958-2

Bricas, N. (2019). Urbanization issues affecting food system sustainability. In Brand, C., Bricas, N., Conaré, D., Daviron, B., Debru, J., Michel, L., & C.-T. Soulard (Eds.), *Designing urban food policies: concepts and approaches* (pp. 1–25). Springer. https://doi.org/10.1007/978-3-030-13958-2

Cayres, D. C. (2015). *Ativismo institucional no coração da Secretaria-Geral da Presidência da República: a Secretaria Nacional de Articulação Social no Governo Dilma Rousseff (2011–2014)* [Doctoral thesis, Universidade Federal de Santa Catarina]. UFSC. https://www.ipea.gov.br/participacao/images/pdfs/tesedomitilacayres.pdf

Daviron, B., Perin, C., & Soulard, C. T. (2017). History of urban food policy in Europe, from the ancient city to the industrial city. In Brand, C., Bricas, N., Conaré, D., Daviron, B., Debru, J., Michel, L., & C.-T. Soulard (Eds.), *Designing urban food policies* (pp. 27–51). Springer. https://doi.org/10.1007/978-3-030-13958-2

Dorigon, C. B. (2019). *Da roça ao restaurante: um estudo sobre redes alimentares de qualidade diferenciada na Serra Gaúcha* [Master's thesis, Universidade Federal do Rio Grande do Sul]. UFRGS Digital Repository. http://hdl.handle.net/10183/197226

Fages, R., & Bricas, N. (2017). *L'alimentation des villes: quels rôles des collectivités du Sud?* AFD. https://hal.archives-ouvertes.fr/hal-01995381

FAO. (2019). *The state of the world's biodiversity for food and agriculture.* FAO Commission on Genetic Resources for Food and Agriculture Assessments. https://doi.org/10.4060/CA3129EN

Fouilleux, E. (2003). *La politique agricole commune et ses réformes.* L'Harmattan.

Fouilleux, E. (2011). Analisar a mudança: políticas públicas e debates num sistema em diferentes níveis de governança. *Estudos, Sociedade e Agricultura, 19*(1), 88–125. https://revistaesa.com/ojs/index.php/esa/article/view/337

Fouilleux, E., & Michel, L. (Eds.). (2020). *Quand l'alimentation se fait politique(s).* Presses Universitaires de Rennes.

Goggins, G. (2018). Developing a sustainable food strategy for large organizations: The importance of context in shaping procurement and consumption practices. *Business Strategy and the Environment, 27*(7), 838–848. https://doi.org/10.1002/bse.2035

Goggins, G., & Rau, H. (2016). Beyond calorie counting: assessing the sustainability of food provided for public consumption. *Journal of Cleaner Production, 112*(1), 257–266. https://doi.org/10.1016/j.jclepro.2015.06.035

Grisa, C. (2018). Mudanças nas políticas públicas para a agricultura familiar no Brasil: novos mediadores para velhos referenciais. *Raízes: Revista De Ciências Sociais e Econômicas, 38*(1), 36–50. https://doi.org/10.37370/raizes.2018.v38.37

Grisa, C., Schneider, S., & França de Vasconcelos, F. (2020). As compras públicas como instrumentos para a construção de sistemas alimentares sustentáveis. In Preiss, P. V., Schneider, S., & G. Coelho De Souza (Eds.), *A contribuição brasileira à segurança alimentar e nutricional sustentável* (pp. 69–90). UFRGS. https://www.ufrgs.br/agrifood/images/2020/07-julho/001115755.pdf

Guéneau, S., Neto, E. J. L., & Braga, C. L. (2019). O processo de construção da Política de Agroecologia e Produção Orgânica do Estado do Maranhão - PEAPOMA. In Sabourin, E. Guéneau, S., Colonna, J., & L. R. Tadeu da Silva (Eds.), *Construção de políticas estaduais de agroecologia e produção orgânica no Brasil: avanços, obstáculos e efeitos das dinâmicas subnacionais* (pp. 197–218). Editora CRV. https://hal.archives-ouvertes.fr/hal-02545007

Hawkes, C., & Halliday, J. (2017). *What makes urban food policy happen? Insights from five case studies.* IPES-Food. https://www.ipes-food.org/_img/upload/files/Cities_full.pdf

HLPE (High Level Panel of Experts on Food Security and Nutrition). (2014). *Food losses and waste in the context of sustainable food systems.* Committee on World Food Security. https://www.fao.org/3/i3901e/i3901e.pdf

IPES Food. (2020). *Covid-19 and the crisis in food systems: symptoms, causes and potential solutions.* https://www.ipes-food.org/_img/upload/files/COVID-19_CommuniqueEN.pdf

Kingdon, J. (2006). Juntando as coisas. In Saraiva, E., & E. Ferrarezi (Eds.), *Políticas públicas. Coletânea – Volume 1* (pp. 225–246). ENAP.

Lotta, G. S., Pires, R. R. C., & de Oliveira, V. E. (2015). Burocratas de médio escalão: novos olhares sobre velhos atores da produção de políticas públicas. In Cavalcante, P. L. C. & G. S. Lotta (Eds.), *Burocracia de médio escalão: perfil, trajetória e atuação* (pp. 23–55). ENAP. https://repositorio.enap.gov.br/handle/1/2063

Lipsky, M. (2019). *Burocracia em nível de rua: dilemas do indivíduo nos serviços públicos* (A. E. Moura da Cunha, Trans.). ENAP. https://repositorio.enap.gov.br/handle/1/4158

Mielitz, C. Dez anos de PAA e a constituição de uma estratégia nacional de segurança alimentar. In Del Grossi, M. E., & D. R. Kroeff (Eds.), *PAA: 10 anos de aquisição de alimentos* (pp. 59–73). Ministério do Desenvolvimento Social e Combate à Fome, Brasil. https://aplicacoes.mds.gov.br/sagirmps/ferramentas/docs/PAA.pdf

Morgan, K., & Sonnino, R. (2008). *The school food revolution: public food and the challenge of sustainable development.* Earthscan.

Morgan, K., & Sonnino, R. (2010). Repensando a alimentação escolar: o poder do prato público. In Assadourian, E. (Ed.), *Estado do mundo: estado do consumo e o consumo sustentável* (C. Strauch, Trans.) (pp. 72–78). Uma Ed.

Niederle, P., & Schubert, M. N. (2020). How does veganism contribute to shape sustainable food systems? Practices, meanings and identities of vegan restaurants in Porto Alegre, Brazil. *Journal of Rural Studies*, *78*, 304–313. https://doi.org/10.1016/j.jrurstud.2020.06.021

Portilho, F. (2020). Ativismo alimentar e consumo político – duas gerações de ativismo alimentar no Brasil. *Redes*, *25*(2), 411–432. https://doi.org/10.17058/redes.v25i2.15088

Portilho, F., Castañeda, M., & Ribeiro de Castro, I. R. (2011). A alimentação no contexto contemporâneo: consumo, ação política e sustentabilidade. *Ciência & Saúde Coletiva*, *16*(1), 99–106. https://doi.org/10.1590/S1413-81232011000100014

Portilho, F., & Micheletti, M. (2018). Politicizing consumption in Latin America. In Boström, M., Micheletti, M., & P. Osterveer (Eds.), *The Oxford Handbook of Political Consumerism* (1st ed.) (pp. 539–557). Oxford University Press. http://www.doi.org/10.1093/oxfordhb/9780190629038.013.24

Porto, S. I. (2014). *Programa de Aquisição de Alimentos (PAA): política pública de fortalecimento da agricultura familiar e da agroecologia no Brasil* [Master's thesis, Universidad Internacional de Andalucía]. UnIA Repository. http://hdl.handle.net/10334/3597

Preiss, P. V. (2017). *As alianças alimentares colaborativas em uma perspectiva internacional: afetos, conhecimento incorporado e ativismo político* [Doctoral thesis, Universidade Federal do Rio Grande do Sul]. UFRGS Digital Repository. http://hdl.handle.net/10183/178604

Procedi, A. (2019). *A trajetória do Programa de Aquisição de Alimentos (PAA) no Rio Grande do Sul: entre a descontinuidade da política pública e a capacidade de ação dos atores sociais* [Unpublished master's thesis, Universidade Federal do Rio Grande do Sul].

Raja, S., Clark, J. K., Freedgood, J., & Hodgson, K. (2018). Reflexive and inclusive: reimagining local government engagement in food systems. *Journal of Agriculture, Food Systems, and Community Development*, *8*(S2), 1–10. https://doi.org/10.5304/jafscd.2018.08B.013

Reece, J. (2018). Seeking food justice and a just city through local action in food systems. *Journal of Agriculture, Food Systems, and Community Development*, *8*(S2), 211–215. https://doi.org/10.5304/jafscd.2018.08B.012

Ribeiro, M. J. A. (2019). *Um alimento político e uma política que alimenta: o activismo do Slow Food no Brasil* [Master's thesis, Universidade Estadual de Montes Claros]. Unimontes. https://www.posgraduacao.unimontes.br/uploads/sites/20/2019/06/Dissertação-de-mestrado-Maria-João-Alves-Ribeiro.pdf

Sabatier, P., & Weible, C. (2007). The advocacy coalition framework: innovation and clarifications. In P. Sabatier (Ed.), *Theories of the policy process* (pp. 189–220). Westview Press. https://doi.org/10.4324/9780367274689

Sabourin, E., Craviotti, C., & Milhorance, C. (2020). The dismantling of family farming policies in Brazil and Argentina. *International Review of Public Policy*, *2*(1), 45–67. https://doi.org/10.4000/irpp.799

Sabourin, E., Guéneau, S., Colonna, J., & Tadeu da Silva, L. R. (2019). Marco teórico e metodológico: a ação pública para a agroecologia. In Sabourin, E., Guéneau, S., Colonna, J., & L. R. Tadeu da Silva (Eds.), *Construção de políticas estaduais de agroecologia e produção orgânica no Brasil: avanços, obstáculos e efeitos das dinâmicas subnacionais* (pp. 17–26). Editora CRV. https://agritrop.cirad.fr/594268/1/ID594268.pdf

Schmitt, C. J., & Barbosa, Y. R. S. (2019). A construção da política estadual de agroecologia e produção orgânica em Minas Gerais: interações Estado-sociiedade na institucionalização da agroecologia. In Sabourin, E., Guéneau, S., Colonna, J., & L. R. Tadeu da Silva (Eds.), *Construção de políticas estaduais de agroecologia e produção orgânica no Brasil: avanços, obstáculos e efeitos das dinâmicas subnacionais* (pp. 75–98). Editora CRV.

Schneider, S., & Gazolla, M. (2017). Cadeias curtas e redes agroalimentares alternativas. In Gazolla, M., & S. Schneider (Eds.), *Cadeias curtas e redes agroalimentares alternativas: negócios e mercados da agricultura familiar* (pp. 9–24). Editora da UFRGS. http://hdl.handle.net/10183/232245

Siliprandi, E., & Cintrão, R. (2013). As mulheres rurais e a diversidade de produtos no Programa de Aquisição de Alimentos. In Del Grossi, M. E., & D. R. Kroeff (Eds.), *PAA: 10 anos de aquisição de alimentos* (pp. 114–151). Ministério do Desenvolvimento Social e Combate à Fome, Brasil. https://aplicacoes.mds.gov.br/sagirmps/ferramentas/docs/PAA.pdf

Silva, M. K., & Oliveira, G. L. (2011). A face oculta(da) dos movimentos sociais: trânsito institucional e intersecção Estado-Movimento - uma análise do movimento de economia solidária no Rio Grande do Sul. *Sociologias, 13*(28), 86–124. https://doi.org/10.1590/S1517-45222011000300005

Soldi, R. (2018). *Sustainable public procurement of food*. European Union. https://cor.europa.eu/en/engage/studies/Documents/sustainable-public-procurement-food.pdf

Sonnino, R. (2019). Translating sustainable diets into practice: the potential of public food procurement. *Redes. Revista do Desenvolvimento Regional, 24*(1), 14–29. https://doi.org/10.17058/redes.v24i1.13036

Sonnino, R., Callenius, C., Lähteenmäki, L., Breda, J., Cahill, J., Caron, P., Damianova, Z., Gurinovic, M. A., Lang, T., Laperriere, A., Mango, C., Ryder, J., Verburg, G., Achterbosch, T., den Boer, A. C. L., Kok, K. P. W., Regeer, B. J., Broerse, J. E. W., Cesuroglu, T., & Gill M. (2020). *Research and innovation supporting the farm to fork strategy of the European Commission* (Policy brief 3). FIT4FOOD2030. https://fit4food2030.eu/wp-content/uploads/2020/04/FIT4FOOD2030-Research-and-Innovation-supporting-the-Farm-to-Fork-Strategy-of-the-European-Commission-Policy-Brief.pdf

Swinburn, B. A., Kraak, V. I., Allender, S., Atkins, V. J., Baker, P. I., Bogard, J. R., Brinsden, H., Calvillo, A., De Schutter, O., Devarajan, R., Ezzati, M., Friel, S., Goenka, S., Hammond, R. A., Hastings, G., Hawkes, C., Herrero, M., Hovmand, P. S., … Dietz, W. H. (2019). The global syndemic of obesity, undernutrition, and climate change: The Lancet Commission report. *The Lancet, 393*(10173), 791–846. https://doi.org/10.1016/S0140-6736(18)32822-8

Teo, C. R. P. A., & Triches, R. M. (Eds.) (2016). *Alimentação escolar: construindo interfaces entre saúde, educação e desenvolvimento*. Argos.

Triches, R. M. (2015). Repensando o mercado da alimentação escolar: novas institucionalidades para o desenvolvimento rural. In Grisa, C., & S. Schneider (Eds.), *Políticas públicas de desenvolvimento rural no Brasil* (pp. 182–200). Editora UFRGS.

Zaneti, T. B. (2017). *Cozinha de raiz: relaciones entre chefs, productores y consumidores a través del uso de productos agroalimentarios únicos en la gastronomía contemporánea* [Doctoral dissertation, Universidade Federal do Rio Grande do Sul]. UFRGS Digital Repository. http://hdl.handle.net/10183/164708

Chapter 17

Family farming food and its relationship with the state: the crystallization of differentiated norms within Argentina's public policy agenda

Clara Craviotti

▸▸ Introduction

The food issue can be seen as a complex web that involves the ways in which food is produced, exchanged and consumed, and the consequences this has on the environment, people's health and development processes. Not only is the issue broad, but so is the variety of actors involved, especially in countries that are producers and exporters of agricultural raw materials and are characterized by profound inequalities in terms of production and access to food.

In recent years, the role of family farming has been highlighted in various forums as a response to the challenges of food security and sovereignty and the conservation of natural resources. The lengthening of food chains has led to the "isolation" of these producers within the agricultural activity, while control of other stages has been taken over by other actors, with a marked distance between producers and consumers (Jacoby et al., 2014). In cases where family production maintains some prominence, not only in terms of production but food processing as well, there are obstacles to gaining access to markets, not only logistical, but also as a result of the demands imposed by health requirements.[1] The latter has become more of an issue in recent decades due to growing concerns about food allergies and intolerances and also due to the emergence of critical events, of which the coronavirus pandemic is a recent example. At the same time, the increasing convergence of quality regulatory systems worldwide has been seen as a key factor in consolidating markets which are functional for large-scale food production (De Haro, 2014).

In this context, this chapter focuses on analyzing the process of approving regulations related to the production of food from family farming in Argentina, initiated towards the end of the 2000s due to the prioritization of this issue as a recipient of public policies. Its development then faced a different political context, due to the

---

1. It is worth specifying that, according to the 2018 National Agricultural Census, only 10% of agricultural holdings in Argentina declare they develop commercial activity on the basis of their production.

redefinition and drastic reduction of actions aimed at family farming (Bertoni & Soverna, 2018; Nogueira et al., 2017). However, the incorporation of a specific article in the Argentine Food Code (CAA) – the country's principal standard regulating the production and commercialization of food – and other related norms signaled a step forward in terms of the visibility of the sector, and indirectly collaborated in the legitimization of a specific way of producing.

Based on the premise that norms are not only technical but also political objects, this chapter analyzes the factors that made the formulation and incorporation of these norms within the CAA possible, the actors that promoted them, their goals and foundations, the main areas of dispute and the agreements reached. It also goes on to examine the extent to which changes in the political context influenced the content and scope of the regulations.

More precisely, it is suggested that these institutional innovations were the product of a set of factors: i) the opening of a "window of opportunity" after the Historical Reparation of Family Farming Law, approved in December 2014, came into force; ii) the continuity of governance schemes involving family farming organizations and key public bodies interested in its promotion, which at certain times managed to form coalitions to support certain policies; and iii) the continuity of certain international committees that supported these actions, as is the case of the REAF (Specialized Meeting on Family Farming of MERCOSUR).

Firstly, however, the political context is thought to have influenced the timing of the gestation process of the standards. Secondly, it also influenced the translation of the initial content of some of them, enabling its scope to be extended "upwards" to cover the stratum of small entrepreneurs, and, at the same time, its restriction in terms of the type of food considered. Thirdly, the vision that guided the development of the standards showed a certain transition with respect to the purpose initially sought, which went from the inclusion of family farming in a broad sense to the achievement of competitiveness by the sector. Consequently, although family farming continued to be a target category for public policies, its place was marginal with respect to the predominant policy aimed at supporting export-oriented agriculture.

This chapter is organized as follows: the theoretical framework and methodology adopted in the paper are synoptically presented before moving on to an analysis of the general evolution of policies towards family farming in the periods considered, identifying the key milestones. The sections that follow outline the central characteristics of the Argentine regulatory framework aimed at ensuring food safety; the process of incorporation of specific family farming-oriented standards, the enabling factors, the actors that promoted their incorporation and their core purposes. The content of the regulations is also addressed, as well as the main disputes and compromises reached. Lastly, the final section of the chapter considers some implications of the analysis in terms of the connection of the policies studied with the different competing *référentiels* and the possible redefinition of the food system.

▸▸ Theoretical-methodological notes

The approach adopted combines contributions from institutionalism focused on actor networks with contributions from the sociology of translation and its complementary, actor-network theory. The first perspective has been prioritized to address the processes of formulation and implementation of public policies that are characterized by their complexity and the participation of different public and private actors (Zurbriggen, 2006). According to this approach, the actors in a network develop more or less stable interactions through which they exchange ideas and resources and negotiate possible solutions to the problems identified. The perspective is useful because it considers their strategies and actions in the network within the broader framework in which the network is embedded, which poses both possibilities and constraints.

Linked to this approach, the concept of advocacy coalitions refers to people who, from a variety of positions, share a belief system and demonstrate a certain degree of coordinated activity over time (Cairney, 2015). In relation to public policies, that belief system can be defined as a *référentiel* (Muller, 2005), which implies the construction of a collective representation of the reality on which action is sought.

It is understood that policy formulation is never a purely technical process but rather involves the elaboration of commitments and the construction of meanings. Thus, disputes arise about the definition of problems, the categories used to describe them, the criteria used for classification and evaluation, and the ideals that guide actions (Lendvai & Stubbs, 2012). From there, the connection can be made between these perspectives and the sociology of translation, in that it emphasizes the fluid nature of political processes and the constant (re)construction of issues, interpretations and the actor networks themselves. The process of translation would be, in Callon's (1986) terms, the very operation of power: to translate is to displace, but also to express in one's own language what others say; if the process is successful, only a single voice is heard as an expression of the network.

In the analysis of this path, it is important to identify the stages (which this author characterizes as problematization, interest, enrolment and mobilization), and the "key actors"; generally, those who have the legitimacy to convene and/or act as translators of the different visions. It is also important to recognize the existence of previous relationships between the actors, as well as their capacities and complementarities (Ghezán et al., 2013). Given that the compromises achieved become supports for the networks formed, reaching consensus implies that the actors' margins of maneuver will be delimited (Callon, 1986). Once consensus is "fixed" in a public policy, it influences the allocation of resources. However, it is important to note that networks possess different degrees of irreversibility. As will be seen below, this point is particularly important in the case of the policies discussed in this chapter.

The incorporation of the sociology of translation and its complementary actor-network theory (Latour, 2008) implies extending the theory of governance networks with the inclusion of non-human actors, such as previously approved norms or those introduced from other contexts. However, in our case we deviate from the principle of symmetry advocated by this theory, which would imply assuming that different categories of actors are equally important.

From a methodological perspective, the preferred theoretical option involves following the actors and analyzing their language as ethnographic data (Lendvai & Stubbs, 2012). Hence the importance of using qualitative research techniques to analyze their discourses and the characteristics of the particular contexts in which they are situated.

In our case, the process started with the timely analysis conducted on the approval of regulations for small, processing facilities at provincial level (Craviotti, 2017), before moving on to focus on the national level, with specific emphasis on the 2014–2019 period. To that end, the meeting minutes of the National Food Commission (32 in total, of which 17 dealt with family farming-related issues) were collected and analyzed, along with complimentary materials corresponding to the minutes, national documents and journalistic notes. Interviews were also conducted with a limited number of key interviewees responding to different areas of government and different institutional periods.

⊳ The political context of the approval of standards for family farming foodstuffs

Institutionalization period of family farming-oriented policies (2006–2015)

In order to understand the more general framework in which the approval of the standards took place, it is important to bear in mind that the category of family farming as a target of specific public policies was constructed and disseminated in Latin America from the beginning of the 2000s onwards. A set of factors facilitated this process, both at international and local level.[2] In the case of Argentina, one decisive factor was the inauguration of a new government in 2003 with a different political orientation from that prevailing during the 1990s, promoting more active state intervention in relation to the economy and a more equitable distribution of income. Thus, the National Forum for Family Farming (FONAF), made up of representatives of state programs and producer organization, was established in 2006, followed by the National Registry of Family Farming (RENAF) in 2007. Other institutional spheres, such as the National Agricultural Technology Institute (INTA) and the National Service of Agrifood Health and Quality (SENASA), established specific areas related to the sector (CIPAF in 2005 and SENAF in 2009).

Throughout this process, the political climate of confrontation with the commodity export sector, prompted by the decision made by the state in 2008 to proceed with its intention to raise taxes for the sector and change its definition (embodied in Resolution 125, later repealed), was significant in triggering the decision to prioritize family farming through new institutional areas. That same year, the Undersecretariat of Rural Development and Family Agriculture was created, later becoming the Secretariat. Similarly, actions were launched to facilitate access to social benefits for family farmers and their insertion into the formal economy through a specific instrument, the Social Agricultural Monotax (*Monotributo Social Agropecuario*).

Analyzed globally, the most intense period in the development of new instruments and specific institutional areas occurred between 2005 and 2010. A subsequent and significant step was taken towards the end of 2014, with the enactment of Law 27.118

---

2. The policy adopted by Brazil was decisive for the establishment of family farming in other Mercosur countries. In this context, its proposal was accepted to create a specific regional body – REAF – made up of States and organizations representing civil society to agree common policy criteria. Prior to this, Argentina had undertaken actions aimed at what was then known as "small production" or "smallholding," which focused on the most undercapitalized segments of the sector. For more details, see Craviotti (2015).

on "Historical Reparation of Family, Peasant and Indigenous Agriculture," which considers family farming as a central actor for regional development. These actions established a favorable framework for initiatives that would again be revisited in a different political context. As in the case of Brazil, the institutional framework opened opportunities for the expression of new actors, ideas and interests in public spaces, serving as an axis of support for those who sought to defend the maintenance of certain policies – even in a different context (Flexor & Grisa, 2016).

It should, however, be noted that in the case of Argentina, the regulation of a law is a fundamental step in its becoming effective and being assigned a specific budget. The family farming law failed to reach that stage before another government coalition came into power, and that had an impact on subsequent actions.

Period of reversal and marginalization of family farming-oriented policies

The Cambiemos government led by Mauricio Macri, who took office in December 2015, was characterized in economic terms by the promotion of pro-market reforms (de Anchorena, 2018). Key among its initiatives, in addition to the strong devaluation of the exchange rate, were its policies of deregulation of the entry and exit of capital flows, high levels of foreign debt and trade liberalization.

In the agricultural sector, tariffs on the export of key commodities (mainly soybeans) were reduced, although the need to promote the country's insertion within the international market through the export of preferably processed foods was emphasized (Lucero & Padín, 2019). The new government introduced significant changes with regard to family farming. Shortly after taking office, the Secretariat of Family Agriculture was merged with the Secretariat for Territorial Coordination and Development, which was then subsequently downgraded to an Undersecretariat. Initially other components of the institutional framework, such as an advisory council composed of government officials and organizational representatives, were maintained, however, the number of meetings held were reduced over time and eventually ceased all together, being replaced by decentralized exchanges with selected groups (Vigil, 2019). The new government drastically cut the number of officials tasked with supporting family farmers, as well as the budgets of several of the programs aimed at the sector (Sabourin et al., 2020).

Other indicators illustrate the qualitative change in the way the sector was viewed. In 2017, the need to update the beneficiaries of the registry of family producers, RENAF, was announced. Organizations reported that they were excluded as registration entities and that new parameters were established, leading to a reduction in the number of registered producers (Montón, 2018; Vigil, 2019). A year later, the need to update the scope of the Agricultural Social Monotax system was decreed. Once again, the consequence was an abrupt reduction in the number of affiliates to the system and the instrument was transferred to the Ministry of Social Development, while the Ministry of Agroindustry ceased to contribute to beneficiary health insurance coverage.

Both official discourses and new programs targeting family farmers and those that remained from previous years showed a tendency to reorient towards the more capitalized segments (Bertoni and Soverna, 2018; Vigil, 2019). Support for family producer organizations was reduced or cut (Vigil, 2019). A business vision, focused on productive and commercial issues displaced the emphasis on organization and access to rights that had predominated during the 2012–2015 period (Bertoni & Soverna, 2018; Nogueira et al., 2017).

However, ambivalently, or perhaps to avoid conflicts arising, part of the institutionality linked to family farming, such as the Historical Reparation Law, was maintained throughout this period, with attempts at regulation that failed to materialize. There was some innovation at national level with the approval of certain instruments that had been set in motion during Cristina Kirchner's second term in office and had not crystallized. One of these was the regulation of the family farming seal, based on a draft presented by Brazil at the REAF. Other instruments that made their way during this period were proposals for articles related to the sector to be incorporated into the Argentine Food Code.

These new articles merit specific analysis, due to the non-linear process followed for their approval, the changes made to their content during the course of their development, and the complexity of the regulatory framework involved, which we will look at in the following section.

▸▸ The characteristics of the process of incorporating new standards into the Argentine Food Code

The CAA is the fundamental standard of the food control system in Argentina. It regulates the hygiene and health standards and commercial identification of food for human consumption, and also applies to imported and exported food. Considered pioneering within Latin America (Ariosti & Carrión, 2014), it was established by Law 18.284 in 1969, and entered into effect in 1971. According to Balbi et al. (2011), the Code incorporates standards agreed within Mercosur that are, in turn, influenced by the Codex Alimentarius, the standards of the European Union and those of the United States Food and Drug Administration (FDA).

The function of the National Food Control System is to ensure compliance with the CAA and contemplate the government agencies in charge. Within the Ministry of Agriculture, SENASA handles the food products specified in Annexes I and II of Decree 815. Within the Ministry of Health, the National Administration of Drugs, Foods and Medical Devices (ANMAT), through the National Food Institute (INAL), regulates the remaining products not specified in that decree. Its role can be seen as more relevant as agrifood chains become more complex: it covers ready-to-eat processed products, packaging and dietary supplements, and alcoholic and non-alcoholic beverages, except for wine.[3]

This multiplicity of agencies has implications at the territoriales level. In the case of the Ministry of Health, ANMAT delegates control to the bromatological authorities of the provinces and municipalities, i.e., it has no territorial presence. On the contrary, in the case of the Ministry of Agriculture, SENASA is a decentralized

---

3. In the case of dairy facilities, both areas of government have joint competence. However, there are variations depending on the destination of the products.

agency with territorial representatives. The fundamental role of this body in the formulation and implementation of health policies is therefore evident, as it is responsible for registering primary production units, intervening in the control of bulk food storage facilities and those involved in processing food of animal origin, enabling their national circulation. Hence the importance of the creation of a specific area related to family farming as part of its structure, in order to provide training and propose regulatory adjustments according to the specific characteristics of these establishments. It is important to note that this was part of a change of vision, promoting a system that contemplated the support of officials and producers and transcended the traditional function of inspection. Reference was also made to this approach in the case of ANMAT.

The CAA is updated on an ongoing basis through joint resolutions signed by the Minister of Agroindustry and the Minister of Health. A whole process exists prior to that, which is coordinated by the National Food Commission (CONAL), with the participation of representatives of both the above ministries, the Secretariat of Domestic Trade, SENASA, ANMAT and the provinces.

Any individual, company or organization may submit proposals to CONAL to modify or add articles to the CAA. The Commission establishes working groups on various topics, in which standards and guidelines are developed and then discussed as a whole. Once approved, these are analyzed by an advisory commission, the CONASE, with the participation of private sector representatives (businessmen and workers from the sector; consumers[4]). The proposals are then subject to public consultation for a month, during which time comments or suggestions can be lodged. It should be said that, in these two cases, the opinions are not binding. Finally, the definitive proposal is submitted to the two ministers (of Agriculture and Health) for signature.

This path (outlined in figure 17.1) implies a certain degree of complexity and sluggishness in the incorporation of new standards, although in recent years the process has been speeded up thanks to the use of digital technologies; the number of annual CONAL meetings has also increased. Moreover, the system tends to favor those actors with most resources (material and symbolic): "if you have someone to present the projects for you, who knows how to put them together, the Commission approves them" (Interview with key interviewee, 2020).

▸▸ Analysis of the process of incorporating Family Farming standards within the CAA

Enabling factors and characteristics of the process

In this section we identify a set of relevant aspects that have influenced the dynamics of the development and incorporation of these standards. Firstly, at international level, the exchange forum facilitated by REAF Mercosur and supported by

---

4. According to Decree 815/88, CONASE is made up of four representatives from the business sector, one of which should be from an SME, two from consumer organizations and one from a worker's organization. The current composition demonstrates the influence acquired by representatives of supermarkets and big industry.

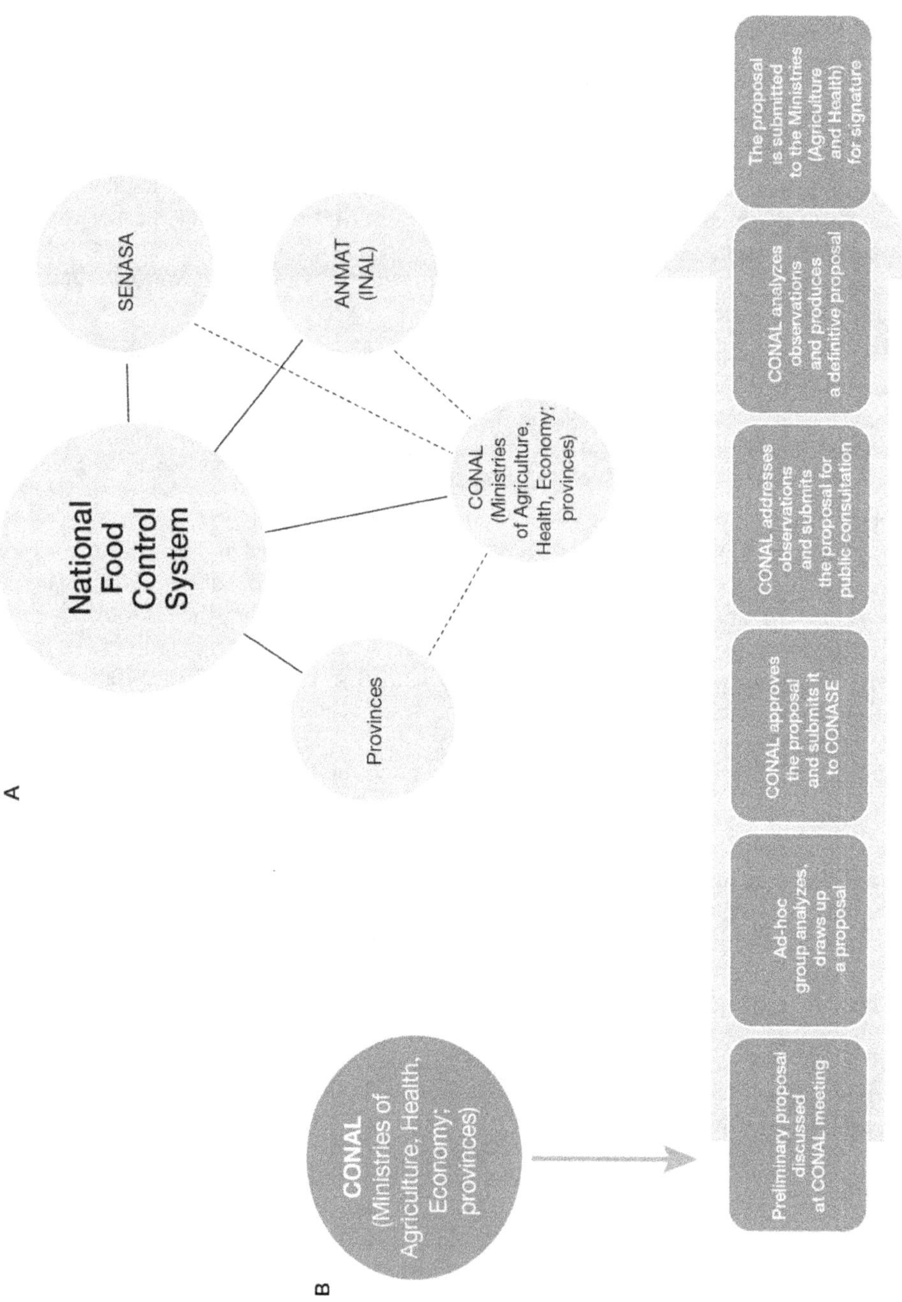

Figure 17.1. National Food Control System and steps for the incorporation of standards within the CAA in Argentina.

(A) The characteristics of the process of incorporating new standards into the Argentine Food Code.
(B) Process for the incorporation of changes within the Argentine Food Code.

international organizations, such as the FAO and IICA, was important to frame and validate the health standards initiatives related to family farming in Argentina, as they form part of the actions initiated by Latin American countries to optimize the coordination and harmonization of guidelines. This exchange began with SENASA's participation in the REAF's technical group on trade facilitation in 2014, continued with a seminar on health issues held in 2016, and then crystallized into a regional network and the holding of a conference on artisanal cheese-making with the participation of international experts from Brazil and Spain. The topic continued with a conference on health issues held in 2019 within the framework of the REAF.

Secondly, the regulatory background of the European Union – considered an influential benchmark with regard to health matters – was considered by key actors at local level in order to legitimize certain concepts, such as "flexibility" in the interpretation of the general requirements of the rules. The National Health Surveillance Agency (ANVISA) measure, RDC No. 49, enacted in Brazil is also noteworthy. This resolution points out the relevance of regularizing small-scale food production with the necessary health safety measures and with a focus on productive inclusion, preserving and valuing traditional food habits (Thomé da Cruz, 2020). It includes guidelines on the "reasonableness" of the requirements and the simplification of procedures for products considered low risk. This aspect was revived in the disputes associated with the Argentine case.[5]

At local level, the regime of production and marketing of handicraft products of Santa Fe (1997), a Misiones province law and an Entre Ríos resolution, both from 2010, represent important regulatory precedents.

One of the relevant factors to the continuity of these initiatives has, even at different political moments, been the governance schemes that integrate representative organizations from the producer sector. At international level, COPROFAM participates in REAF, and, at national level, the family farming-related SENASA area, in which certain sector representatives participate, was established in response to a call on the part of organizations.

As can be seen in table 17.1, the enactment of national regulations on family farming facilities began in 2012, with regulations for slaughterhouses. Two years later, the highest authority of SENASA decided to address the issue of family farming at CONAL, and organized meetings in different regions of the country to promote areas of collaboration.[6]

Based on this initiative, CONAL decided to create a specific working subgroup in 2015, with the mandate to address the proposal submitted by SENASA to incorporate articles on small-scale processing facilities and family farming into the CAA.[7]

---

5. Other international benchmarks mentioned in the interviews were the FAO's work on exemptions for the family farming sector and a US FDA rule regulating on-farm processing.

6. Prior to this, the organization underwent other internal changes, whereby a system of family farming coordination was established, reporting directly to the presidency, as well as points of reference at the territorial level. The membership of its Family Farming Commission was also modified, specifically incorporating family farming organizations (Resolutions 186 and 187).

7. Representatives of the Ministry of Agriculture, INAL, SENASA, INTA, IPAF, INTI, the Prohuerta Program and 14 provinces participated in the subgroup.

Table 17.1. Family farming food processing milestones and regulations

Year	National Milestones	International Milestones	Initiatives for the regulation of family farming-related facilities	
			National	Provincial
2005	CIPAF (INTA)			
2006	FONAF			
2007	RENAF			
2008	Undersecretariat for Rural Development and Family Farming			
2010	Social Agricultural and Livestock Monotax			
2009	SESAF (SENASA)			
2010				Misiones. Provincial register of processing spaces and registration of artisanal food (Law XVII No. 17)
2010				Entre Ríos. Hygiene-health standards and good manufacturing practices for food processing in the framework of micro-enterprises (Resol. 26 of the ICAB)
2012			SENASA: Regulation of mobile slaughterhouse facilities at a fixed point for local supply	
2014	Law 27118 on Family Farming (December)	SENASA's participation in the REAF WG on Trade Facilitation	Creation of Family Farming Coordination in SENASA. Aimed at addressing the subject of family farming in CONAL (August)	

Year	National Milestones	International Milestones	Initiatives for the regulation of family farming-related facilities	
			National	Provincial
2015			Formation of a working subgroup on family farming in CONAL (April). SENASA presentation on the subject (November)	
2015			Initial proposal to add article on small-scale establishment and family farming to the CAA goes through administrative process but does not result in resolution	
2015			Incorporation of Chapter XXIII on animal products into SENASA's regulations for the inspection of products of animal origin (later complemented with requirements for slaughtering and processing facilities for sausage and pork products)	
2016		Regional Seminar "Health and safety of family farming production and food"		
2017		II Regional Seminar "Food safety in peasant and indigenous agriculture"		
2018			Addition of Article 154-quater on establishments that process and/or market food from family farming activities to the CAA	
2018		International conference on the safety of artisan cheeses		
2019		Conference on agrifood health and safety held within the framework of the XXX Specialized Meeting on Family Farming (REAF)		
2019			CONAL approval of the incorporation of Art. 60 bis on artisanal dairies to the CAA	

Having been submitted to the various bodies required for its approval, the proposal reached the public consultation stage, during which some 300 letters of support for the initiative were received. The project was then submitted into the administrative process but was never signed off by the ministers as a resolution.

With a new national government in power, in 2016, CONAL asked the family farming working group to present guidelines to link them to the project (Act 110). While these were presented, the Ministry of Agroindustry suggested the need for a "new wording" of the article, which was presented by its Undersecretariat for Food, in consensus with the Undersecretariat for Dairy and Family Agriculture in 2017 (Minute 118). At a subsequent meeting, INAL presented the proposed resolution. An agreement was made to set the community processing facilities aside and address them at a later stage. The Ministry of Agroindustry contested part of the proposal's content, which left the authorization of foods that were not specified within the standard to the discretion of each health authority. With this option removed, the project moved on through the successive stages, including public consultation. There, this second proposal of the article received 42 notes of support, with Resolution 13 that incorporates Article 154-quater into the CAA being approved in October 2018.

The innovative aspect of this standard was the inclusion of a new category of food factories and stores: those that produce and/or market food produced through family agrifood activity, which are subject to less stringent building standards than the rest of the facilities.[8] The products covered are also specified; with the implicit exclusion of several types of food (meat and most dairy products). That is to say that Article 154-quater represented a step forward in terms of enabling some, although not all, types of facilities to be authorized.

In this second, defining process, the role of the initiator of the proposal (SENASA's Family Farming Coordination), and of those who supported it from the family farming subgroup, as well as that of organizations outside the Commission, is attenuated. In contrast, the Ministry of Agroindustry, in particular its area related to Food, took on a more prominent role, as did the technical secretariats of both ministries, which are the ones that ultimately drafted the fine print of the standard. In this case, the broader institutional context was decisive in explaining the role of the different state actors:

> The Undersecretariat for Family Agriculture (...) actively participated in the first part, frequently taking the lead and proactively moving it within the Ministry of Agroindustry, with a significant degree of reluctance on the part of Food. In the second instance, the Food department was more involved, keen for it to go through, with minimal participation from Family Agriculture, which was involved in a difficult institutional process in which many people were being dismissed and many teams dismantled (Interview, 2020).

However, it is also evident that the bureaucratic mechanism corresponding to the management of the document itself, which proceeded despite the change in government orientations, was also relevant (and could even be seen as a non-human

---

8. They are exempted from certain obligations in terms of facilities (cloakrooms, separate toilets for both sexes; capacity of 15 cubic meters per operator and perimeter fence). The food processing areas can be in the producer's own home, but they must have a separate entrance and be an area exclusively used for processing (i.e., the use of household kitchens is not permitted).

actor, in terms of actor-network theory). "The discussion about the guidelines for that article remained pending. When the guidelines were presented, they said the guidelines would not progress because the articles need to be revised, so the decision was made to pick it up again" (Interview, 2020).

After the enactment of this regulation, the CONAL subgroup in charge of family farming continued to work on developing specific guidelines by product to promote the implementation of Art. 154-quater throughout the territory, and on the determination of guidelines for family farming community facilities. Meanwhile, in September 2018, the Director of Dairy presented a proposal to CONAL to incorporate an article on artisanal dairy facilities into the CAA (Minute 124). An agreement was made to create another working subgroup to work on its development. The proposal presented to CONAL and CONASE went to public consultation and seven notes of support were received with comments. The final draft went through the administrative process, as Art. 60 bis to be incorporated into the CAA (Minute 131). However, as of mid-2020, the respective resolution had not been approved.

Taking stock of the achievements of the new regulations, a key interviewee of the pro-FF coalition reflected that the exceptions in terms of building requirements are not so relevant, but rather that,

> (…) we really had to work on making several concepts more flexible. And not only in relation to those, which in fact are rather ridiculous exceptions, aren't they? But well, they are the ones that ended up going through. And the ones that enabled us to, at least, flag the issue up for discussion. That's how I see the article now. (…) Like a flag, and the organizations saw it as such and celebrated it a lot in that sense too, as if to say, finally! Even though this article doesn't solve much, it sparked another dialogue about artisanal dairy farms; now that's also being considered too, to progress the idea of the dairy-factory, the community facility. It paved the way for other discussions (…) Its most significant achievement is in starting a conversation, a discussion and getting this committee, which is quite conservative, to start to look at a sector for which the rules of an exporting company don't work, and that does not comply with the floor to ceiling stainless steel scenario, that has a different way of producing, that produces in different places, and that clearly represents an inconvenience for an inspection system that is stretched beyond its capacity (…) There has been a revived sense of understanding in relation to culture, tradition and about the fact that the technologicalization of the processes, these advances, do not necessarily have to threaten and disregard something else that has been done for so many years.

However, the official consulted also went on to highlight the fact that several products are not covered by the standard, and the continued existence of problems regarding food control areas at local level (Interview, 2020). In fact, in terms of the practical impact of the incorporation of Art.154-quater to the CAA, the case of the *Red Cañera* network that connects several family producers' organizations in the province of Misiones was disseminated in February 2019, as the first enterprise approved for federal transit based on these criteria.

Another key interviewee agreed that the importance of incorporating an article on family farming into the CAA lies in the fact that "(…) recognizing that there is a certain group of foods that are produced in a different way, is symbolically very valuable (…) If the Food Code says that there are certain foods that are different, it provides the legal basis to create specific public policies" (Interview, 2017).

Analysis of agreements and disputes: recipient subject and valuation languages used; scope of the rules and concessions contemplated

Examining the entire incorporation process for new standards revealed the existence of different coalitions of actors, framed within a wider institutional process. It is, however, also interesting to take a closer look at the modifications that were made to the content of the instruments to get them approved. In the case of Article 154-quater, the different testimonies revealed a strategy that prioritized the standard's approval – otherwise "it would be buried" – with the aim of addressing any unresolved issues at a later date, once certain commitments had been set in stone by the very existence of the standard.

The analysis of journalistic transcripts, resolutions, documents and minutes related to the multiple proceedings in which different actors involved in the approval of these norms participated, provides insight into the different visions that had to be harmonized (or translated, in terms of the theoretical framework adopted) to crystallize into new norms.

As such, we were able to identify two constellations of key concepts within this whole process of approving new standards aimed at family farming: one that emphasizes the need to promote inclusive processes (in a broad sense), taking into account the role of family farming in supplying the domestic market, which proposes a paradigm shift in terms of the type of producer to support, aiming at the construction of a sector that can guarantee food sovereignty. The argument in this context is that consumers need to be provided with guarantees in terms of food safety, but that these guarantees need to be defined together, from the bottom up. The importance of providing technical and financial support to producers' organizations to enable them to get authorized their processing rooms is also stressed. The softening of standards is seen as a central issue to reduce the amount of investment required to adapt facilities or create new ones.

Another constellation of ideas stresses the need to comply with a single safety standard; that differentiated policies is not about relaxing the rules, but on supporting organizations to raise their standards; that the application of good practices is the key to competitiveness and enabling family farming to access domestic and foreign markets in line with the idea that Argentina should become "the supermarket of the world". The inclusion of family farming is seen, in a more restricted sense, as inclusion in formal trade and the regularization of its activities.

Although it does not seem appropriate to see both views as absolute opposites and corresponding to different political eras, in our interpretation, this second view of family farming gained predominance as the implementation of a "pro-market" economic policy strategy progressed at national level. This can be seen in certain, sometimes subtle, redefinitions of the initial proposals.

In this regard, one particular point refers to the subjects and products covered, and its implications in terms of the scope of the rules. In relation to the subjects, the first version of the article to be incorporated into the CAA (154 tris) referred to "small-scale and family farming facilities" and "low-risk foods and processes," produced both in a family and community way. In each case, "health and hygiene requirements and key good practices" are required as part of the processing process. The presentations also stress the need for the ingredients to be self-produced or to come from family farming, for the production processes to be predominantly manual, and the producers registered in the RENAF.

In contrast, in the final version incorporated into the CAA (Article 154-quater), although the existence of the Family Farming Law establishes a need and the importance of this subject for the "food security and sovereignty of the people," its scope refers to "family agrifood activity," excluding any reference to the subject and the scale of production. The same happens with the issue of risk: the products covered by the standard are implicitly low-risk products. As a key interviewee explains,

> (…) it was not the text that we thought was needed…it was a text for negotiation, a text that would make it through the agreements. There was a lot of reluctance to making any kind of exception (…) The question of the subject, whether we identified the subject as family farming or not, or whether we talked about a facility, without saying that it was for a particular sector, that was a whole dispute and that was what caused the first proposal that had been approved to fail (…) In 2015, the Commission approved it, it went to signature, there was a change of administration. And the following year, it was decided that no, that it had to be revised, that what had been approved would not go through as it was, that the undersecretaries were not going to sign it, and so the issue returned to the Commission, and we worked on it again with a different wording. (…) The text talks about family agrifood activity, to avoid saying family farming (Interview, 2020).

This modification implies that the processing facility does not necessarily have to belong to a family farmer registered as such. Moreover, the pro-FF coalition advocated the idea of not specifying products, but rather stipulating processes and putting more emphasis on the handmade and artisanal aspect. In fact, several of the notes from producers who provided feedback during the public consultation stage of the article emphasized that: i) the standard represented progress, but that ii) all products needed to be incorporated, in addition to those considered low risk (with emphasis on fresh cheeses) and that iii) safety was more about the processes than the infrastructure used. This aspect was reiterated during the consultation stage of the artisanal dairy article, where there was also a request to move forward with specific standards for products made with unpasteurized milk.

In this sense, the pro-Food Code coalition (made up of the Food Secretariat, several provinces, and supported by the organizations that make up CONASE) proposed that a list of authorized products should be drawn up at the outset, respecting the "spirit" of the treaty. But, as one interviewee pointed out :

> (…) the drafting of the Code began in 1969, and some of the perspectives considered still reflect that time and space. It doesn't so much look at processes, but it says yes to this and no to that, and it's very restrictive and draws up positive lists. So, let's make a positive list. What happens is that this positive list ends up becoming a prison (…) There were several presentations from INTA, they asked us to extend it – when it went to public consultation – to soft cheeses, to include all the production from goat farmers in the north. When the proposal was submitted, they said the article should be submitted like that or otherwise it would never see the light of day. And we worked on that separately (…) Those of us who were in the group, when they gave us the list,

> the only thing that concerned us was that the main [products] should be included. At the beginning, yerba mate, honey, chocolate, which are low-risk products, were not on the list. (Interview, 2020)

In the specific case of Article 60 bis on artisanal dairy facilities, the initiative for its incorporation into the CAA could, in principle, be interpreted as a consequence from the rule on family farming facilities. It contemplates the same flexibility in terms of facilities as Article 154-quater, to which certain specific requirements are added in terms of processes, such as that the milk be sanitized and subjected to heat treatment (not necessarily pasteurized) and that the direct intervention of the processor be a "substantial" component of the production process. The milk may be the operator's own or from third parties, with a production scale of up to 5000 liters per day and 1000 liters per operator. To gain an understanding of its scope, it is worth noting that the article talks about "the rural and peri-urban population that constitute the broad universe of family agrifood production," without any reference to the Family Farming Law. Neither is artisanal processing associated with this actor, nor does it explicitly consider small dairy farmers that process by-products (scenarios that have been considered as part of some provincial regulations).

Similarly, and in contrast to the situation in 2014–2015, the Family Farming Law could be said to be waning as a device for the agreement of actors, while the more traditional script of the CAA was reaffirmed. In any case, this "translation" was seen by the pro-family farming coalition as the tool that made it possible to enlist reluctant actors within the sector:

> What we tried to do, and I think that was the key to breaking the deadlock, in the first proposal [of Article 154] and in the second one as well, was to talk in the Code's terms (…) There is a kind of culture of maintaining the spirit of the Code. It's very difficult for different proposals to succeed, especially when there are certain interests at stake (…). Nothing to do with the original proposal that included things from other countries and some international organizations, but that could not go through (Interview, 2020).

Therefore, we cannot rule out the adoption of an instrumental approach, which favored focusing on reaching agreements on a small number of "technical" guidelines for inclusion in a national standard – that could eventually be improved on or supplemented by others – rather than prolonging a discussion that would run the risk of being abandoned. The approval of a norm gives legitimacy to the collective effort and enables the actors to continue working.

In addition, and looking at the process as a whole, a certain degree of convergence of key actors can be observed around some common denominators, which can be summarized as follows: the need to consider small-scale producers because they exist, and the fact that it is not enough to pass regulations tailored to their situation but that these should also be complemented by other actions, such as training and access to funding.

▸▸ Conclusions

In this chapter we analyze the process of policy formulation in Argentina that has involved the incorporation of differential criteria for the commercialization of food coming from family farming facilities. The pre-existing normative base in the country is founded on previously established international standards, such as the Codex Alimentarius and other global standards that have been adopted by the dominant players in the system and cement the existence of global markets. These standards not only regulate these markets but also those in the domestic sphere, with a tendency to favor those actors that reach a certain productive scale and have access to both material and symbolic resources.

Underlying this entire process of policy formulation is a series of stages related to the possibility of getting the family farming issue onto the agenda of the relevant institutional departments responsible for reviewing the health standards (CONAL), a task in which some areas of the SENASA have played a central role. As such, they have managed to enlist other actors at national and provincial levels, who have, in turn, acted as spokespersons for others, whose participation is not contemplated by the Commission. The pro-FF coalition had to engage in a negotiation process where disputes about the issue and products covered were resolved. Although this negotiation led to a positive result (the incorporation of an article in the Argentine Food Code), the effective scope of the standard remains an object of dispute that to some extent continues to influence the development of subsequent standards.

Throughout this journey, different translation operations of the initial proposal were carried out, in which certain normative benchmarks from other countries were relegated to second place, although the international interest in the issue channeled through the REAF Mercosur served to ensure the matter remained on the table. As a result of these translation operations, the pro-FF coalition agreed to frame itself within the linguistic parameters of the Argentine Food Code. This is exemplified by the explicitness of the products benefitting from the relaxation of requirements (and implicitly the exclusion of others, such as meat and other products considered high risk). This was due to the more general institutional context given by the change in the orientation of the national government, with emphasis on the promotion of competitiveness and the insertion of Argentina in foreign markets. There was, therefore, a differential weight corresponding to the two coalitions throughout the different stages, where the weakening of the pro-FF coalition was not unrelated to the more general dismantling of the support structures for the sector.

Notwithstanding this, the existence of the Family Farming Law throughout the entire process analyzed, as well as of a collaborative network that was not necessarily visible, together with the bureaucratic mechanism of CONAL's own handling of the issues, made it possible to sustain the proposal to incorporate this standard into the Code, producing a sort of lock-in that ended some discussions. More substantial modifications would probably require a change in the existing institutional mechanism, incorporating actors currently absent within CONAL or CONASE, as well as the reinforcement of strategies to support the sector.

Although these policy issues seem relatively abstract (as some of the actors interviewed pointed out), it is worth noting that they have implications in terms of possible transformations of the food system. The family farming sector supplies the major cities but plays a central role in local supply. However, the entire existing regulatory corpus in Argentina does not consider the possibility of differentiating the procedures aimed at ensuring food safety according to the geographical distance of food processing facilities, and even in trading areas close to them compliance with the most demanding national standards may be required. This latter point is particularly relevant in situations such as those created by the Covid-19 pandemic, where local production became particularly important, in parallel to the health issue and the protocols to be applied.

Safety has been the predominant *référentiel* within all the discussions analyzed, although some actors have tried to link it to that of food sovereignty and security, where family farming is no longer seen as a "problem" but as a possible "solution." Other issues, such as the differential characteristics of these foods by virtue of their artisanal processes and/or the reduced use of chemical inputs (resulting in healthier, not only safer foods) compared to those produced by large companies are still pending; their consideration would undoubtedly give rise to another discussion.

▸▸ References

Ariosti, A., & Olivera Carrión, M. (2014). Food law in Argentina. In Kirchsteiger, E. & T. Baumgartner (Eds.), *Global food legislation: an overview* (pp. 1–21). Wiley.

Bertoni, L., & Soverna, S. (2018, November 25–30). *Evolución de la asistencia técnica y la extensión rural en la Secretaría de Agricultura Familiar, Coordinación y Desarrollo Territorial del Ministerio de Agroindustria de la Argentina (MINAGRO)* [Conference presentation]. X Congreso de la Asociación Latinoamericana de Sociología Rural (ALASRU), Montevideo, Uruguay.

Cairney, P. (2015). Paul A. Sabatier, "An Advocacy Coalition Framework of Policy Change and the Role of Policy-Oriented Learning Therein." In M. Lodge, Page, E. C., & S. J. Balla (Eds.), *The Oxford Handbook of Classics in Public Policy and Administration* (pp. 484–497). Oxford University Press. http://www.doi.org/10.1093/oxfordhb/9780199646135.013.24

Callon, M. (1984). Some elements of a sociology of translation: domestication of scallops and the fishermen of St. Brieuc bay. *The Sociological Review, 32*(S1), 196–223. https://doi.org/10.1111/j.1467-954X.1984.tb00113.x

Craviotti, C. (2015). Regards croisés autour de la légitimation de la catégorie «agriculture familiale» et ses défis conceptuels (A. Ramo, Trans.). *Bulletin de l'Association des Géographes Français, 3*, 322-337. https://doi.org/10.4000/bagf.683

Craviotti, C. (2017). Dilemas en iniciativas de desarrollo orientadas a la agricultura familiar: los productores-elaboradores de quesos en Entre Ríos, Argentina. *Revista de Ciencias Sociales, 30*(41), 199–220.

Cruz Lucero, J., & Padín, J. M. (2019). Hacia una orientación de la política comercial: la construcción del "supermercado del mundo. *Realidad Económica, 48*(325), 91–118. https://ojs.iade.org.ar/index.php/re/article/view/68/16

De Anchorena, B. (2018). Poder empresario y políticas públicas. La captura de las políticas agropecuarias. In García Delgado, D., Ruiz de Ferrier. C., & B. de Anchorena (Eds.), *Elites y captura del Estado. Control y regulación en el neoliberalismo tardío. (2015–2018)* (pp. 183–208). FLACSO. https://www.flacso.org.ar/wp-content/uploads/2018/11/Flacso-Elites-y-captura-del-Estado.pdf

De Haro, Augusto. (2014, October 6–11). *Regulaciones globales y mercados locales. Adaptaciones y controversias* [Conference presentation]. IX Congreso Latinoamericano de Sociología Rural, Mexico City, México

Flexor, G., & Grisa C. (2016). Políticas de seguridad alimentaria y agricultura familiar en Brasil: actores, ideas e instituciones. *América Latina Hoy, 74*, 39–53. https://doi.org/10.14201/alh2016743953

Ghezán, G., Mateos, M., & Cendón, M. L. (2013). Redes y controversias en torno a la valorización de alimentos en el partido de Tandil. *Revista de la Facultad de Agronomía de La Plata, 112*(3), 23–35. http://revista.agro.unlp.edu.ar/index.php/revagro/article/view/160

Jacoby, E., Tirado, C., Diaz, A., Peña, M., Sanches, A., Coloma, M. J., Rapallo, R., Rodríguez, A., Sotomayor, O., Arias, J., & Courtis, C. (2014). *Una mirada integral a las políticas públicas de agricultura familiar, seguridad alimentaria, nutrición, salud pública en las Américas: acercando agendas de trabajo en las Naciones Unidas.* OPS-OMS, FAO, CEPAL, IICA. https://www.fao.org/fileadmin/user_upload/rlc/eventos/231982/doc_20140509_es.pdf

Latour, B. (2008). *Reensamblar lo social. Una introducción a la teoría del actor-red.* Manantial.

Lendvai, N. & Stubbs, P. (2012). Políticas como tradução: situando as políticas sociais transnacionais. *Práxis Educativa, 7*(1), 11-31. https://doi.org/10.5212/PraxEduc.v.7i1.0001

Montón, D. (2018, August 17). *Argentina: La agricultura familiar, entre el vaciamiento y la disputa campesina por la tierra.* Biodiversidad en América Latina. http://www.biodiversidadla.org/Documentos/Argentina_La_Agricultura_Familiar_entre_el_vaciamiento_y_la_disputa_campesina_por_la_tierra

Muller, P. (2005). Esquisse d'une théorie du changement dans l'action publique. Structures, acteurs et cadres cognitifs. *Revue Française de Science Publique, 55*, 155–187. http://www.doi.org/10.3917/rfsp.551.0155

Nogueira, M. E., Urcola, M. A., & Lattuada, M. (2017). La gestión estatal del desarrollo rural y la agricultura familiar en Argentina: estilos de gestión y análisis de coyuntura 2004–2014 y 2015–2017. *Revista Latinoamericana de Estudios Rurales, 2*(4), 25–59. http://www.ceil-conicet.gov.ar/ojs/index.php/revistaalasru/article/view/273

Sabourin, E., Craviotti, C., & Milhorance, C. (2020). The dismantling of family farming policies in Brazil and Argentina. *International Review of Public Policy, 2*(1), 45–67. https://doi.org/10.4000/irpp.799

SENASA. (2014). *Informe técnico sobre los productos de la agricultura familiar en Argentina. Propuesta de trabajo de la Comisión de Agricultura Familiar del SENASA.*

Thomé da Cruz, F. (2020). Agricultura familiar, processamento de alimentos e avanços e retrocessos na regulamentação de alimentos tradicionais e artesanais. *Revista de Economía e Sociologia Rural, 58*(2). https://doi.org/10.1590/1806-9479.2020.190965

Vigil, C. J. (2019). *Agricultura familiar campesina e indígena en la Argentina, 2004–2017.* COPROFAM, FAA.

Zurbriggen, C. (2006). El institucionalismo centrado en los actores: una perspectiva analítica en el estudio de las políticas públicas. *Revista de Ciencia Política, 26*(1), 67–83. https://doi.org/10.4067/S0718-090X2006000100004

Public procurement from Uruguayan family producers and fishers and cross-cutting rural development policies

JUNIOR MIRANDA SCHEUER

▸▸ Introduction

The Oriental Republic of Uruguay, a denomination the country owes to its geographical position to the east of the Uruguay River, borders southern Brazil (the state of Rio Grande do Sul – a dry border) and eastern Argentina (the provinces of Entre Ríos and Corrientes – a border formed by the Uruguay River). To the south of the country lies the River Plate (the estuary of the Uruguay River), which separates the Uruguayan capital of Montevideo from the Argentinian capital of Buenos Aires.

Uruguay is one of the smallest countries in South America. Covering an area of 176,000 km^2, it has 3.5 million inhabitants (5% rural), a Human Development Index (HDI) of 0.808 (very high human development), a GDP of USD 60bn, and a per capita income of USD16,200. It is also seen as a country with an equal society, broadly stable institutions, a high degree of confidence in government, and low levels of corruption, and it has the highest levels of well-being in the region (INE, 2020; UNDP, 2020; World Bank, 2020).

Uruguayan farming, which makes use of just over 16 million hectares, is a driving force of society and the national economy and is characterized by its high food production capacity. According to studies by Mondelli (2014) and Gómez Perazzoli (2019), this small country produces more than domestic demand, with family farming leading the way, and provides food for 20 to 28 million people all over the world.

According to Fernández et al. (2015) and Pereyra (2019), the country currently has sufficient expertise in energy and protein generation (production of 3,206 kcal/inhabitant/day, as opposed to the requirement of 2,180 kcal/inhabitant/day – Torres Ledezma, 2010; FAO, 2020), though this was not always the case. In a brief analysis of food problems, Pereyra (2019) pointed to the fact that one of the main difficulties faced between 1900 and 1990 (economic models that alternated between growth and recession – crises and periods of authoritarian rule [1973/1985] – Lanzaro, 2004) was malnutrition. Following a period of improvement in the *référentiels* (the return of democracy), the economic and financial crisis of 2002 saw the threats of poverty and malnutrition reappear. These gradually receded after 2005 (progressive government) only to resurface with the global recession of 2008/09 (Hristoff & Saravia, 2010). This brief historical review must also consider the effects of the Covid-19 pandemic, though more in terms of access to food than production capacity.

According to the latest reports published by the Food and Agriculture Organization (FAO), the International Fund for Agricultural Development (IFAD), the World Health Organization (WHO), the World Food Program (WFP), and the United Nations Children's Fund (UNICEF), extreme poverty affects approximately 0.3% of the population in Uruguay, poverty 9.4% (mainly in peripheral urban and rural areas), severe food insecurity 7.6%, and moderate to severe food insecurity 25.3%, while the proportion of undernourished people is 2.5% (FAO et al., 2018; FAO et al., 2019).

Obesity (over 18s) is also a cause for concern in Uruguay, having risen from 12.9% of the population in the 1980s to 27.9% in 2016. It is now regarded as the biggest public health problem in the country (consumption of highly processed food stands at more than 15% over the regional average of 129.6 kg per capita year) (FAO et al., 2018; FAO et al., 2019). Access to food in quantity and quality is also revealed as a serious threat to food and nutrition security (Comisión Especial de Población y Desarrollo, 2014; Pereyra, 2019; Torres Ledezma, 2010).

The gradual reduction of poverty and food insecurity and the alternatives designed to mitigate obesity are the result of a set of public policies adopted over time. The most notable of these are as follows (Calanchini et al., 2017; Pereyra, 2019):
– 1920: School Feeding Program, aimed at supplementing and providing food aid to students in rural areas and open-air schools (a glass of milk) and later extended to urban areas (1930);
– 1942: National Food Institute (INDA – current name), which had the role of protecting the most vulnerable members of society and responding to nutrition-related issues;
– 1987: At-Risk Nutrition Program, which promoted health and nutrition in families with children under 18;
– 2005: Creation of the Ministry of Social Development (MIDES) with the aim of promoting social equity and wealth distribution (it later absorbed INDA – action on food and nutrition vulnerability);
– 2005: Food Guide for the Uruguayan Population, reprinted in 2016 and 2019, which aims to inform people about a healthy diet – involving enjoyable, shared meals – with a view to reducing diseases caused by nutrition deficit;
– 2006: 'Tarjeta Uruguay Social' (Social Uruguay Card), aimed at people who have problems securing a basic amount of food and also designed to ensure recipients achieve food and nutrition security;
– 2008: Food and Nutrition Security Observatory (ObSAN), a tool created to provide information on food and nutrition security;
– 2014: Act No. 19.292, dated 16 December 2014, which establishes a market reserve mechanism for agricultural food goods and services from family farming and artisanal fishing.

State procurement in Uruguay is regulated by the Consolidated Accounting and Financial Administration Text (TOCAF – Ministry of Economy and Finance – MEF) and follows a preference scheme, i.e. a device that directly prioritizes certain socioeconomic sectors or companies in order to supply the state through public contracting. There are currently four schemes: Preference for National Industry (goods and services of national origin), Public Procurement Development Program (national suppliers – micro, small and medium-sized enterprises and small agricultural producers), Public Contracting for the Development of the Textile and Clothing Industry (stimulus for the national textile and clothing industry), and Family Farming and Artisanal Fishing (ARCE, n.d.).

It should be pointed out that Uruguayan public procurement does not solely involve foodstuffs from family farming. It is a tool for supplying different products to the state (food, hygiene, office supplies, software, inputs, etc.) and follows the regulations and budgets set out in the TOCAF. The only type of public procurement that has been researched is solely and exclusively food purchases from rural family producers.

Given that rural families are fundamental to food production, food and nutrition security, and rural/economic development (Chiappe et al., 2015; Machado et al., 2018), and that the Public Procurement Act (LCP) is structured around the marketing of family production and artisanal fishers, three questions should be asked: i) Was the framework of the act designed from the perspective of sustainable food production? ii) Which public policies are linked to the institutional market and are capable of promoting sustainable development for rural families? iii) Does this set of policies shape and/or promote sustainable food systems for family producers?

The aim of the research was thus to understand the operational framework of the Public Procurement Act and to ascertain how it interacts with other government policies promoting the sustainable rural development of the food production of Uruguayan family producers and artisanal fishers.

The research, which was qualitative in nature, was based on a review of the LCP's normative texts and the regulatory decree, relating it to an understanding of the institutional framing of the agenda, the structure of the problem that the LCP sought to minimize/remedy, the objective of public procurement, the target audience for which the action was created, the general composition of the LCP's structure, the marketing data collated to that point, the setting of food prices, and public government transparency regarding the implementation of the LCP.

For the second part of the objective, we analyzed the state framework to identify the main public policies that in some way interact with the LCP and that permeate, as far as possible, sustainable food systems. As well as highlighting procedures, we examined the state of the art in related themes. The linking, interaction and/or set of policies is a departure from the concept of policy mix, i.e. the ability of a set of public instruments to achieve a certain end (Milhorance et al., 2019; Rogge & Reichardt, 2013). In this case, the "end" relates to assessment of the LCP, and other policies, as mechanisms for the sustainable promotion of food by rural households.

Aside from the introduction of the topic, the objective, and the research methodology, this chapter comprises three other sections:

The first of them presents the operational framework of the LCP in terms of agenda, problem, objective, target audience, structuring, marketing, pricing, and public transparency.

The second section discusses a mix of policies that are linked to the LCP, such as the definition of family producer and artisanal fisher, the National Agroecology Plan, food and nutrition security, the School Feeding Program, food safety, the Family Producer Seal, Rural Development Forums, and other public policies.

Finally, we provide some passing thoughts on the LCP and the policy mix.

▸▸ Understanding the operational framework of the LCP

The public procurement agenda

Uruguay's Public Procurement Act has antecedents in the policies developed by neighboring countries. A concept discussed by Solanas (2009), "Mercosurization" refers to the dissemination of cognitive, normative, organizational and strategic processes instigated by the political structures and public policies of the members of the Southern Common Market (MERCOSUR) in other countries in the region.

According to De Torres et al. (2018), the agenda around institutional markets and the sale of family farming produce (among other public policies – Grisa & Niederle, 2018) was made possible by the Specialized Meeting on Family Farming (REAF – a political dialogue between MERCOSUR governments and organizations), and a line of research conducted by the Cooperative Program for the Development of AgriFood and Agro-Industrial Technology in the Southern Cone. The issues presented at the REAF have become regulatory instruments (proposals and recommendations) for differentiated public policies for family farming, which may or may not be taken up by member states or parties (Grisa & Niederle, 2018). In 2010, state procurement policies were highlighted in the REAF and Brazil, a pioneer in the field with the Food Procurement Program (2003) and the National School Feeding Program (introduced in the 1950s and reformulated in 2009), with the experience acquired being passed on in exchanges with government officials and civil society in Argentina, Chile, Ecuador, Paraguay and Uruguay (REAF, 2014).

At a local level, policies relating to rural families are disseminated through the General Directorate of Rural Development (DGDR) of the Ministry of Livestock, Agriculture and Fishing (MGAP) and social organizations such as the National Commission for Rural Development and the Association of Rural Women of Uruguay (De Torres et al., 2018).

Grisa and Niederle (2018, p. 18) argue that the Uruguayan agenda was implemented through dialogues promoted mainly by the REAF and the "(…) debate on public procurement in the region (…) led by Brazil [ideological leadership – Ramos, 2019], on the basis of the [Food Procurement Program and the National School Feeding Program] raised (…) in line with the needs of each country and the political agenda (…)".

The agenda in question was shaped by the Uruguayan executive branch in a pilot project implemented between 2009 and 2010 as part of the Rural Uruguay Program (Chiappe et al., 2015). The experiences acquired (local and international) enabled an interinstitutional team (the MIDES and the MGAP – with the DGDR playing a leading role) of the executive branch to draft a bill (Machado et al., 2018), with discussions then being taken up by the legislative branch of the Population and Development Commission. This culminated in 2014 in the passing of the Public Procurement Act (Act No. 19.292) (Parlamento del Uruguay, 2014).

The structure of the LCP problem

Article 1 of Act No. 19.292 and regulatory decree No. 86, dated 27 February 2015, state that the production of family farming and artisanal fisheries is of general interest. The official presentations of the LCP contain no specific explanation of the problems that the act seeks to address.

In addressing the problem (based on other studies) there are two points that stand out: production capacity and marketing. With regard to the first point, family producers account for 62% of agricultural establishments in the country (25,285 farms across 15% of the area farmed) and are responsible for 88% of pork production, 84% of poultry, 80% of vegetables, 78% of fodder seeds, 73% of milk, 69% of viniculture, 66% of sheep, 59% of cereal seeds, 58% of cattle, and 38% of fruit (Leporati et al., 2014; MGAP, 2011; Sganga et al., 2016), among others. According to De Torres et al. (2018), much of the food consumed in Uruguay originates from family farming. It can thus be said that while there is production competence, family farmers are hindered or restricted when it comes to selling their produce/putting it on the market. The second point addresses the structural problem of family production, taking into account its limited competitiveness in relation to non-family producers and the ability to negotiate with buyers, with marketing to intermediaries seen as a solution to this (Machado et al., 2018). The consequence of this state of affairs is a perception of prices below the market level (less than half the retail price) and limited access to certain points of sale (Mila, 2015).

In looking at the two points together, it could be inferred that policy makers identified the commercial limitations of family producers and the payment of low prices for their production as a problem. The LCP plays its part here by facilitating the flow of production and allocating part of family farming production to state-run institutional markets (short supply chains), reducing the effects of commercial oligopolies (Chiappe et al., 2015) on price formation and market ownership.

Purpose of the LCP

The stated objective of the LCP is to enable family producers and artisanal fishers to flourish by allowing them to access better prices for what they produce (reduction of marketing intermediaries) and through the creation of a state-wide market reserve (guaranteed sales to the state as part of short supply chains). Such measures allow for planning in the production system (production/marketing) and closer organizational ties with the target audience (social capital) (Centro de Información Oficial, 2014).

Although the LCP aims to promote family producers and artisanal fishers, it is understood that its main objective (interpretation and disclosure of arguments throughout the text) is to ensure the supply of foodstuffs to the state through the procurement of family production, which has certain priorities in terms of marketing.

Policy makers have not, therefore, determined the mechanisms or types of specific support for promoting family production. Article 3 of the LCP explicitly states that the market is reserved for the target audience but adds that this reserve will be maintained unless there is supply (Centro de Información Oficial, 2014). This passage exempts the government from committing itself to promoting family production through specific public policies or a set of policies (policy mix), including food and nutrition security, production sustainability (organic or agroecological), technical support, training, rural credit, etc.

The decree states, however, that the MIDES is responsible for promoting and providing support to socially vulnerable family producers and artisanal fishers (a specific group) (Centro de Información Oficial, 2015), but does not specify any actions or link them to other government policies and/or programs. This somewhat confused role assigned to the MIDES is disproportionate to its capabilities. It is the MGAP/DGDR that has the tools (mainly financial and human) to develop sustainable agrifood systems for family farmers.

Target audience of the LCP

By way of context, Uruguay refers to the family unit as "family producers", unlike Brazil and other countries in the region, which use the term "family farming" (*campesina*). This terminology (widely recognized in institutional regulations) is often used erroneously by government agencies and even academics.

To avoid confusion and conceptual frustrations, we define family producers and family fishery producers in line with the text of MGAP resolution No. 1.013, dated 11 November 2016 (revised version of the first resolution of 2008). "Family producers" are understood as being natural persons with direct responsibility for the management of an agricultural establishment and/or production activity allowing for the hiring of up to two wage earners (or 500 hours of work per year), with a land area not exceeding 500 hectares, the family residence located on the property or nearby (within 50 km), and whose additional property income cannot exceed a monthly average of USD500 (value set out in 14 Benefits and Contributions Thresholds – BPC) (MGAP, 2016a).

Family fishing producers are natural persons directly responsible for artisanal fishing activities (vessels under 10 tons and in possession of a valid fishing permit) or for fishing from land (permit for fishing from the banks of waterways), who do not have more than one permit (artisanal fishing or fishing from land), most of whose family income comes from fishing, and who do not hire more than three non-family wage earners or the equivalent of an annual workload of 1,250 hours (MGAP, 2016a).

The LCP target audience is composed solely of family farmers and artisanal fishers with links to a company (registered, development and agricultural), association (agricultural) or cooperative (production, social and agricultural), composed of at least five producers, with at least 70% of them of family origin. Such organizations are known as Qualified Organizations (OHs) and such groups of producers must apply to the DGDR to register an OH in the National Registry of Qualified Organizations (RENAOH), which grants the right to sell part of what they produce to the state (Centro de Información Oficial, 2014).

Groups of family producers and/or artisanal fishers that are organized (OH) and registered (RENAOH) are required to report the type of agricultural production they engage in and their expected potential, the seasonality of their produce, and compliance with Best Agricultural Practices. Organizations are asked to register in the Single Registry of Providers to the State (Centro de Información Oficial, 2015).

It is noteworthy that the LCP requires the target audience (family producers and artisanal fishers) to become organizations, or, in Mila's interpretation (2015), collectives. This requirement encourages the creation of associative and cooperative ties between participating groups, resulting in the construction of social capital, the creation of social actors with "powers" to demand new avenues and/or amendments to the LCP (though there is no means for this in the LCP, it is present in the Rural Development Forums as a cross-cutting element – subheading "3.7. Rural Development Forums") and, according to Machado et al. (2018), in the organization of families as a strategy for economic and rural development.

Although the target audience is outlined, the LCP and the decree do not set out criteria for prioritizing the potential beneficiaries of public procurement. In Article 5 of the LCP (Centro de Información Oficial, 2014, p. 1), it is stated that "(...) at least 70% [of the OH] must be made up of family farmers and/or artisanal fishers". This means to say that there is legal margin in the LCP for the inclusion of agricultural producers who are not "family farmers", as defined in MGAP resolution No. 1.013 (MGAP, 2016a).

The LCP thus sets out a margin of 30% (excluding the 70% made up by family producers) of producers not defined as family producers in the OH. The regulatory decree of the LCP (Centro de Información Oficial, 2015) establishes the target audience as *family producers* (our emphasis) and artisanal fishers duly registered and accredited in the Register of Family Producers and in the Register of Family Fishing Producers (DGDR tools for the recognition of family status – see subheading "3.1. Definition of family farmers/producers and artisanal fishers"). Furthermore, one of RENAOH's functions is specifically to "(...) carry out the necessary checks so that the true 'beneficiaries' of the system are mostly family producers" (Mila, 2015, p. 14).

In line with analysis and in an update of the LCP in 2018 (Act No. 19.685, dated 17 October 2018 – Centro de Información Oficial, 2018a), an indirect prioritization criterion of the target audience can be inferred through the encouragement given to women to take part in institutional markets (see "2.5. Operational framework of the LCP"), provided that the female audience performs operational functions in the organization and in the production system itself.

The lack of an specific prioritization criterion in the LCP may be attributable to the intention of the LCP to largely benefit family producers (and some non-family producers), regardless of their socioeconomic situation. This hypothesis is unlikely, owing to the LCP's objective of not giving due prominence to the sustainable rural development of family producers, with the main role involving the supply of food to state institutions.

In addition, the budget and/or funding for public procurement (conditional on the procurement mechanism and the prices detailed in the TOCAF) are not set out primarily for the target audience. In other words, the LCP does not have a budget of its own for the procurement of food from family producers but has funding for state procurement in general.

Operational framework of the LCP

The LCP is essentially structured in two ways: centralized and decentralized procurement. In centralized mode, purchases are made by the Centralized Procurement Unit (UCA) of the MEF, with a partial market reserve of at least 30% of the procurement of foodstuffs from family producers and artisanal fishers (in relation to the funding sum). The UCA (the centralizing mechanism) issues an annual public call for tender for the procurement of products (in this case, foodstuffs) to supply state bodies that wish to be included in this mode (which is autonomous in terms of decision making) (Centro de Información Oficial, 2014).

In the decentralized mode, purchases are made directly from the target audience (family producers and artisanal fishers). The market reserve reaches 100% (of funding), regardless of the tool used (tender or direct purchase) and pricing (Centro de Información Oficial, 2014). In conjunction to both modes, Act No. 19.685 established a minimum market reserve for Gender-Equal Qualified Organizations (OH+G) of 50% in both centralized and decentralized procurement processes (Centro de Información Oficial, 2018a).

Purchases are made through one of the following mechanisms (depending on their cost): public tender, shortened tender (common and extended), direct purchase (common and extended) and purchase by exception. Procurement is managed (own procedures and operations) by the State Contracting and Procurement Agency (ARCE – which enjoys technical autonomy) or the UCA (ARCE, 2019; Chiappe et al., 2015; Mila, 2015).

State procurement is coordinated through the joint intra- and inter-institutional efforts of the MGAP and DGDR, responsible for the OH and RENAOH and for support/information on the origin of agricultural products; the Directorate General of Farming Establishments (support/information on bee products and fresh and/or processed fruit and vegetables); the Agricultural Observatory (which determines foodstuff prices); the National Directorate of Aquatic Resources (support/information on products from artisanal fisheries); the MGAP and the MIDES (intra-institutional communication of the LCP and development of the capabilities of OHs); the MEF (a body represented by the UCA); and the ARCE (support/information and processing procedures and procurement), with each performing particular activities and allowing the LCP to function.

The implementation of state procurement follows the top-down concept (vertical governance), with actors from the institutional hierarchy to the target audience determining and rolling out LCP decisions and actions (Scheuer & Vassallo, 2019). Although the participation of civil society (Carvalho, 2011) in the LCP (previously established rules and resources – TOCAF) involves friction, the Rural Development Forums (subheading "3.7. Rural Development Forums") provide a platform for debates that can, in theory, lead to structural changes in the LCP.

Executive, legislative and judicial branches, bodies with functional autonomy, decentralized services, autonomous entities and state governments are required to follow the aforementioned modes (Centro de Información Oficial, 2014) (figure 18.1).

A broad section of the Uruguayan state is required to comply with the LCP's guidelines. According to a survey conducted by Chiappe et al. (2015), the largest volume of foodstuffs were requested by the Ministry of National Defense (Army, Navy and Air Force); the Ministry of the Interior (prisons – around 10,000 inmates – police hospitals, the fire service, central offices, National Police School, Republican Guard); the State Health Services Administration (approximately 37 hospitals); the National Administration of Public Education (around 255,000 students benefit from the School Feeding Program); and the National Food Institute (meals for more than 10,000 people and 49,000 children up to the age of three), among other institutions.

Marketing and pricing

Family production marketable through the LCP is divided into four groups: i) natural produce; ii) artisanal processed products; iii) agroindustrial processed products; iv) artisanal fishing products. With regard to processed food, it should be pointed out that the LCP states that the source or intermediate product and processed or final product must originate from the same establishment belonging to the OH (Centro de Información Oficial, 2014).

As regards the processed foods group, Article 4 of the LCP regulatory decree provides for a compromise with the provisions of Article 3 of the Act: "(…) at least half of the supply must come from family producers belonging to the Qualified Organization" (Centro de Información Oficial, 2015, p. 2). This passage allows for reinterpretation. In other words, it does not make the source of the food clear, whether it originates entirely from family producers (the act) or only 50% of it originates from them (the decree).

The safety of foodstuffs is achieved through the health and food standards set out in the Best Agricultural Practices. HOs must fulfil this requirement if they are to register as suppliers to the state. However, there are no restrictions in the LCP and the decree – beyond the aforementioned groups – as to which types of products may be sold. The types of food are only specified at the request stage by the UCA and ACCE, which somewhat undermines the LCP's objective of allowing family farmers to plan their production system, while also disregarding issues linked to sustainable food systems.

Food prices, which are slightly above market prices, are usually set in line with the prices published in the Agricultural Observatory. For fruits, vegetables and eggs, the pricing of products cannot exceed 40% of the pricing published by the Observatory and, for those not included, the respective amount is limited to the average price of the Consumer Price Index of the National Institute of Statistics (a public body). Should it not be possible to set amounts using these tools, the MGAP (executive branch) will set the best price in line with circumstances. In exceptional situations of high food price rises (above 25%), the state reserves the right to suspend public procurement in part or in full (Centro de Información Oficial, 2014).

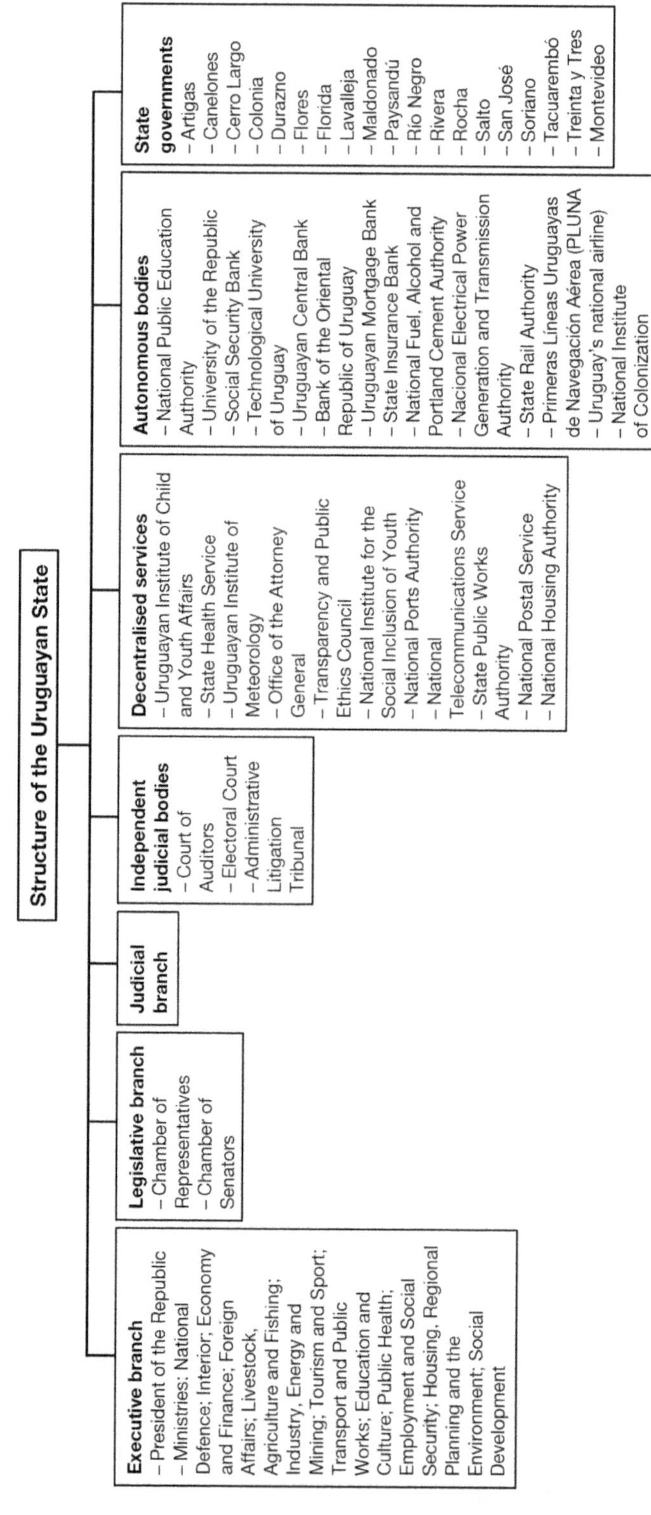

Figure 18.1. Structure of the Uruguayan State.

There are currently 15 OHs. The food they sell to the state is of agricultural origin (bee products, fruit, vegetables, fodder, cereals, grains, mushrooms, pulses, and eggs), processed food and beverages (fresh and processed red meat, fresh and processed white meat, cold cuts and sausages, frozen and/or processed food, canned and dried food, milk and dairy products, non-alcoholic beverages, alcoholic beverages), and fish (unprocessed fish and seafood). In geographical terms, access to public procurement is afforded to family producers and artisanal fishers in the states of Canelones, Flores, Florida, Montevideo, Paysandú, Salto and Soriano (37% of the national territory) and in some municipalities (DGDR, n.d.).

Public transparency

Act No. 18.381, dated 17 October 2008, addresses the right of access to public information. Article 1 of the Act sets out the objective of "(...) promoting transparency in the administrative functions of all public bodies, whether state bodies or not, and guaranteeing people's fundamental right of access to public information". Article 2 states that public access shall be curtailed for "(...) restricted or confidential information" (Centro de Información Oficial, 2008b).

The importance of this act is consistent with a governance problem found in the LCP. Both the LCP and the decree affirm the transparency of procurement-related data, prices obtained from family producers and artisanal fishers, quantity/volume, and suppliers, among other data.

In more specific terms, insufficient data and (public) microdata was identified to ascertain the number of producers taking part in the LCP, their geographical distribution, end consumers, the prices applied in procurement from family producers and artisanal fishers, and the volume of food sold, etc., which renders the task of conducting a preliminary study of the LCP problematic. Furthermore, this information cannot be described as being "restricted or confidential" in nature.

Research conducted by Chiappe et al. (2015, p. 3) also revealed the problem of the public transparency of data, with the authors stating that in "(...) various state bodies [carrying out procurement] have no documents that may be consulted in order to ascertain sales volume, (...) and their estimation on the basis of certain considerations and assumptions, applied to partial data".

▶▶ Interaction of the LCP with other policies

The state, through policy makers and the proactive participation of society, has the prerogative of providing and promoting mechanisms that enhance its ability to implement public policies in a progressive manner (Kattel & Lembert, 2010). In addition, there is also a need to understand the issues under discussion and to develop and link existing public policies; in other words, a policy mix of actions that are guided by one main purpose, particularly in the creation of policies regarding sustainable food production, the rural development of family producers and artisanal fishers, and sustainable procurement by the state.

Taking the description of the operational structure of the LCP as a starting point, a limitation was observed in relation to the promotion of the sustainable rural development of family producers and artisanal fishers ("end"). The limitations of the LCP refer mainly to the weakness of the objective, the prominent role of food supply to the state, the absence of sectoral support, the interrelationship with other public policies (with the exception of the characterization of family producer and artisanal fisher), and the lack of any strong connection between the financial resources made available for public food procurement and rural families.

This section presents the possible links between the LCP and other government policies that directly or indirectly promote sustainable food systems. We thus highlight the definition of family farmer/producer and artisanal fisher (institutional recognition of rural families), the National Agroecological Production Plan (sustainable food systems), food and nutrition security (social regulatory framework), the School Feeding Program (access to food), food safety (quality controls), the Family Producer Seal (recognition of rural families), Rural Development Forums (decentralization and participation in public policies), among other public policies (sectoral and cross-cutting initiatives).

Definition of family farmer/producer and artisanal fisher

As explained in subheading "2.4. LCP Target Audience", the LCP is integrated with the MGAP resolution defining what is a family farmer/producer. In other words, only family producers and artisanal fishers (unless otherwise stated) who meet the specific requirements set out in the ministerial resolution can benefit from the LCP.

In addition to the defining of what is a family farmer/producer (and not artisanal fishers, who were included in the MGAP resolution following their inclusion in the LCP – Machado et al., 2018), the MGAP resolution also establishes a registry (compulsory for the everyone in this category) with the purpose of identifying the potential target audience and implementing separate public policies (cross-cutting, complementary, and specific – De Torres et al., 2018 – from the MGAP to other entities) for the sector, including the LCP.

National Agroecological Production Plan

As a result of popular and academic pressure, the government was compelled to enact Act No. 19.717, dated December 21, 2018. The Agroecology Act aims to strengthen food sovereignty and security by protecting the environment (production, distribution and consumption of agroecological foodstuffs – natural or processed), creating benefits that enhance the quality of life of Uruguayan people (Centro de Información Oficial, 2018b).

The law established an Honorary Commission (under the auspices of the DGDR/MGAP) with the aim of drafting the National Agroecological Production Plan and coordinating (implementation) and monitoring (execution) its operation, promoting discussions on and consideration of agroecology, creating theme-based commissions, and bringing different institutions and the government into the development of the agroecology plan (Centro de Información Oficial, 2018b). The document, which has yet to be finalized, is in the public consultation phase. Funding for the act's implementation has not been included in the government's budget for the 2020-2024 period.

There are essentially four aspects to the potential link between the National Agroecological Production Plan and the LCP. Firstly, both government tools are aimed at the same target audience, i.e. family farmers/producers (previously defined in the MGAP resolution – "3.1. Definition of family farmers/producers and artisanal fishers"). Secondly, the drafting of the plan should encompass promotion and ease of access to differentiated, local and neighborhood markets (short supply chains) and especially those markets already in place (for example, the LCP). Thirdly, it deals with mechanisms for the sustainable development of family production, which are overlooked in the LCP. Finally, it provides for the consolidation of food sovereignty and food security by means of an accessible supply of safe food products (see "3.5. Food safety"), impacting on the quality of life of the people of Uruguay.

Food and nutrition security

The regulatory framework pertaining to food and nutrition security is still negotiating though the bureaucratic procedures of the Uruguayan legislature. The "Right to Food, Food and Nutrition Security of the Population" project came up for discussion for the first time in the Chamber of Deputies' Special Population and Development Commission in June 2015. There has been no further movement since 19 February 2020, when its status was recorded as "archived at the end of the legislature" (Parlamento del Uruguay, 2020).

The project aims to "(…) establish a legal *référentiel* (…)" that allows the state to develop permanent public policies and strategies of national importance for the population's right to food and food and nutrition security, with a view to ensuring a healthy lifestyle and diet. The text also refers to a sustainable production system and the availability and accessibility of food produced by family producers and artisanal fishers for the population (Cámara de Representantes, 2015, p. 5), issues that are linked to both agroecology and the LCP.

The legislative dispute over the framework in question has not curtailed the actions of the Uruguayan state. Implicitly, the issue has been addressed since the 1920s, through the School Feeding Program, while Article 72 of the Constitution of the Republic implies the inherent right to access to food (Centro de Información Oficial, 1967). The study conducted by Calanchini et al. (2017) provides an in-depth regulatory review of the state's responsibilities with regard to food and nutrition security. This text is recommended for further consideration of the issue.

More specifically, the purpose of the INDA and ObSAN is to provide the state, policy makers and society (Fernández et al., 2015) with detailed information on the subject and shape projects and/or policies for overcoming obstacles related to poverty and food and nutrition insecurity. The INDA promotes food-support initiatives (food packages) for the socioeconomically vulnerable through the purchase of food products and donations (a useful tool for linking with the LCP).

School Feeding Program

In the same vein as the issue of food and nutrition security addressed in this introduction, the program (under the auspices of the Board of Elementary and Primary Education – National Administration of Public Education) is designed to ensure that students in the public and private network eat sufficiently nutritional diets, the aim being to improve their education and reduce food-risk situations (DGEIP, 2020). The aforementioned program is part of Act No. 19.140, dated October 11, 2013 (Centro de Información Oficial, 2013), which safeguards the health of school-children (public and private) by promoting healthy eating habits and a nutritious diet and thereby prevent overweight, obesity, and other related diseases. It puts the emphasis on the consumption of natural foods that are minimally processed and prepared in accordance with specific recommendations.

The survey revealed an interrelationship between the LCP and the School Feeding Program, discussed under subheading "2.5. Operational framework of the LCP", and Act No. 19.140 (The procurement of food from family producers – Chiappe et al., 2015). However, owing to the difficulty in obtaining microdata and to the public transparency of the LCP, drawing a direct parallel between these government instruments is not an easy task.

Food safety

Created in 2015, the Directorate General of Food Safety Control forms part of the MGAP (there had been related initiatives prior to its creation). It was set up with the objective of planning, formulating, coordinating and implementing policies at three levels: health barriers, biosecurity, and food safety (MGAP, 2017).

The food safety plan provides for the creation of health standards for food and feed, the adoption of quality procedures (Best Practices for Production, Processing and Storage), the encouraging of agricultural producers to participate in quality and safety procedures, and the certification of organic produce, etc. (MGAP, n.d.-b). Among other precepts of food safety (observed in the LCP), there is an emphasis on promoting the safety and quality of foodstuffs for domestic consumption (including public procurement), which in turn promotes the supply to the population of products of recognized quality (national and international standards) and which are detailed in Best Agricultural Practices (for the production of fruit and vegetables).

Best Agricultural Practices are a set of technical procedures relating to the production system (producer, production, plant material, soil and water management, disease control, pest management, etc.), harvesting (the stages before and during harvest), post-harvesting (treatment, handling, packaging, storage and transport), water management, pesticide management, general equipment, animal management, waste management, worker management, and general operations and traceability (MGAP, 2016b).

The Family Producer Seal

The DGDR/MGAP and the Inter-American Institute for Cooperation on Agriculture (IICA – also discussed in the REAF) carried out studies and engaged in activities on the project "Towards the construction of a family farming seal in Uruguay". Essentially, it seeks to enhance the value of family production through a seal that certifies and denotes food produced by family producers and artisanal fishers (MGAP, n.d.-a). According to De Torres et al. (2018), the project's main target audience is rural families producing vegetables, fruit, and small animals (poultry and pigs).

As well as a kind of "trademark", the seal affords a competitive advantage (over other companies) that adds value to produce and provides a range of information to the consumer (production processes and traceability), thus ensuring wider access to formal markets and also within the LCP.

Rural Development Forums

Created as part of Act No. 18.126, dated 15 May 2007, Rural Development Forums reflect the complete decentralization of public rural development policies. The regulatory hierarchy provided for by the Act is composed of the Rural Development Forums (local level), which is subject to the State Agricultural Council (regional), which is answerable to the National Agricultural Council (national), a triumvirate supported by the Agricultural Council – overseen by the MGAP (Centro de Información Oficial, 2007).

The objective of the Rural Development Forums is to encourage social agricultural sectors to become fully involved in and participate in the debate on public policies as they relate to the rural world (demands and concerns beyond production – De Torres et al., 2018). As part of a staggered process, arguments are forwarded to the State Agricultural Council, which is responsible for identifying and putting forward local concerns and disseminating and assisting with MGAP initiatives. The National Agricultural Council receives regional recommendations and advises the MGAP on the creation of national agricultural policies aimed at rural development. Finally, the Agricultural Council helps develop, link and coordinate agricultural policies with the MGAP's decentralized initiatives and encourages all regions to participate and become fully involved (Centro de Información Oficial, 2007).

As seen, the forums encourage the emergence of new, bottom-up social actors (horizontal governance), who make their demands to the state itself, especially to the MGAP (De Torres et al., 2018). Although the LCP does not establish a direct relationship with the decentralization provided for in Act No. 18.126, some family producers are regarded as new social actors because they set up OHs and also participate in Rural Development Forums, as is the case with family producers in the states of Maldonado and Florida. In other words, the LCP provides an opportunity for the formation of new actors and the Forums give them a platform for discussing public policies (a participative and intersectoral role – Sabourin & Arbeletche, 2018).

Other public policies

Other public policies for which the government is responsible (state capacity) and which are linked to the LCP are the Rural Microcredit Program, Technical Support and Rural Extension (ATER), "Somos Mulheres Rurais" (We Are Rural Women), 'Somos Daqui' (We Are From Here), the Regional Planning and Sustainable Development Act, and the Agricultural Insurance Law.

Coordinated by the DGDR and the Uruguayan Foundation for Cooperation and Mutually Supportive Development, the Rural Microcredit Program provides short-term financing to family producers who do not have access – due to a lack of financial means – to the mechanisms available on the formal credit market. The purpose of the program is to answer their financial needs and promote local production. It is a mechanism that to some extent addresses the absence of such an instrument in the LCP and which supports the structure of the production system of the LCP's target audience.

In addition to rural credit, Technical Support and Rural Extension is a fundamental tool for the sharing of technical and unspoken knowledge and the optimization and promotion of sustainable food systems for family producers and artisanal fishers (MGAP, n.d.-a).

Like the LCP, the Somos Mulheres Rurais program promotes OH+Gs. It also allocates resources to rural women to help bring about social inclusion and sustainable rural development. The Somos Daqui program is a similar initiative and provides funding for rural youth projects aimed at rural and non-rural development.

Act No. 18.308, dated 18 June 2008, concerns regional planning and sustainable development through cross-cutting actions aimed at enhancing the quality of life of Uruguayan people, social integration, and the democratic and sustainable use and enjoyment of natural and cultural resources (Centro de Información Oficial, 2008a). This piece of legislation complements the Agroecology Act and is linked to the LCP in terms of non-compliance with the sustainable agricultural production system.

Finally, the Agricultural Insurance Act (Act No. 19.678, dated 26 October 2018) supports family and non-family producers in the event of adverse weather (catastrophe or disaster) and health events, providing economic-productive stability for its beneficiaries (Bado Cardozo, 2019) and the continuation of the family production model.

▸▸ Passing thoughts

Uruguay's LCP arose out of regional discussions promoted by the REAF, with the Uruguayan state adapting it to the socioeconomic context of the country. Unlike Brazil, where state procurement programs for family farming have a strong slant towards rural development and sustainable food systems, the Uruguayan LCP instead gives the state a central role in the supply of food products, with preference given to family producers and artisanal fishers.

With regard to the problem, the LCP does not specifically determine what it intends to address/remedy, but it can be inferred that public procurement allows family producers and artisanal fishers to sell part of their production to the state (short supply chains) and to obtain better prices. This aspect has in all probability contributed to the setting of an objective that does not seek the sustainable rural development of beneficiaries (as explained above), despite the state's demand that rural families receive encouragement.

Though the target audience is stipulated (family producers and artisanal fishers), non-family producers are included as beneficiaries of the LCP, undermining the Act's objective of helping rural families. Furthermore, there is no explicit prioritization criterion for potential beneficiaries, which leaves access open to all family (and non-family) producers, regardless of their socioeconomic status.

The LCP is vague in other ways and/or has other limitations, such as the source of processed food (whole or partial) and difficulties faced by rural families with planning the production system. Nor does the LCP promote sustainable (organic or agroecological) food production or provide a price differentiation mechanism for family producers (FS) that operate in a sustainable way. Although a market reserve is determined for these types of food, there are no resources for procurement from the target audience, not to mention the lack of public transparency of the mechanisms/results employed/obtained. The LCP is not widely implemented across the country and there are few HOs in relation to the production capacity of rural families.

An exception to all this is the requirement for rural families to organize themselves into HOs in order to participate in state procurement, forming social capital and social actors (although no direct participation mechanism is discussed). The promotion of gender equality is another extremely positive aspect of the LCP, bringing women into the production cycle and the organization of OHs. In addition to these two factors, it is also worth highlighting the setting of food prices above market prices.

The arguments raised about the structural weaknesses of state procurement in promoting the development of sustainable food systems can be remedied through linkage with other policies, in other words, a policy mix that encompasses a set of actions and dynamics that optimize the sustainable development of family producers and artisanal fishers.

The study provides examples of some policies: the LCP is heavily dependent on the definition and registration of family producers, which directs the potential target audience towards state procurement; the Agroecology Plan addresses the gap in the LCP on sustainable food production; in the LCP there is no relationship with food and nutrition security (nor is there a government definition), though the state does have solid tools (INDA and ObSAN) for developing the subject and linking it to family production; the interaction of the LCP with the School Feeding Program in the supply of food for rural families; food safety, which determines quality parameters for the production and sale of food; the Family Producer Seal project, which is designed to make family production competitive; the Rural Development Forums, which influence debate on public policies (including the LCP) for sustainable rural development; as well as the Rural Microcredit Program (credit for production), ATER (technical support), Somos Mulheres Rurais (female empowerment), Somos Daqui (support for young people), the Regional Planning and Sustainable Development Act (sustainable planning), and the Agricultural Insurance Act (continuation of the family model).

The policy mix in question consists of actions ranging from institutional recognition of rural families to sustainable food systems, linking with a social regulatory framework for food and nutrition security, influencing access to food for schoolchildren, establishing quality controls on food production, recognizing rural families in the shape of a seal, the decentralization of and public participation in the debate on public policies, as well as other sectoral and cross-cutting tools. This path addresses (or would address) the gaps in the LCP and enables the "end" with regard to the sustainable rural development of food by family producers and artisanal fishers.

⏵ References

ARCE (Agencia Reguladora de Compras Estatales). (n.d.). *Regímenes de preferencia*. República Oriental del Uruguay. Retrieved March 9, 2020 from https://www.gub.uy/agencia-compras-contrataciones-estado/politicas-y-gestion/planes/regimenes-preferencia

ARCE. (2019). *Manual de Contratación Pública* (version 11). República Oriental del Uruguay. https://www.comprasestatales.gub.uy/ManualesDeUsuarios/manual-procedimiento-compras/Compraporexcepcion.html.

Bado Cardozo, V. S. (2019). Ámbito de aplicación y eficacia de la Ley de Contratos de Seguros uruguaya. *Revista de la Facultad de Derecho, 47*, Article e20194702. http://dx.doi.org/10.22187/rfd2019n47a2

Calanchini, J., Borche, A., & Canclini, G. (2017). *El derecho a la alimentación en el marco normativo de Uruguay 1985–2014*. Observatorio del Derecho a la Alimentación, Facultad de Derecho (UDELAR). https://www.fder.edu.uy/sites/default/files/2018-02/EL%20DERECHO%20A%20LA%20ALIMENTACIÓN%20EN%20URUGUAY%20%20(2017).pdf

Cámara de Representantes. (2015). *Comisión Especial de Población y Desarrollo* (carpeta n.° 198). Parlamento de la República Oriental del Uruguay.

Carvalho, P. D. (2011). *Ação coletiva transnacional e Mercosul: organizações da sociedade civil do Brasil e do Paraguai na construção da reunião especializada sobre agricultura familiar (REAF)* [Master's thesis, Universidade de Brasília]. UnB Repository. http://repositorio.unb.br/bitstream/10482/8836/1/2011_PriscilaDelgadoDeCarvalho.pdf

Centro de Información Oficial. (1967). *Constitución de la República*. Dirección Nacional de Impresiones y Publicaciones Oficiales, Uruguay. https://www.impo.com.uy/bases/constitucion/1967-1967

Centro de Información Oficial. (2007). *Ley no. 18.126, del 12 de mayo de 2007*. Dirección Nacional de Impresiones y Publicaciones Oficiales, Uruguay. https://www.impo.com.uy/bases/leyes/18126-2007/8

Centro de Información Oficial. (2008a). *Ley no. 18.308, del 18 de junio de 2008*. Dirección Nacional de Impresiones y Publicaciones Oficiales, Uruguay. https://www.impo.com.uy/bases/leyes/18308-2008

Centro de Información Oficial. (2008b). *Ley no. 18.381, del 17 de octubre de 2008*. Dirección Nacional de Impresiones y Publicaciones Oficiales, Uruguay. https://www.impo.com.uy/bases/leyes/18381-2008/9

Centro de Información Oficial. (2013). *Ley no. 19.140, del 11 de octubre de 2013*. Dirección Nacional de Impresiones y Publicaciones Oficiales, Uruguay. https://www.impo.com.uy/bases/leyes/19140-2013

Centro de Información Oficial. (2014). *Ley no. 19.292, del 16 de diciembre de 2014*. Dirección Nacional de Impresiones y Publicaciones Oficiales, Uruguay. https://www.impo.com.uy/bases/leyes/19292-2014

Centro de Información Oficial. (2015). *Decreto no. 86/015, del 27 de febrero de 2015*. Dirección Nacional de Impresiones y Publicaciones Oficiales, Uruguay https://www.impo.com.uy/bases/decretos/86-2015

Centro de Información Oficial. (2018a). *Ley no. 19.685, del 26 de octubre de 2018*. Dirección Nacional de Impresiones y Publicaciones Oficiales, Uruguay. https://www.impo.com.uy/bases/leyes/19685-2018/2

Centro de Información Oficial. (2018b). *Ley no. 19.717, del 21 de diciembre de 2018*. Dirección Nacional de Impresiones y Publicaciones Oficiales, Uruguay. https://www.impo.com.uy/bases/leyes/19717-2018

Chiappe, M., Oyhantçabal, G., Pizzolón, A., & Rodríguez, L. (2015). *Informe final: desarrollo rural y aplicación de políticas de compras públicas a productos alimenticios provenientes de la agricultura familiar.* FAO, Facultad de Agronomía de la Universidad de la Republica.

Comisión Especial de Población y Desarrollo. (2014). *Producción familiar agropecuaria y pesca artesanal*. Cámara de Representantes, Parlamento de la República Oriental del Uruguay. https://parlamento.gub.uy/documentosyleyes/ficha-asunto/120934/tramite

De Torres, M. F., Arbeletche, P., & Sabourin, E. (2018). Agricultura familiar en Uruguay: reconocimiento y políticas públicas. *Raízes: Revista de Ciências Sociais e Econômicas, 38*(1), 116–128. https://doi.org/10.37370/raizes.2018.v38.42

DGDR (Dirección General del Desarrollo Rural). (n.d.). *Organizaciones autorizadas a operar*. Ministerio de Ganadería, Agricultura y Pesca, Uruguay. Retrieved March 9, 2020 from http://www.mgap.gub.uy/organizaciones/organizacionesmostrar.aspx

DGEIP (Dirección General de Educación Inicial y Primaria). (2020). *Programa de Alimentación Escolar*. Administración Nacional de Educación Pública, Uruguay. https://www.dgeip.edu.uy/finalidad-del-pae/

FAO. (2020). *Perfil nacional de seguridad alimentaria y nutricional Uruguay*. FAO, ALADI, ECLAC.

FAO, IFAD, UNICEF, WFP, & WHO. (2019). *El estado de la seguridad alimentaria y la nutrición en el mundo 2019. Protegerse frente a la desaceleración y el debilitamiento de la economía* [*The state of food security and nutrition in the world 2019. Safeguarding against economic slowdowns and downturns*]. FAO. https://www.fao.org/3/ca5162es/ca5162es.pdf

FAO, OPS, WFP, & UNICEF. (2018). *Panorama de la seguridad alimentaria y nutricional en América Latina y el Caribe 2018*. https://www.fao.org/3/CA2127ES/CA2127ES.pdf

Fernández, V., Arciet, J., Bove, I., Rodríguez, S., Camacho, S., Curutchet, M. R., Magnani, D., & Sampayo, V. (2015). *Derecho a la alimentación, seguridad alimentaria y nutricional: logros y desafíos de Uruguay*. Facultad de Ciencias de la Universidad de la República Uruguay. https://www.gub.uy/ministerio-desarrollo-social/comunicacion/publicaciones/derecho-alimentacion-seguridad-alimentaria-nutricionallogros-desafios

Gómez Perazzoli, A. (2019). Uruguay: país productor de alimentos para un sistema alimentario disfuncional. *Agrociencia (Uruguay)*, *23*(1), Article e70. http://dx.doi.org/10.31285/agro.23.1.8

Grisa, C., & Niederle, P. (2018). Difusão, convergência e tradução nas políticas de compras públicas da agricultura familiar no âmbito da REAF Mercosul. *Revista Latinoamericana de Políticas y Acción Pública*, *5*(2), 9–30. https://repositorio.flacsoandes.edu.ec/bitstream/10469/14612/2/RFLACSO-MP5%282%29.pdf

Hristoff, A., & Saravia, L. (2010). *Situación de la seguridad alimentaria y nutricional en Uruguay 2009*. Observatorio de Seguridad Alimentaria y Nutricional Uruguay, Instituto Nacional de Alimentación. http://dspace.mides.gub.uy:8080/xmlui/handle/123456789/566

INE (Instituto Nacional de Estadística). (2020). *Demografía y estadísticas sociales*. http://www.ine.gub.uy/web/guest/censos-2011

Kattel, R., & Lember, V. (2010). Public procurement as an industrial policy tool: an option for developing countries? *Journal of Public Procurement*, *10*(3), 368–404. https://doi.org/10.1108/JOPP-10-03-2010-B003

Lanzaro, J. (2004). Fundamentos de la democracia pluralista y estructura política del Estado en el Uruguay. *Revista Uruguaya de Ciencia Política*, *14*, 103–135. https://hdl.handle.net/20.500.12008/6914

Leporati, M., Salcedo, S., Jara, B., Boero, V., & Muñoz, M. (2014). La agricultura familiar en cifras. In Salcedo, S., & L. Guzmán (Eds.), *Agricultura familiar en América Latina y el Caribe: recomendaciones de política* (pp. 35–56). FAO. https://www.fao.org/3/i3788s/i3788s.pdf

Machado, A., Pizzolon, A., & Vaz Tourem, J. (2018). Compras institucionales: estado del arte del proceso uruguayo. In Perez-Cassarino, J., Triches, R. M., Baccarin, J. G., & C. R. Paz Arruda Teo (Eds.), *Abastecimento alimentar: redes alternativas e mercados institucionais* (pp. 121–136). UFFS, UNICV. https://doi.org/10.7476/9788564905726

MGAP (Ministerio de Ganadería, Agricultura y Pesca). (n.d.-a). *Dirección General de Desarrollo Rural*. República Oriental del Uruguay. Retrieved March 18, 2020 from https://www.gub.uy/ministerio-ganaderia-agricultura-pesca/dgdr

MGAP. (n.d.-b). *División inocuidad y calidad de alimentos*. Retrieved April 18, 2020 from https://www.gub.uy/ministerio-ganaderia-agricultura-pesca/institucional/estructura-del-organismo/division-inocuidad-calidad-alimentos

MGAP. (2011). *Censo general agropecuario 2011: resultados definitivos*. República Oriental del Uruguay. https://www.gub.uy/ministerio-ganaderia-agricultura-pesca/sites/ministerio-ganaderia-agricultura-pesca/files/2020-02/censo2011.pdf

MGAP. (2016a, November 11). *Resolución No. 1.013/016 - Definición del productor familiar agropecuario*. República Oriental del Uruguay. https://www.gub.uy/ministerio-ganaderia-agricultura-pesca/institucional/normativa/resolucion-1013016-definicion-del-productor-familiar-agropecuario

MGAP. (2016b, December 6). *Guía de buenas prácticas agrícolas para la producción de frutas y hortalizas frescas en Uruguay.* República Oriental del Uruguay. https://www.gub.uy/ministerio-ganaderia-agricultura-pesca/comunicacion/publicaciones/guia-buenas-practicas-agricolas-para-produccion-frutas-hortalizas

MGAP. (2017). *Uruguay agrointeligente: los desafíos para un desarrollo sostenible.* República Oriental del Uruguay. https://www.gub.uy/ministerio-ganaderia-agricultura-pesca/sites/ministerio-ganaderia-agricultura-pesca/files/2019-12/libro%20completo%20con%20hipervinculos.pdf

Mila, F. (2015, November 11–12). *La agricultura familiar como proveedora de las compras públicas de productos agrícolas y alimentos* [Workshop presentation]. Programa Regional FIDA MERCOSUR, Montevideo, Uruguay.

Milhorance, C., Sabourin, E., & Mendes, P. (2019). *Implementation and coordination of climate change adaptation policies in Bahia and Pernambuco semi-arid regions* (Working Papers Series no. 3). UNB, CIRAD.

Mondelli, M. (2014, September 2). *Lineamientos estratégicos para el desarrollo competitivo agroexportador* [Conference presentation]. II Foro Cooperativo Agropecuario CAF 2014, Montevideo, Uruguay. http://www.caf.org.uy/site/wp-content/uploads/2014/09/Documento-PDF-4.pdf

Parlamento del Uruguay. (2014). *Produccion familiar agropecuaria y pesca artesanal. Interes general. Declaración.* https://parlamento.gub.uy/documentosyleyes/ficha-asunto/120934/tramite

Parlamento del Uruguay. (2020). *Alimentacion, seguridad y soberania alimentaria. Derecho. Ley marco. Aprobacion.* https://parlamento.gub.uy/documentosyleyes/ficha-asunto/124753/ficha_completa

Pereyra, I. (2019). Políticas públicas sobre nutrición en Uruguay y la autonomía de las personas. *Revista Cubana de Salud Pública, 45*(1), Article e1238. https://www.scielosp.org/article/rcsp/2019.v45n1/e1238

Ramos, Á. (2019). *Estado de las políticas diferenciadas para la agricultura familiar campesina y indígena en siete países de América Latina.* COPROFAM.

REAF (Reunión Especializada sobre Agricultura Familiar). (2014). *10 años cambiando realidades.*

Rogge, K. S., & Reichardt, K. (2013). *Towards a more comprehensive policy mix conceptualization for environmental technological change: a literature synthesis* (Working Paper Sustainability and Innovation, no. S3/2013). Fraunhofer Institute for Systems and Innovation Research.

Sabourin E., & Arbeletche, P. (2018). *Rapport final du projet "Comprendre le fonctionnement et les effets de dispositifs participatifs innovants à l'interface entre élevage et environnement. Le cas des Forums de Développement Rural en Uruguay".* Projet ECOS, CIRAD, UDELAR. https://agritrop.cirad.fr/586173/1/Rapport%20final%20Projet%20MESAS%20Uruguay%20ECOS-UDELAR%202017.pdf

Scheuer, J. M., & Vassallo, M. (2019). Análise do Programa Nacional de Fortalecimento da Agricultura Familiar no município gaúcho de Roque Gonzales, Brasil. *Revista Geográfica Acadêmica, 13*(1), 40–61. https://revista.ufrr.br/rga/article/view/5376

Sganga, F., Cabrera, C., Gonzalez, M., & Rodriguez, S. (2016). *Producción familiar agropecuaria uruguaya y sus productores familiares a partir de los datos del Censo General Agropecuario y el Registro de Productores Familiares.* Ministerio de Ganadería, Agricultura y Pesca, Oficina de Programación y Política Agropecuaria.

Solanas, F. (2009). El impacto del impacto del MERCOSUR en la educación superior: un análisis desde la "Mercosurización" de las políticas públicas. *Education Policy Analysis Archives/Archivos Analíticos de Políticas Educativas, 17,* 1–18. http://www.redalyc.org/articulo.oa?id=275019727017

Torres Ledezma, C. (2010). *Informe final de la consultoria realizada en el marco del proyecto Fortalecimiento del Observatorio de Seguridad Alimentaria y Nutricional.* Observatorio de Seguridad Alimentaria y Nutricional, Uruguay.

UNDP (United Nations Development Programme). (2020). *El PNUD en Uruguay.* https://www.uy.undp.org/content/uruguay/es/home/about-us.html

World Bank. (2020). *Uruguay: panorama general* [*Uruguay: overview*]. https://www.worldbank.org/en/country/uruguay/overview

Soy on one side, livestock on the other: the food issue with regard to indigenous people and public policies in Paraguay

SILVIA APARECIDA ZIMMERMANN,
DIANA JAZMIN BRITEZ COHENE, NOELIA RIQUELME

▸▸ Introduction

Paraguay has slightly more than seven million inhabitants, where around 1.6 million people (23.5%) live in households whose per capita income is lower than the cost of a basic food basket, and are therefore considered poor (DGEEC, 2020). Around 284,000 people, or 4% of the country's population is in extreme poverty, where their monthly per capita income does even cover the cost of a minimum food basket. The highest proportion of those living in poverty and extreme poverty is found in rural areas, where 33% of the population is considered poor. Of the inhabitants considered to be in extreme poverty, 73% are also in rural areas (DGEEC, 2020).

The importance of the country's agricultural sector (crops, livestock and forestry) is significant: the sector contributes an estimated 25% of GDP (IMAS, 2019). Data from the 2017 and 2018 Permanent Household Surveys show that 38% of the country's total population lives in rural areas and 37% of the total employed population also lives in rural areas (DGEEC, 2020). According to the national production trade union, "agriculture, the economy and society are still very closely tied, especially in Paraguay, whose production system is sustained by agricultural crops and their respective chains, where rural areas are predominant" (UGP, 2015, p. 3).

As a result, although Paraguay has other important activities, the country maintains agrarian characteristics and supports agricultural and livestock production, with substantial inequality within rural areas in terms of income, land tenure, access to basic services and public policies (sectoral and non-sectoral), among other problems that contribute to the social vulnerability of the rural population (Wesz Jr., Zimmermann & Ríos, 2018; Riquelme, 2016). In addition, in recent years peasant and indigenous territories as well as food production have declined, and every year the country "increasingly depends on [food] imports to meet the demand for staple foods, which were previously provided by family farming" (Villalba, 2019, p. 7). Paraguay maintains food sovereignty only in its cassava and banana production. Likewise, family farming in Paraguay "accounts for 91% of the total number of agricultural holdings, while in terms of surface area, it totals only 6%" (Riquelme, 2016, p. 26).

The presence of indigenous people, especially the Guaraní, marks Paraguayan history and determines the agrifood system of farmers, who then migrate to the cities with their production, gathering and consumption practices (Doughman, 2011). A "people" is understood to be "a group of people who are characterized by their own culture and social life. In Paraguay, as in other places in America, during colonial times they were called a nation, because they had been born in a single territory, generally had their own language, and followed shared customs" (DGEEC, 2014, p. 129). In Paraguay, the areas held by indigenous people are gradually being reduced, which compromises not only their forms of production, gathering and consumption, but also those of the country's entire population.

This chapter aims to present the productive, gathering and consumption reality of indigenous people in Paraguay, their agrifood system, and the national public policy initiatives for food sovereignty and security that attempt to guarantee the *tekoporã*, or "good life" in Guaraní, of indigenous people. The three dimensions of public policies – polity-politics-policy (Secchi, 2015; Souza, 2006) – are analysed. Polity deals with the institutional dimension of politics (i.e., the institutions and rules that establish how power is distributed and exerted). Politics deals with the conflicts and behavior of the actors (i.e., who are the political, state, economic and social actors involved?). Policy (or policies) refers to public policies, their operating rules, objectives, results and government actions (Jaime et al., 2013).

The methodology used for the analysis included a literature review; consultation of official documents and analysis of statistical data; and semi-structured interviews, conducted between November 2019 and August 2020 via WhatsApp and Skype, with political actors[1] involved in implementing and studying food sovereignty and food security policies in Paraguay. For the qualitative study, interviewees were questioned about the policies they consider most relevant for food sovereignty and food security in their country, conflicts and progress. Seven experts were interviewed, including three with technical responsibility at the FAO; three representatives of organized civil society (national peasants' movement of Paraguay – MCP; the National Coordinator of Organizations of Rural and Indigenous Women Workers – CONAMURI; and the national women farmers' organization – ONAC); and one government representative (Senator Hugo Richer, member of the Parliamentary Front against Hunger).

After a brief introduction, this chapter discusses the data from the surveys of indigenous people in Paraguay, with a particular focus on data on the food system. The second part delves into the pressure of agribusiness on the sustainable food systems of indigenous people. Third, the establishment of the National Plan for Indigenous Peoples is described, where these peoples are recognized in national policy, while various segments of civil society have begun mobilizing for the development of public policy, a process that has been paralyzed by the Covid-19 pandemic.

---

1. We are grateful to the interviewees for their time and the support of the National Council of Scientific and Technological Development (CNPq) as part of the public call MCTI/CNPq nº 01/2016 and the support of the Federal University for Latin American Integration through the calls PRPPG/UNILA no. 110/2018, PRPPG/UNILA no. 137/2018 and PRPPG/ no. 80/2019.

▸▸ The indigenous peoples of Paraguay: the food issue

It is clear that the indigenous population of Paraguay has increased compared to previous censuses (table 19.1; figure 19.1). The data presented below comes from the third national census of indigenous people of 2012, from the General Directorate of Statistics, Surveys and Censuses (DGEEC), which produced several documents providing information that could shed light on the situation of the indigenous people of Paraguay and the sustainable agrifood systems where they live.

Table 19.1. National and indigenous population by census year, 1981–2012, in Paraguay

Population	Census Year			
	1981	**1992**	**2002**	**2012**
Total	2,954,171	4,152,588	5,63,198	6,435,218
Indigenous	38,703	49,487	89,169	117,150
% Indigenous	1.3	1.2	1.7	1.8

Source: DGEEC (2014, p. 49).

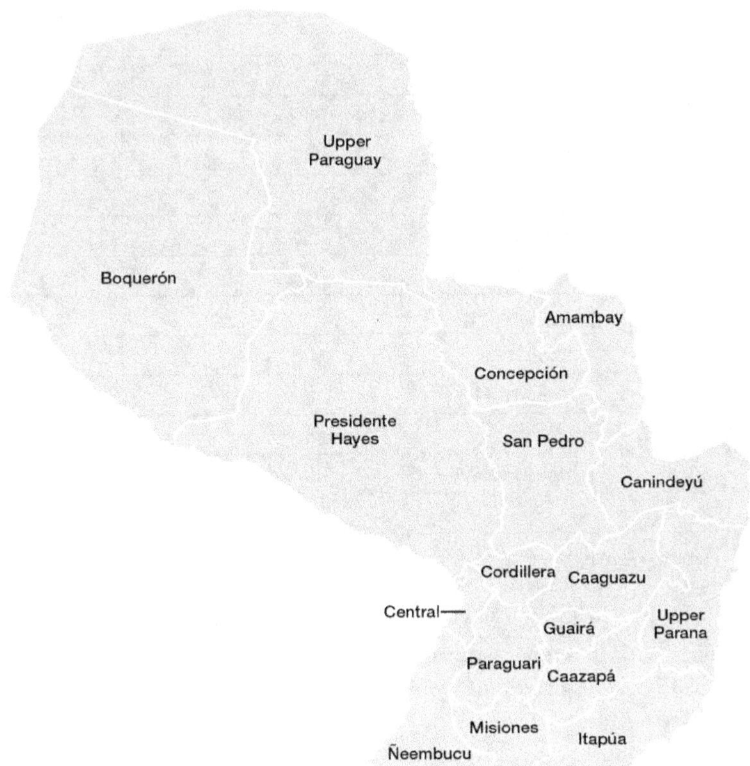

Figure 19.1. Departmental distribution of indigenous communities (%) in Paraguay, 2012. Source: DGEEC (2015, p. 34).

*The map does not show villages or neighborhoods, only communities.

The survey counted "493 [indigenous] communities and 218 villages or neighborhoods, totaling 711 communities, villages or neighborhoods, distributed in 19 towns, belonging to five linguistic groups, inhabiting 13 departments of the country and in Asunción. The people with the largest number of communities is the Mbyá Guaraní, and the one with the smallest number is the Tomárãho people" (DGEEC, 2015, p. 15). No indigenous communities were found in the Cordillera, Paraguarí, Misiones or Ñeembucú departments. The largest number of communities are in the eastern region, in the departments of Canindeyú and Caaguazú, with 106 (22%) and 59 (12%) communities, respectively. In the western region, the departments of Presidente Hayes (50 communities, 10%) and Boquerón (46 communities, 9%) had the largest communities (table 19.2).

Table 19.2. Departmental distribution of indigenous communities (%) in Paraguay, 2012

Region	Department	Indigenous Communities	
		Number	**%**
Eastern	Asunción	1	0.2
	Concepción	20	4.1
	San Pedro	28	5.7
	Guairá	8	1.6
	Caaguazú	59	12
	Caazapá	29	5.9
	Itapúa	31	6.3
	Upper Paraná	38	7.7
	Central	6	1.2
	Amambay	45	9.1
	Canindeyú	106	21.5
Western	President Hayes	50	10.1
	Boquerón	46	9.3
	Upper Paraguay	26	5.3
Total		**493**	**100%**

Source: DGEEC (2015, p. 34).

The vast majority of indigenous peoples live in rural areas (91%, or 103,396 indigenous people), with a smaller number in urban areas (9%, or 9,858 indigenous people) (table 19.3).[2] Approximately 60% of those who live in urban areas are in the Boquerón department and 19% in the Central department, with the rest distributed across the remaining departments. The majority of indigenous people who live in rural areas are in the departments of Presidente Hayes (24%), Boquerón (18%), Canindeyú (13%) and Amambay (11%) (DGEEC, 2015).

---

2. The information only includes data from the national population and household census of indigenous peoples, with a total of 113,254 indigenous people living in urban and rural areas (DGEEC, 2015).

Table 19.3. Indigenous population by urban and rural areas and sex, by people and age groups in Paraguay, 2012

Age groups (years)	Total			Areas and sex					
				Urban			Rural		
	Total	Men	Women	Total	Men	Women	Total	Men	Women
Total	113,254	58,563	54,691	9,858	5,042	4,816	103,396	53,521	49,875
0-4	16,516	8,443	8,073	1,378	718	660	15,138	7,725	7,413
5-9	16,330	8,362	7,968	1,302	663	639	15,028	7,699	7,329
10-14	14,704	7,633	7,071	1,177	620	557	13,527	7,013	6,514
15-19	13,777	6,891	6,886	1,205	616	589	12,572	6,275	6,297
20-24	11,024	5,702	5,322	1,015	513	502	10,009	5,189	4,820
25-29	8,200	4,276	3,924	784	403	381	7,416	3,873	3,543
30-34	6,687	3,472	3,215	721	350	371	5,966	3,122	2,844
35-39	5,944	3,070	2,874	585	300	285	5,359	2,770	2,589
40-44	4,665	2,491	2,174	404	208	196	4,261	2,283	1,978
45-49	4,038	2,156	1,882	365	184	181	3,673	1,972	1,701
50-54	3,407	1,824	1,583	314	158	156	3,093	1,666	1,427
55-59	2,429	1,293	1,136	181	98	83	2,248	1,195	1,053
60-64	1,922	1,003	919	152	72	80	1,770	931	839
65-69	1,576	856	720	117	55	62	1,459	801	658
70-74	929	482	447	72	37	35	857	445	412
75-79	645	370	275	54	32	22	591	338	253
80-84	292	158	134	20	11	9	272	147	125
85-89	118	55	63	9	2	7	109	53	56
>90	51	26	25	3	2	1	48	24	24

Source: DGEEC (2014, p. 243).

Men make up a slight majority (52%) of the population compared to women (48%), and most of the population is very young, with around 54% of the indigenous people surveyed under 19 years of age (figure 19.2). Around 81% of the economic activities carried out by Paraguayan indigenous people are related to the primary sector; the main activities are crop and livestock production, hunting, fishing and gathering. Approximately 10% of indigenous people work in the tertiary sector (provision of services), and 7% in the secondary sector (industries and construction) (DGEEC, 2014, p. 74).

Hunting and fishing activities are found in 655 (92%) of the indigenous communities, with fishing being the most common, followed by hunting of armadillos, lizards and deer, along with forest birds, wild pigs, coatis, capybaras, crocodiles, monkeys, ostriches, rabbits, guinea pigs, tapirs, anteaters, tortoises, tigers and others. Indigenous peoples also raise livestock, with the activity found in 693 (97%) communities. Chickens, ducks and guineafowl were raised in 643 (93%) communities; other animals included dogs, cattle, pigs, goats, horses, sheep, mules and donkeys and oxen.

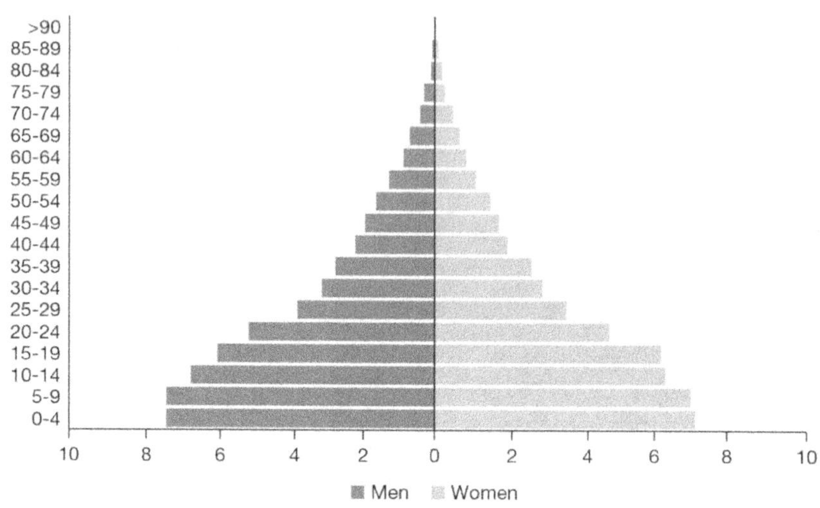

Figure 19.2. Indigenous population by age group and sex in Paraguay, 2012. Source: DGEEC (2014, p. 53).

A total of 630 (87%) indigenous communities gathered food from forests, fields or other places for their food supply, including wild honey (78%), guavira (41%), carob (31%), green beans (29%) and coconuts (28%). Other foods included guembe, tuna, heart de palm, wild mango, yvaviju, chaga mushrooms, dwarf yatay palm, chili pepper, Brazilian cherry, among others (DGEEC, 2015). The practice of food gathering is historical among the Guaraní people, who have always lived in a symbiotic relationship with nature:

> The Guaranís had already developed a food system that was extremely well suited to the ecosystem of the northern part of the Río de la Plata basin. They produced more than enough food to meet their needs (Meliá, 1988). At the same time, theirs was a civilization of leisure: their food production and gathering was very efficient, allowing them sufficient free time for physical activities, games, arts, family and educating their children and young people, and developing a deep knowledge of the human body and nature (Bertoni, 2009). According to Bertoni, the Guaraní diet was almost vegetarian; given their agricultural achievements, their extensive knowledge of various wild fruits, and the abundance of fish in the rivers and streams, they did not overly depend on hunting (Bertoni, 2009) (...) Guaraní agriculture showed good agroforestry management, with its impact offset by changing crops periodically, opening new areas by slashing and burning, and leaving the old areas to recover and grow back into forests. Their agricultural production was extremely diversified; they had more than 24 species of manioc, 21 species of sweet potato and 13 species of corn. These base crops were accompanied by numerous species of squash, legumes including various beans, and peanuts, which the Guaraní were the first to master (Noelli, 1994). (Doughman, 2011, p. 81)

Crafts are an important part of the life of the communities, villages and barrios; around 535 (75%) produce crafts, where women do most of this type of work (68%). Among the materials used are vase plant (53%), wood (49%), wool (39%), elephant ears (36%) and palm leaf (35%). Other materials include takuapi, seeds, takuara, cotton, feathers, mud, wild animal skins and giant bulrush. Some materials are produced in the communities and others are collected from nature, which means that sustainable production practices must be ensured around the indigenous communities so that the forests remain diversified (DGEEC, 2015).

The census identified a total of 28,926 indigenous dwellings (table 19.4); most are private, while eight are collective dwellings (religious communities, hospitals or health centers, educational boarding schools, etc.). A total of 27,905 of the individual indigenous dwellings are occupied, while for the others, people may be temporarily absent or the buildings are undergoing repairs or construction, or are abandoned (DGEEC, 2014). In these dwellings, the indigenous people were asked if the members carried out subsistence activities. Of those who responded, hunting, fishing, food gathering and cultivation for their own consumption were important for a significant portion of the population. Respondents mentioned product sales most with regard to growing crops.

Table 19.4. Dwellings occupied by inhabitants who have at least one household member engaged in subsistence activities in Paraguay, by activity, 2012

Type of livelihood activity	Total	Destination of the Activity			
		Sales	Consumption	Sales and Consumption	Not reported
Hunting	13,090	336	11,098	1,455	201
Fishing	15,104	304	13,690	958	152
Food gathering from the forest, field or other places	14,632	330	12,406	1,612	284
Cultivation of crops	18,356	547	13,843	3,772	194

Source: DGEEC (2014, p. 161).

Table 19.5. Dwellings occupied by inhabitants by means of obtaining food supplies in Paraguay, 2012

Place where food is obtained	Total	Forma de obtención de Provista						
		Cash	Credit	Voucher	In exchange for work	Bartering	Other	Not reported
Country total	**27905**	**23,016**	**2,163**	**453**	**1,086**	**103**	**283**	**801**
Supermarket/ small grocery/ stand	15,265	13,471	1,010	246	423	24	43	48
Middlemen	2,096	1,768	218	–	83	5	15	7
Warehouse	1,004	557	163	4	223	48	4	5
Indigenous grocery	6,036	4,923	662	92	241	14	54	50
Cooperative	2,369	2,132	80	100	39	–	8	10
Other	391	120	23	7	56	11	158	16
Not reported	744	45	7	4	21	1	1	665

Source: DGEEC (2014, p. 156).

Other census data on food dynamics of the indigenous people is related to the way they obtain supplies (table 19.5). Most make cash purchases of foods in super-markets, small grocery stores or stands, followed by indigenous grocery stores, cooperatives and warehouses/middlemen.

Over the years, it is clear that indigenous populations in Paraguay have declined and that their forms of food production, gathering and consumption have changed. However, what the following data shows is that a large part of these changes resulted from a change in the Paraguayan agrifood system due to the expansion of agribusiness.

▸▸ The pressure of agribusiness on the sustainable food systems of indigenous people

The food of indigenous people is marked by gathering, hunting and fishing, due to a historical experience of living in harmony with nature. However, the advance of agribusiness in Paraguay has altered this reality. The census shows that indig-enous communities face similar problems as Paraguayan peasant family farmers, since they have problems with land tenure, access to natural resources, access to infrastructure (roads, housing, potable water, health services, etc.), among others. Census respondents mentioned difficulties associated with land tenure and recogni-tion of rights, land titles, extension, use, the institutions that grant authorizations, official processes to legalize their situations, renting of land to third parties, and natural resources. The most frequently cited problems were a significant reduction of wild animals, spraying of agricultural chemicals by neighbours and contamination of waterways (DGEEC, 2015).

Over the years there has been a redistribution of the indigenous population by region in Paraguay (figure 19.3). In 1981 almost two-thirds of the indigenous people lived in the western region, but today more than 50% live in the eastern region, as can be seen in the figure below.

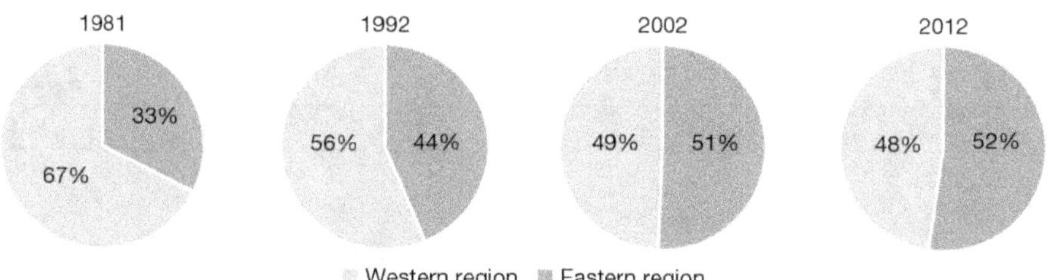

Figure 19.3. Distribution of the indigenous population by region (%) in Paraguay, according to census year for the 1981–2012 period. Source: DGEEC (2014, p. 54).

To a certain extent, the redistribution of indigenous people occurred under the pressure of large producers, who plan land use and manage machinery. Most of these producers are foreigners, and especially Brazilians, Germans, Japanese and Mennonites. The Mennonites arrived in the western region in the 1930s and later also acquired land in the eastern region.

In 1947, the Mennonites arrived in the area. They bought large tracts of land (also belonging to the aforementioned industrial Paraguay), including those occupied by the indigenous people. (…) Over time, they took over more and more land, until they reached the land on which our communities were still living. Until about six years ago, the situation was very difficult. We received an offer from the Mennonites themselves to rent our land, and we decided to rent 15 of the 30 hectares we have. Today, soy and wheat plantations surround our houses; and when they (the tenants, who are foreign entrepreneurs) spray, they affect us, killing our animals (chickens) and we start to feel unhealthy… (…) Meeting our needs, mainly for our sons and daughters, forces us to rent our land to the Mennonites. We know that, in time, they may take our land. But we have no other alternative if we want to eat. The money we receive is not enough…. The land is very good, very fertile, of a red clay color. We want to work on a production plan, to try to definitively recover our land and stop renting it (Report by an *apud* indigenous person, Baquero & Gómez, 2012, p. 422).

The reports point to the difficulties that indigenous people suffer in guaranteeing that their sustainable food systems are maintained. As with peasant family farmers, indigenous people are also pressured by foreign land acquisitions, whether by purchase, undue appropriation by entrepreneurs or renting. Furthermore, the penetration of foreign companies makes the situation more critical, and especially those located in the eastern region, in the departments bordering Argentina and Brazil, which have fertile land of high quality and constitute environmental spaces rich in basic natural resources: water and mountains. (Baquero & Gómez, 2012).

The problems related to land tenure in indigenous communities differ. Of the 493 indigenous communities, 143 (29%) indicate problems related to land tenure, with most (66%) in the eastern region compared to the western region (34%) (CENSO, 2015, p. 41). Among indigenous communities with issues, 47 (33%) declare having problems related to provision or renting land to third parties, 42 (29%) faced undue appropriation by business owners, 31 (22%) mentioned invasion by rural inhabitants, 23 (16%) superposition of legal titles, among other difficulties. In the eastern region, the majority of problems are found in the departments of Canindeyú, Caaguazú and Amambay, and in the western region, in the departments of Presidente Hayes and Boquerón.

The indigenous census recorded that the renting or provision of land is an important source of income for 182 indigenous communities (approximately 37%), where 95 communities rented to Paraguayans, 70 to Brazilians and 28 to others. Land that is rented or provided to third parties is used for extensive crop farming (93 communities), grazing (58 communities) and extraction of wood and coal production (12 communities). In the eastern region, most land is used for extensive crop farming, while grazing is more common in the northern part of the eastern region and central part of the western region. These divisions coincide with the dynamics of soy production and livestock grazing, and the regions where agribusiness is expanding (Wesz Jr., 2020; Silveira et al., 2019; Schmalko & Sarta, 2018). Not only is there pressure from grazing, but also pressure from agricultural products, and especially soy. Table 19.6 shows the increase between 2002 and 2017 of livestock production, while crop production is shown in table 19.7 for 2017.

Table 19.6. Numbers of livestock production in 2002 and 2017 in Paraguay

Department	Livestock production		Increase	Increase (%)
	2002	**2017**	**2002-2017**	**2002-2017**
Concepción	651,631	1,158,562	506,931	78
San Pedro	1,082,258	1,319,848	237,590	22
Guairá	195,297	150,923	–44,374	–23
Caaguazú	596,781	552,926	–43,855	–7
Caazapá	294,950	327,899	32,949	11
Itapúa	484,768	436,357	–48,411	–10
Upper Paraná	392,751	203,418	–189,333	–48
Misiones	461,154	489,705	28,551	6
Paraguari	418,617	471,508	52,891	13
Ñeembucu	398,768	582,099	183,331	46
Cordillera	231,757	254,156	22,399	10
Central	79,362	57,604	–21,758	–27
Amambay	590,544	970,061	379,517	64
Canindeyú	571,758	713,660	141,902	25
President Hayes	1,695,542	2,525,699	830,157	49
Boquerón	1,033,356	1,949,310	915,954	89
Upper Paraguay	198,887	1,659,916	1,461,029	735
Total	**9,378,181**	**13,823,651**	**4,445,470**	**47**

Source: Adapted from Schmalko and Sarta (2018, p. 59).

Table 19.7. Land area occupied by crop production in 2017 in Paraguay

Department	Soybean (ha)	Corn (ha)	Wheat (ha)	Rice (ha)	Sugarcane (ha)	Sunflower (ha)	Total
Concepción	38,860	7,223	–	50	344	44	46,521
San Pedro	325,397	103,539	11,050	3,610	2,508	1,465	447,569
Guairá	12,603	4,749	3,400	39	44,693	161	65,645
Caaguazú	444,938	146,442	85,000	492	17,444	7,909	702,225
Caazapá	168,045	56,598	55,250	16,017	5,930	2,929	304,769
Itapúa	603,521	70,352	183,600	41,511	654	14,646	914,284
Upper Paraná	926,158	269,632	127,500	810	325	8,788	1,333,213
Misiones	35,137	4,749	8,500	63,364	2,262	513	114,525
Paraguari	48	4,839	–	1,794	25,468	–	32,149
Ñeembucu	–	4,839	–	8,185	108	37	13,169
Cordillera	–	3,958	–	3,063	7,291	–	14,312
Central	–	341	–	798	3,215	–	4,354
Amambay	169,955	45,545	10,200	176	1,475	220	227,571
Canindeyú	650,220	216,695	25,500	93	6,215	4,189	902,912
President Hayes	–	361	–	–	39	–	400
Boquerón	4,774	8	–	–	–	–	4,782
Upper Paraguay	344	22	–	–	–	–	366
Total	**3,380,000**	**939,892**	**510,000**	**140,002**	**117,971**	**40,901**	**5,128,766**

Source: Adapted from Schmalko and Sarta (2018, p. 43).

The concentration of extensive crop farming, the production of pasture for livestock, as well as the extraction of wood in and around the indigenous communities' lands has generated innumerable difficulties. As natural resources are affected, so are the food sources of the indigenous peoples due to the importance that hunting, fishing and gathering play in these communities. The census shows that of the total number of indigenous communities surveyed, 254 (52%) communities declared some difficulty with regard to natural resources. The most frequently reported difficulty is the significant decrease of wild animals (134 communities), followed by agrochemical spraying (109 communities) and contamination of waterways (101 communities). The problems related to agrochemical spraying and waterway contamination are mainly found in the eastern region, which coincides with the regions with the highest production of annual crops of soybeans, corn, wheat, rice, sugar cane and sunflower (Schmalko & Sarta, 2018).

▶▶ Indigenous people on the government agenda: the establishment of the National Program for Indigenous Peoples (PNPI)

Paraguay's 1992 constitution marked a new stage because it recognized Guaraní as the official language of the country, established a series of recognitions and a legal framework of national and international laws regarding the protection of indigenous rights in Paraguay (Bote, 2019). For Bote, in the history of Paraguay, there were few moments of recognition of indigenous peoples in national politics until this period. One rare example was the creation of the Department of Indigenous Affairs in 1958, which occurred due to defense-related issues of the Cold War, where "it was necessary to control the movements in the jungle areas where the opposing forces were hiding" (Ibid., p. 62). Bote also noted that the most important achievement during president Stroessner's time was the Statute of the Indigenous Communities, which recognized the legal personality of the communities and their land regime, and considered the Customary Law, although besides the "unclear and in some cases, contradictory provisions of this law, its most evident lack was the absence of a sanction mechanism for non-compliance" (Ibid., p. 64).

In the 2000s, after efforts by the Ministry of Public Health and Social Welfare, representatives of the indigenous people, civil society organizations, cooperation agencies and the Interinstitutional Indigenous Health Committee, the National Indigenous Health Policy was created through Resolution No. 143/2008, coordinated by the Directorate General of Assistance to Vulnerable Groups. In 2009, the Ministry of Public Health validated the National Indigenous Health Policy, and made the Directorate General of Health Service Development and the Directorate of Primary Health Care responsible for the monitoring and execution of the policy (MPS, 2009).

The policy mentions the health problems of indigenous people due to the changes in the agrifood systems in Paraguay. For example, the massive spraying and indiscriminate use of pesticides in the country have caused acute and chronic poisoning, while the practice "also affects their subsistence, since they drastically reduce their sources of food and natural medicine" (MPS, 2009, p. 16). The policy also emphasizes a concern with "changes in their lifestyles, reduction of their habitat, migration to urban areas, changes in eating habits, etc." (Ibid., p. 16), which can lead to high blood pressure and diabetes, among other illnesses, and that "the primary occupation of indigenous people is aggravated today by high deforestation" resulting in "the destruction of the area from which they obtained the majority of their food" (Ibid., p. 17). For the reasons described above, among the specific objectives of the National Indigenous Health Policy, we point out the specific objective of "creating alliances with involved institutions to carry out actions related to food security" (Ibid., p. 25). Thus, the National Indigenous Health Policy continues to be an important action that also deals with food security.

In recent years, the government has promoted the creation of a National Program for Indigenous Peoples (PNPI), which aims to establish guidelines for action around seven thematic areas (land, legal framework, housing, education, production systems, food security and uncontacted peoples) to create a national policy developed by different indigenous leaders who have a deep understanding of the socioeconomic reality of their community (Bote, 2019).

Taking into account this process and the reported reality, in 2014 the Inter-Institutional Committee on Food Security and Nutrition was created, headed by the Paraguayan Indigenous Institute. According to the report of one of the FAO representatives interviewed for this research: "(…) between 2012 and 2013 a consultancy [at the FAO] was developed where participatory dialogues were held with representatives of indigenous people and institutions to work on a food security diagnosis and proposal (…)"

The result of the process, which took more than a year of work, was a diagnosis of the situation of indigenous peoples in Paraguay (MAG, 2014). This diagnosis was carried out within the framework of the TCP/RLA/3403 Project "Food Security and Nutrition Policies and Indigenous People in Paraguay," coordinated by the Ministry of Agriculture and Livestock, through its Department of Technical Assistance to Indigenous Communities under the Directorate of Agricultural Extension, the support of the Paraguayan Indigenous Institute (INDI), and the cooperation of the FAO. As stated by the FAO representative in Paraguay: "The result of this process, which took more than a year of work, was a diagnostic document about the food security of indigenous people in Paraguay, and a proposal to work on a Food and Nutritional Security Plan. (…) This first process was released by the Ministry of Agriculture through this department of Technical Assistance to Indigenous Communities" (Interviewee 1, June 13, 2020).

The diagnosis concluded that public policies have a low impact on indigenous people and lack differentiated strategies. There is also a dispersion of interinstitutional coordination bodies in the indigenous context, and the level of institutionalization is low. Indigenous participation is low in influential forums for public policy, compounded by the existing weakness of indigenous coordination. The problem persists with regard to the recovery, assurance and possession of ancestral territories, as well as long-standing structural issues and an ignorance about the culture and worldview of these people around food security (MAG, 2014, p. 15). The diagnosis was the first work specifically carried out with indigenous people, and was conducted between 2012 and 2013.

In 2015, the follow-up of the diagnosis took place with the leadership of INDI and its Resolution 302/2015, which created the Inter-Institutional Committee for Food

Security and Nutrition. Its purpose is to strengthen the INDI by coordinating inter-institutional actions with indigenous people in the framework of the Food Security Program in an advisory capacity and to follow up on the commitments made by public and private institutions and indigenous organizations in this area (STP, 2016, p. 47). In the process of building the Committee, representatives of departmental governments were incorporated. The changes in the presidency of INDI contributed to the discontinuity of the meetings and decisions for the creation of the plan:

> In the second stage, now with the leadership of INDI, to strengthen this forum for coordination. We worked… it was not easy to establish a common forum and work on the objective, the regulations of this group. We needed several forums for dialogue and discussion, and we ended up having institutionalized this interinstitutional committee, where local and departmental governments were also included in this process (…) And we ended this process with the institutionalization of this interinsti-tutional committee in the INDI. Since then, the INDI president has changed, there have been several sticking points on standby until now (Interviewee 1, June 13, 2020).

The Committee is comprised of an executive component and two indigenous organizations. In 2018, the Committee was responsible for the creation of the "Protocol for the process of consultation and free, prior and informed consent of indigenous people living in Paraguay," which basically guarantees the right of indigenous people to be consulted about any project that could affect their traditional lands, territories, natural resources and livelihoods. The protocol also guarantees the right of the affected indigenous peoples to give their free, prior and informed consent to the proposal, as well as to decide if they want to participate in the consultations or to terminate the consultations at any time. The Paraguayan state is required to hold the consultation (Paraguay, Decree No. 1039/2018). Thus:

> Specific actions were carried out with this committee (…) because this committee is presided over by an executive coordinator, who nominated INDI and two indigenous organizations alongside it, which are responsible for the executive coordination of this committee (…) through this body. The executive coordination was the one that also supported the reactivation of the process of creating the protocol for consultation and free, prior and informed consent. This had already been under discussion for several years with the indigenous people and the executive coordination reactivated and able to revalidate the protocol to presented it to the executive, which was then approved. Today, after several back and forths, we already have Decree 1039 for the consultation and free, prior and informed consent (Interviewee 1, June 13, 2020).

Since 2015, Paraguay has been subject to the commitment to carry out informal free consultation for the creation of the National Plan for Indigenous Peoples (PNPI) of Paraguay, which was carried out between 2016 and 2019. The process of building the PNPI was coordinated by the INDI with a lead team that includes the Technical Secretariat for Economic and Social Development Planning. The Directorate of Indigenous Education of the MEC, the Department of Technical Assistance to Indigenous Communities of the Directorate of Agrarian Extension of the Ministry of Agriculture and Livestock, the Directorate General of Human Rights of the Ministry of Foreign Affairs, the National Secretariat of Culture, the Guarani Feder-ation, the Federation for the Self-determination of Indigenous People, and the FAO were also involved in the discussions.

The plan, which has a 2020 timeline that is almost ready, is the result of a participatory construction in which approximately 600 community leaders recognized by the INDI and representing the indigenous people of the departments took part (STP, 2019). The plan presents thematic issues defined by the lead team, based on the particular contexts of each area, such as land, territory, natural resources, spirituality, food security, craftwork, women, identity, housing, children, health, education and youth, among others. In the words of the FAO representative, it is understood that:

> The plan has four strategic focus areas. The first focus area deals with conceptual issues, on the worldview of indigenous people, where we touch on quite specific themes that can strengthen the identity of indigenous people. In which we speak of participation, free determination, identity, culture, worldview and above all what is essential for them, which are natural resources, land and territory, truth. The second focus area is based on social and economic rights, where we have all the thematic issues related to basic services: health, education, food security, roads, electricity (...) During this consultation process, we worked on understanding their vision to be able to build the plan with them and put forward the main issues. We developed the thematic focus areas with them (...) The first point with them was what was their feeling, their worldview with regard to this thematic focus area (...) and after this analysis, after they told us how they see this thematic focus areas, we tried to get a deeper understanding of the current situation that they are going through in their communities with a broader vision in the department and at the national level. After that, we had a proposal for improvement. We also discussed with them from their worldview, how they see, what we should do, what they should also do to improve the situation in which they are in terms of this thematic focus area. (...) Then, we had another axis that is basically directed at the population... so, for those with special needs, such as women, children, elderly people, children living on the street. (...) Each thematic focus area has general and specific objectives, central measures, and as I said, we are seeing apart from the institutions which existing initiatives can contribute and which initiatives/programs have to be established in order to be able to comply with the central measures, really (Interviewee 1, June 13, 2020).

While Covid-19 paralyzed the final implementation of the plan, one interviewee said that it was possible that it would be finalized by August. Despite not having a specific public policy, the indigenous peoples of Paraguay have access to policies that guarantee their sovereignty and food security, although it is fragile as mentioned in the diagnosis (MAG, 2014). Accordingly:

> On the one hand, this is the budget issue. However, more than anything, it was due to the political instability that we had in the institution, in the governing body, the INDI. Because we had leadership changes and when this happened, it took us time to be able to bring the new leader up to speed, and then when we started to start up, leadership changed again. So this process was not continuous, it was really a challenging process, and today we already have the draft plan of indigenous peoples that in April we had to approve with the leaders that were designated in each territory, and well, because of the pandemic we could not develop that forum to do so and have the plan validated by the indigenous peoples and then request an extension by the executive. Today we are hoping that we can create this forum, we are assessing, analyzing alternatives to be able to get the approvals so that we don't miss out again this year and still not have the plan. And this plan is the most historic and the biggest thing that we have here at a national level for the indigenous peoples. So, this is a broad process that we are reaching towards the end (Interviewee 1, June 13, 2020).

Another initiative emerged between 2015 and 2018. The Joint Programme on Food Security and Nutrition (PC-SAN) was created and implemented with the financial support of the Spanish Agency for International Development Cooperation and six other United Nations agencies, funds and programs, under the leadership of the Pan American Health Organization/World Health Organization, the FAO, the World Food Programme, the United Nations Children's Fund, UN Women and the Office of the United Nations High Commissioner for Human Rights. On the government side, nine public institutions were involved, under the leadership of the Technical Secretariat for Economic and Social Development Planning, the Paraguayan Institute of Indigenous People, the Ministry of Agriculture and Livestock, the Ministry of Education and Science, the Ministry of Public Health and Social Welfare, the Secretariat of Environment, the Secretariat of Social Action, the Ministry of Women, the Secretariat of National Emergency and the support of the governors and municipalities of the PC-SAN's area of influence. Around 3,000 indigenous families and 600 rural families benefitted, with Spain contributing USD 1.5 million in the period from 2015 to 2018 (AECID, 2018; STP, 2018).

The PC-SAN aimed to protect the vulnerable population, ensure food and nutritional security and reduce child and maternal malnutrition in the departments of Presidente Hayes, Caazapá and Caaguazú. The program's main contributions were: i) coordination of actions between the USF and regional health centres to regularly deliver milk to communities; ii) production of participatory diagnostics focused on sustainable lifestyles of those involved; iii) promotion of food production, with provision of seeds and technical support to participating indigenous and rural families; iv) installation of school gardens, orchards and demonstration plots and teaching of sustainable management practices; v) promotion of the commercialization of products (agricultural and handicrafts) and income generation, the creation of fairs and teaching about the processes of commercialization, from product presentation to marketing and linking with local markets through the fairs; vi) income generation capacity building through leadership training for women to promote social and financial inclusion; vii) promotion of "Semilla Róga" seed banks, to ensure the survival of native species and seeds for the next season; viii) development of public spaces for safe water and hygiene adapted to rural population and indigenous communities; ix) training and technical inputs for honey gathering; and x) participatory production of intercultural training and food security materials with indigenous peoples (STP, 2018). With the Covid-19 pandemic, indigenous people have become even more fragile, especially those who live with food insecurity. They were included in the beneficiaries of the Ñangareko Food Security Program (*ñangareko* means "take care of" in Guaraní). The program consists of monetary transfers of 230,000 Paraguayan guaraní (USD 35[3]), which can be used only for the purchase of food and hygiene products or direct delivery of food. Data from the National Emergency Secretariat, on April 4, 2020, show that more than 100,000 kilograms of food were distributed to approximately 6,000 indigenous families in the towns of Fuerte Olimpo, Carmelo Peralta, Puerto Sastre and Isla Margarita in the department of Alto Paraguay (AIP, 2020).

---

3. Dollar amounts are converted based on the February 16, 2021 quotation table on the Banco Central del Paraguay website. Available at: https://www.bcp.gov.py/webapps/web/cotizacion/monedas

Table 19.8. The dimensions of the Policy for Indigenous Peoples in relation to food sovereignty and security in Paraguay, 2020

Policy dimensions		
Policy	*Polity*	*Politics*
National Indigenous Health Policy (2009 to present)	Resolution No. 143/2008: the Ministry of Public Health validated the National Indigenous Health Policy General Directorate for the Development of Health Services and the Directorate of Primary Health Care. Mesa Interinstitucional de Salud Indígena	Ministry of Public Health and Social Welfare Organization of civil society Federation for the Self-determination of Indigenous People (FAPI) National Coordinator of Pastoral Indigenous – CONAPI Cooperation agencies Directorate General for Assistance to Vulnerable Groups
National Indigenous Peoples Plan (PNPI)	Interinstitutional Commission on Food and Nutritional Security (SAN) Paraguayan Indigenous Institute (INDI) Ministry of Agriculture and Livestock Department of Technical Assistance to Indigenous Communities Directorate of Agricultural Extension Resolution 302/2015 (INDI) Decree No. 1039/2018 – Protocol for the process of consultation and free, prior and informed consent with indigenous peoples living in Paraguay.	United Nations Food and Agriculture Organization (FAO) 600 community leaders recognized by the INDI
Joint Program for Food Security and Nutrition (PC-SAN) (2015–2018)	Technical Secretariat for Economic and Social Development Planning Paraguayan Indigenous Institute (INDI) Ministry of Agriculture and Livestock Ministry of Education and Science Ministry of Public Health and Social Welfare Secretariat of Environment Secretariat of Social Action Ministry of Women National Emergency Secretariat Governments and municipalities in the area of influence of the PC-SAN (in the Presidente Hayes, Caazapá and Caaguazú departments)	Spanish Agency for International Development Cooperation (AECID) Pan American Health Organization/World Health Organization (PAHO/WHO) United Nations Food and Agriculture Organization (FAO) World Food Programme (WFP) United Nations Children's Fund (UNICEF) UN Women Office of the United Nations High Commissioner for Human Rights (OHCHR)

Within public policies for indigenous people (table 19.8), the actions of indigenous organizations in Paraguay stand out. Palau et al. (2018, pp. 94-95) indicate that the organizations ask "for respect of their territories, compliance with legal requirements, public policies for communities, and that the National Indigenous Institute (INDI) serve as the representative body for request for social movements to the Paraguayan state administered by an indigenous representative." The same study touches on the complex relationship between the Paraguayan state and indigenous people, above all due to the colonial heritage that historically excluded indigenous people from territories and public policies. Demands for public policies from two associations for the country's indigenous people (the Mesa de Articulación Indígena y Mujeres Indígenas del Paraguay and Indigenous Women of Paraguay), report the stalled efforts of Territorios ("territories"), an INDI administered by an indigenous representative, the non-compliance with Convention 169 of the International Labour Organization and the United Nations Declaration on the right to prior consultation, but above all, the lack of public policies for indigenous communities.

In recent years, many of the organizations were mobilized for the approval of Act No. 6286/2019 on "Defense, Restoration and Promotion of Family Farming," approved and published on May 25, 2019. The law's first article recognizes the responsibility of the state in the:

> (...) recovery, preservation and stimulation of the economy; social protection and improvement of the quality of life of rural families and indigenous people, so that their economic and productive endeavors develop with dignity through the implementation of programs that facilitate access to land, housing, public services, communications and transportation; training and generation of science and technology for the countryside, price stabilization mechanisms, markets, as well as adequate technical and financial assistance to the entire production chain linked to rural family farming. (Paraguay, Act No. 6286/2019)

Having a law does not guarantee that the Paraguayan state will direct its actions to farmers and indigenous people, but it is a political instrument for social organizations to "fight for rural reform that democratizes access to land and recovers its role as a producer of healthy food for the population" (Riquelme, 2020, p. 40).

▶▶ Final considerations

Although indigenous people, especially the Guaraní people, have a long history in Paraguay, despite the fact that their recognition in public policies is very recent. Moreover, indigenous people face various structural problems in their everyday lives, where "poverty, exclusion and disenfranchisement are issues that seriously threaten the development of these populations, keeping them enclosed in a vicious cycle that is very difficult to escape" (Rojas & Bote, 2019, p. 83).

The creation and execution of the census of indigenous peoples was fundamental for the visibility of indigenous people in Paraguay, and the recent construction of the PNPI demonstrates the mobilization of civil society involved in the planning of the plan, which exerts pressure to include indigenous people in the government agenda. With regard to progress, the interviews underlined a decline in investment from the INDI. Rojas and Bote (2019, p. 84) point out that in "2019 the Paraguayan Indigenous Institute's budget fell by 35% compared to the prior year."

The food system and habits are a reflection of the ways in which they are organized. In today's world, the expansion of agribusiness – mainly soy production and livestock farming – has forced indigenous people to abandon their territory and move to the cities. New intensive production practices (use of modified seeds, fertilizers and pesticides) and occupation of the territory increase deforestation, which has an enormous impact on the indigenous people, their food system, their production practices, food gathering, production of handicrafts and food consumption. For Rojas Brítez (2012, p. 14), "it is the dispossession of the Guaraní people from their ancestral lands that marks the accelerated destructuring of their *tekoporã*. It is now what defines their sociocultural profile."

It is therefore necessary to reflect on the permanence of indigenous people, their knowledge, and their sustainable contribution to the Paraguayan population:

> With so little, you do so much with them. You do not need many resources. Often, your support, your presence, your accompaniment can do a lot for them. That is what I always say, they are so few here in Paraguay, in terms of population percentage, and the fact that they are as they are is because in reality at the moment of planning, at the moment of implementing actions, we are not really trying to improve their living conditions (…) There is no recipe, (…) you may have a base to start from, here the important thing is to listen and build on that together. There is a lot to do for them. And Paraguay has much to give to indigenous people" (Interviewee 1, June 13, 2020).

The Federation for the Self-Determination of Indigenous Peoples (FAPI) in its Five-Year Strategic Plan 2017–2021 states:

> In recent decades, the pressure for land and territories of indigenous people and communities has increased, leading to dispossessions and violations of collective rights, usurpations and evictions of communities from their traditional lands for the subsequent dismantling and cultivation of soybeans or other cereals in the eastern region and cattle farming in the western region. A very worrying factor to highlight in the framework of the accelerated expansion of the agricultural frontier is the attempt to modify Act No. 904/81, which guarantees the right to land for indigenous people, although its parameters do not sufficiently contemplate the cultural aspects in terms of the quality and quantity necessary for the cultural reproduction of indigenous people (FAPI, 2017, p. 09).

In Paraguay, a visible process of "disenfranchisement of the rights of indigenous people over their lands, territories and resources, vital for their survival and dignity, which generates conflicts and violations of human rights" (Guereña & Rojas Villagra, 2016, p. 83) in rural areas to contact the population while changing its food system, its forms of food production, gathering and consumption. It was observed that the process does not only affect indigenous people in the country, but also the entire population (we recommend reading of the chapters 5 and 10 of this book). In this sense, one cannot fall into the trap of thinking that the PNPI will have strength by itself. Thus, in addition to ensuring that the PNPI is implemented, other processes in Paraguay must be rethought, such as rural development, the advance of agribusiness, and the various forms of production and exploitation of nature. Otherwise, the soy-and-livestock diptych will continue to suffocate the agrifood system of the country, the indigenous people and also of the entire Paraguayan nation.

▸▸ References

ABC Color. (2018, September 3). *Recortaron 24.000 millones al Indi*. Editorial AZETA. https://www.abc.com.py/edicion-impresa/locales/recortaron-24000-millones-al-indi-1737063.html

AECID. (2018, July 30). *Programa Conjunto de Seguridad Alimentaria y Nutricional*. https://www.aecid.org.py/programa-conjunto-de-seguridad-alimentaria-y-nutricional/

Agencia de Información Paraguaya (AIP). (2020, April 4). *Gobierno benefició hasta la fecha a 41.662 familias por medio de Ñangareko*. Ministerio de Tecnologías de la Información y Comunicación. https://www.ip.gov.py/ip/gobierno-beneficio-hasta-la-fecha-a-41-662-familias-por-medio-de-nangareko/

Baquero, F. S., & Gómez, S. (2012). *Dinámicas del mercado de la tierra en América Latina y el Caribe: concentración y extranjerización*. FAO. https://www.fao.org/3/i2547s/i2547s.pdf

Da Silveira, C. V., Da Silveira, G. S., Kukiel, E. D. G., & Vieira, R. M. (2019, May 27–31). *Dinâmica regional da economia paraguaia: o caso da soja e carne* [Conference presentation]. XVIII ENANPUR, Natal, Brazil. http://anpur.org.br/xviiienanpur/anaisadmin/capapdf.php?reqid=1031

Dirección General De Estadística, Encuestas y Censos. (2014). *Pueblos indígenas en el Paraguay. Resultados finales 2012 de población y viviendas*. Gobierno Nacional de Paraguay. https://www.ine.gov.py/Publicaciones/Biblioteca/indigena2012/Pueblos%20indigenas%20en%20el%20Paraguay%20Resultados%20Finales%20de%20Poblacion%20y%20Viviendas%202012.pdf

Dirección General De Estadística, Encuestas y Censos. (2015). *Censo de comunidades de los pueblos indígenas, resultados finales 2012*. Gobierno Nacional de Paraguay. https://www.ine.gov.py/Publicaciones/Biblioteca/comunidad%20indigena/Censo%20de%20Comunidades%20de%20los%20Pueblos%20Indigenas%20Resultados%20Finales%202012.pdf

Dirección General de Estadística, Encuestas y Censos. (2020). *Principales resultados de pobreza monetaria y distribución de ingreso – 2019*. Gobierno Nacional de Paraguay. https://www.ine.gov.py/Publicaciones/Biblioteca/documento/5781_Pobreza%20Monetaria%202019_Boletin.pdf

Doughman, R. (2011). *La chipa y la soja: la pugna gastro-política en la frontera agroexportadora del este paraguayo*. BASE Investigaciones Sociales. http://www.baseis.org.py/adjuntos/La%20chipa%20y%20la%20soja.pdf

Federación por la Autodeterminación de los Pueblos Indígenas (FAPI). (2017). *Plan Estratégico Quinquenio 2017–2021*. https://www.fapi.org.py/wp-content/uploads/2020/07/Plan-Estratégico-FAPI.pdf

Gómez Bote, J. M. (2019). *La política indigenista en el Paraguay desde 1992 hasta nuestros días: Integración y desafíos de las comunidades nativas actuales* [Master's thesis, University of Bergen]. Bergen Open Research Archive. https://hdl.handle.net/1956/20215

Guereña, A., & Rojas Villagra, L. (2016). *Yvy jára: Los dueños de la tierra en Paraguay*. OXFAM.

Imas, V. J. (Ed.). (2019). *Seguridad y soberanía alimentaria en Paraguay. Sistema de indicadores y línea de base*. Centro de Análisis y Difusión de la Economía Paraguaya. https://www.conacyt.gov.py/sites/default/files/upload_editores/u294/seguridad_soberania_alimentaria_cadep.pdf

Jaime, F. M., Dufour, G., Alessandro, M., & Amaya, P. (2013). *Introducción al análisis de políticas públicas* (1st ed.) Universidad Nacional Arturo Jauretche. https://www.unaj.edu.ar/wp-content/uploads/2017/02/Pol%C3%ADticas-públicas2013.pdf

Ministerio de Agricultura y Ganadería, Paraguay. (2014). *Informe Nacional – Diagnóstico y propuestas para el desarrollo de una política pública de seguridad alimentaria y nutricional de los pueblos indígenas en Paraguay*. FAO. https://www.fao.org/3/i3863s/i3863s.pdf

Ministerio de Educación y Ciencias, Paraguay. (2018). *Decreto nº 1039, Protocolo para el proceso de consulta y consentimiento libre, previo e informado con los pueblos indígenas que habitan en el Paraguay*. https://www.mspbs.gov.py/dependencias/portal/adjunto/7fdaf1-DECRETO1039Protocolodeconsultayconsentlibreprevioeinformadopueblosindigenas.pdf

Ministerio de Salud Pública, Paraguay. (2009). *Política nacional de salud indígena*. https://www.mspbs.gov.py/dependencias/portal/adjunto/a25a59-PoliticaNacionaldeSaludIndigenaParaguay.pdf

Palau, M., Coronel, C., Irala, A., & Yuste, J. C. (2018). *Canalización de demandas de los Movimientos Sociales al Estado paraguayo*. Base Investigaciones Sociales.

Riquelme, Q. (2016). *Agricultura familiar campesina en el Paraguay. Notas preliminares para su caracterización y propuestas de desarrollo rural.* Centro de Análisis y Difusión de la Economía Paraguaya.

Riquelme, Q. (2020). *Ley de la Agricultura Familiar Campesina en Paraguay, Ley n°6286: logro y desafío para las organizaciones campesinas e indígenas.* Centro de Documentación y Estudios. https://www.cde.org.py/wp-content/uploads/2020/05/Estudio-Ley-6286-Paraguay-1.pdf

Rojas Brítez, G. (2012). *Los pueblos guaraníes en Paraguay: una aproximación socio-histórica a los efectos del desarrollo dependiente* (Documento de trabajo no. 13). Centro de Estudios y Educación Popular Germinal. http://germinal.pyglobal.com/pdf/documento_trabajo_13.pdf

Schmalko, C. A., & Sarta, A. M. (2018). *Mapeando el agronegocio en el Paraguay.* Base Investigaciones Sociales.

Secchi, L. (2014). *Políticas públicas. Conceptos, esquemas de análise, casos práticos* (2nd ed.). Cengage Learning.

Secretaria Técnica de Planificación del Desarrollo Económico y Social, Paraguay. (2016). *Gobierno Aperto Paraguay. Informe final de autoevaluación del 2do plan de gobierno abierto. Julio 2014 a Junio 2016.* https://informacionpublica.paraguay.gov.py/public/10697123-AutoevaluacinfinalSegundoPA-GApdf-AutoevaluacinfinalSegundoPAGA.pdf

Secretaría Técnica de Planificación del Desarrollo Económico y Social, Paraguay. (2018). *Apuntes de experiencias y lecciones aprendidas: Esta es la historia que construimos en el periodo 2013–2018.* https://www.stp.gov.py/v1/wp-content/uploads/2020/03/STP_-Apuntes-de-experiencias-y-lecciones-aprendidas-2013-2018.pdf

Secretaría Técnica de Planificación del Desarrollo Económico y Social, Paraguay. (2019, September 26). *Avanza construcción del Plan Nacional de Pueblos Indígenas en Paraguay.* https://www.stp. gov.py/v1/avanza-construccion-del-plan-nacional-de-pueblos-indigenas-en-paraguay/content/ uploads/2014/12/pnd2030.pdf

Souza, C. (2006). Políticas públicas: uma revisão na literatura. *Revista Sociologias, 8*(16), 20–45. https://doi.org/10.1590/S1517-45222006000200003

Unión de Gremios de la Producción. (2015). *Agricultura y desarrollo en Paraguay.*

Villalba, A. M. (2019). *Políticas públicas diferenciadas para la agricultura familiar campesina. Paraguay (2004–2017).* CLAEH.

Wesz, Jr., V. J. (2020). A rentabilidade dos produtores de soja no Paraguai: concentração e exclusão. *Estudos Sociedade e Agricultura, 28*(1), 156–179. https://doi.org/10.36920/esa-v28n1-7

Wesz, Jr., V. J., Zimmermann, S. A., & Carreras Rios, F. D. (2018). La institucionalización de políticas públicas para la agricultura familiar en Paraguay. *Raízes: Revista de Ciências Sociais e Econômicas, 38*(1), 80–97. https://doi.org/10.37370/raizes.2018.v38.40

List of Spanish or Portuguese acronyms

ABC: Low-Carbon Agriculture Program, Brazil

ACHIPIA: Chilean Food Safety and Quality Agency

ACIP: Peasant, Indigenous and Popular Unity Committee

ADEPA: National Cotton Producer Association, Bolivia

ALBA: Bolivarian Alliance for the Peoples of Our America

ANAPO: Association of Oilseed and Wheat Producers, Bolivia

ANMAT: National Administration of Drugs, Foods and Medical Devices, Argentina

APEGA: The Peruvian Gastronomic Society

ARP: Paraguayan Rural Association

CAA: Paraguay Agricultural Advisory Council

CAA: Argentine Food Code

CAN: Paraguay National Agricultural Census

CAO: Eastern Agricultural Chamber, Bolivia

CAP: Paraguayan Agricultural Coordinator

CDSs: Sandinista Defense Committees

CELAC: Community of Latin American and Caribbean States

CIMMYT: International Maize and Wheat Improvement Center

CLA: Causal Layered Analysis

CNAFCI: Paraguayan National Committee for Family and Indigenous Agriculture

CNC: National Peasant Confederation, Mexico

CNI: National Intersectional Coordinator

CONACYT: National Science and Technology Council, Paraguay

CONAL: National Food Commission, Argentina

CONAN: National Council for Food and Nutrition, Bolivia

CONASUPO: National Company of Popular Subsistence, Mexico

CONICYT: National Commission for Scientific and Technological Research, Chile

COPLAMAR: National Commission for Deprived Zones and Marginalized Groups

CORA: Agrarian Reform Corporation

CORFO: Chilean economic development agency

CPE: Political Constitution of the State, Bolivia

DGEEC: General Directorate of Statistics, Surveys and Censuses, Paraguay

ECDBC: Colombian Low-Carbon Development Strategy

ECLAC: Economic Commission for Latin America and the Caribbean

EDBC: Low Carbon Development Strategy, Nicaragua

EINCV: Nutritional intervention strategy through the life cycle to prevent obesity and other noncommunicable diseases, Chile

EMBRAPA: Brazilian Agricultural Research Corporation

ENABAS: The Nicaraguan Basic Food Company

FAO: UN Food and Agriculture Organization

FECOPROD: Paraguayan Federation of Production Cooperatives

Fegasacruz: Santa Cruz Livestock Farming Federation, Bolivia

FEPRINCO: Paraguayan Federation of Production, Industry and Commerce

FF: Family Farming

FIA: Foundation for Agrarian Innovation, Chile

FIP: Paraguayan Federation of Industrialists

FLAMA: Latin American Federation of Supply Markets

FNC: National Peasant Federation

FNS: Food and nutrition security

FONAF: National Forum for Family Farming, Argentina

FONDEF: Scientific and Technological Development Support Fund, Chile

FSLN: Sandinista National Liberation Front, Nicaragua

GAP: Good agricultural practice

GHI: Global Hunger Index

GISAMAC: Intersectoral Group for Health, Food, Environment and Competitiveness, Mexico

HIPC: Heavily Indebted Poor Countries

ICESCR: International Covenant on Economic, Social and Cultural Rights

IDR: Nicaraguan Rural Development Institute

IFAD: International Fund for Agricultural Development

IICA: Inter-American Institute for Cooperation on Agriculture

IMF: International Monetary Fund

INAFOR: Nicaraguan National Forestry Institute

INAL: National Food Institute, Argentina

INAN: Paraguay National Institute of Food and Nutrition

INC: National Institute of Culture of Peru

INCAFE: Nicaraguan Coffee Institute

INCAP: Institute of Nutrition of Central America and Panama

INDAP: Institute of Agricultural Development, Chile

INDAP: Agricultural Development Institute

INDERT: National Institute for Rural and Land Development, Paraguay

INDI: Paraguayan Indigenous Institute

INTA: National Agricultural Technology Institute, Argentina

INTA: Nicaraguan Agricultural Technology Institute

INTA: Institute of Nutrition and Food Technology, Chile

IPHAN: National Historical and Artistic Heritage Institute

IPSA: Institute of Agricultural Protection and Health

IPTA: Paraguayan Agricultural Technology Institute
JUNAEB: National Board of Student Aid and Scholarships, Chile
JUNJI: National Board of Children's Gardens, Chile
LCP: Public Procurement Act, Uruguay
MACA: Bolivian Ministry of Peasant and Agricultural Affairs
MAG: Ministry of Agriculture and Livestock, Nicaragua
MANA: Food and Nutrition Improvement Program of Antioquia, Colombia
MARENA: Ministry of Environment and Natural Resources, Nicaragua
MAS: Movement for Socialism–Political Instrument for the Sovereignty of the Peoples
MasAgro: Sustainable Modernization of Traditional Agriculture Program, Mexico
MCP: Paraguayan Peasants' Movement
MDS: Ministry of Social Development, Argentina
MEFCCA: Ministry of Family, Community, Cooperative and Associative Economy, Nicaragua
MERCOSUR: The Southern Common Market
MESA: Extraordinary Ministry of Food Security and Fight against Hunger, Brazil
MIDINRA: Ministry of Agricultural Development and Agrarian Reform
MINED: Nicaraguan Ministry of Education
MNR: Revolutionary Nationalist Movement, Bolivia
NAFTA: North American Free Trade Agreement
NAMA: Nationally Appropriate Mitigation Action, Colombia
NI: New Institutionalism
PAA: Food Acquisition Program, Brazil
PACAM: Complementary Feeding Program for Older Adults, Chile
PAE: School Feeding Program, Chile
PAHO: Pan American Health Organization
PAN: National Action Party
PANF: Family Food and Nutrition Program, Mexico
PANI: Comprehensive Nutritional Food Program, Paraguay
PASAF: Program on Healthy Eating and Physical Activity, Chile
PESA: Special Program for Food Security, Mexico
PESA: Strategic Food Security Project
PIDER: Public Investment for Rural Development Program, Mexico
PINE: Comprehensive School Nutrition Program, Nicaragua
PLANAL: National Plan for Food and Nutrition Sovereignty and Security, Paraguay
PMDC: Multisectoral Zero Malnutrition Program, Bolivia
PNA: Nicaraguan National Food Program
PNA: National Food Program, Nicaragua
PNAC: National Supplemental Feeding Program, Chile
PNAE: National School Feeding Program, Brazil
PNAIR: Nicaragauan National Rural Agro-Industry Program
PNAN: National Food and Nutrition Plan
PNAS: National Plan for a Healthy Argentina
PNDH: National Human Development Plan

PNF: Nicaraguan National Forestry Program

PNPI: National Program for Indigenous Peoples, Paraguay

PNSA: National Food Security Plan, Argentina

PNSAN: National Food and Nutrition Security Policy, Nicaragua

PP-AL: Public Policy and Rural Development in Latin America network

PPA: Food Production Program, Nicaragua

PRI: Institutional Revolutionary Party

PROAN: National Assistance Program on Food and Nutrition

Progresa: Education, Health and Food Program, Mexico

PRONAF: National Programme for the Strengthening of Family Farming, Brazil

PRONAL: National Food Program, Mexico

PRONASE: National Seed Production Agency; Mexico

PRONASOL: National Solidarity Program, Mexico

PRSP: Poverty Reduction Strategy Paper

PSSAN: Municipal Public Policy Proposal for Food and Nutritional Security and Sovereignty

REAF: Specialized Meeting on Family Farming of MERCOSUR

REAF: Specialized Meeting on Family Farming of MERCOSUR

RENAF: National Registry of Family Farming, Argentina

SADER: Secretariat of Agriculture and Rural Development

SADER: Mexican Secretariat for Rural Development

SAM: Mexican Food System

SDGs: Sustainable Development Goals

SEDESOL: Mexican Secretariat of Social Development

SEGALMEX: Mexican Food Security Agency

SEGOB: Mexican Secretariat for Home Affairs

SENASA: National Service of Agrifood Health and Quality, Argentina

SICA: Central American Integration System

SINAREFI: National System of Plant Genetic Resources for Food and Agriculture

SME: small- and medium-sized enterprise

SNA: National Society of Agriculture, Chile

SNPCC: National System of Production, Consumption and Commerce

SNSSAN: National System for Food and Nutrition Sovereignty and Security, Paraguay

SOFOFA: Society of Industrial Development, Chile

SSN: Social Safety Net

TCOs: Native Community Lands

TIPNIS: Isiboro Sécure National Park and Indigenous Territory

UIP: Paraguayan Industrial Union

UNASUR: Union of South American Nations

USDA: United States Department of Agriculture

WFP: World Food Programme

WTO: World Trade Organization

About the authors

Ana María Costa. Professor-researcher at the Department of Social Sciences, Faculty of Agricultural Sciences, Universidad Nacional de Mar del Plata, Argentina. amariacosta.27@gmail.com

Carlos Urrutia. Bachelor's degree in Economics from the Universidad Nacional Mayor de San Marcos. Research assistant, Institute of Peruvian Studies (IEP). currutia@iep.org.pe

Carolina Trivelli. Senior researcher at the Institute of Peruvian Studies (IEP), consultant for the FAO and other international organizations. Former Minister of Development and Social Inclusion of Peru. trivelli@iep.org.pe

Catia Grisa. Professor of the postgraduate programs in rural development (PGDR) and regional dynamics and development (PGDREDES) at the Universidade Federal de Rio Grande do Sul (UFRGS), Brazil. catiagrisaufrgs@gmail.com

Cecilia Inés Aranguren. Researcher in rural economy and sociology, Agricultural Experimental Station in Balcarce (Instituto Nacional de Tecnología Agropecuaria – INTA). aranguren.cecilia@inta.gob.ar

Clara Craviotti. Researcher at the National Scientific and Technical Research Council (CONICET), Argentina; professor of the Agricultural Social Studies Master's Program at the Latin American Faculty of Social Sciences (FLACSO). c.craviotti@conicet.gov.ar

Daniel Campos R.D. Professor and researcher at the Instituto de Posgrado en Desarrollo (IPD) and Sociedad de Estudios Rurales y Cultura Popular (SER), Paraguay. danielcampos@ser.org.py

Diana Jazmin Britez Cohene. Student in rural development and food security at the Universidade Federal da Integraçao Latino-Americana (UNILA), Brazil. jazminbritez2010@hotmail.com

Diego S. Taraborrelli. Researcher at the Economy and Foresight Research Center (CIEP–INTA, Argentina). Coordinator of the Platform for Contributions to the Development and Management of Public Policies. taraborrelli.diego@inta.gob.ar

Fernanda França de Vasconcellos. PhD student in rural development at the Universidade Federal de Rio Grande do Sul (UFRGS). fernandacfv@yahoo.com.br

Francisco Javier Chavarría Arauz. Tenured professor at the Universidad Nacional Autónoma de Nicaragua, Regional Multidisciplinary Faculty of Matagalpa. Involved in research in the field of management of environmental and natural resources in agricultural ecosystems in central Nicaragua. fcha29@yahoo.com

Graciela Borrás. Researcher in economics and rural sociology, Agricultural Experimental Station in Balcarce (Instituto Nacional de Tecnología Agropecuaria – INTA). Professor, Faculty of Psychology, Universidad Nacional de Mar del Plata, Argentina. borras.graciela@inta.gob.ar

Geneviève Cortes. Professor of geography at Paul-Valéry University Montpellier 3, France. Researcher, Joint Research Unit UMR-ART-Dev (Actors, Resources and Territories in Development), Montpellier, France. genevieve.cortes@univ-montp3.fr

Guy Henry. Bioeconomist and regional delegate for Latin America, French agricultural research and international cooperation organization working for the sustainable development of tropical and Mediterranean regions (CIRAD). guy.henry@cirad.fr

Héctor Ávila-Sánchez. Researcher at the Regional Center for Multidisciplinary Research and professor of undergraduate and graduate geography, Universidad Nacional Autónoma de México. ahector@unam.mx

Ignacio Alonso. Researcher at the Economy and Foresight Research Center (CIEP–INTA, Argentina). alonso.ignacio@inta.gob.ar

Jairo Rojas Meza. Tenured professor at the Universidad Nacional Autónoma de Nicaragua, Regional Multidisciplinary Faculty of Matagalpa. Coordinator of the PhD program in sustainable rural territorial development. President of the board of directors of the Territorial Development Research and Training Institute (INFODET). infodetjairo@gmail.com

Janaína Deane de Abreu Sá Diniz. Professor of the postgraduate programs in environment and rural development (PPG-Mader) and for sustainability with traditional peoples and territories (PPG-PCTs) at the Universidade de Brasília (UnB), Brazil. janadiniz@unb.br

Jean-François Le Coq. Researcher at the French agricultural research and international cooperation organization working for the sustainable development of tropical and Mediterranean regions (CIRAD), member of the Joint Research Unit UMR-ART-Dev (Actors, Resources and Territories in Development), and visiting researcher at Alliance of Bioversity International and CIAT, Cali, Colombia. jflecoq@cirad.fr; jf.lecoq@cgiar.org

Jessica Pereira Garcia. Master's degree in environment and rural development from the Universidade de Brasília (UnB), Brazil. jessicapg15@gmail.com

Jorge Albarracin. Researcher-professor of the postgraduate degree program in development sciences (CIDES), Universidad Mayor de San Andrés (UMSA). Coordinator of the PhD program in rural development sciences, Bolivia. jalbarracindeker@gmail.com; jorgealbarracin@cides.edu.bo

José Anibal Quintero Hernández. Researcher-professor in rural development, Faculty of Social Sciences of the Institución Universitaria Colegio Mayor de Antioquia, Colombia. jose.quintero@colmayor.edu.co; joseanibalq@gmail.com.

Junior Miranda Scheuer. Assistent profesor at the Faculty of Agronomy, Universidad de la República (UDELAR), Montevideo, Uruguay. scheuerjr@gmail.com

María C. Benavidez. Researcher and director of the Sociedad de Estudios Rurales y Cultura Popular (SER), Paraguay. Specialist in gender and development. celsycampos@ser.org.py

Maria Mercedes Patrouilleau. Researcher at the Economy and Foresight Research Center (CIEP–INTA, Argentina). Advisor for the Programa Argentina Futura, Chief Cabinet Secretary. Coordinator of the degree program in Foresight studies and futures analysis at CIDES-UMSA, Bolivia. patrouilleau.mercedes@gmail.com

Mário Avila. Professor of the postgraduate programs in environment and rural development (MADER) and public management at the Universidade de Brasília (UnB), Brazil. unbavila@gmail.com

Mauro Guilherme Maidana Capelari. Professor of the postgraduate program in sustainable development (PPGCDS) and member of the Laboratory for Policy and Sustainability (PoliS) at the Center for Sustainable Development (CDS) at the Universidade de Brasília, Brazil (CDS). mauro.capelari@unb.br

Michel Leporati Néron. Director of innovation and technological transfer at the University of Talca, Chile. Director of CERES BCA biosafety services and food quality. mleporati@utalca.cl; michel.leporati@ceresbca.cl

Noelia Riquelme. Graduated with a degree in rural development and food security from the Universidade Federal da Integraçao Latino-Americana (UNILA). Former program assistant at FAO Paraguay. *(In memoriam)*.

Paulo Niederle. Professor of the postgraduate programs in sociology (PPGS) and rural development (PGDR) de la Universidade Federal de Rio Grande do Sul (UFRGS). Researcher at the National Scientific and Technological Research Council (CNPq), Brazil. pauloniederle@gmail.com

Pedro Pablo Benavídez. National director of agricultural technology transfer at INTA–Nicaragua. Professor for the PhD program for management and territorial development and doctoral committee member of the Universidad Católica del Trópico Seco de Estelí (UCATSE). Professor for the PhD program in management and quality of scientific research – third cohort, Universidad Nacional Autónoma de Nicaragua Managua. pebena@yahoo.com

Rafael Cabral. Master's student in the postgraduate program in environment and rural development (PPGMADER) at the Universidade de Brasília (UnB). Economist at the Ministry of Agriculture, Livestock and Food Supply (MAPA), Brazil. rafaelcab@gmail.com.

Ruby Elisabeth Castellanos Peñaloza. Professional at the Santiago de Cali municipality's Department of Public Health, profesor and researcher of the nutrition and diet program of the nstitución Universitaria Escuela Nacional del Deporte, Colombia, ruby.castellanos@cali.gov.co

Sandrine Freguin-Gresh. Researcher, French agricultural research and international cooperation organization working for the sustainable development of tropical and Mediterranean regions (CIRAD), Joint Research Unit UMR ART-Dev (Actors, Resources and Territories in Development), Montpellier, France. freguin@cirad.fr

Sara Rankin-Cortázar. Research associate, Food Environment and Consumer Behavior Team, Alliance of Bioversity International and CIAT, Cali, Colombia. s.rankin@cgiar.org

Silvia Aparecida Zimmermann. Professor at the Universidade Federal da Integraçao Latino-Americana (UNILA), researcher of the Observatorio de Políticas Públicas para la Agricultura (OPPA/UFRRJ) and Observatorio de las Agriculturas Familiares Latinoamericanas (AFLA/UNILA/UFRGS). Foz do Iguaçu, Brazil. silvia.zimmermann@unila.edu.br

Stéphane Guéneau. Researcher at the French agricultural research and international cooperation organization working for the sustainable development of tropical and Mediterranean regions (CIRAD), Joint Research Unit UMR MoISA (Montpellier Interdisciplinary center on Sustainable Agrifood systems), Montpellier, France. stephane.gueneau@cirad.fr

Susana Silvia Brieva. Professor and researcher at the Department of Social Sciences of the Universidad Nacional de Mar del Plata (UNMDP). Director of the graduate program in economics and territorial development (FCA-UNMDP), Argentina. sbrieva@mdp.edu.ar

Tainá Bacellar Zaneti. Professor at the Centre of Excellence in Tourism (CET), Universidade da Brasília (UnB), Brazil. tainazaneti@gmail.com

Cover illustration: Mexico street market – Fresh vegetables in San Cristobal de las Casas, Chiapas, Mexico © James Kelley, stock.adobe.com

Edition : James Calder, Lindsay Higgins, Teri Jones-Villeneuve, Abisola Persem and Claire Walker

Layout: Hélène Bonnet, Studio 9

Print deposit on publication: February 2022

Imprimé pour vous par Books on Demand (Allemagne)